Kinematics and Robot Design V, KaRD2022

Kinematics and Robot Design V, KaRD2022

Editor

Raffaele Di Gregorio

MDPI • Basel • Beijing • Wuhan • Barcelona • Belgrade • Manchester • Tokyo • Cluj • Tianjin

Editor
Raffaele Di Gregorio
University of Ferrara (UNIFE)
Ferrara, Italy

Editorial Office
MDPI
St. Alban-Anlage 66
4052 Basel, Switzerland

This is a reprint of articles from the Special Issue published online in the open access journal *Robotics* (ISSN 2218-6581) (available at: https://www.mdpi.com/journal/robotics/special_issues/VXGEC1WKX7).

For citation purposes, cite each article independently as indicated on the article page online and as indicated below:

Lastname, A.A.; Lastname, B.B. Article Title. *Journal Name* **Year**, *Volume Number*, Page Range.

ISBN 978-3-0365-8076-0 (Hbk)
ISBN 978-3-0365-8077-7 (PDF)
doi.org/10.3390/books978-3-0365-8077-7

Contents

About the Editor

Raffaele Di Gregorio

Raffaele Di Gregorio is currently full professor of machine mechanics at Engineering Dept. of Ferrara University, Italy. He received a M.Sc. degree in nuclear engineering, a M.Sc. degree in mechanical engineering and a M.Sc. degree in automotive engineering from Polytechnic University of Turin in 1982, 1985 and 1988, respectively, and a Ph.D. degree in applied mechanics from University of Bologna, in 1992.

In 1983, he spent one year at FIAT Research Center, Orbassano (IT) with an ATA grant.

Since 1984 till 1992, he has been officer of Italian Army's Technical Corp at Military School of Turin and, at STAVECO of Bologna. In 1993, he taught at ITIS O. Belluzzi of Bologna. Since 1994, he has joined the Engineering Dept. of Ferrara University.

His research interests include kinematics and dynamics of mechanisms and machines, biomechanics, robotics, vibration mechanics and vehicle mechanics.

He has been "observer/member" of IFToMM TC "Computational Kinematics" since 2007 and "member" of IFToMM PC "Standardization of Terminology" since 2009. He is an ASME member and has been "general member" of the ASME Mechanisms and Robotics Committee since 2007 till 2012; then, he has continued to serve this Committee as past member.

He has been author/co-author of about 90 technical papers published in refereed international journals and as many papers published in conference proceedings, and actively collaborates with many international journals of the field.

Preface

Kinematics plays a central role in machine design since all functional aspects of a machine need tools that belong to this discipline for the analysis and/or adjustment of specific requirements. The design of robotic/automatic systems confirms this fact. Indeed, topics such as the analysis and synthesis of mechanisms, robot modelling and simulation, robot control, mobility and singularity analysis, performance measures, accuracy analysis, path planning and obstacle avoidance, collaborative robotics, the identification of novel manipulator architectures, metamorphic mechanisms, compliant mechanism analysis and synthesis, micro/nano-manipulator design, origami-based robotics, medical and rehabilitation robotics, and bioinspired robotics, all deal with kinematics. Therefore, the social impact that mechatronics has in delineating future perspectives of human welfare must be partly tributed to kinematics.

Kinematics and Robot Design V (KaRD2022) is the 5th edition of the KaRD series, hosted by MDPI's *Robotics*. The KaRD series of open access Special Issues started in 2018 and is now an open environment where researchers can present their works and discuss all topics focused on the many ways in which kinematics is involved in the design of robotic/automatic systems using supplementary multimedia materials, which are uploadable during the submission. All the papers are peer-reviewed as soon as they are submitted and, if accepted, they are immediately published in MDPI's *Robotics* and appear on the website of the KaRD issue. All the papers of each KaRD edition are also collected into freely downloadable e-books. In short, the KaRD series is an "agora", where researchers efficiently exchange their experiences.

In the KaRD series, as in all the well-established serial international conferences/publications, the activity of the guest editor is supervised/supported by a scientific committee that is composed of the following members, who come from all over the world. Their service is gratefully acknowledged: Massimo Callegari (Polytechnic University of Marche, Italy), Juan Antonio Carretero (University of New Brunswick, Canada), Yan Chen (Tianjin University, China), Daniel Condurache ("Gheorghe Asachi" Technical University of Iași, Romania), Xilun Ding (Beijing University of Aeronautics & Astronautics, China), Mary Frecker (Penn State - College of Engineering, USA), Clement Gosselin (Laval University, Canada), Just Herder (TU Deft, Netherlands), Larry Howell (Brigham Young University, USA), Xianwen Kong (Heriot-Watt University, UK), Pierre Larochelle (South Dakota School of Mines & Technology, USA), Giovanni Legnani (University of Brescia, Italy), Haitao Liu (Tianjin University, China), Daniel Martins (Universidade Federal de Santa Catarina, Brazil), Andreas Mueller (Johannes Kepler Universität, Austria), Andrew Murray (University of Dayton, USA), Leila Notash (Queen's University, Canada), Matteo Palpacelli (Polytechnic University of Marche, Italy), Alba Perez (Universitat Politècnica de Catalunya, Barcelona, Spain), Victor Petuya (University of the Basque Country, Spain), José Maria Rico Martinez (Universidad de Guanajuato, Mexico), Nina Robson (California State University, Fullerton, USA), Jon M. Selig (London South Bank University, UK), Bruno Siciliano (University of Naples Federico II, Italy), Tao Sun (Tianjin University, China), Yukio Takeda (Tokyo Institute of Technology, Japan), Federico Thomas (Institute of Industrial Robotics, Spain), Volkert Van Der Wijk (TU Deft, The Netherlands).

All those that operate in the KaRD series editorial project believe in and aim to achieve a peaceful world where people and nations respect each other and cooperate to make the world a better place, where nobody is left aside. Unfortunately, these goals are still far from being reached and disseminating open science is our contribution to this aim.

KaRD2022 (https://www.mdpi.com/journal/robotics/special_issues/VXGEC1WKX7), after a rigorous peer-review process, accepted 13 papers. The accepted papers cover some theoretical and many design/applicative aspects. This book collects the thirteen papers published in KaRD2022. The book is organized as follows. The first three papers (1–3) address relevant theoretical aspects. The next paper (4) presents the State-of-the-Art on gripper design, and the successive two papers (5, 6) propose a novel gripper and a novel end effector, respectively, designed for specific applications. Then, the next three papers (7–9) address problems related to robot control, and the two successive papers (10, 11) propose novel robot architectures and a study of their kinematics. Finally, the last two papers (12, 13) deal with multi-leg mobile robots.

Raffaele Di Gregorio
Editor

1. Sean Mather and Arthur Erdman, Reformulation of Theories of Kinematic Synthesis for Planar Dyads and Triads. *Robotics* **2023**, *12*, 22, https://doi.org/10.3390/robotics12010022
2. Burkhard Corves and Amir Shahidi, Kinematic Graph for Motion Planning of Robotic Manipulators. *Robotics* **2022**, *11*, 105, https://doi.org/10.3390/robotics11050105
3. Takashi Kusaka and Takayuki Tanaka, Partial Lagrangian for Efficient Extension and Reconstruction of Multi-DOF Systems and Efficient Analysis Using Automatic Differentiation. *Robotics* **2022**, *11*, 149, https://doi.org/10.3390/robotics11060149
4. Jaime Hernandez, Md Samiul Haque Sunny, Javier Sanjuan, Ivan Rulik, Md Ishrak Islam Zarif, Sheikh Iqbal Ahamed, Helal Uddin Ahmed and Mohammad H Rahman, Current Designs of Robotic Arm Grippers: A Comprehensive Systematic Review. *Robotics* **2023**, *12*, 5, https://doi.org/10.3390/robotics12010005
5. Alexey M. Romanov, Ntmitrii Gyrichidi and Mikhail P. Romanov, A Novel Gripper with Integrated Rotary Unit and Force Control for Pick and Place Applications. *Robotics* **2022**, *11*, 155, https://doi.org/10.3390/robotics11060155
6. Md Mahafuzur Rahaman Khan, Asif Al Zubayer Swapnil, Tanvir Ahmed, Md Mahbubur Rahman, Md Rasedul Islam, Brahim Brahmi, Raouf Fareh and Mohammad Habibur Rahman, Development of an End-Effector Type Therapeutic Robot with Sliding Mode Control for Upper-Limb Rehabilitation. *Robotics* **2022**, *11*, 98, https://doi.org/10.3390/robotics11050098
7. Jan Reinhold, Henry Baumann and Thomas Meurer, Constrained-Differential-Kinematics-Decomposition-Based NMPC for Online Manipulator Control with Low Computational Costs. *Robotics* **2023**, *12*, 7, https://doi.org/10.3390/robotics12010007
8. Luca Bruzzone and Shahab Edin Nodehi, Application of Half-Derivative Damping to Cartesian Space Position Control of a SCARA-like Manipulator. *Robotics* **2022**, *11*, 152, https://doi.org/10.3390/robotics11060152
9. Henrique Simas and Raffaele Di Gregorio, Collision Avoidance for Redundant 7-DOF Robots Using a Critically Damped Dynamic Approach. *Robotics* 2022, 11, 93, https://doi.org/10.3390/robotics11050093
10. Luca Carbonari, Matteo-Claudio Palpacelli and Massimo Callegari, Inverse Kinematics of a Class of 6R Collaborative Robots with Non-Spherical Wrist. *Robotics* **2023**, *12*, 36, https://doi.org/10.3390/robotics12020036
11. Jessica Villalobos, Irma Y. Sanchez and Fernando Martell, Singularity Analysis and Complete Methods to Compute the Inverse Kinematics for a 6-DOF UR/TM-Type Robot. *Robotics* **2022**, *11*, 137, https://doi.org/10.3390/robotics11060137
12. Hongxi Zhu and Ulrike Thomas, Mechanical Design of a Biped Robot FORREST and an Extended Capture-Point-Based Walking Pattern Generator. *Robotics* **2023**, *12*, 82, https://doi.org/10.3390/robotics12030082
13. Md. Hasibur Rahman, Saadia Binte Alam, Trisha Das Mou, Mohammad Faisal Uddin and Mahady Hasan, A Dynamic Approach to Low-Cost Design, Development, and Computational Simulation of a 12DoF Quadruped Robot. *Robotics* **2023**, *12*, 28, https://doi.org/10.3390/robotics12010028

robotics

MDPI

Review

Reformulation of Theories of Kinematic Synthesis for Planar Dyads and Triads

Sean Mather * and Arthur Erdman

Department of Mechanical Engineering, University of Minnesota, 111 Church Street SE, Minneapolis, MN 55455, USA
* Correspondence: mathe587@umn.edu

Abstract: Methods for solving planar dyads and triads in kinematic synthesis are scattered throughout the literature. A review of and a new compilation of the complex number synthesis method for planar dyads and triads is presented. The motivation of this paper is to formulate uniform solution procedures, pointing out the commonalities of various approaches and emphasizing a consistent method for synthesizing mechanisms defined by specified precision positions. Particular emphasis is given to the solution method using compatibility linkages. The textbook *Advanced Mechanism Design Vol II* by Erdman and Sandor (1984) only includes a small portion of the available information on this method, and several researchers have added to the basic knowledge in the years since. In some cases, the approach and nomenclature were not consistent, yielding a need to describe and chart a generic formulation and solution procedure for dyads/triads using compatibility linkages and solution structures. The present method offers benefits for solving for exact dyad/triad solutions for complex multiloop mechanisms and could be a promising tool for reducing the computational load of finding complex mechanisms, and for visualizing properties of the solution space.

Keywords: mechanism synthesis; compatibility linkages; Burmester curves

Citation: Mather, S.; Erdman, A. Reformulation of Theories of Kinematic Synthesis for Planar Dyads and Triads. *Robotics* **2023**, *12*, 22. https://doi.org/10.3390/robotics12010022

Academic Editor: Raffaele Di Gregorio

Received: 15 November 2022
Revised: 21 December 2022
Accepted: 29 January 2023
Published: 1 February 2023

1. Introduction

The goal of synthesizing linkages and mechanisms to perform a particular task is a centuries-old practice. One famous example, called the South Pointing Chariot, was purportedly created by Chinese engineer Ma Jun (c. 200–265). As its name implies, a clever gear system driven by a chariot's wheels forces a statue on the back of the chariot to continually point south. This was true no matter how many turns the chariot took, provided both of its wheels rolled without slipping. This was a valuable navigational tool that significantly pre-dated the invention of the conventional magnetic compass [1,2]. However, for many centuries to follow, no formal or systematic methodology for synthesizing new mechanisms was developed.

Professor Robert Willis articulated this problem in the preface of his 1841 text "Principles of Mechanism" when he said, "By some strange chance, very few have attempted to give a scientific form to the ... results of mechanism; for it cannot be said that the few and simple ... examples in books of mechanics, are to be regarded as even forming a foundation ... that will enable us either to reduce the movements and actions of a complex engine to system or to give answers to the questions that naturally arise upon considering such engines" [3].

In the remainder of the text, Willis laid a foundation for later work and a challenge for mathematicians and engineers to create mathematical synthesis techniques. This call was taken up by the likes of Franz Reuleaux, James Watt, Ludwig Burmester, and Ferdinand Freudenstein. Each of them made a unique contribution to the field, such that by the early-mid 1900s, a mathematical basis for solving mechanism problems had been established.

The generations of kinematicians that followed more thoroughly fleshed out the techniques formulated by these early researchers, and developed more methods, such as complex number and continuation methods. As the field has continued to expand, few centralized solution methodologies have arisen, but rather a collection of largely unique approaches that are mostly specific to the type of linkage topology.

There are many distinct ways to define problems, and, consequently, many distinct methods to solve them. Some define precision positions (x, y coordinates and relative angle) that a coupling link must pass through, some a path a single point must pass through, and others seek to generate vast sets of possible mechanism solutions through continuation methods [4–8]. Recent studies in kinematic synthesis have primarily emphasized algorithmic approaches.

For example, Purwar and Deshpande investigated a machine-earning approach to kinematic synthesis, with the intent of mitigating the solution mechanisms' sensitivity to the initial conditions [9]. In another paper, Baskar and Bandyopadhyay discuss an algorithm aimed at reducing the computational load of calculating the finite roots of large systems of polynomial equations, a problem that arose in kinematic synthesis as the mathematical method of polynomial continuation was implemented [10]. Ref. [11] demonstrates a procedure for synthesizing RR, PR or RP dyads, but using a blend of exact and approximate positions.

While evidently valuable, this paper leans away from these algorithm driven approaches in favor of more classical synthesis approaches that focus on directly solving the kinematic equations. Countless complex planar mechanisms can be formed by a combination of dyads and triads, which can be viewed as kinematic building blocks. As two examples, consider the multilink mechanisms shown in Figures 1 and 2. The first deploys the footrest of a chair, while the second moves the leading-edge flap of a wing into its working position. Both mechanisms are composed of multiple dyad and triad chains. Rather than attempting to develop a custom kinematic synthesis process for every complex linkage, a uniform strategy is preferred.

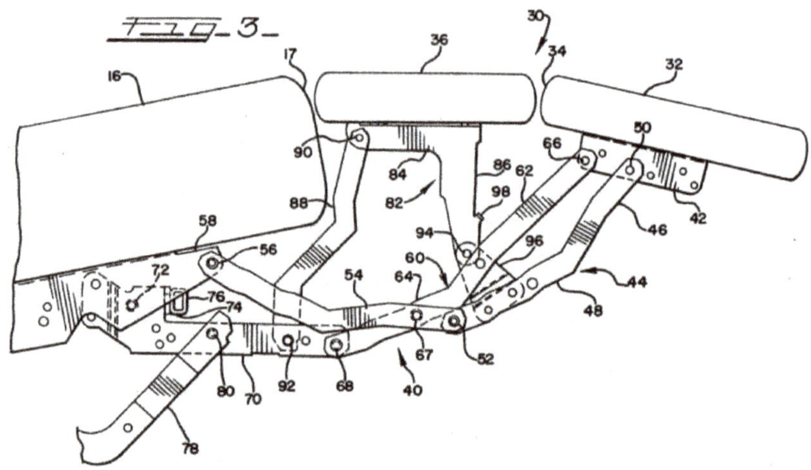

Figure 1. Patent figure of a chair with a deployable footrest [12].

Owing to some inherent properties of mechanisms and machines formed by links and joints, kinematic synthesis methods found in the literature share certain underlying mathematical principles that make finding one or more solutions possible. There exist a few analytical approaches to solving triad synthesis synthesis problems, some of which are analyzed in Reference [13], including a unique approach coined the "relative precision position approach for triad synthesis (p. 433)." Here, emphasis is placed on the solution method called the "compatibility linkage" for problems defined by precision positions. This method was first introduced by Sandor and Freudenstein [14] and summarized in Hartenberg, and Denavit [15] and later in Erdman and Sandor [4]. Further contributions building on the foundation established by Sandor and Freudenstein were made by Chuen-Sen Lin and other authors [16–19].

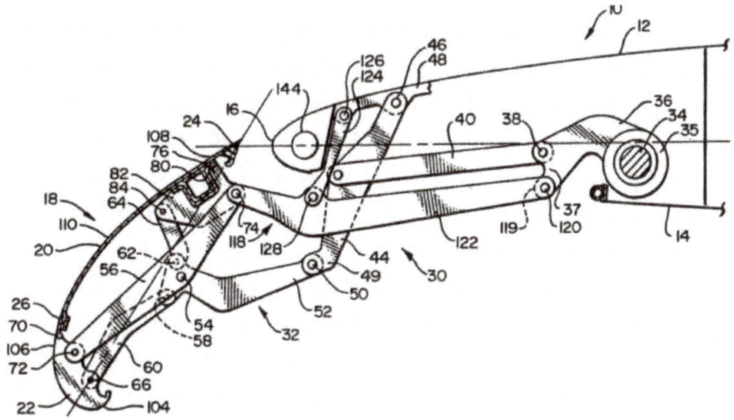

Figure 2. Patent figure of a leading edge flap of a wing [20].

2. Body

2.1. Precision Position Solution Methods

The starting point for using the complex number method for solving kinematic synthesis problems (defined by precision positions) is modelling the linkage mechanism using a number of "standard form equations" [4]. The equations are slightly different for a dyad and a triad. A dyad represents two links in the mechanism and has the form:

$$\mathbf{W}\left(e^{i\beta_j}-1\right)+\mathbf{Z}\left(e^{i\alpha_j}-1\right)=\boldsymbol{\delta}_j \tag{1}$$

A triad represents three links in relative motion and has the form:

$$\mathbf{W}\left(e^{i\alpha_j}-1\right)+\mathbf{V}\left(e^{i\beta_j}-1\right)+\mathbf{Z}\left(e^{i\gamma_j}-1\right)=\boldsymbol{\delta}_j-\boldsymbol{h}_j \tag{2}$$

Note that **W**, **Z**, **V**, $\boldsymbol{\delta}_j$ and **h** are vectors defined with complex numbers. Links in the mechanism that are not binary may be defined by more than one dyad or triad loop. These equations are illustrated in Figure 3a,b.

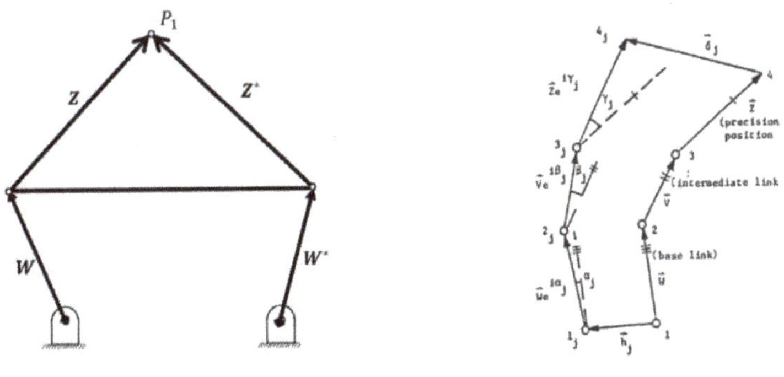

(**a**) (**b**)

Figure 3. (**a**) The standard form depiction of a four-bar comprised two dyads. (**b**) The standard form depiction of a triad is shown in two positions. Reprinted with permission from ref. [21]. Copyright 1987 Chuen-Sen Lin.

Each of the terms on the left-hand side in the standard form equations represent a link (**W**, **V**, or **Z**) in an assembled mechanism. They are multiplied by rotational operators which represent the rotations of the link from the starting position to each prescribed position. The term $\boldsymbol{\delta}_j$ represents the vector between the precision point in the i-th and j-th position (i.e., $\boldsymbol{\delta}_2$ goes between P1 and P2). In most cases, the terms $\boldsymbol{\delta}_j$ and α_j are prescribed in the problem, and β_j is taken as a free choice. As the number of precision positions increases, the number of free choices that can be made decreases until there are no free choices left. As seen in Table 1, for problems involving a dyad in two or three positions, the number of free choices is such that the standard form equations can be solved with a linear solution, either directly or by Cramer's rule for a dyad in three positions. However, in the four-precision position case, there are three vector equations (six scalar equations) which must be solved simultaneously, but seven unknowns. As a result, a nonlinear solution method must be used. This is where the method of compatibility linkages is so useful.

Table 1. Maximum number of solutions for an unknown dyad/triad when δ_j and α_j are prescribed in the equation $\boldsymbol{W}\left(e^{-i\beta_j}-1\right)+\boldsymbol{Z}\left(e^{-i\alpha_j}-1\right)=\boldsymbol{\delta}_j$ for dyads, and $\boldsymbol{W}\left(e^{-i\alpha_j}-1\right)+\boldsymbol{V}\left(e^{-i\beta_j}-1\right)+\boldsymbol{Z}\left(e^{-i\gamma_j}-1\right)=\boldsymbol{\delta}_j-\boldsymbol{h}_j$ for triads.

Dyad/ Triad	Number of Positions	Number of Scalar Equations	Number of Scalar Unknowns	Number of Free Choices	Number of Solutions
Dyad	2	2	5	3	$O(\infty^3)$
Dyad	3	4	6	2	$O(\infty^2)$
Dyad	4	6	7	1	$O(\infty^1)$
Dyad	5	8	8	0	Finite *
Triad	2	2	7	5	$O(\infty^5)$
Triad	3	4	8	4	$O(\infty^4)$
Triad	4	6	9	3	$O(\infty^3)$
Triad	5	8	10	2	$O(\infty^2)$
Triad	6	10	11	1	$O(\infty^1)$
Triad	7	12	12	0	Finite *

* Solutions come in sets of 0, 2, or 4 for dyads, and sets of 0, 2, 4, or 6 for triads [4] (p. 94), [21] (p. 21).

Unlike the standard form dyad equation, the triad equation also includes the vector term h. This term adjusts the tail end of the vector chain, allowing for the solution method applicability even in completely ungrounded triad cases. The dyad equation can be modified to include the term h if required.

Figure 4 illustrates how a linkage system can be viewed as combinations of dyads and triads. Even though this six-bar is far less complex that the mechanisms shown in Figures 1 and 2, the following process can be applied to mechanisms with more loops and links in a similar way. For example, although a Stephenson III six-bar is shown, the other six-bar chains can be placed in the dyad-triad standard form as reported by Lonn [22].

The six-bar shown in Figure 4 has three loops, one dyad and two triads. They are defined by Equations (3)–(5) [4]. The first loop equation describes a dyad, while the next two describe triad loops.

$$\mathbf{Z}_1\left(e^{i\varphi_j}-1\right)+\mathbf{Z}_2\left(e^{i\gamma_j}-1\right)=\boldsymbol{\delta}_j \tag{3}$$

$$\mathbf{Z}_5\left(e^{i\psi_j}-1\right)+\mathbf{Z}_4\left(e^{i\beta_j}-1\right)-\mathbf{Z}_3\left(e^{i\gamma_j}-1\right)=\boldsymbol{\delta}_j \tag{4}$$

$$\mathbf{Z}_6\left(e^{i\theta_j}-1\right)+\mathbf{Z}_7\left(e^{i\beta_j}-1\right)-\mathbf{Z}_3\left(e^{i\gamma_j}-1\right)=\boldsymbol{\delta}_j \tag{5}$$

Using free choices of link vectors, this mechanism can be solved with three dyads [4] (p. 113). Ref. [4] has other examples of assigning dyad and triad standard form modelling to multiloop mechanisms including an eight bar with four triads and geared mechanisms. Once a linkage system is modelled with combinations of dyad and triad standard form equations, the compatibility linkage solution process is used to reveal the potential solution space.

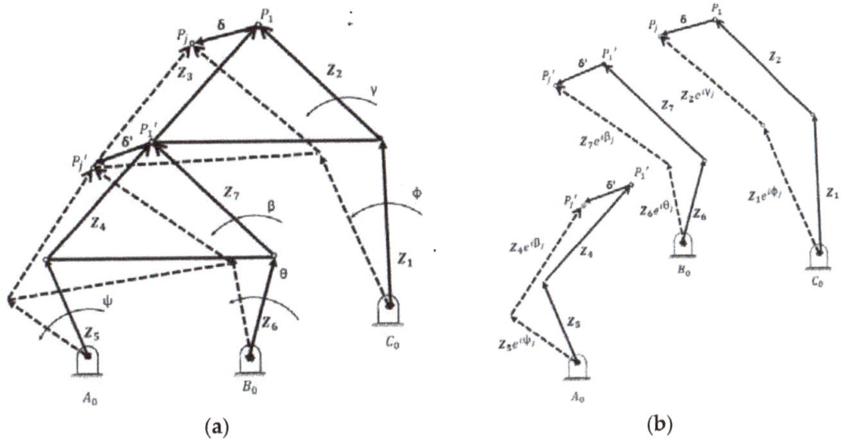

Figure 4. (a) A multiloop mechanism shown in an assembled form. (b) The same multiloop mechanism decomposed into three components, two triads and one dyad.

A practical mechanism which reveals these loops can be seen in Figure 5a, with the loops shown in Figure 5b. The mechanism guides a rotor on a drone from a vertical position into a horizontal position, allowing for the same rotor to produce vertical or forward thrust. This setup would allow for the drone to takeoff vertically but fly in a typical "fixed-wing" configuration once in the air, improving its efficiency. This particular example is very challenging due to significant constraints on both the ground and moving pivots. The ground pivots must both be within the nacelle, and the moving pivots are very close to the link holding the propellor. In addition, the mechanism must deploy smoothly without exhibiting poor transmission angles. The dimensional synthesis resulted in the Z values shown in Table 2.

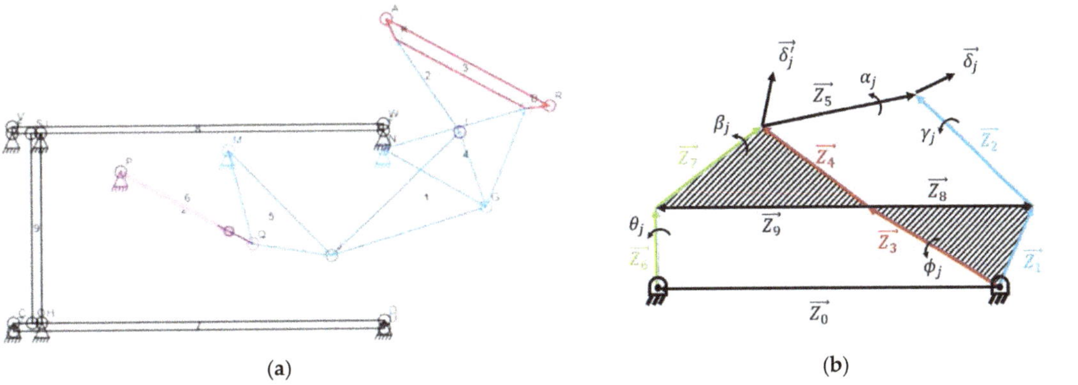

Figure 5. (a) The synthesized Watt 1 multiloop mechanism for guiding a drone rotor out of the nacelle and into a vertical position [23,24]. (b) The breakdown of the vector loops comprising the Watt 1 mechanism [23].

Table 2. Synthesized Z-Vectors.

Link	Vector (PP1)
Z_0	−3.377 + 0i
Z_1	0.405 − 3.178i
Z_2	0 + 3.178i
Z_3	−1.522 − 2.037i
Z_4	−0.436 + 1.620i
Z_5	2.364 + 0.416i
Z_6	−1.560 − 2.773i
Z_7	2.978 + 2.357i
Z_8	1.927 − 1.141i
Z_9	−3.170 − 0.684i

A proof-of-concept prototype was assembled, seen in Figure 6.

(a) (b)

Figure 6. (**a**) A 3D printed prototype in the initial "fixed-wing" position [23]. (**b**) The prototype in the open, vertical liftoff configuration.

2.2. *Compatibility Linkage Solution Procedure, Dyad for 4 Precision Points*

The purpose of the compatibility linkage is to find "compatible values" of several unknown variables in a set of nonlinear synthesis equations that are compatible with the known or specified variables. This method results in a closed-form solution to these equations. The compatibility linkage technique, which was introduced by Sandor and Freudenstein [14], takes advantage of insights gained by graphical and analytical precision position methods—both approaches provide keys to generating solutions for triads and dyads.

As with other precision position methods, it is assumed that the designer has either determined or measured the required x and y locations, and angles, of the precision point in each position, meaning that $\delta_{i\text{-}j}$ and $\alpha_{i\text{-}j}$ are known. Depending on the number of positions being considered for a particular problem, the designer may have additional free choices to make, but the change in position coordinates and angle between positions should always be known.

The compatibility linkage general solution procedure will be emphasized and illustrated with a planar dyad. The first step is to translate the known information into the standard form vector equations [4,12] (see Equations (1) and (2)). As seen in Figure 3, each vector **W**, **Z**, (and for triads, **V**), represents a link in a dyad or triad. Note that the two links **Z** and **Z′** shown in Figure 3a do not represent two unique links, but rather two vectors embedded in the same link.

The number of standard-form equations should be one less than the number of precision positions selected in the problem. The only terms that change in each of these equations are the angles β_j, α_j, γ_j, and the vector $\boldsymbol{\delta}_j$ for each additional position j. These equations are then translated into a matrix form, which looks like this:

$$\begin{bmatrix} e^{i\beta_2}-1 & e^{i\alpha_2}-1 \\ e^{i\beta_3}-1 & e^{i\alpha_3}-1 \\ e^{i\beta_4}-1 & e^{i\alpha_4}-1 \end{bmatrix} \begin{bmatrix} W \\ Z \end{bmatrix} = \begin{bmatrix} \delta_2 \\ \delta_3 \\ \delta_4 \end{bmatrix} \tag{6}$$

[4] (p. 180).

This equation will look roughly the same for a triad, except a column is added for **V** in the first matrix, and **V** is added to the column vector **WVZ**. An augmented matrix can be formed by adding the column vector $\delta_{2\text{-}4}$ to the matrix on the left-hand side. A known property of this type of system is that a solution only exists if the rank of the augmented matrix is two (for a dyad in four positions), with rank referring to the number of linearly independent rows in the matrix. The rank can be most easily checked by finding the determinant of the matrix. For square matrices such as the augmented matrix under consideration, if the determinant equals zero, the rank of the matrix is based on the non-zero cofactor (also called minor in math) of the maximum possible order [25]. The following expressions are derived from these properties.

$$Det\ M = Det \begin{bmatrix} e^{i\beta_2}-1 & e^{i\alpha_2}-1 & \delta_2 \\ e^{i\beta_3}-1 & e^{i\alpha_3}-1 & \delta_3 \\ e^{i\beta_4}-1 & e^{i\alpha_4}-1 & \delta_4 \end{bmatrix} = 0 \tag{7}$$

[4] (p. 181).

This determinant can be written into the following expression, known as the compatibility equation:

$$\boldsymbol{\Delta}_2 e^{i\beta_2} + \boldsymbol{\Delta}_3 e^{i\beta_3} + \boldsymbol{\Delta}_4 e^{i\beta_4} + \boldsymbol{\Delta}_1 = 0 \tag{8}$$

[4] (p. 181).
where each vector $\boldsymbol{\Delta}_{2\text{-}4}$ is the cofactor matrix associated with the corresponding β value in the augmented matrix. The cofactors are found by eliminating the row and column containing each value of β, as follows (e.g., $\boldsymbol{\Delta}_2$):

$$\begin{bmatrix} \cancel{e^{i\beta_2}-1} & \cancel{e^{i\alpha_2}-1} & \cancel{\delta_2} \\ \cancel{e^{i\beta_3}-1} & \begin{bmatrix} e^{i\alpha_3}-1 & \delta_3 \\ e^{i\alpha_4}-1 & \delta_4 \end{bmatrix} \\ \cancel{e^{i\beta_4}-1} & \end{bmatrix} \tag{9}$$

The cofactor matrix formed from what remains after eliminating this row and column is marked in the gray box. For a dyad in four positions, the cofactor matrices are given as follows:

$$\boldsymbol{\Delta}_2 = \begin{vmatrix} e^{i\alpha_3}-1 & \delta_3 \\ e^{i\alpha_4}-1 & \delta_4 \end{vmatrix} \tag{10}$$

$$\boldsymbol{\Delta}_3 = -\begin{vmatrix} e^{i\alpha_2}-1 & \delta_2 \\ e^{i\alpha_4}-1 & \delta_4 \end{vmatrix} \tag{11}$$

$$\boldsymbol{\Delta}_4 = \begin{vmatrix} e^{i\alpha_2}-1 & \delta_2 \\ e^{i\alpha_3}-1 & \delta_3 \end{vmatrix} \tag{12}$$

[4] (p. 181).

Note that for dyads, the cofactor matrices will always be 2×2, while for triads, the cofactor matrices will be 3×3. Additionally, each of the vectors $\boldsymbol{\Delta}_{2\text{-}4}$ is a matrix signified with vertical lines rather than conventional matrix brackets. This is a mathematical shorthand representing a determinant, meaning that each of these terms (once the determinant is evaluated) is a vector with a magnitude and direction. The term $\mathbf{\Delta_1}$ is unique from the others, defined by the following expression:

$$\mathbf{\Delta_1} = -\mathbf{\Delta_2} - \mathbf{\Delta_3} - \mathbf{\Delta_4} \tag{13}$$

[4] (p. 181).

As brilliantly noted in [14], this equation can be viewed as a four link mechanism in its starting position—thus named the compatibility linkage. Equation (13) is the equation of closure where $\mathbf{\Delta_1}$ is the fixed link and the rest of the vectors close the loop by connecting the chain's head to its tail. Plotting each of the four above vectors without applying any of the beta rotation angles mathematically will resemble Figure 7 (for a dyad in four positions, see Table 3 for more configurations).

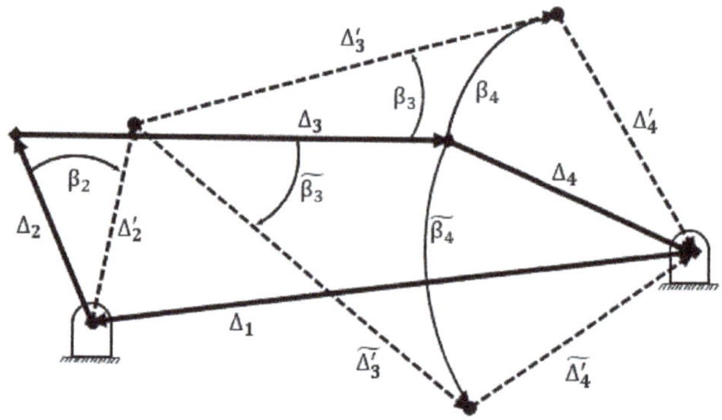

Figure 7. The compatibility linkage for a dyad in four precision positions. The resultant position of each link is shown after applying a rotation of β_2. There are two combinations of the links $\mathbf{\Delta_3}$ and $\mathbf{\Delta_4}$ which close the linkage. Both represent a viable solution to the original problem for the given value of β_2.

Table 3. Summary of the solution process for compatibility linkages for dyads and triads.

Dyad/ Triad	Number of Positions	Number of Links	Number of Cofactor Matrices	Procedure	Compatibility Linkage BKC
Dyad	4	Four-bar compatibility linkage	4; $\mathbf{\Delta_1}, \mathbf{\Delta_2}, \mathbf{\Delta_3}, \mathbf{\Delta_4}$	1. Write standard form equations, put them in an augmented matrix form, write the cofactor matrices 2. Create $\mathbf{\Delta_1}$ as the sum of $\mathbf{\Delta_{2\text{–}4}}$, then draw the linkage. 3.Rotate $\mathbf{\Delta_2}$ by β_2, rotate $\mathbf{\Delta_3}$ and $\mathbf{\Delta_4}$ to close the loop. 4. Read off the angular displacement of $\mathbf{\Delta_3}$ and $\mathbf{\Delta_4}$ to get β_3 and β_4.	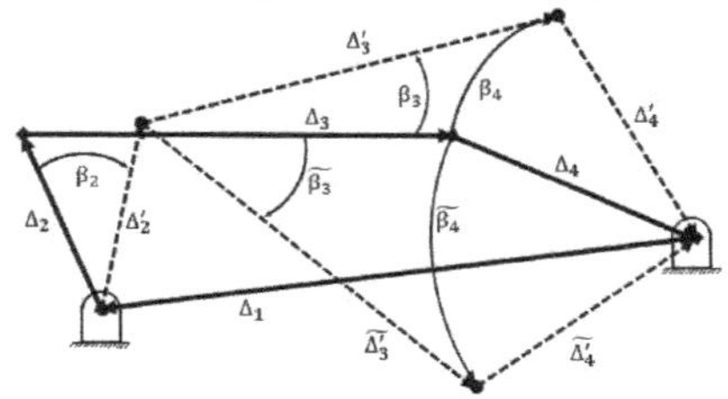
Dyad	5	Seven—bar structure, six-bar compatibility linkage	7; $\mathbf{\Delta_1}, \mathbf{\Delta_2}, \mathbf{\Delta_3}, \mathbf{\Delta_4}, \mathbf{\Delta_1}', \mathbf{\Delta_2}', \mathbf{\Delta_3}'$	1. Write standard form equations, put them in an augmented matrix form, write the cofactor matrices. There are two sets, $\mathbf{\Delta}$ and $\mathbf{\Delta}'$. 2. Using a consistent scale, plot both complete loops, with the base of the $\mathbf{\Delta_1}$ links at the same x, y position 3. Rotate either loop until $\mathbf{\Delta_2}$ and $\mathbf{\Delta_2}'$ are colinear. 4. Form a parallelogram about the $\mathbf{\Delta_3}$ links, creating point D. 5. Remove the newly formed link DE to find the final compatibility linkage. 6. Rotate $\mathbf{\Delta_2}$ by β_2, adjusting other links accordingly. Positions where BE and CD are parallel represent solutions.	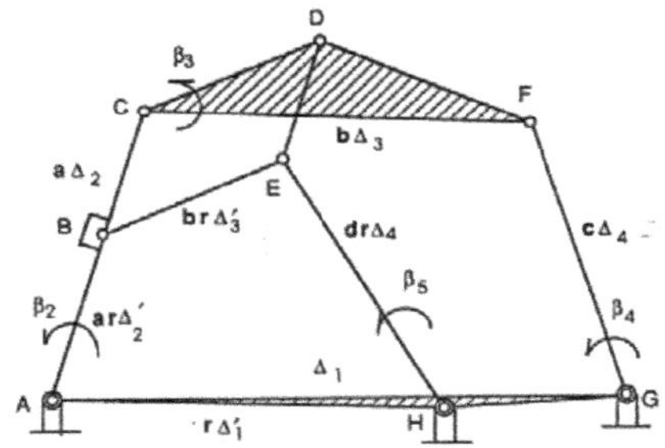 [21] (p. 109)
Triad	5	Five-bar 2-DOF compatibility linkage	5; $\mathbf{\Delta_1}, \mathbf{\Delta_2}, \mathbf{\Delta_3}, \mathbf{\Delta_4}, \mathbf{\Delta_5}$	Repeat steps 1–4 of the dyad in 4PP. The difference is that the loop has five links instead of four, so the designer will need to set two free choices, typically the angles of $\mathbf{\Delta_2}$ and $\mathbf{\Delta_5}$.	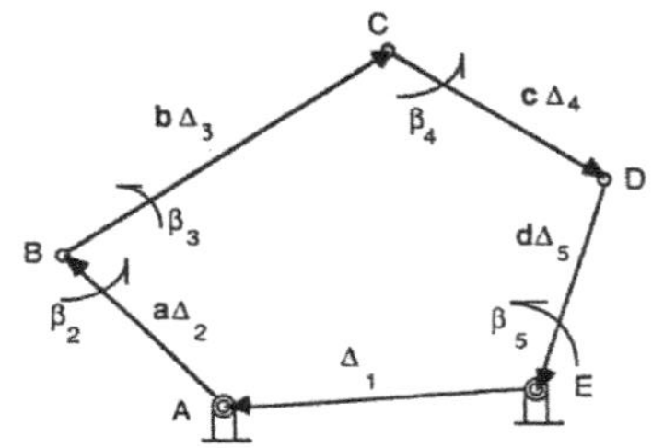 [21] (p. 113)

Table 3. *Cont.*

Dyad/ Triad	Number of Positions	Number of Links	Number of Cofactor Matrices	Procedure	Compatibility Linkage BKC
Triad	6	Eleven-bar structure, ten-bar compatibility linkage	9; $\mathbf{\Delta_1, \Delta_2, \Delta_3, \Delta_4, \Delta_5, \Delta_1', \Delta_2', \Delta_3', \Delta_4'}$	Repeat steps 1–6 for the dyad in five positions. In the example at right, $\mathbf{\Delta_3}$ is used as the ground pivot. There are two parallelogram loops to form in step four. One about $\mathbf{\Delta_4}$, forming BCEF, and one about $\mathbf{\Delta_1}$, forming GIJK. Remove a link from either parallelogram to convert to the compatibility linkage.	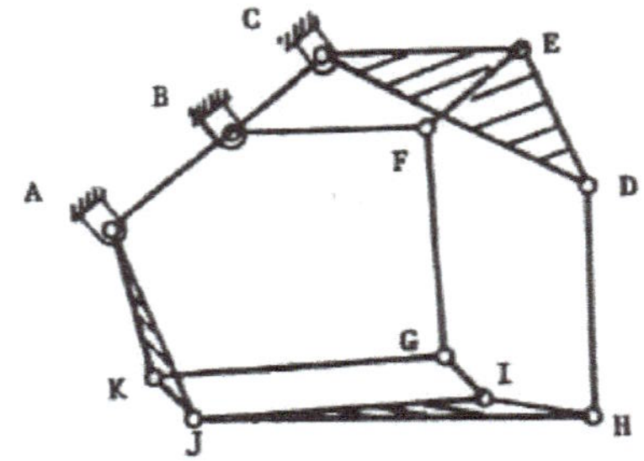 [21] (p. 35)
Triad	7	Fifteen-bar structure, fourteen-bar compatibility linkage	13; $\mathbf{\Delta_1, \Delta_2, \Delta_3, \Delta_4, \Delta_5, \Delta_1', \Delta_2', \Delta_3', \Delta_4', \Delta_1'', \Delta_2'', \Delta_3'', \Delta_4''}$	Repeat steps 1–6 for the dyad in five positions. There are several parallelogram loops to form in step four. Loops should be formed between each layer of the linkage. Here, they are formed about $\mathbf{\Delta_4}$ and $\mathbf{\Delta_1}$, and the layers of link $\mathbf{\Delta_2}$ are fused to form a single link. Remove a link from any parallelogram to convert to the compatibility linkage *.	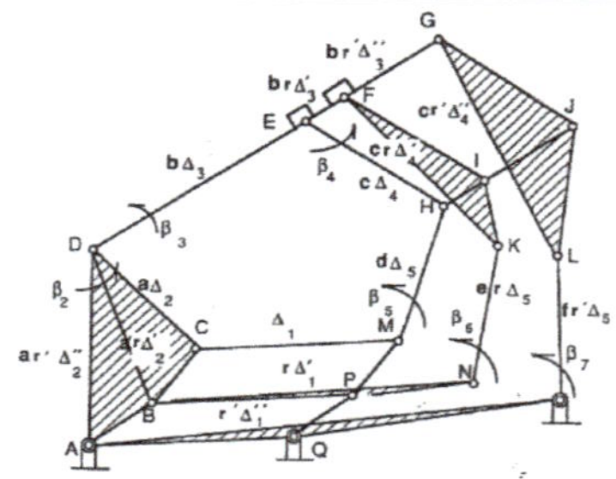 [21] (p. 117)

* Additional information regarding the solution procedure for a triad in six and seven precision positions is shown in Appendix A, including a simplified form of the six-precision position compatibility linkage.

One frequent point of confusion is that the compatibility linkage is related to the actual solution dyad. This is not the case. Rather, the constructed compatibility linkage is only the tool to allow the user to find compatible solutions for the unknown angles in the standard form equations.

Once the linkage is assembled, the user applies a rotation of β_2 to link $\mathbf{\Delta_2}$. Consequently, the links for $\mathbf{\Delta_3}$ and $\mathbf{\Delta_4}$ need to translate and rotate by some amount to keep the loop closed, as $\mathbf{\Delta_1}$ is considered ground, and does not move. Once solved, the displacement angles of links $\mathbf{\Delta_3}$ and $\mathbf{\Delta_4}$ represent the solution values of β_3 and β_4. These values are then plugged into the original standard form dyad equations. With β_3 and β_4 identified, solving for the vectors W and Z using standard linear algebra techniques is possible.

While the problems may be a bit more complex, increasing the number of precision positions or transitioning from a dyad to a triad changes very little about the underlying methodology for compatibility linkages. While this paper will not emphasize quadriads, it is even possible to apply the method of compatibility linkages to solving four link chains [21]! Here, each higher-order case up to seven precision positions will be briefly examined, highlighting the key differences of each from the dyad in four positions explained above. See Table 3 for a summary of these cases and see Appendix A for a detailed solution procedure of the triad in six and seven positions.

2.3. *Dyad in 5 Precision Positions*

Moving from four to five positions is likely the biggest single jump in complexity for solving problems using the method of compatibility equations. This is because there is no longer a single compatibility equation, but rather two. The compatibility equations for a dyad with five prescribed positions (no free choices) are:

$$\begin{vmatrix} e^{i\beta_2}-1 & e^{i\alpha_2}-1 & \delta_2 \\ e^{i\beta_3}-1 & e^{i\alpha_3}-1 & \delta_3 \\ e^{i\beta_4}-1 & e^{i\alpha_4}-1 & \delta_4 \end{vmatrix} = 0 \tag{14}$$

and

$$\begin{vmatrix} e^{i\beta_2}-1 & e^{i\alpha_2}-1 & \delta_2 \\ e^{i\beta_3}-1 & e^{i\alpha_3}-1 & \delta_3 \\ e^{i\beta_5}-1 & e^{i\alpha_5}-1 & \delta_5 \end{vmatrix} = 0 \tag{15}$$

[4] (p. 201).

Or, in their simplified form:

$$\mathbf{\Delta_2}e^{i\beta_2} + \mathbf{\Delta_3}e^{i\beta_3} + \mathbf{\Delta_4}e^{i\beta_4} + \mathbf{\Delta_1} = 0 \tag{16}$$

$$\mathbf{\Delta'_2}e^{i\beta_2} + \mathbf{\Delta'_3}e^{i\beta_3} + \mathbf{\Delta_4}e^{i\beta_5} + \mathbf{\Delta'_1} = 0 \tag{17}$$

[21] (p. 107) where

$$\mathbf{\Delta_1} = -\mathbf{\Delta_2} - \mathbf{\Delta_3} - \mathbf{\Delta_4} \tag{18}$$

$$\mathbf{\Delta'_1} = -\mathbf{\Delta'_2} - \mathbf{\Delta'_3} - \mathbf{\Delta_4} \tag{19}$$

The $\mathbf{\Delta'}$ terms are formed in the same way as the $\mathbf{\Delta}$ terms (cofactors of the augmented matrix), but they are taken from the second matrix. These equations must be fulfilled simultaneously to find a valid solution for W and Z. Previously, finding the solution to these compatibility equations would have required using a technique known as "Sylvester's Dyalitic Eliminant". While this method worked, the process is computationally involved and mathematically rigorous. Using the method of compatibility linkages described below allows the designer to avoid this complexity while being able to visualize the solution process.

To form a solution structure, identify each of the two independent four-bar loops formed by the $\mathbf{\Delta}$ terms. The first loop includes $\mathbf{\Delta_1}$ through $\mathbf{\Delta_4}$, while the second is formed from $\mathbf{\Delta_1}'$, $\mathbf{\Delta_2}'$, $\mathbf{\Delta_3}'$, and $\mathbf{\Delta_4}$ ($\mathbf{\Delta_4}$ used twice because $\mathbf{\Delta_4}'$ is identical to it, either notation is acceptable). Using a consistent scale, line up these two four bars such that the tails of

Δ_2 and Δ_2' share the same x, y coordinate, and rotate either four-bar linkage (keeping all internal angles the same) such that Δ_2 and Δ_2' have the same angular direction. The result should now resemble Figure 8, with Δ_1 (A-G-H) and Δ_2 (A-B-C) appearing as ternary links. There are two distinct four-bar chains, or loops, between them.

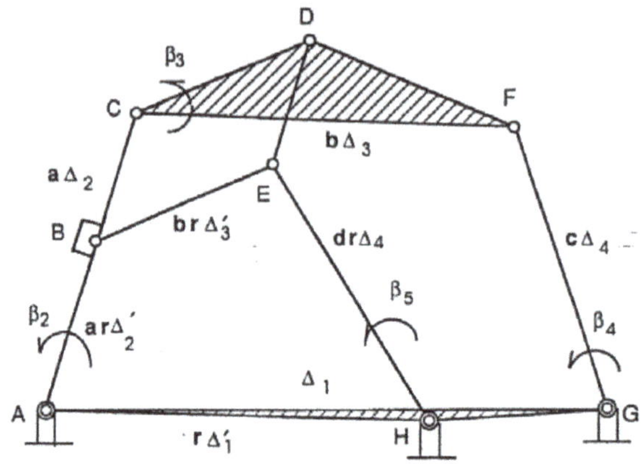

Figure 8. The compatibility linkage for a dyad in five precision positions. Linkage positions in which the lines CD and BE are parallel represent solutions to the synthesis problem [21] (p. 109).

The final key step shown in Figure 8 is adding point D. This point is found by creating the parallelogram BCDE, turning link CF (corresponding to Δ_3) into a ternary link. The completed mechanism is known as the solution structure. Once identified, the link DE may be removed, but point D will remain as a reference. After removing link DE, the mechanism transforms from a zero degree-of-freedom structure to a one degree-of-freedom linkage; this is the final compatibility linkage.

To find solutions, rotate the (now ternary) link Δ_2. The rotation of Δ_2 is the only input required to fully define the system, so each other link is determined once the angle of Δ_2 is set. As this mechanism moves, at any position where the links CD and BE are parallel, the linkage represents a compatible solution to the original problem. The exception is the first position, as CD and BE will always be parallel initially, by definition.

For each unique parallel position, the displacement angles of the links correspond to the angles $\beta_{2\text{–}5}$. Specifically, $\angle\Delta_3 = \beta_3$, $\angle\Delta_4$ (outer loop)= β_4, and $\angle\Delta_4$ (inner, Δ' loop)= β_5. These compatible angles are then inserted back into the standard form dyad equations. With four vector equations and two vector unknowns, the equation can now be solved for W and Z via a linear solution. The number of geometric inversions of the compatibility linkage corresponds to the number of solution sets to the compatibility equations. The term geometric inversion refers to the number of unique mechanisms that can be created by changing which link is fixed, meaning distinct inversions do not have unique angular displacements, just different grounded links. In this case, that means there are six sets of unique combinations of $\beta_{2\text{–}5}$ which fulfill the original compatibility equations. However, two of these solutions correspond to the slider and concurrency special points. As a result, only up to four dyad solutions exist—that is, there are zero, two or four viable solutions for each choice of independent variable x.

2.4. Triad in 5 Precision Positions

As a designer transitions from synthesizing a dyad to a triad, the underlying solution procedure will remain the same, but a few key steps will either change or be added. First, the standard triad equation has an additional term, $e^{i\gamma_j}$-1, associated with the Z vector. See Figure 3b depicting the vector form of a triad to see where this term fits in the vector chain. Similarly, an intermediate vector V has been added, increasing the number of links in the chain from two to three. The meaning of some of the angles have changed as well. The angle α no longer describes the angle of the coupler link, but rather the angle of the vector **W**. β is now assigned to the intermediate link **V** and will continue to be selected as the free choice for these problems. The new angle, γ, replaces α as the angle describing the coupler's rotation. It is important to note that using triads instead of dyads for five precision positions does not increase the number of loops in the compatibility linkage—there is still only a single loop—but the triad does increase the number of terms that must be identified. All values of γ must also be prescribed along with all the information that was prescribed for a dyad in five positions. This volume of free choices enables a designer to make many decisions about their desired mechanism, but this can also be overwhelming due to the vast potential solution space. For the triad in five positions, there are four simultaneous vector equations, and one compatibility equation:

$$\mathbf{\Delta_2}e^{i\beta_2} + \mathbf{\Delta_3}e^{i\beta_3} + \mathbf{\Delta_4}e^{i\beta_4} + \mathbf{\Delta_5}e^{i\beta_5} + \mathbf{\Delta_1} = 0 \tag{20}$$

[21] (p. 111)

$$\mathbf{\Delta_1} = -\mathbf{\Delta_2} - \mathbf{\Delta_3} - \mathbf{\Delta_4} - \mathbf{\Delta_5} \tag{21}$$

As with the compatibility linkage for a dyad in four positions, only one compatibility equation exists for a triad in five positions. As a result, the compatibility equation only has a single loop. However, one significant difference between the two is the additional link in the five-bar compatibility equation. This results in a solution structure with two degrees of freedom rather than one. However, this challenge can be avoided by giving the designer a second free choice. Typically, these free choices are chosen as β_2 and β_5, though any other combination of two angles is also valid. Once these free choices are made, the solution procedure is the same as the dyad in four positions, as all that is left is a geometrically deterministic triangle. The remaining link positions and angles can be solved by using the law of cosines. Table 3 and Figure 9 represent summaries of the dyad and triad solution procedures using the compatibility linkage approach. The similarities across these cases are indicated, perhaps suggesting a future software kinematic synthesis package. One example in this direction was achieved by Chase [26], although there was limited use of this software at that time.

A full explanation of the solution procedure using the compatibility linkage for a triad in six and seven precision positions can be found in Appendix A. While kinematic chains above dyads and triads will not be discussed in detail here, Lin demonstrated the general solution procedure for the compatibility linkage of a quadriad [21]. Theoretically, even higher-order chains also ought to be solvable by compatibility linkages. However, it becomes increasingly difficult to fathom a sufficiently complex yet practical mechanism synthesis problem that would justify their use. Even so, chains incorporating five or more vectors/links, and their potential applications, remain a possible area for further study.

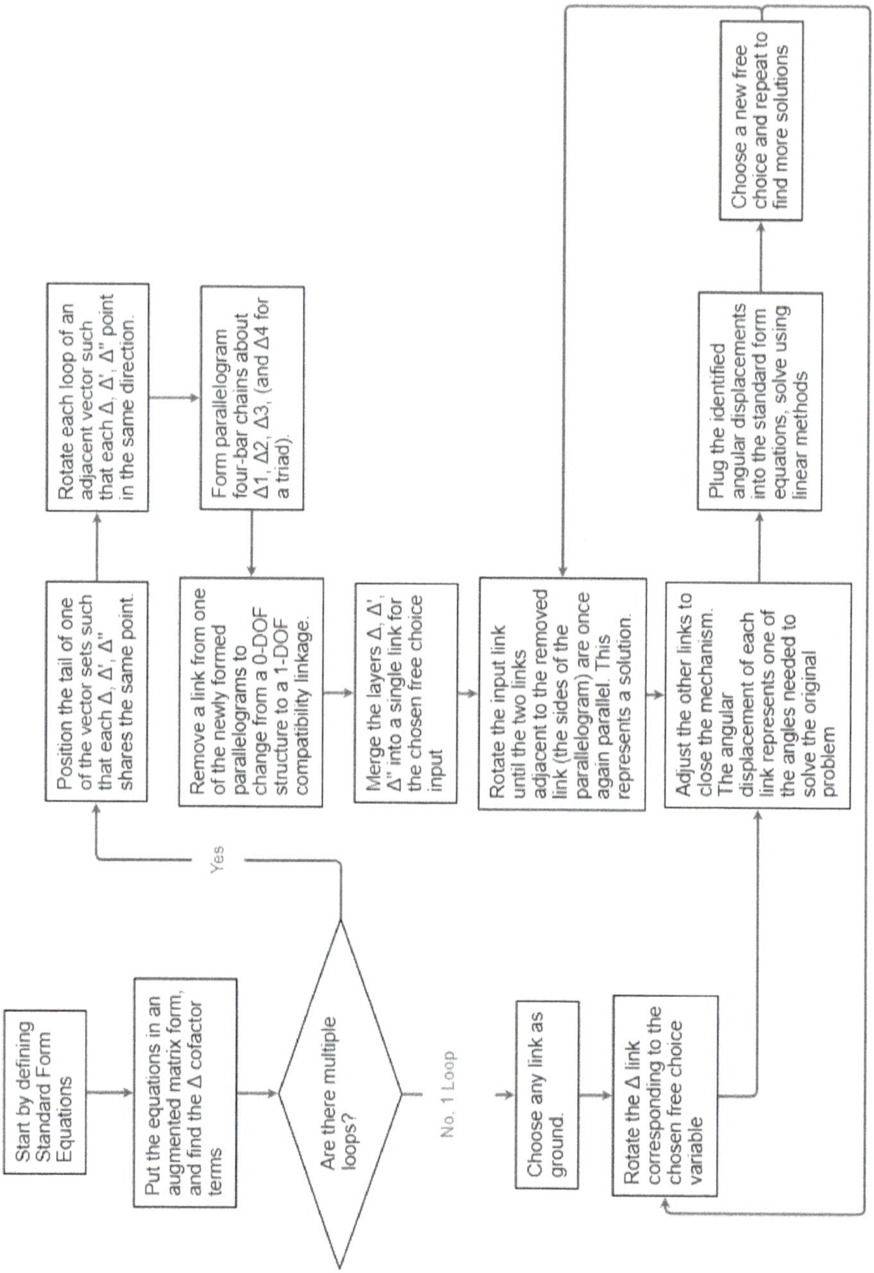

Figure 9. General solution procedure flow chart.

2.5. General Solution Procedure

A flow chart is provided in Figure 9 depicting the general solution procedure using the method of compatibility linkages. Inspiration for the chart comes from [21] (pp. 140–143).

To solidify the general solution procedure, the authors find it prudent to provide the following numerical example, a dyad in five precision positions. The problem is defined by the precision positions stated in Table 4.

Table 4. Precision positions and alpha angles.

Position	Coordinates	Alpha (deg)
1	−21.700 + 22.035i	0
2	−11.883 + 12.018i	19.684
3	−5.973 + 11.767i	30.751
4	4.744 + 10.766i	59.361
5	5.345 + 12.318i	84.696

Using Equations (7)–(13) and (19), the vectors representing the links in the compatibility linkage are found, shown in Table 5. These vectors form two four-bar loops, shown in Figure 10. The end of vector $\mathbf{\Delta_1}$ is chosen as the shared point between the two loops.

Table 5. Direction and magnitude of each $\mathbf{\Delta}$ vector.

Delta Link	Vector Coordinates
$\mathbf{\Delta_1}$	−1.2736 + 0.1990i
$\mathbf{\Delta_2}$	0.9214 − 3.4612i
$\mathbf{\Delta_3}$	1.5541 + 3.7924i
$\mathbf{\Delta_4}$	−1.2018 − 0.5302i
$\mathbf{\Delta_1}'$	−3.3851 + 1.1261i
$\mathbf{\Delta_2}'$	5.2149 − 9.7840i
$\mathbf{\Delta_3}'$	−0.6280 + 9.1881i

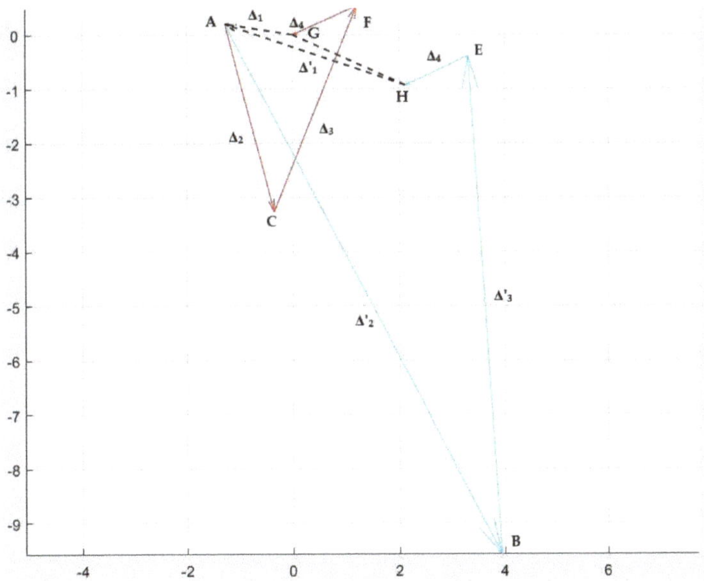

Figure 10. The unmodified plot of the $\mathbf{\Delta}$ vectors for a dyad in five precision positions.

At this point, the links $\mathbf{\Delta_2}$ and $\mathbf{\Delta_2}'$ are chosen as the input which will drive the compatibility linkage. As a result, all of the $\mathbf{\Delta}'$ links are rotated about the head of vector $\mathbf{\Delta_1}$ to align $\mathbf{\Delta_2}$ and $\mathbf{\Delta_2}'$ so that they are colinear, a rotation of −13.15 degrees (CW). Additionally, a parallelogram is formed by drawing a vector from the end of $\mathbf{\Delta_3}'$ in the direction of $\mathbf{\Delta_2}$.

This vector has length $\mathbf{\Delta_2} - \mathbf{\Delta_2}'$=1.9307 − 7.2528i. After applying these changes, the compatibility linkage takes the form shown in Figure 11.

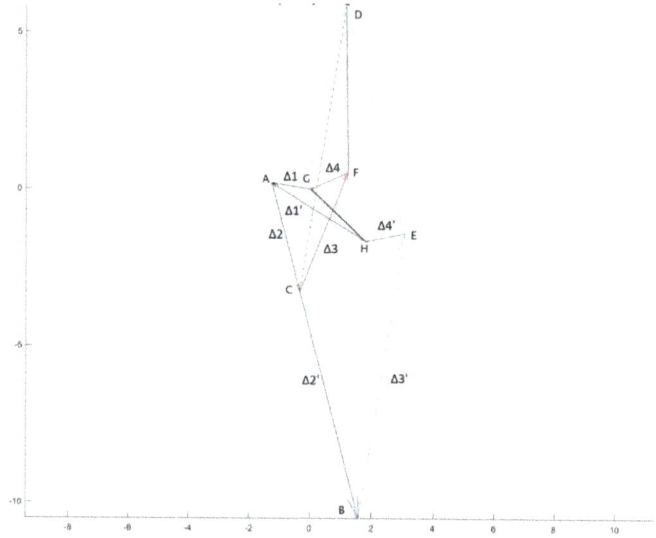

Figure 11. The solution structure for a dyad in five precision positions.

Figure 11 is the solution structure representing this problem. Removing the link DE forms the compatibility linkage. To use it, the designer can directly begin rotating the linkage to try to find solutions. However, a useful intermediate step is to find the range of acceptable β_2 values for which the compatibility linkage closes. This range is found by drawing a circle with radius $\mathbf{\Delta_2}$ around the tip of $\mathbf{\Delta_1}$, as well as a circle of radius $|\mathbf{\Delta_2}| + |\mathbf{\Delta_3}| + |\mathbf{\Delta_4}|$ around the tail of $\mathbf{\Delta_1}$. Repeat this process for the $\mathbf{\Delta}'$ loop. Here the range of the $\mathbf{\Delta}$ loop is more limiting. The range of $\mathbf{\Delta_2}$ is shown in Figure 12, with the circle's intersections denoting the limits of $\mathbf{\Delta_2}$.

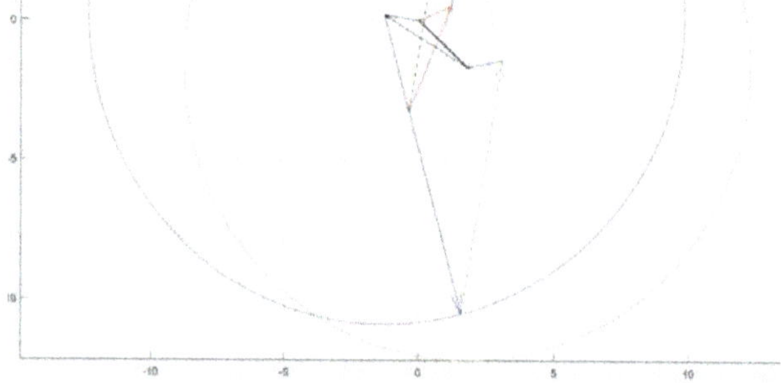

Figure 12. Procedure for finding the range of β_2. Here the two intersections of the circles represent the upper and lower bound of the angle.

From these circle intersections, it is possible to calculate an upper and lower bound of β_2 as 116 degrees above the initial position (CCW), and 27 degrees below the initial position (CW). The mechanism is rotated over this range, and any positions where links CD and BE are parallel to each other is recorded. In this problem, there are two such positions, shown in Figure 13a,b.

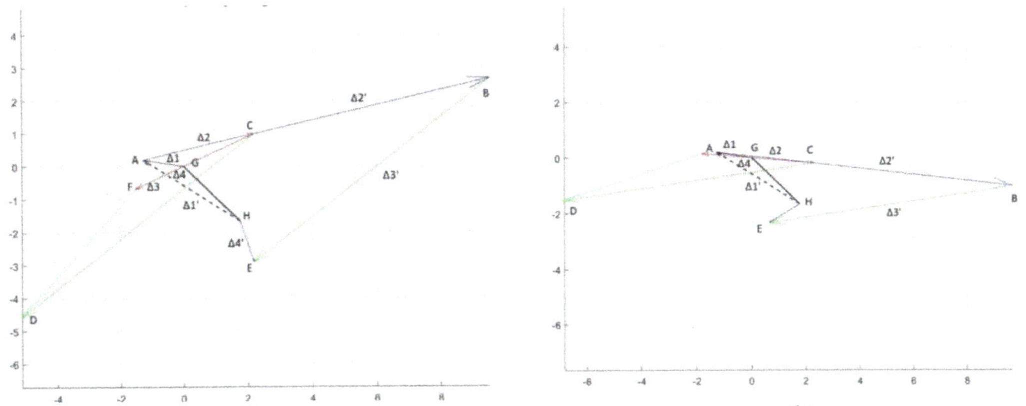

(**a**) (**b**)

Figure 13. (**a**) Compatibility linkage in solution position one. (**b**) Compatibility linkage in solution position two.

In each of these two compatibility linkage positions, the links are measured to determine their angular displacement relative to the initial position. From these displacements, two dyads are found, corresponding to two solution positions, by plugging the values back into the standard form equations and finding a linear solution. These two dyads are plotted using the software Lincages in Figure 14 [26,27].

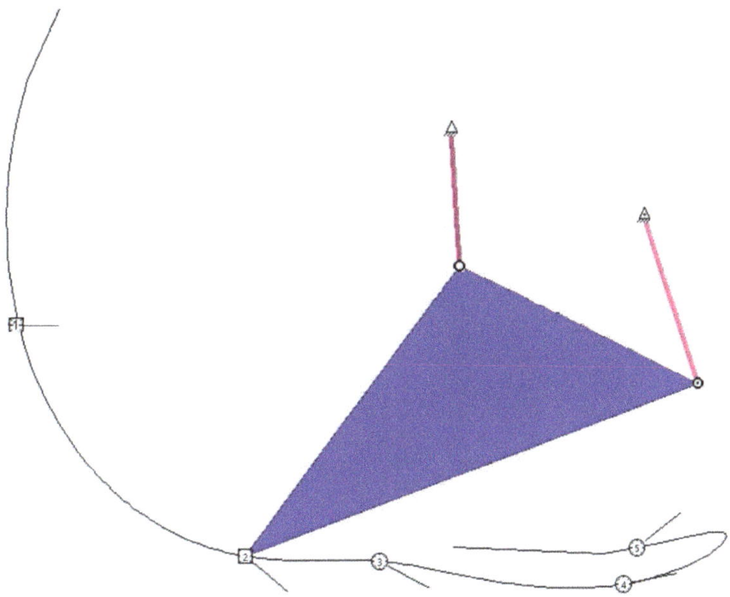

Figure 14. Final solution linkage visualized in the Lincages software. The triangles represent the ground pivots of this mechanism.

2.6. Special Cases

As with other synthesis methods, there are several special cases when solving problems using the compatibility linkage. A few of the most common will be emphasized here, and Figure A4 shows a more complete table of special cases.

The first set of special cases occurs when the free choice angle β_2 is equal to α_2, γ_2, or 0. This results in Equation (2) taking the form shown in Equation (22) (for $\beta_2 = \alpha_2$). Each of these cases is resolved by the solution containing a slider.

$$\mathbf{W}\left(e^{i\alpha_j} - 1\right) + \mathbf{V}\left((p_j)e^{i\alpha_j} - 1\right) + \mathbf{Z}\left(e^{i\gamma_j} - 1\right) = \boldsymbol{\delta}_j - \boldsymbol{h}_j \tag{22}$$

In this solution case, all angle variables are predefined prior to performing any calculations. The only scalar term which is not defined is p_j. p_j is called the stretch factor, and p_2 is the free choice for this problem.

The second set of special cases is also caused by other angular similarities. They are: one link with no angular displacement (i.e., $\alpha_j = 0$), two links with the same angular displacement (i.e., $\alpha_j = \gamma_j$), and multiple links with no angular displacement (i.e., $\alpha_j = \gamma_j = 0$). Each of these special cases is resolved through some combination of sliders, with the exception of $\gamma_j = 0$, for which no solutions exist. For a full depiction of the special cases for triads, see Figure A4.

2.7. Advantages of the Compatibility Linkage Method

By analyzing a compatibility linkage for range of rotation of the "input link" of the compatibility linkage, some interesting properties of possible solution mechanisms are revealed. Frequently, the link $\boldsymbol{\Delta}_2$ will have a finite rotational range, meaning that only values of β_2 falling in the acceptable range can produce solutions. This is quite useful, as previously, the range of acceptable β_2 (free choice) values would have been found through an exhaustive search. Through the method of compatibility linkages, the designer can clearly identify the upper and lower bounds of β_2 based on how far link $\boldsymbol{\Delta}_2$ in the compatibility linkage will rotate in either direction from its starting position. For example, a crank-rocker type compatibility linkage will give β_2 a range that allows any value to be used as a free choice. In contrast, a double-rocker compatibility linkage will restrict the range of β_2 [17–19]. In the latter case, one can expect solutions only for a limited range of β_2, clockwise or counterclockwise—thus larger values of $\pm\beta_2$ are rare. This is a quite useful design rule.

Applying the Grashof theory to a compatibility linkage reveals some interesting behavior. Depending on the type of mechanism formed by the compatibility equations (Grashof vs non-Grashof, crank-rocker, double-rocker, etc.), solution regions may emerge. If a compatibility linkage has more than one branch (e.g., more than one unique configuration, such that reaching the second configuration requires temporarily removing at least one pin joint), then there will correspondingly be multiple sets of β_2 values that produce viable solutions. This can be seen in reference [18] (p. 4), depicting the Center-point Burmester curve for a double-rocker compatibility linkage.

A non-Grashof triple-rocker mechanism, on the other hand, has a single circuit. As a result, it will have continuous solutions throughout its full potential range of motion of the input angle. However, this will still not cover a full 360 degrees, as rockers are inherently limited in this regard.

J.A. Schaaf and J.A. Lammers furthered this research, identifying fourteen specific classes of compatibility linkages and their corresponding center-point curve shapes [19]. These fourteen groups are divided into three categories; Grashof, non-Grashof, and change-point mechanisms. Within each of these groups, depending on which link $\boldsymbol{\Delta}_{1\text{–}4}$ is the shortest, the general shape of the center-point curve can be determined. See their paper for a full list of these categories [19]. While this theory has presently only been applied to the compatibility linkage of a dyad in four precision positions, there is reason to believe that the same line of analysis may yield similar findings for the triad in five precision positions, and perhaps even higher numbers of precision positions.

Applying the Grashof criteria to the compatibility linkage is not the same as applying the same criteria to the finished solution. Its use for the compatibility linkage reveals interesting information about potential solution regions in which, for any value of β_2, a solution exists, or regions where no dyad/triad solutions exist. However, mechanisms produced from the compatibility linkage approach may still be subject to circuit, branch, and order defects. Additionally, they may have poor force properties or low transmission angles.

2.8. Defects

The compatibility linkage is useful in that it reveals numerous prospective solutions, but the designer will still need to determine if a candidate mechanism found by this method meets their requirements and that it does not exhibit defects, such as the combination of dyads selected not reaching all design positions on one circuit of the mechanism. Chase and Mirth detailed an effective procedure for identifying and addressing these defects [28]. Whether there is a relationship between the Grashof type of compatibility linkage and any or all these properties remains a possible area for research. Similarly, applying the Grashof criteria to the higher-order compatibility linkages could be further investigated. Investigating the circuit defects of a compatibility linkage will reveal unique solution regions as there are gaps where no solutions exist for a particular value of β_2.

2.9. Eight or More Precision Positions

Cases that would require more than seven precision positions are less common in industry, as usually, a less complex solution method can produce a satisfactory mechanism design. However, a few options are available if a designer wishes to move beyond this limit. First, Chuen-Sen Lin derived compatibility linkage solutions for quadriads in up to nine positions. The solution structures produced for these mechanisms are quite complex but are solved in largely the same way as the dyads and triads. See his work from the University of Minnesota [21] or the subsequent work he and his students completed at the University of Alaska Fairbanks [29,30].

2.10. Connections to General Burmester Theory

Burmester Theory underpins many of the precision position synthesis techniques in the field of mechanisms. The theory largely revolves around the position of the poles for a particular moving plane. They are found by identifying the intersection of the perpendicular bisectors between two positions for two arbitrary points on the moving plane. In four positions, the center-point curve passes through the six standard poles, while the circle-point curve passes through the poles $\mathbf{P_{12}}$, $\mathbf{P_{13}}$, $\mathbf{P_{14}}$, and the image poles $\mathbf{P_{23}}'$, $\mathbf{P_{24}}'$ and $\mathbf{P_{34}}'$.

In addition to the poles, points called "opposite pole quadrilaterals", or "Π-points", are found by identifying the intersection of lines passing through each pair of non-adjacent poles. There are twelve of these points in four positions. The first six are shown in Equation (23), each of which intersects the center-point curve, much like the natural poles.

$$\begin{array}{ll} \Pi_{12} = \overline{P_{13}P_{23}}\ x\ \overline{P_{13}P_{23}} & \Pi_{13} = \overline{P_{12}P_{23}}\ x\ \overline{P_{14}P_{34}} \\ \Pi_{14} = \overline{P_{12}P_{24}}\ x\ \overline{P_{13}P_{34}} & \Pi_{23} = \overline{P_{12}P_{13}}\ x\ \overline{P_{24}P_{34}} \\ \Pi_{24} = \overline{P_{12}P_{14}}\ x\ \overline{P_{23}P_{34}} & \Pi_{34} = \overline{P_{13}P_{14}}\ x\ \overline{P_{23}P_{24}} \end{array} \tag{23}$$

[31,32] (pp. 25–26).

The next six are formed from some combination of the image poles and are shown in Equation (24). The circle point curve passes through each of these points in addition to the poles listed above.

$$
\begin{array}{ll}
\Pi^1_{12} = \overline{P_{13}P^1_{23}}\, x\, \overline{P_{14}P^1_{24}} & \Pi^1_{13} = \overline{P_{12}P^1_{23}}\, x\, \overline{P_{14}P^1_{34}} \\
\Pi^1_{14} = \overline{P_{12}P^1_{24}}\, x\, \overline{P_{13}P^1_{34}} & \Pi^1_{23} = \overline{P_{12}P_{13}}\, x\, \overline{P^1_{24}P^1_{34}} \\
\Pi^1_{24} = \overline{P_{12}P_{14}}\, x\, \overline{P^1_{23}P^1_{34}} & \Pi^1_{34} = \overline{P_{13}P_{14}}\, x\, \overline{P^1_{23}P^1_{24}}
\end{array}
\quad (24)
$$

[31,32] (pp. 25–26).

Using the full collection of these points, an initial depiction of both Burmester curves can be drawn. This visual tool may lend exceptional value. A sample plot of the Burmester curves for a dyad can be seen in Figure 15:

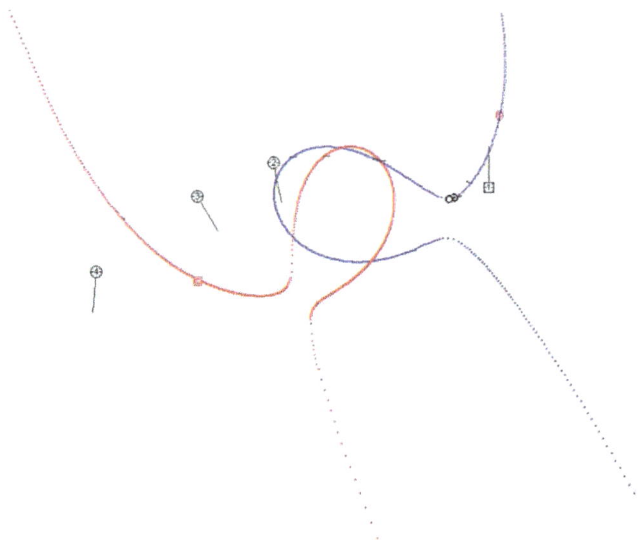

Figure 15. An example of the center-point and circle-point Burmester curves for a dyad in four positions [32].

Similarly, the Burmester Curve of Triad is shown in Figure 16:

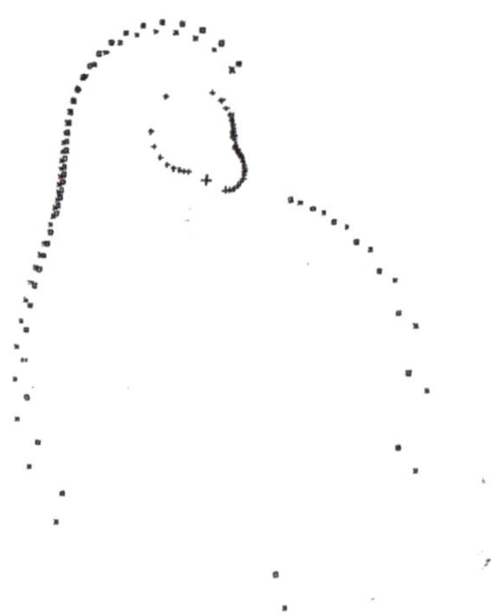

Figure 16. The three Burmester curves for a triad in six precision positions. There are two circle-point curves (corresponding to the two moving pivots) and one center-point curve [21] (p. 86).

Each point on the Burmester curve is part of a 'Burmester Point Pair', or a Burmester point set in the case of the triad. These pairs represent a corresponding moving pivot location and ground pivot location. The method of compatibility linkages is an extremely effective tool for finding these curves—for each value of the free choice that is valid, a new pair of points in the curves is generated.

2.11. Closing Thoughts

The potential applications of kinematic synthesis through the compatibility linkage are intriguing, and there are still opportunities for further investigation. Throughout this paper β_2 has been used as the free choice, and the compatibility linkage has been applied to find the values of the angles $\beta_{i\text{-}j}$. For dyads, this paper assumes the designer wants to solve the problem for motion generation. However, by forming new cofactor matrices about α rather than β, the compatibility linkage could be used for path with prescribed timing problems as well. The LINCAGES software package utilized this realization [33–35]. Similarly for triads, it should be possible to use the compatibility linkage to solve the standard form equation for any of the angles β, α or γ, with the only procedural change being rewriting the cofactor matrices. This would allow the designer to solve triad synthesis problems defined for motion generation or for path with prescribed timing. There are typically two path generators for each motion generator due to cognates [4]. As a result, further investigation into the unique properties of the compatibility linkage and their cognates is warranted.

Schaaf and Lammers investigated the compatibility linkage of a dyad in four positions and found that its Grashof type played a distinctive role in determining the shape of the Burmester curves [19]. Inspired by their findings, we speculate that the compatibility linkage of the triad for five precision positions will exhibit similar properties. The triad has a five-bar compatibility linkage with two degrees of freedom, but it only has a single loop. Additionally, after setting the angle of one of the free choices (e.g., β_5), the rest of the linkage behaves like a four-bar, and the second free choice can be rotated through all its values (e.g., β_2). This likely means that the findings of Schaaf and Lammers are applicable to the triad, and that for each free choice of β_5 a new center-point curve could be generated which resembles the corresponding class of dyads.

In addition to the applications for multiple prescribed position synthesis, we speculate that the compatibility linkage can be utilized for mixed position-velocity synthesis as well. Using the standard form equations mixed position-velocity synthesis is already possible. In a two-precision position problem, for example, a designer may choose to include a third equation describing the velocity of the precision point in the first position. The standard form equation can be rewritten as:

$$\mathbf{W}\left(e^{i\dot{\beta}_j}-1\right)+\mathbf{Z}\left(e^{i\dot{\alpha}_j}-1\right)=\boldsymbol{V}_j \tag{25}$$

Here $\dot{\beta}_j$ and $\dot{\alpha}_j$ represent the angular velocities of their respective links, and V_j is the velocity vector of the precision point. To evaluate this expression using a compatibility linkage, the cofactors would need to be rewritten, but after making that change the general solution procedure should flow in exactly the same way [4,36,37].

Multiple researchers have demonstrated using multiply separated positions to synthesize a path generation mechanism by using derivative equations [4] (pp. 239–245), [38]. The resultant tracer point curve closely resembles a prescribed function for a significant range of that function. However, to achieve this result for a problem defined by a single position and its four derivatives, Sylvester's dialytic eliminant was employed. In the same way as before, we speculate that this method can be avoided by employing the compatibility linkage method. This would only require rewriting the cofactor matrices with appropriate derivatives.

3. Conclusions

In this paper, the compatibility linkage approach to kinematic synthesis has been shown to have broad applicability to a large range of linkage mechanisms. Computationally, it is a simplified way of traversing the solution space of potential dyads and triads. Similarly, the method adds the ability to visualize spatial properties of the solution space which are not easily identified through other methods. The process of using the assembled compatibility linkage to find solutions flows naturally from common mechanism analysis techniques, meaning that once completed, the compatibility linkage can be used to generate numerous potential solutions. Furthermore, while the scope of the method presented in this paper is already quite unifying, there remain many opportunities to expand the method even further through additional research.

Author Contributions: Both authors contributed equally to all aspects of this paper. All authors have read and agreed to the published version of the manuscript.

Funding: This research received no external funding.

Data Availability Statement: Not applicable.

Acknowledgments: The authors appreciate the contributions of previous researchers whose work is referenced in this paper, notably Chuen-Sen Lin, and colleagues at the University of Minnesota, including Tom Chase and James Van de Ven.

Conflicts of Interest: The authors declare no conflict of interest. There was no funding provided or influence for the design of the study; in the collection, analyses, or interpretation of data; in the writing of the manuscript; or in the decision to publish the results.

Appendix A. Detailed Solution Procedure for a Triad in Six and Seven Precision Positions Using the Method of Compatibility Linkages

Triad in six precision positions:
Displacement equations:

$$\begin{bmatrix} e^{i\alpha_2}-1 & e^{i\beta_2}-1 & e^{i\gamma_2}-1 & \boldsymbol{\delta_2}-\boldsymbol{h_2} \\ e^{i\alpha_3}-1 & e^{i\beta_3}-1 & e^{i\gamma_3}-1 & \boldsymbol{\delta_3}-\boldsymbol{h_3} \\ e^{i\alpha_4}-1 & e^{i\beta_4}-1 & e^{i\gamma_4}-1 & \boldsymbol{\delta_4}-\boldsymbol{h_4} \\ e^{i\alpha_5}-1 & e^{i\beta_5}-1 & e^{i\gamma_5}-1 & \boldsymbol{\delta_5}-\boldsymbol{h_5} \\ e^{i\alpha_6}-1 & e^{i\beta_6}-1 & e^{i\gamma_6}-1 & \boldsymbol{\delta_6}-\boldsymbol{h_6} \end{bmatrix} \begin{bmatrix} \vec{W} \\ \vec{V} \\ \vec{Z} \\ -1 \end{bmatrix} = \vec{0} \tag{A1}$$

[21] (p. 22).

As with a dyad in five positions, there are two compatibility equations to solve simultaneously from this matrix. They are:

$$\boldsymbol{\Delta_2}e^{i\beta_2} + \boldsymbol{\Delta_3}e^{i\beta_3} + \boldsymbol{\Delta_4}e^{i\beta_4} + \boldsymbol{\Delta_5}e^{i\beta_5} + \boldsymbol{\Delta_1} = 0 \tag{A2}$$

And:

$$\boldsymbol{\Delta'_2}e^{i\beta_2} + \boldsymbol{\Delta'_3}e^{i\beta_3} + \boldsymbol{\Delta'_4}e^{i\beta_4} + \boldsymbol{\Delta_5}e^{i\beta_6} + \boldsymbol{\Delta'_1} = 0 \tag{A3}$$

[21] (p. 23).

Once the $\boldsymbol{\Delta}$ terms have been defined, the compatibility linkage is drawn. As before, align the tails of vectors $\boldsymbol{\Delta_2}$ and $\boldsymbol{\Delta_2}'$, and rotate the $\boldsymbol{\Delta}'$ loop (keeping inter-link angular relations constant) so that $\boldsymbol{\Delta_2}$ and $\boldsymbol{\Delta_2}'$ point in the same direction. In this case, there are two parallelogram loops to form. The first is the loop BCEF, and the second is GIJK, as seen below.

Note that Figure A1 was generated slightly differently from the previous compatibility linkages. Here, $\boldsymbol{\Delta_3}$ and $\boldsymbol{\Delta_3}'$ were chosen as the ground pivots, and these links were overlaid with each other, with the head rather than the tail of $\boldsymbol{\Delta_2}$ taken as the common point between the two loops. As a result, the parallelogram BCEF is formed about the link $\boldsymbol{\Delta_4}$ and $\boldsymbol{\Delta_4}'$ ($\boldsymbol{\Delta_4}$ is extended to form the ternary link CDE), while the parallelogram GIJK was formed about links Δ_1 and Δ_1' (Δ_1 is extended to form the ternary link HIJ). Finally, a ternary link AJK is formed incorporating Δ_2 and Δ_2' (connecting the points J and K). A critical observation to make regarding these parallelogram loops is that each was made possible by existing relationships between the Δ and Δ' terms. Δ_2 and Δ_2' share the same angular displacement, even prior to being connected. The same is true of Δ_3 and Δ_3', and Δ_4 and Δ_4'. This means that combining these links is an unnecessary but helpful simplification of the compatibility linkage, as this new form above requires only a single input to fully determine the rest of the mechanism. This is a significant advantage, as each independent five-bar loop had two degrees of freedom, making the problem more complex.

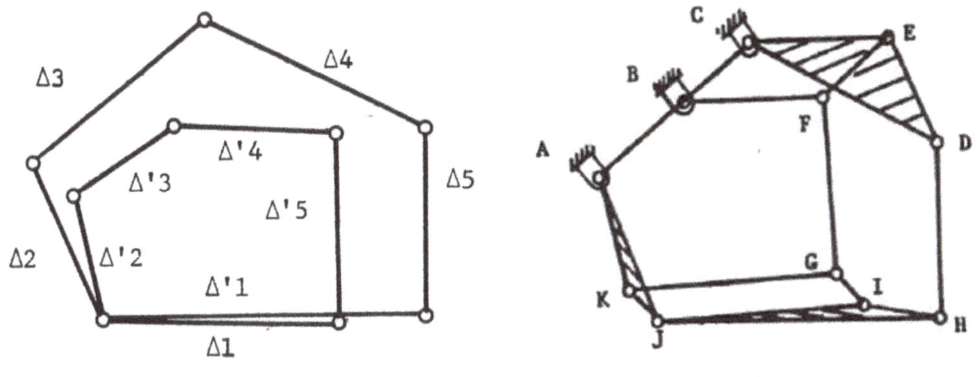

(a) (b)

Figure A1. (**a**) The two loops of the triad in six positions shown without modification; (**b**) The compatibility linkage of a triad in six positions after aligning Δ_3 and forming parallelograms [21] (pp. 31–35).

The compatibility linkage of the triad in six positions can be used to find solutions in the form described above. However, the linkage is unique in that it has an additional layer of possible simplification that the designer can take advantage of. As was discussed earlier, the number of unique geometric inversions of the compatibility linkage corresponds to the number of unique solutions to the original problem. In this case, though, once the free-choice angle β_2 is selected and implemented, the relative positions of links AJK and HIJ remain consistent regardless of which geometric inversion is considered. As a result, several links can be eliminated. The pivots A, G, and H can be considered as a single ternary link, reducing the ten-bar linkage to a seven-bar zero-DOF structure [17] (p. 36). After implementing each of these steps, the new structure looks like this (Figure A2):

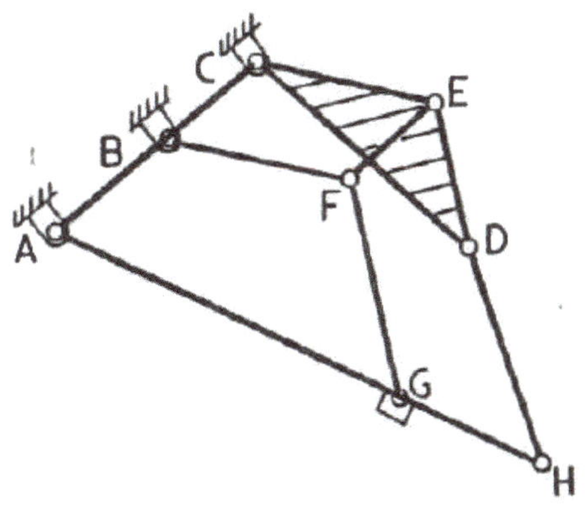

Figure A2. The simplified form of the compatibility linkage for a triad in six positions [21] (p. 38).

The final positions of each of the points in this compatibility linkage are shown in Table A1. To find solutions, remove the link EF, creating a Watt-II solution linkage. For positions (apart from the starting position) in which links, CE and BF are parallel, the mechanism represents a solution to the original problem.

Table A1. Simplified compatibility linkage global point positions (triad, 6PP).

Point	Position in Plane
A	(0,0)
B	$\Delta_3{}'$
C	Δ_3
D	$\Delta_3 + \Delta_4$
E	$\Delta_3 + \Delta_4{}'$
F	$\Delta_3{}' + \Delta_4{}'$
G	$-\Delta_2{}' + \Delta_1$
H	$-\Delta_2 + \Delta_1$

The solution values are then taken from the angular displacements of links AH, BF, HD, and GF, which correspond to the values of -β_3, β_4-β_3, β_5-β_3, and β_6-β_3, respectively. These relationships are shown in Table A2. While making these simplifications does take more time initially, the payoff is substantial. Analysis of a six-bar mechanism is easier than a ten-bar linkage, not to mention that the Watt-type mechanisms are much more thoroughly covered in the literature. A designer working through this process will find the following relations:

Table A2. Selected Links in the simplified compatibility linkage and their corresponding beta values.

Link	Angular Displacement
AH	-β_3
BF *	β_4-β_3
HD	β_5-β_3
GF	β_6-β_3

* BF or CE can be used, as they share the same angular displacement.

The attentive reader may note that the angle β_2 is neglected in Tables A1 and A2, and it no longer plays a role as a free choice or as a solution angular displacement. Fortunately, once the beta values have been calculated, not all of them must be used. Only three of the beta values need to be incorporated to calculate the value of each unique solution for the triad. Chuen-Sen Lin shows an example, here taken as Equation (A4), using β_2, β_3, and β_4.

$$\begin{bmatrix} \vec{W} \\ \vec{V} \\ \vec{Z} \end{bmatrix} = \begin{bmatrix} e^{i\alpha_2}-1 & e^{i\beta_2}-1 & e^{i\gamma_2}-1 \\ e^{i\alpha_3}-1 & e^{i\beta_3}-1 & e^{i\gamma_3}-1 \\ e^{i\alpha_4}-1 & e^{i\beta_4}-1 & e^{i\gamma_4}-1 \end{bmatrix}^{-1} \begin{bmatrix} \delta_2 - h_2 \\ \delta_3 - h_3 \\ \delta_4 - h_4 \end{bmatrix} \quad \text{(A4)}$$

[21] (p. 39).

The solutions to this reduced system of standard form equations are the final solutions to the precision position problem.

Triad in seven precision positions:

A triad in seven precision positions is by far the most mathematically complex of the compatibility linkage types listed here. As with the transition from four to five positions for a dyad, the transition from six to seven positions for a triad reduces the number of potential solutions from an infinite number (based on the infinite number of potential free choice values) to a finite value, Solutions come in sets of 0, 2, 4, or 6 depending on the intersections of the Burmester curves. As a result, most authors recommend refraining from using seven positions when synthesizing triads. The benefits of having a free choice almost always outweigh the cost of giving up a precision position. With that said, should a designer choose to proceed with the compatibility linkage approach to a triad in seven positions, they'll find that there are three compatibility equations that must be fulfilled. These equations are shown below, with loop one:

$$\boldsymbol{\Delta}_2 e^{i\beta_2} + \boldsymbol{\Delta}_3 e^{i\beta_3} + \boldsymbol{\Delta}_4 e^{i\beta_4} + \boldsymbol{\Delta}_5 e^{i\beta_5} + \boldsymbol{\Delta}_1 = 0 \tag{A5}$$

Loop Two:

$$\boldsymbol{\Delta'}_2 e^{i\beta_2} + \boldsymbol{\Delta'}_3 e^{i\beta_3} + \boldsymbol{\Delta'}_4 e^{i\beta_4} + \boldsymbol{\Delta}_5 e^{i\beta_6} + \boldsymbol{\Delta'}_1 = 0 \tag{A6}$$

Loop Three:

$$\boldsymbol{\Delta''}_2 e^{i\beta_2} + \boldsymbol{\Delta''}_3 e^{i\beta_3} + \boldsymbol{\Delta''}_4 e^{i\beta_4} + \boldsymbol{\Delta}_5 e^{i\beta_7} + \boldsymbol{\Delta''}_1 = 0 \tag{A7}$$

[21] (p. 115).
where,

$$\boldsymbol{\Delta}_1 = -\boldsymbol{\Delta}_2 - \boldsymbol{\Delta}_3 - \boldsymbol{\Delta}_4 - \boldsymbol{\Delta}_5 \tag{A8}$$

$$\boldsymbol{\Delta'}_1 = -\boldsymbol{\Delta'}_2 - \boldsymbol{\Delta'}_3 - \boldsymbol{\Delta'}_4 - \boldsymbol{\Delta}_5 \tag{A9}$$

$$\boldsymbol{\Delta''}_1 = -\boldsymbol{\Delta''}_2 - \boldsymbol{\Delta''}_3 - \boldsymbol{\Delta''}_4 - \boldsymbol{\Delta}_5 \tag{A10}$$

Conveniently, as with the previous analyses of triads, each of these compatibility equations represents a five-bar chain in the compatibility linkage. As a result, the setup of the linkage, in this case, will feel quite familiar. As before, the angular displacements of some equivalent $\boldsymbol{\Delta}$ terms are equal. Now, however, a third link is added to each set, yielding ($\boldsymbol{\Delta}_2$, $\boldsymbol{\Delta}_2'$, $\boldsymbol{\Delta}_2''$), ($\boldsymbol{\Delta}_3$, $\boldsymbol{\Delta}_3'$, $\boldsymbol{\Delta}_3''$), and ($\boldsymbol{\Delta}_4$, $\boldsymbol{\Delta}_4'$, $\boldsymbol{\Delta}_4''$). In response to some stimulus in the mechanism, each of these links in the compatibility linkage will have the same displacement angle as the others in its set. This allows for the forming of several sets of parallelograms between the different chains formed from the compatibility equations. The chains start out in the form shown in Figure A3.

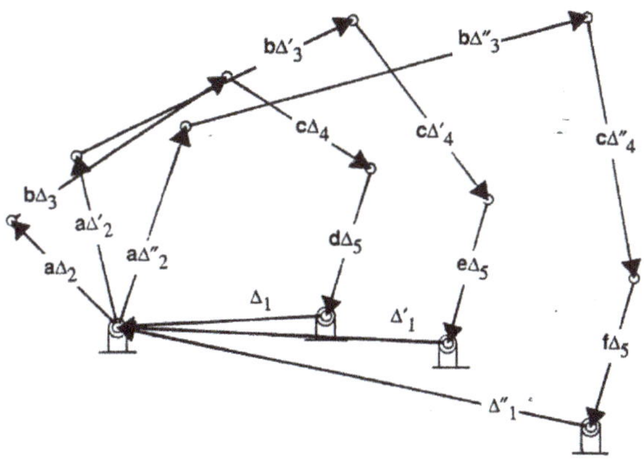

Figure A3. The three loops of the triad compatibility linkage for seven precision positions prior to modification. The end of link $\boldsymbol{\Delta}_1$ is selected as the common point [21] (p. 116).

After rotating the links to create alignment and creating the appropriate parallelograms between each layer of the loop, the linkage takes the form shown in Table 2. Links IJ, IH, MP, and PQ have all been created to form parallelogram loops. These parallel motion structures are created around links $\boldsymbol{\Delta}_1$, $\boldsymbol{\Delta}_2$, $\boldsymbol{\Delta}_3$, and $\boldsymbol{\Delta}_4$, which were each of the links that had identical angular displacement relationships. As with the lower-order structures, the designer may use this linkage by removing any of these created links, changing the chain from a 15-bar zero-degree-of-freedom structure to a 14-bar one-degree-of-freedom linkage. From here, the designer must identify the mechanism positions (outside of the starting position) in which the sides of the parallelogram from which the link was removed are parallel to each other. At these positions, the angular displacements of links CD, DE, EH, HM, KN, and LO from their starting positions correspond to β_2-β_7, respectively [21] (p. 115). Further simplifications of the compatibility linkage for a triad in seven precision positions remain an area for further study.

Appendix B. Special Cases of the Triad Compatibility Linkage

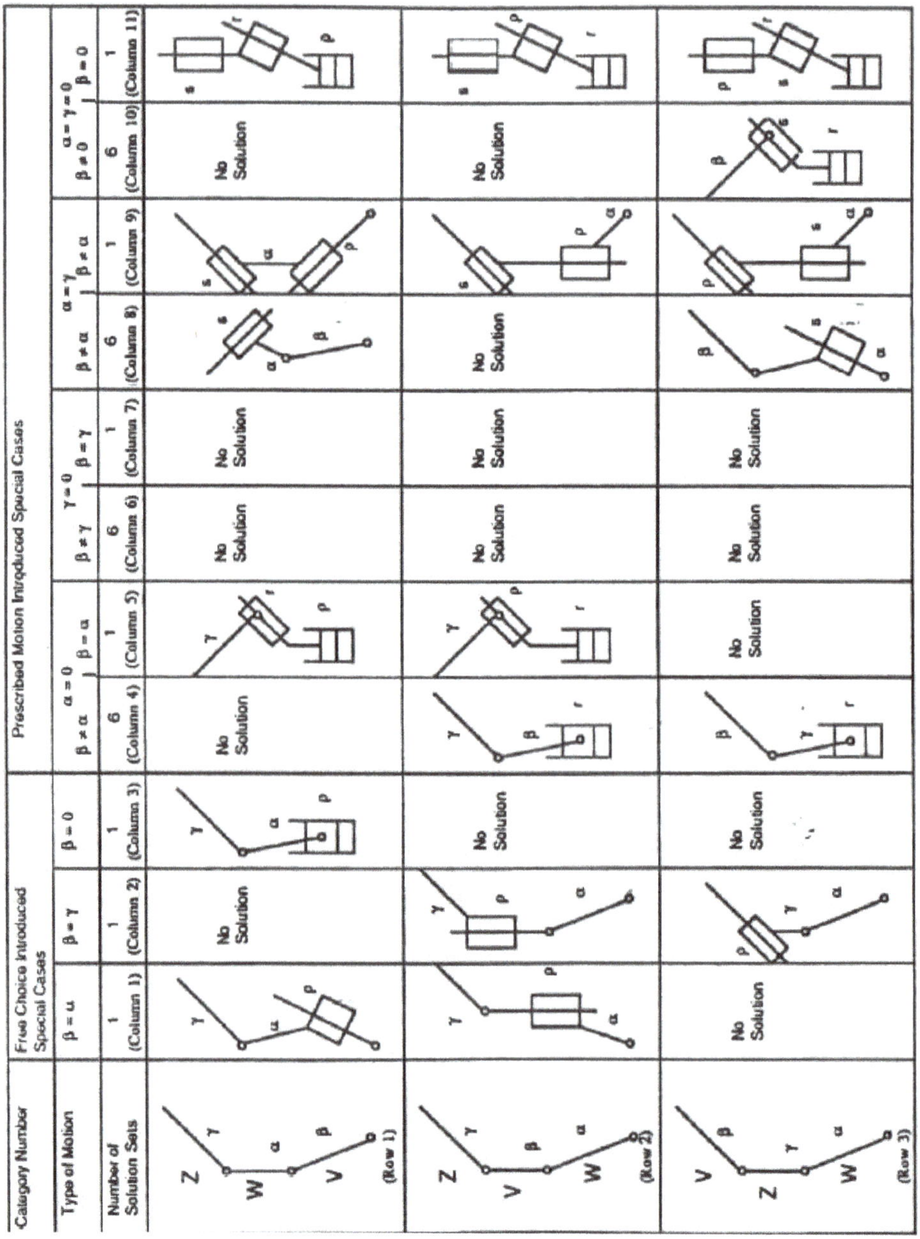

Figure A4. Special cases of the triad compatibility linkage [21] (p. 50).

References

1. Needham, J. *Science and Civilisation in China—Volume 4: Physics and Physical Technology*; Cambridge University Press: Cambridge, UK, 1971. Available online: https://archive.org/details/principlesmecha02willgoog (accessed on 7 November 2022).
2. Compass. Available online: https://education.nationalgeographic.org/resource/compass (accessed on 7 November 2022).
3. Willis, R. *Principles of Mechanism*; Nabu Press: Charleston, SC, USA, 2010. Available online: https://archive.org/details/principlesmecha02willgoog/page/n8/mode/2up (accessed on 7 November 2022).
4. Sandor, G.N.; Erdman, A.G. *Advanced Mechanism Design: Analysis and Synthesis*; Pearson: New York, NY, USA, 1984; Volume 2, ISBN 978-0130114372.
5. Wampler, C.W.; Morgan, A.P.; Sommese, A.J. Numerical Continuation Methods for Solving Polynomial Systems Arising in Kinematics. *J. Mech. Des.* **1990**, *112*, 59–68. [CrossRef]
6. Subbian, T.; Flugrad, D.R. Four-Bar Path Generation Synthesis by a Continuation Method. *J. Mech. Des.* **1991**, *113*, 63–69. [CrossRef]
7. Subbian, T.; Flugrad, D.R. Six and Seven Position Triad Synthesis Using Continuation Methods. *J. Mech. Des.* **1994**, *116*, 660–665. [CrossRef]
8. Pathak, V.K.; Singh, R.; Sharma, A.; Kumar, R.; Chakraborty, D. A Historical Review on the Computational Techniques for Mechanism Synthesis: Developments Up to 2022. *Arch. Comput. Methods Eng.* **2022**. [CrossRef]
9. Deshpande, S.; Purwar, A. Computational creativity via assisted variational synthesis of mechanisms using deep generative models. *J. Mech. Des. Trans. ASME* **2019**, *141*, 121402. [CrossRef]
10. Baskar, A.; Bandyopadhyay, S. An algorithm to compute the finite roots of large systems of polynomial equations arising in kinematic synthesis. *Mech. Mach. Theory* **2019**, *133*, 493–513. [CrossRef]
11. Zhao, P.; Ge, X.; Zi, B.; Ge, Q.J. Planar linkage synthesis for mixed exact and approximated motion realization via kinematic mapping. *J. Mech. Robot.* **2016**, *8*, 051004. [CrossRef]
12. Re, F.M.; Williams, J.O. Chaise Lounge Recliner Chair. U.S. Patent 5,090,768A, 17 July 1991.
13. Chase, T.R.; Erdman, A.G.; Riley, D.R. Triad Synthesis for up to Five Design Positions With Application to the Design of Arbitrary Planar Mechanisms. *J. Mech. Transm. Autom. Des.* **1987**, *109*, 426–434. [CrossRef]
14. Freudenstein, F.; Sandor, G.N. Synthesis of Path-Generating Mechanisms by Means of a Programmed Digital Computer. *J. Eng. Ind.* **1959**, *81*, 159–167. [CrossRef]
15. Hartenberg, R.; Denavit, J. *Kinematic Synthesis of Linkages*; McGraw-Hill: New York, NY, USA, 1964. Available online: https://idoc.pub/documents/r-hartenberg-j-denavit-kinematic-synthesis-of-linkages-1964pdf-jlk92zy1r745 (accessed on 10 November 2022).
16. Hayes, M.J.D.; Zsombor-Murray, P.J. Solving the Burmester Problem Using Kinematic Mapping. In Proceedings of the ASME 2002 International Design Engineering Technical Conferences and Computers and Information in Engineering Conference. Volume 5: 27th Biennial Mechanisms and Robotics Conference, Montreal, QC, Canada, 29 September–2 October 2002; pp. 1439–1446. [CrossRef]
17. Myszka, D.H.; Murray, A.P. Identifying sets of four and five positions that generate distinctive center-point curves. In Proceedings of the ASME 2009 International Design Engineering Technical Conferences and Computers and Information in Engineering Conference, San Diego, CA, USA, 30 August–2 September 2009; pp. 463–473. [CrossRef]
18. Chase, T.R.; Erdman, A.G.; Riley, D.R. Improved Centerpoint Curve Generation Techniques for Four-Precision Position Synthesis Using the Complex Number Approach. *J. Mech. Trans. Autom.* **1985**, *107*, 370–376. [CrossRef]
19. Schaaf, J.A.; Lammers, J.A. Geometric Characteristics of the Center-Point Curve Based on the Kinematics of the Compatibility Linkage. In Proceedings of the ASME 1992 Design Technical Conferences. 22nd Biennial Mechanisms Conference: Mechanism Design and Synthesis, Scottsdale, AZ, USA, 13–16 September 1992; pp. 475–480. [CrossRef]
20. Sakurai, S. Three-Position Variable Camber Krueger Leading Edge Flap. U.S. Patent 5,158,252A, 24 October 1991.
21. Lin, C.S. Dimensional Synthesis of a Planar Triad for Six Precision Positions. Ph.D. Thesis, University of Minnesota, Twin Cities, MN, USA, 1987.
22. Erdman, A.G.; Lonn, D. A Unified Synthesis of Planar Six-Bar Mechanisms Using Burmester Theory. In Proceedings of the Fourth World Congress on the Theory of Machines and Mechanisms, Newcastle Upon Tyne, UK, 8–12 September 1975; pp. 867–872.
23. Berneman, C.; Bican, L.; Howard, M.; Oladepo, I. Kinematic Synthesis of a Drone Tilt Mechanism. *Univ. Minn. Twin Cities Adv. Mech. Des.* **2022**, *1*, 1–29.
24. Rector, D.M. Linkage Mechanism Designer and Simulator. 2019. Available online: www.linkagesimulator.com (accessed on 6 September 2022).
25. Rank of a Matrix. Available online: https://www.cuemath.com/algebra/rank-of-a-matrix/ (accessed on 17 October 2022).
26. Erdman, A.G.; Chase, T.R. New Software Synthesizes Complex Mechanisms. *Mach. Des.* **1985**, *57*, 107–113.
27. Yu, N.; Erdman, A.G.; Byers, B.P. LINCAGES 2000: Latest Developments and Case Study. In Proceedings of the ASME 2002 International Design Engineering Technical Conferences and Computers and Information in Engineering Conference. Volume 5: 27th Biennial Mechanisms and Robotics Conference, Montreal, QC, Canada, 29 September–2 October 2002; pp. 1421–1425. [CrossRef]
28. Mirth, J.A.; Chase, T.R. Circuit Rectification for Four Precision Position Synthesis of Four-Bar and Watt Six-Bar Linkages. *J. Mech. Des.* **1995**, *117*, 612–619. [CrossRef]

29. Lin, C.-S.; Erdman, A.G.; Jia, B.-P. Use of Compatibility Linkages and Solution Structures in the Dimensional Synthesis of Mechanism Components. *Mech. Mach. Theory* **1996**, *31*, 619–635. [CrossRef]
30. Lu, X. A New Method to Group the Solutions from Dimensional Synthesis of Planar Triads for Six Prescribed Precision Positions. Master's Thesis, University of Alaska, Fairbanks, AK, USA, 1992; p. 118.
31. Mlinar, J.R.; Erdman, A.G. An Introduction to Burmester Field Theory. *J. Mech. Des.* **2000**, *122*, 25–30. [CrossRef]
32. Mlinar, J.R. An Examination of the Features of the Burmester Field and the Linear Solution Geometry of Dyads and Triads. Ph.D. Thesis, University of Minnesota, Twin Cities, MN, USA, 1997.
33. Peterson, R.; Logan, L.; Erdman, A.G.; Riley, D.R. Three Precision Point Synthesis of a Four Bar Linkage: An Example Using the LINCAGES-4 Program. In Proceedings of the Computers in Engineering 1988 Conference, San Francisco, CA, USA, 31 July–4 August 1988; pp. 91–96.
34. Erdman, A.G.; Gustafson, J.E. LINCAGES: Linkage Interactive Computer Analysis and Graphically Enhanced Synthesis Package. *Am. Soc. Mech. Eng. (Pap.)* **1977**, *77*, 77-DET-5.
35. Nelson, L.; Erdman, A.G. Recent Enhancements to the LINCAGES-6 Synthesis Package, including Circuit Rectification. In Proceedings of the ASME 1994 Design Technical Conferences collocated with the ASME 1994 International Computers in Engineering Conference and Exhibition and the ASME 1994 8th Annual Database Symposium. 23rd Biennial Mechanisms Conference: Mechanism Synthesis and Analysis, Minneapolis, MN, USA, 11–14 September 1994; pp. 263–271. [CrossRef]
36. Holte, J.E.; Chase, T.R.; Erdman, A.G. Approximate Velocities in Mixed Exact-Approximate Position Synthesis of Planar Mechanisms. *J. Mech. Des.* **1999**, *123*, 388–394. [CrossRef]
37. Holte, J.E.; Chase, T.R.; Erdman, A.G. Mixed Exact-Approximate Position Synthesis of Planar. *J. Mech. Des.* **2000**, *122*, 278–286. [CrossRef]
38. Tesar, D.; Sparks, J.W. Multiply Separated Position Synthesis. *ASME Paper* **1968**, *3*, 68-MECH-66. [CrossRef]

robotics

MDPI

Article

Kinematic Graph for Motion Planning of Robotic Manipulators

Burkhard Corves and Amir Shahidi *

Institute of Mechanism Theory, Machine Dynamics and Robotics (IGMR), RWTH Aachen University, 52062 Aachen, Germany
* Correspondence: shahidi@igmr.rwth-aachen.de

Abstract: We introduce a kinematic graph in this article. A kinematic graph results from structuring the data obtained from the sampling method for sampling-based motion planning algorithms in robotics with the motivation to adapt the method to the positioning problem of robotic manipulators. The term kinematic graph emphasises the fact that any path computed by sampling-based motion planning algorithms using a kinematic graph is guaranteed to correspond to a feasible motion for the positioning of the robotic manipulator. We propose methods to combine the information from the configuration and task spaces of the robotic manipulators to cluster the samples. The kinematic graph is the result of this systematic clustering and a tremendous reduction in the size of the problem. Hence, using a kinematic graph, it is possible to effectively employ sampling-based motion planning algorithms for robotic manipulators, where the problem is defined in higher dimensions than those for which these algorithms were developed. Other barriers that hindered adequate utilisation of such algorithms for robotic manipulators with articulated arms, such as the non-injective surjection of the forward kinematic function, are also addressed in the structure of the kinematic graph.

Keywords: motion planning; sampling-based motion planning algorithms; heuristic search; robotic manipulators; open-chain mechanisms

Citation: Corves, B.; Shahidi, A. Kinematic Graph for Motion Planning of Robotic Manipulators. *Robotics* **2022**, *11*, 105. https://doi.org/10.3390/robotics11050105

Academic Editor: Raffaele Di Gregorio

Received: 19 August 2022
Accepted: 28 September 2022
Published: 5 October 2022

1. Introduction

Robotic applications are fundamentally characterised by the planned motion of the systems. Therefore, the study of motion planning algorithms has been active in the robotic community since the early stages of robotic research. The problem of motion planning has been dealt with both analytical and sampling-based planning approaches. Analytical motion planning addresses the problem of the motion within both aspects of geometrical and temporal transition [1]. This can further be seen as a one- or multi-dimensional problem. In the field of robotics, a one-dimensional problem is suitable for planning in the space of generalised joint coordinates of the robot, where the multi-dimensional problem deals with the applications of motion planning in $\mathcal{T}$-space of the robot. Analytical planning algorithms have proven to be applicable for optimising the motion of the system in continuous space. However, their success is particularly subject to preliminary parametrisation of the problem. Moreover, in practical cases, where the task of the manipulator is defined in $\mathcal{T}$-space of the robot, they are dependent on inverse kinematic algorithms, as the robotic systems are actually controlled in the space of generalised joint coordinates. Furthermore, checking the collision states in $\mathcal{O}$-space ($\mathcal{O}$-space refers to the spatial volume occupied by the robot in all its feasible configurations) demands knowledge of the overall configuration of the system.

The planning algorithms primarily rely on the $\mathcal{C}$-space of the robotic manipulators [2,3], as per definition, the position of every point on the structure and the entire configuration of the robot can be represented as a point $q \in \mathcal{C}$-space. Conventionally, motion planning algorithms, such as optimisation-based planning (e.g., potential fields [4]) and combinatorial planning [5] (e.g., cell decomposition [6]), are conducted on an exact and explicit descriptions of $\mathcal{C}$-space. This explicit formulation is, however, computationally expensive and sophisticated mathematical operations are required to compute a plan. In addition, it renders the scalability of such algorithms to higher dimensions impractical. Moreover, the transformation of obstacles into $\mathcal{C}$-space can be very complicated and challenging [7]. Hence, these algorithms demand complex collision detection [8].

Sampling-based planning algorithm [9] circumvent the expenses of planning in explicit descriptions of $\mathcal{C}$-space by conducting simple BOOLEAN tests on samples drawn from $\mathcal{C}$-space to perform collision detection. The admissible sets of the configurations, that is, $\boldsymbol{q}$ belonging to $\mathcal{C}_{\text{free}}$ that results in the connection of the initial state (configuration) of the robot to a final state (configuration), that satisfies the goal posture conditions, will result in the path ($\tau\colon\ [0, 1] \rightarrow \mathcal{C}_{\text{free}}$). To evaluate the performance of the sampling-based planning algorithm a common theoretical evaluation criteria is the *completeness* of the algorithm. Complete algorithms report if a solution exists in a finite amount of time and return one if there exists one [5], which can only be fulfilled by combinatorial algorithms, because they rely on an exact description of the $\mathcal{C}$-space. A weaker definition applies for sampling-based planning algorithm, called *resolution completeness*, that states that the algorithm reports the existence of the solution depending on the sampling resolution. There exists yet another definition, *probabilistic completeness*, that states that the probability of finding the solution, if there exists one, approaches to one.

The sampling of $\mathcal{C}$-space can be performed through *probabilistic* or *deterministic* (*regular* and *irregular*) techniques [5]. Due to the discrete character of the finite samples, to evaluate the quality for representation of $\mathcal{C}$-space, that is, the coverage of $\mathcal{C}$-space, there is yet another criterion to be considered, namely *denseness*. We handle the subject in detail in Section 1.3.2. The samples drawn via the above-mentioned methods construct a graph $\mathfrak{G}(\mathfrak{V}, \mathfrak{E})$, with the vertices $\mathfrak{V}$ being the drawn samples and the edges $\mathfrak{E}$ being the connectivity information of the samples. Neighbouring and parenting methods should be defined while constructing the graph.

Obviously, the state of research in the field of sampling-based planning algorithm has focused mainly on the $\mathcal{C}$-space of the robot. However, the tasks of the robotic systems are defined in the $\mathcal{T}$-space, being the configuration space of the task at hand. Hence, in practical robotic problems, finding a feasible path in the $\mathcal{T}$-space of the robotic manipulator is preferable. Furthermore, even though sampling-based planning algorithm are of a discrete nature and an explicit transformation of the obstacles from $\mathcal{O}$-space into $\mathcal{C}$-space is not necessary, BOOLEAN collision checks should be carried out by calling the Forward Kinematics (FK) (see Section 3.4). In addition, FK is generally surjective but not injective, and thus a pure planning in $\mathcal{T}$-space cannot guarantee a configurational collision-free motion of the robot in $\mathcal{O}$-space.

Motivation of developing the kinematic graph is, to the extent of its contribution, to elevate the shortcomings of sampling-based planning algorithms for mechanisms with open-chain topology, such as robotic manipulators with articulated arms, and make efficient application of such planners in higher dimensions possible. The scope of this study limits itself to that of the positioning problem of the manipulators, more precisely, the positioning problem of the regional structure of the decoupled structures (see Section 1.3). In the following, our motivation to develop the Kinematic Graph (KG) is detailed and the state-of-the-art developments with similar motivations, i.e., the planners that utilise the $\mathcal{T}$-space information in planning, will be presented. Afterwards, the scope of this work will be discussed using three aspects. First, we introduce the manipulator structures that can benefit best from the structure of the proposed graph. Nevertheless, this structural spectrum is not exhaustive and is limited to cases of the most practical manipulator structures. Then, the subproblem of the sampling-based planning algorithm will be introduced and the ones that are extended based on our proposed approach will be identified. Finally, we summarise and conclude with our contributions.

1.1. Motivation

There are powerful and efficient methods developed for heuristic-based (and incremental) search algorithms for efficient motion planning in lower dimensions. Examples of such methods are those that deal with applications of mobile robotics in dynamic environments, for instance navigation [10], and dynamic A* [11]. A majority of the sampling-based planning algorithms do not use these informed search algorithms, and sacrifice cost minimisation in favour of high-speed planning. Additionally, they are basically relying on random sampling of the $\mathcal{C}$-space. The main deficit of most developed sampling-based planning algorithm is, however, that they rely merely on the $\mathcal{C}$-space of the system.

In the literature, mainly the high dimensionality of the sampling-based planning algorithm for robotic manipulators is mentioned as the basic hurdle of the extension of these planners for the case studies of manipulation. This applies, of course, as these algorithms are developed just taking the $\mathcal{C}$-space of the systems into account and completely ignore the essential differences of the cardinalities of the $\mathcal{C}$-space and $\mathcal{T}$-space of the robotic manipulators. The tasks of the manipulators are logically defined in their $\mathcal{T}$-space, and hence, this space is best suitable to perform planning and deal with planning-related aspects such as collision avoidance. Our hypothesis is that *respecting the differences of the cardinalities of $\mathcal{C}$-space and $\mathcal{T}$-space and overlaying the information from these spaces will result in a tremendous reduction in the size of the problem and thus make sampling-based planning algorithm an appropriate method for efficient motion planning for manipulators with open-chain topology*. The results from Section 3.2 prove the correctness of this hypothesis.

The restriction that the planning space, namely the $\mathcal{C}$-space, taking into account the non-injective surjection of the FK, imposes to sampling-based planning algorithm for the case of manipulation problems has not been given appropriate consideration when the planning needs to be carried out in $\mathcal{T}$-space. Strictly speaking, due to the non-injective surjection of the forward kinematics function ($\mathscr{K} : \mathcal{C} \rightarrow \mathcal{O}$), a pure planning in $\mathcal{T}$-space can guarantee neither a collision-free motion of the robot nor a feasible motion in consideration of actuator limits. Hence, it is generally desirable to avoid using analytical Inverse Kinematics (IK). Some approaches, however, attempt to find a roadmap in the layers of multiple answers to IK by approximating the search space in these layers and performing minimisation on distances between the intended (given) path of end-effector and the answers to FK [12]. The alternatives of using the analytical IK are the numerical solutions to IK and kinematic control schemes that prove to be very efficient [13–15]. For the former case, the chosen solver and sensitivity of the solver to the initial guess ($q \in \mathcal{C}$) should be given special attention, and in the latter case, an efficient geometrical modelling of the system is of great importance. Nevertheless, the motion to be fed into the numerical IK, the kinematic control schemes, is to be planned purely in the $\mathcal{T}$-space of the manipulator, yet not directly consisting of the information from $\mathcal{C}$-space. There are also alternatives to IK, e.g., pseudo-IK [16], which rely on the answers from IK numerical solutions. The approach presented in [16] also attempts to minimise the number of discontinuities in path when mapping from $\mathcal{T}$-space to $\mathcal{C}$-space. The motivation behind the development of the KG is to enable feasible planning in terms of both $\mathcal{C}$-space and $\mathcal{T}$-space motion, i.e., *for any path that is generated using this graph, it is guaranteed that a motion for positioning can be computed that is kinematically feasible for the robotic manipulator to execute, i.e., no discontinuities in the path occur*.

Another motivation for the development of the KG is to exploit the advantages of the heuristic-based search algorithms to perform efficient motion planning for the manipulators, on the one hand, and to enable "natural" motion of the robot and to increase the repeatability of solutions with similar initial postures and goals, on the other hand. The matter is stated in Section 1.3.2 and discussed in detail in Section 3.

1.2. Similar Works

There have been recent attempts to utilise $\mathcal{T}$-space information in sampling-based planning problems to achieve more practical motion. Berenson et al. [17] defined the goal region in the $\mathcal{O}$-space of the robot, where the goal posture is defined, rather than in its $\mathcal{C}$-space. Two approaches based on rapidly-exploring random trees (RRT) [18] are implemented: one using JACOBIAN-based gradient descent toward this goal region and the other based on a bi-directional RRT.

Cohen et al. [19] compute the heuristics in EUCLIDEAN space of $\mathcal{T}$-space using breadth-first search. This approach to computation of the heuristics has proven to also be beneficial for avoiding collisions in cluttered environments. In this approach, the heuristics are computed via an additional simplified search in a reduced dimension of the $\mathcal{T}$-space of the system.

Rickert et al. [20] present an approach called an exploring/exploiting tree that makes an adjustment in sampling according to the information form $\mathcal{O}$-space. This way, a balance is realised between exploitation (enhancing existing solutions) and exploration (searching for new solutions). The completeness of the planner is, however, traded off for computational efficiency.

To perform palletising through robotic systems, Scheurer and Zimmermann [21] decompose the $\mathcal{O}$-space into cylindrical cells and search for the collision-free path through these cells. The path in $\mathcal{C}$-space of the robot is regenerated using IK. It seems that the feasibility of a $\mathcal{O}$-space path is assumed.

A hierarchical path planner based on an exact representation of $\mathcal{O}$-space for collision avoidance in conjunction with exploration on $\mathcal{C}$-space was developed by Mesesan et al. [22].

In addition, Ref. [23] introduces a hierarchical structure for encoding configuration-to-workspace mapping information for collision checking of an enormous number of samples during operation, enabling real-time path planning for robots.

1.3. Scope

1.3.1. Manipulator Structure

Within the scope of this article, we study the case of a motion planning problem for robotic manipulators where the planning problem addresses either the positioning of the end-effector of the mechanism or the positioning of a *point of interest* on the mechanism. The former case is common for planar robotic manipulators or the mechanisms that are employed to perform gantry-like manipulation, such as a Scara robot. The latter case is common for manipulators that demonstrate decoupled structures. For the case of such manipulators, it is a common practice in robotics to consider the positioning (the task of the regional structure, i.e., the articulated arm) and orientation (the task of the local structure, i.e., the wrist) problems separately but in conjunction with each other. With this background in mind, the algorithm presented in this article addresses the problem of path planning for regional structure. Strictly speaking, we handle the problem of the positioning of the point of interest, but we do not treat the orientation problem in this article and leave the detailed elaboration of this problem to our future work. We call the point of interest for the manipulators that demonstrate decoupled structures the centre of the wrist or the centre of the local structure and symbolise it with $\mathrm{P_W}$. In the balance of this article, without loss of generality, we consider the $\mathrm{P_W}$ as the point of interest, because always a wrist can be added to manipulators built for positioning purposes. For the simulations and discussion on the results of the planned motion, we use the structure of the open-chain robotic manipulator designed at IGMR, named IGOR (blog.rwth-aachen.de/robotik/en/igor, visited on 19 August 2022).

The manipulators demonstrating the decoupled structures can be identified easily based on the DENAVIT–HARTENBERG (DH) parameters [24]. There are different interpretations and extensions to DH parameters, e.g., [25,26], or some attempts to make them "less ambiguous", e.g., [27]. Here, we refer to the classical form of the DH parameters. That is, the links for a multibody system with open-chain involving n bodies are numbered $0, 1, \ldots, n$; joint i connecting bodies $i-1$ and i; and the coordinate system i is attached to link $i-1$. Overviews of the modelling of the multibody systems can be found in, e.g., [28,29]. As a common definition of robotic manipulators with decoupled structures (for simplicity in the indexing, and without loss of generality, let us consider the robotic manipulators that are designed to perform six-dimensional (6D) tasks and have six bodies connected with six joints), the ones are meant that demonstrate $a_4 = a_5 = d_4 = 0$ ([30], Section 4.4). These systems actually do have a decoupled structure with a spherical wrist (see, e.g., Figure 1a), and it is possible to develop analytical IK for such structures. Nonetheless, it is possible to extend the application of the KG for manipulators with DH parameters $a_4 = a_5 = 0$, i.e., the ones that do not have spherical wrists (see, e.g., Figure 1b), but the trace of the end-effector positions is left on the surface of a torus-shaped manifold centred at the origin of the fifth coordinate system and has a radius equal to d_4 and thus will be the point of interest. In this case, the centre of the surface will be the point of interest P_w.

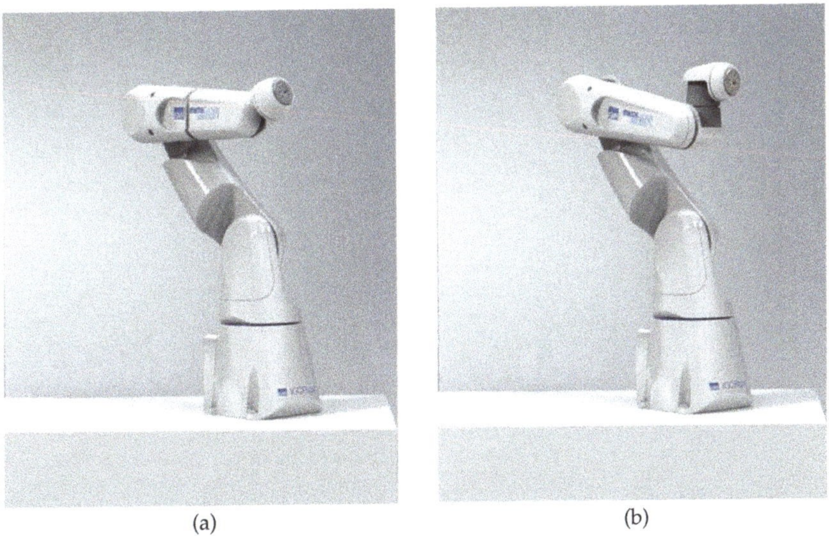

Figure 1. IGOR: the open-chain robotic manipulator designed at IGMR (**a**) with spherical wrist; (**b**) with non-spherical wrist.

1.3.2. Graph Structure

Any sampling-based planning algorithm is composed of six main subproblem sequences as follows: sampling strategy, calculation of metrics, finding neighbours, parent allocation, strategies for collision detections, and exploration strategy itself [31].

To perform sampling, we use the same procedure as that for the conventional deterministic regular sampling of the $\mathcal{C}$-space. This specific sampling strategy is chosen here, motivated by the fact that the closure of the *dense* regular deterministic samples yield $\mathcal{C}$-space, as each limit sample (interior $\cup$ boundary samples) in this set represents an adherent point (or closure point). *Denseness* implies that the samples can get arbitrarily close to any configuration of the mechanism [5]. To sample the $\mathcal{T}$-space, we use convex sampling of the enclosing environment of the manipulator. The vertices of the KG, however, are not composed merely of any of these samples, but the combinations of information from these samples (for details see Section 2).

Each vertex on the graph denotes a state (or configuration) of the manipulator. Thus, the metrics calculation refers to the spatial and/or temporal distances or any other kinematic, kinetostatic, or dynamic effort (such as transition in potential energy of the system, change in manipulability, etc.) that is induced to the system through this state transition.

Neighbouring strategy for the KG is originated in a neighbouring strategy of the samples from $\mathcal{C}$-space, but in consideration of the vertices of the graph (for details see Section 2.1). The parent assignment strategies are left to the exploration algorithm, e.g., here, the back propagation of the A*.

A discrete search algorithm, such as DIJKSTRA's algorithm [32], A* algorithm [33] or their modern versions, any-time repairing (ARA*) [34], lifelong planning A* (LPA*) [35] or dynamic A* (D*) [11,36] for deterministic, and rapidly exploring dense-trees (RRT/RDT) [18], and probabilistic roadmap planners (PRM) [37] for probabilistic methods, realises the exploration strategy of the algorithm in the graph. In this article, we pursue the use of a heuristic search algorithms on the generated graph, as they provide theoretical guarantees such as completeness and optimality of the delivered solutions. Moreover, based on the heuristic function the number of evaluations is limited, as the most effective next actions are chosen, and thus, the amount of time the algorithm needs to compute a plan will be reduced [38]. Henceforth, for the exploration strategy itself, we use the A* search algorithm [33], and no extension will be proposed to the search algorithm. However, this introduces another subproblem that needs to be taken into account: the introduction of an informative heuristic. Definition of a powerful and informative heuristic can have a tremendous effect on the quality and predictability or repeatability of the motion (see Sections 3 and 4).

1.3.3. Contribution

The contribution of this article is the presentation of the novel structure of the graph to be fed to the search algorithm of the sampling-based planning algorithm. We present the idea of combining information from both the $\mathcal{C}$-space and $\mathcal{T}$-space of the robot for the construction of a graph structure dubbed *Kinematic Graph* $\mathfrak{G}_k\,(\mathfrak{V}_k,\,\mathfrak{E}_k)$, where the vertices $\mathfrak{V}_k$ inherit the information from both $\mathcal{C}$-space and $\mathcal{T}$-space, and edges $\mathfrak{E}_k$ are originated from connectivity information from $\mathcal{C}$-space. The KG is to be constructed a priori. This graph that is developed specifically for mechanisms with open-chain topology, is proven to keep the promises in Section 1.1. The most important problems are enabling of the efficient and complete employment of the sampling-based planning algorithm in higher dimensions, and the guarantee of the feasibility of the planned motion for these mechanisms. The efficiency of the KG over the $\mathcal{C}$-space-based graphs are presented in detail in Section 3. To the extent of our knowledge, no graph structure with such premises based on the combination of the information from the $\mathcal{C}$-space and the $\mathcal{T}$-space of the robotic manipulators has been presented as of yet. This article extends our conference manuscript [39] by presenting the detailed algorithm of the construction of the KG which was visually presented and extensive evaluation of the performance of KG. In this vein, the comparison between KG and traditional $\mathcal{C}$-space-based graphs is discussed in detail in Section 3. To facilitate the visualisation of the motion of the mechanisms using KG, the results are explained in detail using a two-DoF mechanism. The extension of the result to spatial mechanisms is discussed in the Applications. Moreover, we present approaches to compute the costs and heuristics and perform collision avoidance and practical illustration of the implementation of the KG.

2. Materials and Methods

Kinematic graph $\mathfrak{G}_k\,(\mathfrak{V}_k,\,\mathfrak{E}_k)$ promotes ideas to meet the challenges of sampling-based planning algorithm for open-chain mechanisms due to the non-injective surjection of FK by introducing spatial information from $\mathcal{T}$-space directly into the $\mathfrak{V}_k$. Therefore, both $\mathcal{C}$-space and $\mathcal{T}$-space of the robot should be sampled. In the following, the algorithm to construct the KG is elaborated in detail.

2.1. Kinematic Graph—The Algorithm

Algorithm 1 requires the kinematical model of the mechanism. The forward kinematic function $\mathscr{K}$ should be provided such that it solves the positioning problem of the point of interest of the manipulator (end-effector or P_w). Furthermore, the sampling resolution of the $\mathcal{C}$-space and $\mathcal{T}$-space should be provided. As will be elaborated throughout this Section, these parameters can be considered to be the regulating parameters for the construction of the KG. To construct the KG, samples are to be drawn from both the $\mathcal{C}$-space and the $\mathcal{T}$-space of the robot. Although the same terminology (sampling) is used for both spaces, the sampling procedure and concept for these spaces are different from each other, to be discussed in the following. Finally, the reach of the P_w should also be determined.

Algorithm 1 Construction of kinematic graph $\mathfrak{G}_k\,(\mathfrak{V}_k, \mathfrak{E}_k)$.

```
input: 𝒦, C-space, q, C_res, T_res, and r
output: 𝔊_k (𝔙_k, 𝔈_k)
N = SampleConfigSpace(q, C_res)                              ▷ See, e.g., List 2
for n ∈ N do
    n.pos ← 𝒦(n.q)                                           ▷ 𝒦 : C → O
V = DiscretiseEnvironment(r, T_res)                          ▷ See, e.g., List 2
V = ApproximateTaskSpace(V, N, T_res)
𝔙_k = ConstructVertices(V)
ConstructkinematicGraph(𝔙_k)
procedure ApproximateTaskSpace(V, N, T_res)
    repeat
        n = leastn(N)
        for v ∈ V do
            if ||n.pos − v.cent||_∞ ≤ T_res/2 then
                v.N_v ← v.N_v ∪ {n}
    until N = ∅
    return V = {v ∈ V | v.N_v ≠ ∅}
procedure ConstructVertices(V)
    c_idx ← 0
    repeat
        v = leastv(V)
        repeat
            n = (v.N_v).randpop()
            n.c_idx ← c_idx
            C ← ∅
            C ← C ∪ {n}
            C.c_idx ← c_idx
            C.v.cent ← v.cent
            O ← neighbours(n) ∩ v.N_v
            repeat
                n = O.randpop()
                n.c_idx ← c_idx
                C ← C ∪ {n}
                O ← (O ∪ (neighbours(n) ∩ v.N_v)) \ C
            until O = ∅
            c_idx ← c_idx + 1
            𝔙_k ← 𝔙_k ∪ {C}
        until v.N_v = ∅
    until V = ∅
    return 𝔙_k
procedure ConstructKinematicGraph(𝔙_k)
    repeat
        C = 𝔙_k.randpop()
        addVertex(𝔊_k, C)
        repeat
            for n ∈ neighbours(N_C.randpop()) do
                C' = c ∈ 𝔙_k : c.c_idx = n.c_idx
                if C' ≠ C then
                    addEdge(𝔊_k, 𝔢(C, C'))
        until N_C = ∅
    until 𝔙_k = ∅
    return 𝔊_k (𝔙_k, 𝔈_k)
```

First, the $\mathcal{C}$-space should be sampled. As stated in Section 1.3, we use the conventional deterministic regular sampling of the $\mathcal{C}$-space. To perform the sampling, as described in **procedure** *SampleConfigSpace* in List 2, a function (here `meshgridx(•)`) is to be defined such that it generates a regular grid from the elements of the vectors of an input matrix. The vectors of this matrix should contain the sequence of numbers ($\in \mathbb{R}$) representing the discretisation of jth the generalised coordinate of $\boldsymbol{q}$ with the resolution of $\mathcal{C}_{\text{res}}$. The drawn samples will be treated as *objects* and are called *nodes* in the following. The closure of these nodes yields the set N. Each node n is identified with an index n_{idx} and stores, at this stage, the generalized coordinates of $\mathcal{C}$-space, i.e., $\boldsymbol{q}$ of the sample it represents. For these nodes, a neighbouring strategy should then be determined. After samples are drawn from $\mathcal{C}$-space, the spatial coordinate of P_w should be stored in each n by applying the forward kinematics function $\mathscr{K}$ (see Algorithm 1, Line 2).

List 2 List of the auxiliary procedures and functions utilised in Algorithm 1.

procedure *SampleConfigSpace*(q, $\mathcal{C}_{res}$)
  $J_{\mathcal{C}} \leftarrow [\,]$
  for $j \in q$ **do**
    j = `discretise`(j_{min}, j_{max}, $\mathcal{C}_{res}$)
    $J_{\mathcal{C}}$.`append`(j^T)
  return N = `meshgridx`($J_{\mathcal{C}}$)

`discretise(•)`: returns a sequence of numbers between $\bullet_1$ and $\bullet_2$ with resolution $\bullet_3$ as a vector.

`meshgridx(•)`: returns the set N via a deterministic regular grid from input matrix •.

procedure *DiscretiseEnvironment*(r, $\mathcal{T}_{res}$)
  V = `meshgridw`($r + \epsilon, \mathcal{T}_{res}$)
  for v ∈ V **do**
    $v.N_v \leftarrow \emptyset$
  return V

`meshgridw(•)`: returns the set V via a convex discretisation of $\mathcal{T}$-space with resolution •.

`leastn(N)`: returns the node n ∈ N with smallest EUCLIDEAN norm of n.pos and removes it from N.

`leastv(V)`: returns the voxel v ∈ V with smallest EUCLIDEAN norm of v.cent and removes it from V.

`•.randpop()`: returns a randomly selected member from the set • and removes it from the set •.

`neighbours(n)`: returns a set containing the neighbour nodes of node n, based on neighbouring strategy in $\mathcal{C}$.

`addVertex`($\mathfrak{G}$, $\mathfrak{v}$): adds the vertex $\mathfrak{v}$ to the graph $\mathfrak{G}$.

`addEdge`($\mathfrak{G}$, $\mathfrak{e}(\mathfrak{v}_1, \mathfrak{v}_2)$): adds the edge $\mathfrak{e}$ between the vertices $\mathfrak{v}_1$ and $\mathfrak{v}_2$ in the graph $\mathfrak{G}$.

After sampling the $\mathcal{C}$-space, the $\mathcal{O}$-space should be sampled. The cardinal character of this space, in conjunction with its relationship with $\mathcal{C}$-space via $\mathscr{K}$, motivates the sampling via convex discretisation of $\mathcal{O}$-space (see **procedure** *DiscretiseEnvironment* in List 2). Therefore, a function `meshgridw(•)` should be defined that basically works with the same logic as that of function `meshgridx(•)` and returns the closure of the *voxels* objects v in a set V. Voxels are determined with their *centroids*. Note, however, that these voxels should be guaranteed to spatially enclose the reach r of the point P_W. Hence, in case of the regular cubic discretisation of $\mathcal{O}$-space, a value ϵ should be added to r, i.e., the discretisation of $\mathcal{O}$-space should enclose the sphere with radius $r + \epsilon$. As a final touch for the preparation of the voxels, to each voxel v, a set $v.N_v$ is allocated that contains the transformed nodes from $\mathcal{C}$-space ($\mathscr{K} : \mathcal{C} \rightarrow \mathcal{O}$) and is initialised to the empty set $\emptyset$.

Next, the $\mathcal{T}$-space of the mechanism can be approximated as the set of voxels that are occupied with nodes, based on information from n.pos (see **procedure** *ApproximateTaskSpace* in Algorithm 1). To determine the nodes that belong to each voxel, an assignment strategy should be determined. For voxels of a cubical shape, infinity norm ($\|\bullet\|_\infty$) can be used (see Algorithm 1 Line 11f). Ties can be broken arbitrarily. Here, according to the definition of the function `leastn`(N) in List 2, ties are broken in favour of the voxel with a smaller EUCLIDEAN norm of the centroid of the voxel v.cent. The closure of the nodes in each voxel forms the set $\mathrm{v.N_v}$ and is to be added as an attribute to the voxel object. Finally, the set of voxels V can be reduced to the set of occupied voxels, i.e.,

$$\mathrm{V} = \{\mathrm{v} \in \mathrm{V} \,|\, \mathrm{v.N_v} \neq \emptyset\}. \tag{1}$$

Now, the closure of V forms the approximated $\mathcal{T}$-space.

Construction of the vertices of the $\mathfrak{G}_k$, i.e., $\mathfrak{V}_k$

Whereas the representation of the nodes in $\mathcal{C}$-space has a "nice" regular distribution, the distribution of the transformed nodes in $\mathcal{T}$-space is rather disarranged, for $\mathscr{K}$ is generally a nonlinear function. Moreover, the non-injective surjection of $\mathscr{K}$ transforms the nodes from "different neighbourhoods" of $\mathcal{C}$-space to the "same neighbourhood" (respectively, same spot) in $\mathcal{T}$-space. The concept behind the KG is to find these "clusters" based on information from both $\mathcal{C}$-space and $\mathcal{T}$-space and store them as the vertices of the $\mathfrak{G}_k$, i.e., $\mathfrak{V}_k$, and connect them to each other via information from $\mathcal{C}$-space and store the connectivity information as the edges of the $\mathfrak{G}_k$, i.e., $\mathfrak{E}_k$). First, we start with finding the clusters in each voxel. The **procedure** *ConstructVertices* in Algorithm 1 illustrates the logic of clustering. The logic is as follows: *in each voxel ($\mathcal{T}$-space information) find and cluster all the nodes that are continuously connected to each other in $\mathcal{C}$-space ($\mathcal{C}$-space information), that is, you can traverse between them continuously in $\mathcal{C}$-space.* The clustering of the samples from $\mathcal{C}$-space is valid, because (i) the joints of the robot are assumed to move continuously and (ii) KG is constructed a priori, hence the entire $\mathcal{C}$-space is assumed to be $\mathcal{C}_{\text{free}}$ (see Section 3.4 for performing collision avoidance in KG). To ease the bookkeeping of the clusters and later the procedure of finding edges between the clusters, we use an index $\mathrm{c_{idx}}$ to be assigned to each cluster and to each node that belong to these clusters. Then, we iterate through the nodes of each voxel in the set V. This is elaborated in the second **repeat** loop in the **procedure** *ConstructVertices* in Algorithm 1. For each voxel, we initialise a cluster with $\emptyset$ and add the set $\{\mathrm{c_{idx}}, \mathrm{v.cent}\}$ to it. Then, we draw/remove a random node from the voxel. This will be the initial stage of gathering the nodes in this cluster. This node is added to the cluster, along with all the $\mathcal{C}$-space neighbours of this node that also belong to $\mathrm{v.N_v}$, to a temporary set called the open set O, of the nodes for clustering, i.e.,

$$\mathrm{O} \leftarrow \mathtt{neighbours}(\mathrm{n}) \cap \mathrm{v.N_v}. \tag{2}$$

Then, we repeat the same process, but this time over the elements of O, while skipping the addition of the already existing nodes in this cluster, i.e.,

$$\mathrm{O} \leftarrow (\mathrm{O} \cup (\mathtt{neighbours}(\mathrm{n}) \cap \mathrm{v.N_v})) \setminus \mathrm{C}, \tag{3}$$

until there are no more neighbour nodes to be found in this voxel (see the third **repeat** loop in the **procedure** *ConstructVertices* in Algorithm 1). Note that the implementation of **procedure** *ConstructVertices* in Algorithm 1 is basically the breadth-first search (BFS), with the termination condition that the nodes should be neighbour of each other in the $\mathcal{C}$-space. Then, we add the cluster (which is a set of nodes on its own) to the set of $\mathfrak{V}_k$ and repeat the process for the remaining nodes in the voxel, until it is empty. Then, we iterate over all voxels of V.

Observe the following important properties:

- The KG abstracts the samples from $\mathcal{C}$-space to the clusters of the samples from $\mathcal{C}$-space that reach the voxels in $\mathcal{T}$-space with different configurations;
- Each cluster belongs to merely one voxel in $\mathcal{T}$-space.

Note that the essential limitation of the sampling-based planning algorithm applies. Generally, the collisions can effectively be checked only for the available clusters of the samples, and not for the path segments connecting them to each other. There are, however, effective methods developed to check the collision in path segments (see, e.g., [5], Section 5.3.4).

Construction of the edges of the $\mathfrak{G}_k$, i.e., $\mathfrak{E}_k$

The final step to be taken for construction of $\mathfrak{G}_k$ is finding the connectivity between clusters and form the $\mathfrak{E}_k$. The logic is illustrated in **procedure** *ConstructKinematicGraph* in Algorithm 1. The logic is as follows: *for each cluster, find all the clusters in $\mathfrak{V}_k$ that contain at least one node that is a neighbour of a node in the cluster in $\mathcal{C}$-space ($\mathcal{C}$-space information).* This test can be performed easily using the index c_{idx} that has been stored in the clusters and nodes during the construction of the vertices. Thus, the neighbour clusters of a cluster are those that have an identical c_{idx} to the nodes in the clusters, i.e.,

$$\mathrm{C}' = \mathrm{c} \in \mathfrak{V}_k : \mathrm{c.c}_{idx} = \mathrm{n.c}_{idx}. \tag{4}$$

If this is a different cluster, add an edge to $\mathfrak{E}_k$, connecting C and C′, while avoiding duplications (see **procedure** *ConstructKinematicGraph* in Algorithm 1, Line 38f and function `addEdge`($\mathfrak{G}_k, \mathfrak{e}(\mathrm{C}, \mathrm{C}')$) in List 2). Iterate over the clusters of $\mathfrak{V}_k$ and the nodes or neighbour nodes in the cluster returned by function `neighbours`(n) (see List 2).

3. Discussion

Evaluation of the kinematic graph will be performed in this section for both the KG on its own and the performance of the planned motion based on the path generated using the KG on planar and spatial mechanisms. The software was developed in Python using the graph-tool python library [40]. We determine what sort of cost and heuristic functions can be utilised based on the structure of the vertices and the edges of the KG, $\mathfrak{G}_k\,(\mathfrak{V}_k, \mathfrak{E}_k)$.

3.1. *Shape of the Kinematic Graph*

The KG contains the information from both $\mathcal{C}$-space and $\mathcal{T}$-space, and thus can be plotted in both spaces. An exemplarily sketch of the KG is depicted in Figure 5 in Section 3.4. Figure 5a demonstrates the plot of the graph based on the average CARTESIAN coordinates of the P_w in $\mathcal{T}$-space. To make the overlaying vertices $\mathfrak{V}_k$ (clusters) visible, a small offset is imposed. Figure 5b demonstrates the KG in $\mathcal{C}$-space, which can also be interpreted as the "unfolded" version of the KG in $\mathcal{T}$-space. An overview of the colour coding related to the search algorithm is presented in Section 3.4.

3.2. *Computational Complexity*

Theoretical worst-time computational complexity, with the assumption that the algorithm has to explore the entire graph to find the optimal solution, of the search algorithms is $O(b^d)$. In this expression, b is the *branching factor* and is determined by the average number of neighbours, or successors, of a vertex in the graph in which the search is performed, and d is the *solution depth,* that is, the shortest path between the start and the goal vertices. This notation may seem to be not of much use though, except that, due to the exponential relation between b and d, it is clear that solving problems with significant b and d values, as in the case of sampling-based planning algorithm for robotic manipulators with articulated arms, becomes computationally intangible. This fact is the most referenced barrier of utilisation of such algorithms for manipulation problems (see Section 1.1). This worst-time computational complexity provides us, however, with some intuition to evaluate the KG.

3.2.1. Branching Factor

Based on the MOORE neighbourhood strategy, the number of neighbours of a state in a two-dimensional (2D) problem, e.g., the positioning problem of a planar mechanism or a holonomic mobile robot, is eight. This is implied from the fact that the state of each dimension can remain unchanged or can move in positive and negative directions (restrictions apply at the boundaries). Hence, the simple formula of

$$b = |succ\_states|^{dim} - 1, \tag{5}$$

where $|succ\_states|$ is the number of the successor states of each one-dimensional action that can be derived to evaluate the average number of neighbours of a state (for a 2D problem: $b = 3^2 - 1 = 8$).

Obviously, the branching factor increases exponentially. For a three-dimensional (3D) problem, the branching factor will be $b = 3^3 - 1 = 26$. In the case of 6D manipulation problems, the branching factor will be $b = 3^6 - 1 = 726$. The essential effect of dividing the problem into positioning and orientation and applying the sampling-based planning algorithm for the positioning problem, can now be inferred.

3.2.2. Solution Depth

The significant reduction in the dimension of the problem based on the clustering can basically be deduced from Algorithm 1. In this Section, we attempt to present a quantification of this result. Let us consider a planar two-DoF mechanism. Although the topology of the $\mathcal{C}$-space of this mechanism is of non-EUCLIDEAN shape (it has the shape of a torus), it can be parametrised and represented as a 2D EUCLIDEAN space, with axes of θ_1 and θ_2. The construction of the KG can be regulated based on two parameters: $\mathcal{C}_{\text{res}}$ and $\mathcal{T}_{\text{res}}$. For the sake of simplicity, let us consider the same limits for the joints of the mechanism, $\{\theta_1, \theta_2\} \in \{[-\pi, \pi]^{\text{T}}, [-\pi, \pi]^{\text{T}}\}$. A specific number of joint values ($|jv|$) will be generated, based on deterministic regular discretisation of the joint range and $\mathcal{C}_{\text{res}}$. For instance, for $\mathcal{C}_{\text{res}} = \pi$, $|jv|_{\{\mathcal{C}_{\text{res}}=\pi\}} = 3$. Let us consider three different $\mathcal{C}_{\text{res}} = 2.0°$, $1.0°$, and $0.5°$. Then, $|jv|_{\{\mathcal{C}_{\text{res}}=2.°\}} = 181$, $|jv|_{\{\mathcal{C}_{\text{res}}=1.°\}} = 361$, and $|jv|_{\{\mathcal{C}_{\text{res}}=.5°\}} = 721$.

The first step of the construction of the KG is generating a $\mathcal{C}$-space graph, $\mathfrak{G}_{\mathcal{C}}\ (\mathfrak{V}_{\mathcal{C}}, \mathfrak{E}_{\mathcal{C}})$. Based on the MOORE neighbourhood strategy, the number of vertices ($|\mathfrak{V}_{\mathcal{C}}|$) and edges ($|\mathfrak{E}_{\mathcal{C}}|$) of this graph can be computed as follows (note that this is physically an undirected graph) [41]:

$$|\mathfrak{V}_{\mathcal{C}}| = |jv|^2, \tag{6}$$

and,

$$|\mathfrak{E}_{\mathcal{C}}| = 4|jv|^2 - 6|jv| + 2. \tag{7}$$

Thus, $|\mathfrak{V}_{\mathcal{C}}|_{\{\mathcal{C}_{\text{res}}=2.°\}} = 32{,}761$, $|\mathfrak{V}_{\mathcal{C}}|_{\{\mathcal{C}_{\text{res}}=1.°\}} = 130{,}321$, and $|\mathfrak{V}_{\mathcal{C}}|_{\{\mathcal{C}_{\text{res}}=.5°\}} = 519{,}841$. Also, $|\mathfrak{E}_{\mathcal{C}}|_{\{\mathcal{C}_{\text{res}}=2.°\}} = 129{,}960$, $|\mathfrak{E}_{\mathcal{C}}|_{\{\mathcal{C}_{\text{res}}=1.°\}} = 519{,}120$, and $|\mathfrak{E}_{\mathcal{C}}|_{\{\mathcal{C}_{\text{res}}=.5°\}} = 20{,}570{,}440$.

Now we continue to generate the $\mathfrak{G}_{\text{k}}\ (\mathfrak{V}_{\text{k}}, \mathfrak{E}_{\text{k}})$ by regulating the second parameter, $\mathcal{T}_{\text{res}}$. We consider four different $\mathcal{T}_{\text{res}} = 0.1$m (coarse) , 0.05m, 0.03m, and 0.01m (fine). The number of the vertices and edges, $|\mathfrak{V}_{\text{k}}|$ and $|\mathfrak{E}_{\text{k}}|$ respectively, generated via Algorithm 1 and the relation between the number of the dimensions of $\mathfrak{G}_{\text{k}}$ and the underlying $\mathfrak{G}_{\mathcal{C}}$ are summarised in Table 1. For the examples in the following sections, the applied resolutions for contraction of the KG are $\mathcal{C}_{\text{res}} = 1.$ and $\mathcal{C}_{\text{res}} = 0.05$.

For most (almost all well-determined) pairs of $\mathcal{C}_{\text{res}}$ and $\mathcal{T}_{\text{res}}$ the values of the third and fifth columns are rather small, suggesting that the KG, $\mathfrak{G}_{\text{k}}$, is much smaller than its underlying joint space graph, $\mathfrak{G}_{\mathcal{C}}$. This is, however, not a surprising result, and merely a quantification of the logic of construction of the KG. This also has a tremendous effect on the reduction in the storage size of the KG, enabling us to store more information, such as the ones that are necessary for direct collision checks in $\mathcal{T}$-space and hence preventing overhead computation time for calling $\mathscr{K}$ functions in online applications.

Table 1. Quantified evaluation of the size of the KG.

$\mathcal{C}_{res} \mid \mathcal{T}_{res}$	$\lvert\mathfrak{V}_k\rvert$	$\frac{\lvert\mathfrak{V}_k\rvert}{\lvert\mathfrak{V}_\mathcal{C}\rvert}$	$\lvert\mathfrak{E}_k\rvert$	$\frac{\lvert\mathfrak{E}_k\rvert}{\lvert\mathfrak{E}_\mathcal{C}\rvert}$
2. ∣ 0.1	581	0.0177	1745	0.0134
2. ∣ 0.05	2441	0.0745	7289	0.0561
2. ∣ 0.03	6579	0.2008	23,153	0.1782
2. ∣ 0.01	30,561	0.9328	120,303	0.9257
1. ∣ 0.1	601	0.0046	1760	0.0034
1. ∣ 0.05	2389	0.0183	7134	0.0137
1. ∣ 0.03	6534	0.0501	19,597	0.0377
1. ∣ 0.01	64,348	0.4937	237,516	0.4575
0.5 ∣ 0.1	577	0.0011	1704	8.3e-5
0.5 ∣ 0.05	2353	0.0045	7031	0.0003
0.5 ∣ 0.03	6510	0.0125	19,635	0.0009
0.5 ∣ 0.01	57,827	0.1112	183,843	0.0089

3.3. Cost and Heuristic Functions

To calculate the costs of the transitions between different states C and C′ (where C′ is a successor of C, i.e., C′ ∈ succ(C)) and the heuristic functions (which estimate the cost of the shortest path from each state to reach the goal), metric functions are to be determined (see Section 1.3.2) that are quantitatively dependent on the states themselves. Logically, the first pair of choices to be stored in each state are the averaged values of the CARTESIAN coordinates of the P_w

$$\mathrm{C.pos} = \overline{\mathrm{N_C.pos}}, \tag{8}$$

and joint states

$$\mathrm{C.}q = \overline{q(\mathrm{N_C})}. \tag{9}$$

Moreover, the average CARTESIAN coordinates of any critical point on the mechanism can also be stored, as an example of such, the CARTESIAN coordinates of the elbow of the articulated arm. This specific information will help twofold: (i) checking the collision states directly in $\mathcal{T}$-space (see Section 3.4) and (ii) determining the influence of the arm orientation during the planning for the orientation problem of the wrist.

The above local information can be utilised to determine the cost and heuristic functions based on different distance metrics, such as the EUCLIDEAN norm in the corresponding spaces. In addition to this local information, any other indicator based on kinetic, kinematic, or kinetostatic performance criteria can be stored in the KG to achieve a desired motion. It is important to note that the heuristic-based search algorithms require admissible and consistent heuristic functions. This is fulfilled when the amount of the heuristic function in each state does not overestimate the cost of the shortest path from each state to reach the goal and consequently amounts to zero at the goal state.

The commonly used evaluation criteria in the literature are predominantly devoted to the kinetostatic performance values, specifically the manipulability of the robotic manipulators (see, e.g., [30], Section 5.8). These criteria are fundamentally functions of the JACOBIAN matrix of the manipulators, J, and can generally be split into positional and orientational parts. The geometrical shape of these manipulabilities are ellipses (2D) or ellipsoids (3D). The direction and the length of the principal semi-axes of these ellipses/ellipsoids evaluate the quality of velocity transmission in the corresponding directions. Hence, this provides us with a good measurement for evaluate the distance to the singularities (where the area/volume vanishes) and the directions that lead to them. There can be different measures defined for quantifying the manipulability ellipsoids based on the singular values of

J, i.e., the eigenvalues of JJ^{T}. The computationally favourable index is the area/volume of the ellipses/ellipsoids, which amounts to

$$\mu = \sqrt{\left|JJ^{\mathrm{T}}\right|}, \tag{10}$$

where $|\bullet|$ refers to the determinant of the matrix $\bullet$ ([14], Section 5.4). Henceforth, in the balance of this article we refer to this measure when the manipulability of mechanism is mentioned.

Based on different metrics that one may use in the planning process, distinct evolution of configurational motion is expected for the motions with similar start configurations to the goal posture of the end-effector. Furthermore, the computation time, i.e., the amount of the exploration of the search space to find the optimal motion, is primarily dependent on the selected criteria. Figure 2 demonstrate four different motions of a two-DoF planar mechanism with similar start configurations and the goal postures of the end-effector. The trace of the path of the end-effector ($\tau : [0, 1] \rightarrow \mathcal{T}$) is demonstrated with solid lines evolving from green to grey. Obviously, these heuristic function fulfil the conditions of the admissibility of the heuristic functions for the heuristic-based search algorithms in the corresponding spaces in which they are defined.

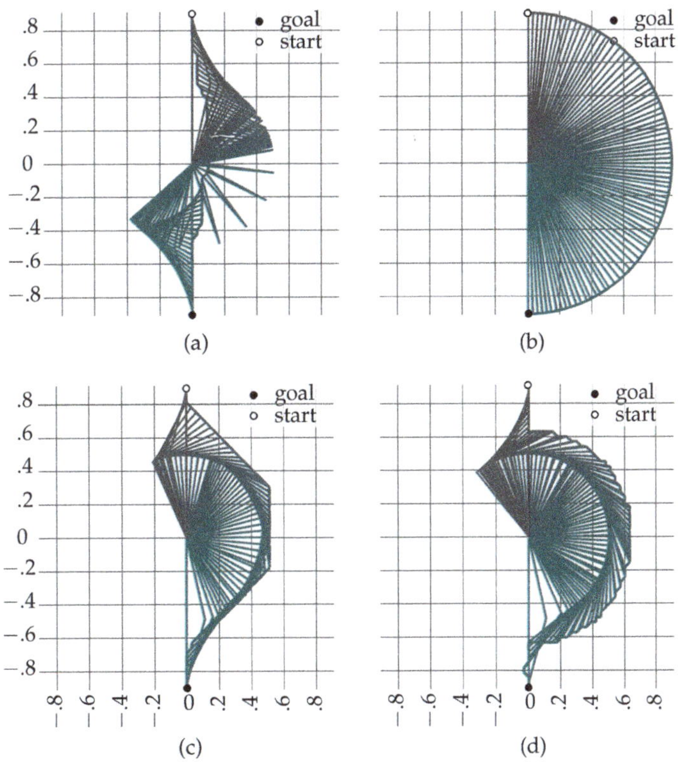

Figure 2. The motion of a two-DoF planar mechanism. (**a**) Cost and heuristic functions: distance in $\mathcal{T}$-space. (**b**) Cost and heuristic functions: distance in $\mathcal{C}$-space. (**c**,**d**) Cost and heuristic functions: combination of the distance in $\mathcal{T}$-space and the linear manipulability of the mechanism.

When the costs and heuristic functions are set to be the distances in $\mathcal{C}$-space, merely 8.36% of the search space is explored. This amount rises to 17.34% for the case of $\mathcal{T}$-space exploration. It should be noted that, as mentioned in Section 1.1, it is desired not to call on the IK. However, this is necessary to be able to calculate the $\mathcal{C}$-based heuristics in this scenario. Here, in this example and the example shown in Figures 4b and 6b, we use the $\mathcal{C}$-based heuristics just for demonstrative purposes. Nevertheless, it is observable in the literature that these heuristics has been utilised often. This is not of a great importance in practical scenarios, however, because the KG enables us to use powerful heuristics based on $\mathcal{T}$-space and other performance criteria of the mechanisms.

Now, let us consider a combinatorial pair of costs and heuristics: distance travelled in $\mathcal{T}$-space and the manipulability of the mechanism. Here, special attention should be devoted to this combination. The main task of the heuristic function is guiding the search towards the goal posture of the end-effector, where the function's value is zero. This seems to be trivial in the context of heuristic functions that are a function of the distances travelled in $\mathcal{C}$-space and $\mathcal{T}$-space: the functions sink towards the goals, and the distance yet to be travelled to reach the goal at the goal posture of the end-effector is zero. The distributions of the manipulability in $\mathcal{T}$-space are, however, geometrically in hyperbolic shapes of different dimensions. Hence, a simple distance function leads the search either to the portions of the $\mathcal{T}$-space with higher manipulabilities or even to singularities. Nonetheless, a combination is very beneficial. For instance, we can exploit the distance function in $\mathcal{T}$-space to provide information on the direction of the exploration towards the goal and ensure the admissibility of the heuristic function and a reformed manipulability function to guide the search towards the portions of the $\mathcal{T}$-space with higher manipulabilities. Examples of such are

$$\mathscr{C}(\mathrm{C}, \mathrm{C}') = \|\mathrm{C.pos} - \mathrm{C}'.\mathrm{pos}\|_2^2(\mu_{max} - \mu(\mathrm{C}'.\boldsymbol{q})), \tag{11}$$

and

$$\mathscr{H}(\mathrm{C}) = \|\mathrm{P}_{\mathrm{W}g}.\mathrm{pos} - \mathrm{C.pos}\|_2^2(\mu_{max} - \mu(\mathrm{C}.\boldsymbol{q})), \tag{12}$$

with $\mathrm{P}_{\mathrm{W}g}$.pos and $\|\bullet\|_2$ representing the goal postures of the points P_W and l_2 EUCLIDEAN norm, respectively. Alternatively, the inverse of the maximum manipulability (of course not at and near the configuration singularities) can be used. The flow of the heuristic for the combinatorial case of Figure 2d is illustrated in Figure 3. The arrows and the dotted line on the diagrams show, qualitatively sketched, the path of the end-effector (in the $\mathcal{C}$-space and the $\mathcal{T}$-space, respectively) that the planner outputs. Note that the amount of heuristic at the goal posture of the end-effector is equal to zero. In this case, 20.91% of the search space was explored. Qualitatively, however, this motion is the most "natural" motion generated amongst the previous motions.

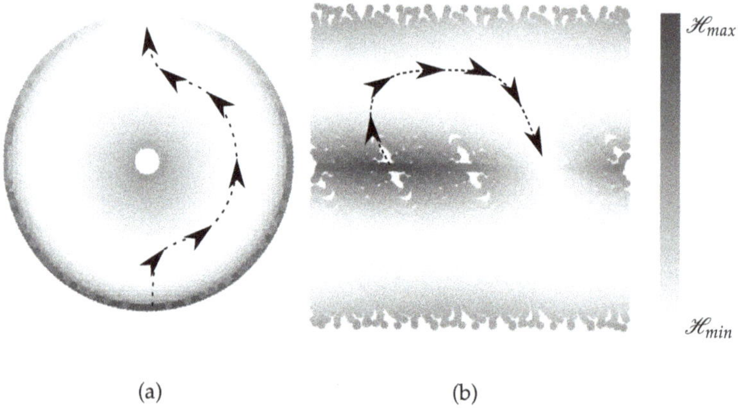

Figure 3. The flow of heuristic as combination of distance in $\mathcal{T}$-space and linear manipulability of two-DoF mechanism. (**a**) $\mathcal{T}$-space representation. (**b**) $\mathcal{C}$-space representation.

3.4. Collision Avoidance

In the conventional approaches of sampling-based planning algorithm, explicit transformations of the obstacles from $\mathcal{O}$-space to $\mathcal{C}$-space are not performed. Instead, BOOLEAN checks should be conducted to examine whether a specific configuration causes any collision in $\mathcal{O}$-space. Hence, a call to $\mathscr{K}$ is inevitable. Geometrical relations are followed to determine the spatial occupation of the bodies of the mechanism.

When planning using the KG, we are able to store the necessary information of critical points of the mechanism (see Section 3.3). The construction of the KG is conducted a priori, and thus, in online applications the BOOLEAN collision checks reduce to the geometrical calculations to determine the spatial occupation of the mechanism that can be directly performed in $\mathcal{O}$-space. Therefore, planning using KG outperforms conventional planning in $\mathcal{C}$-space in terms of computational efficiency for collision detection in online single-query applications.

Figure 4 demonstrates cases where there is a $\mathcal{O}$-space obstacle in the $\mathcal{T}$-space of the mechanism. For this scenario, it is *not* possible to plan a feasible motion using purely the $\mathcal{T}$-space information. The meaning of the colours are the same as those in Section 3.3.

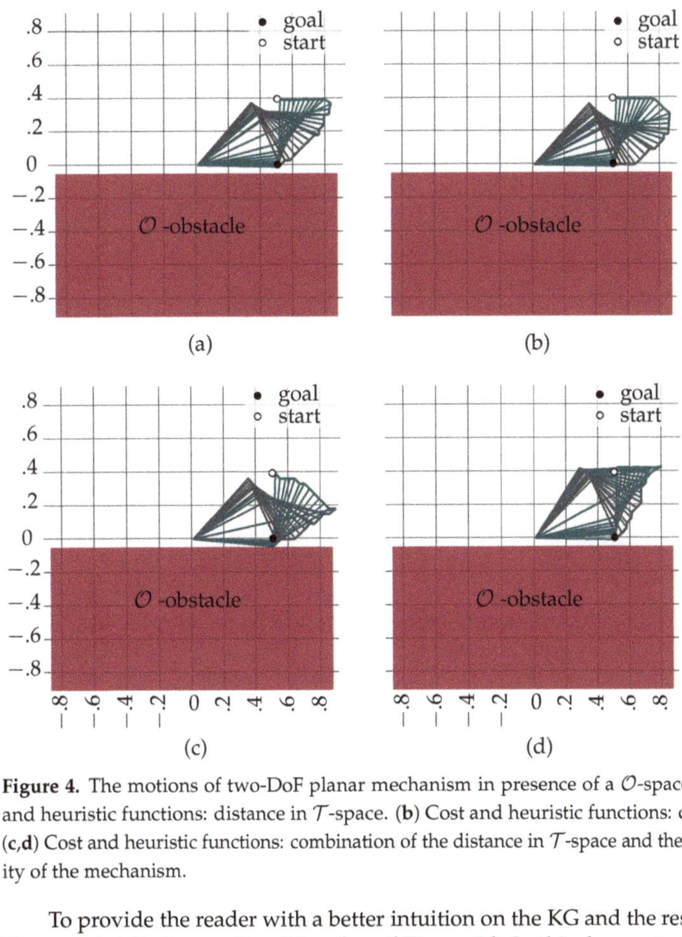

Figure 4. The motions of two-DoF planar mechanism in presence of a $\mathcal{O}$-space obstacle; (**a**) Cost and heuristic functions: distance in $\mathcal{T}$-space. (**b**) Cost and heuristic functions: distance in $\mathcal{C}$-space. (**c**,**d**) Cost and heuristic functions: combination of the distance in $\mathcal{T}$-space and the linear manipulability of the mechanism.

To provide the reader with a better intuition on the KG and the result of the search, Figure 5 details the results of the plan of Figure 4d. In this demonstration, the obstacle is mapped entirely into the KG, and the collision check does not follow the instruction described above for online single-query applications. The black vertices are those that cause collision with obstacles, the green vertices are those that are expanded during the search, and the yellow vertices are those that are met (are in priority queue) but not expanded. The paths are shown in red.

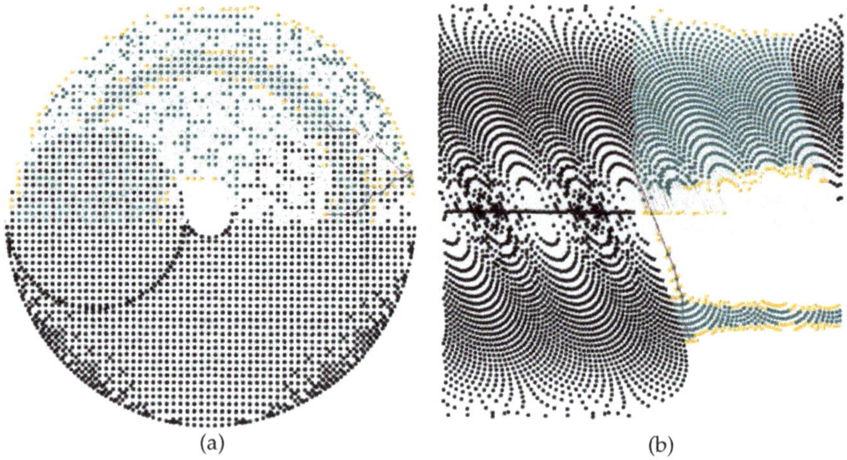

Figure 5. Demonstration of the complete KG $\mathfrak{G}_k$ ($\mathfrak{V}_k$, $\mathfrak{E}_k$), involving search results for the example of Figure 4d. (**a**) Demonstration of the KG in $\mathcal{T}$-space (for the vertices with the same CARTESIAN coordinate a small offset is imposed to avoid complete overlap). (**b**) Demonstration of the KG in $\mathcal{C}$-space.

Close investigation of these figures reveal the properties of the KG detailed in the balance of this article. As an example, observe the vertices with the same $\mathcal{T}$-space CARTESIAN coordinate (i.e., the overlay in $\mathcal{T}$-space), of which some may cause collision (with configurational interpretation of "elbow-down" in this case) and some may lie on the path (with configurational interpretation of "elbow-up" in this case).

3.5. *Limitation of Kinematic Graph*

3.5.1. "Holes" in the Kinematic Graph

The construction of the KG is subject to the regulation of two parameters: $\mathcal{C}_{res}$ and $\mathcal{T}_{res}$. Whereas $\mathcal{C}_{res}$ determines with which and how many, if any, configurations the structure reaches the spatial regions of the $\mathcal{T}$-space, $\mathcal{T}_{res}$ determines the size of the clusters. If there are no configurations that reach a specific segment of the $\mathcal{T}$-space where voxels are generated based on $\mathcal{T}_{res}$, these voxels remain empty, which results in generation of "holes" in the $\mathcal{T}$-space representation of the KG. This phenomenon occurs basically when the $\mathcal{C}$-space is sampled "coarsely" and the $\mathcal{T}$-space is sampled "finely". This can lead to a problem where no start index can be found in the KG in the case that the point P_w lies on a "hole" at the start configuration of the mechanism because this voxel does not belong to the approximated $\mathcal{T}$-space, i.e., no index c_{idx} can be found to start the search.

3.5.2. Sparsity in Configuration Space

A feasible path is the one that can be traced by the mechanism, that is, the one that takes the physical limitation of the actuators of the mechanism into consideration. The path planner, hence, should generate paths that correspond to configurational executable motions of the mechanism from the start configuration to the goal posture of the end-effector. Due to the clustering process, the $\mathcal{C}$-space representation of the KG can be sparse in the vicinity of the configuration singularities of the mechanism. This phenomenon can be comprehended by inspection of Figure 5b. The reason for this phenomenon is the small change in the position of point P_w in the $\mathcal{C}$-space in the vicinity of the singularities. As the clustering was performed based on the movements encoded in the $\mathcal{T}$-space, if a path traverses in the vicinity of a configuration singularity, or passes through it, moving from one cluster to the next, it demands large steps in the $\mathcal{C}$-space. This may lead to violation of the physical limitation of the actuators of the mechanism. It is worthwhile to mention that this is an essential limitation due to the configuration singularity of the mechanism and can be mitigated by performing interpolation on the generated path segments in post-processing steps. Besides that, collision avoidance in the path segments should be given due attention (see, e.g., [5], Section 5.3.4).

3.5.3. Completeness

The method developed in this article is primarily suitable for the positioning problem of robotic manipulators. The consideration of the problem as a decentralised problem for positioning and orientation is given, based on the discussions of Section 1.1. As discussed in Section 3.2.1, this has a significant effect on the simplification of the problem. However, when solving the problem for the whole manipulator (regional and local structure), the completeness of the algorithm can be guaranteed merely for the positioning problem. This is, however, a theoretical limitation, and not a practical one [20].

4. Applications

Application of the KG will be extended in this section to the more practical cases of robotic manipulators, specifically the ones introduces in Section 1.3. We mention different practical issues that the reader may face during the implementation of the KG and convey the experienced best practices to facilitate the implementation of KG.

4.1. Implementation for Spatial Robotic Manipulators

Let us start with examples that demonstrate the application of the KG for spatial manipulators. For the examples in this section, we again consider the same cost and heuristic functions introduced in Section 3.3.

In the examples, we simulate a relatively cluttered environment to challenge the search algorithm (see Figure 6a). The meaning of the colours are the same as those in Section 3.3. A quick comparison of the results of these experiments reveal some similarities with those demonstrated in Figure 2. For instance, when the costs and heuristics are set to be the minimum distance travelled in $\mathcal{C}$-space, the manipulator tends to move across the borders of its $\mathcal{O}$-space, minimizing the movements in $\mathcal{C}$-space while sacrificing the manipulability. The behaviour in the case of minimum distance travelled in $\mathcal{T}$-space is also comparable to that of the 2D scenario, moving in the vicinity of the base of the manipulator. Additionally, the combination of cost and heuristic functions based on $\mathcal{T}$-space information and manipulability of the manipulator results in the most "natural" and "predictable" behaviour of the manipulator.

4.2. Best Practices for Implementation of the Kinematic Graph

4.2.1. Finding the Start Index in the Kinematic Graph

In the initial configuration ($\boldsymbol{q}_i$), the CARTESIAN position of point $\mathrm{P_W}$ may not be of much use to determine the start index in KG, as there may be several vertices (clusters) at the same position (due to reachability of the position with several configurations). In this case, the start index $\mathrm{c_{idx}}$ is the answer to the optimisation function

$$\mathrm{c_{idx}} = \underset{\mathrm{c_{idx}}}{\arg\min}(\boldsymbol{q}_i - \mathrm{C}.\boldsymbol{q}). \tag{13}$$

4.2.2. Finding the Goal Vertex When Planning in Configuration Space

If the CARTESIAN posture of the goal ($\mathrm{P_{W}}_g$.pos) is determined in $\mathcal{T}$-space but the attempt is to plan a path in the $\mathcal{C}$-space, the voxel v in which the goal posture finds itself can be found via

$$\left\|\mathrm{P_{W}}_g.\mathrm{pos} - \mathrm{v.cent}\right\|_\infty \leq \frac{\mathcal{T}_{\mathrm{res}}}{2}. \tag{14}$$

The vertices with the same v.cent are the potential goal vertices in the KG, and the one with the least EUCLIDEAN norm of the distance from $\boldsymbol{q}_i$ to C.$\boldsymbol{q}$ (minimal geodesic on $\mathcal{C}$) is the actual goal vertex in the $\mathcal{C}$-space, if the minimal geodesic is collision-free. This may be helpful when the calculation of heuristic functions from $\mathcal{C}$-space is desirable.

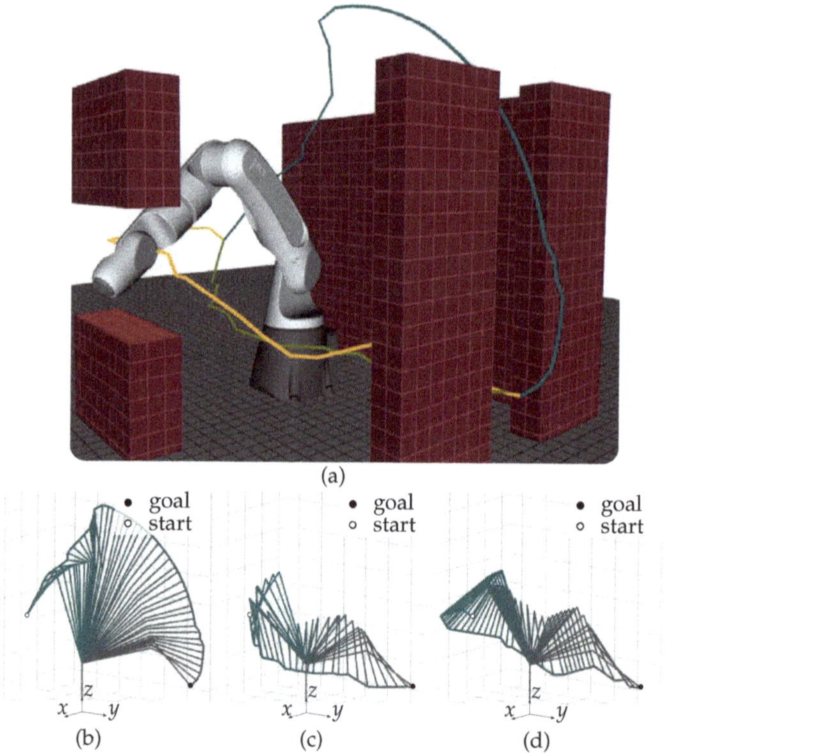

Figure 6. Examples of implementation of the sampling-based planning algorithm for the spatial robotic manipulator IGOR. (**a**) The oblique view of the experimental set up. (**b**) Cost and heuristic functions: distance in $\mathcal{C}$-space. Amount of exploration: 22.66% of the KG (the output path of the P_w in the $\mathcal{T}$-space is shown in Figure (**a**) in dark green). (**c**) Cost and heuristic functions: distance in $\mathcal{T}$-space. Amount of exploration: 4.69% of the KG (the output path of the P_w in the $\mathcal{T}$-space is shown in Figure (**a**) in light green); (**d**) Combination of distance in $\mathcal{T}$-space and linear manipulabilities of the mechanism based on (12). Amount of exploration: 11.3% of the KG (the output path of the P_w in the $\mathcal{T}$-space is shown in Figure (**a**) in orange).

4.2.3. Defining the Stop Criteria When Planning Explicitly in Task Space

The planning procedure returns a path when some condition at the goal posture of the end-effector is satisfied. If the goal posture is explicitly given as P_{wg}.pos, then the computation of the path is concluded when

$$\left\|\text{C.v.cent} - P_{wg}.\text{pos}\right\|_2^2 \leq \delta, \tag{15}$$

with δ being a small positional tolerance. This is the case for the mechanisms designed for positioning tasks.

4.2.4. Defining the Stop Criteria When Planning Implicitly in Task Space

When the goal posture is not given explicitly but the complete posture of the end-effector is, then two conditions should be satisfied:

- The position of the end-effector (*ee*.pos) should be reachable from P_{W_g}.pos, i.e., *ee*.pos should be on the surface of the sphere/torus-shaped manifold covering point P_W;
- The collision of the wrist with the articulated arm should be addressed. This check can be performed using the position of the elbow.

4.2.5. Reaching the Goal

When the goal posture of the manipulator is given explicitly in the $\mathcal{T}$-space, it is not likely that the P_{W_g}.pos coincides with a vertex of the KG. Nevertheless, the output can be considered as a "perfect" initial guess for the numerical solution of the goal posture. A better practice is, however, using kinematic control loops (see, e.g., [15]). This is done by feeding the post processed path from the sampling-based planning algorithm with the KG by combining it with the motion plan of the wrist into the kinematic control loop.

5. Conclusions

In this article, we have presented a detailed introduction to the structure of novel graph dubbed *Kinematic Graph KG*. We have analysed the performance of the KG and have shown that the KG holds the premises arisen from the motivation of developing it, including, but not limited to, the following:

- Any path that is generated using the KG is guaranteed to correspond to a feasible motion that is kinematically and configurationally feasible for the robotic manipulator to execute (however, the issues of collision avoidance in path segments should be considered). That is, planning using the KG is not affected by the hindrances due to the non-injective surjection of the forward kinematics function for mechanisms with open-chain topology, such as robotic manipulators with articulated arms;
- Using the KG, it is possible to effectively employ sampling-based planning algorithm for robotic manipulators, i.e., the problem of higher dimensions;
- Using the KG, it is possible to employ cost and heuristic functions for heuristic search algorithms from the combination of the information from $\mathcal{C}$-space and $\mathcal{T}$-space of the robotic manipulators.

Up to now, we have employed cost and heuristic functions based on the distances in EUCLIDEAN spaces of $\mathcal{C}$-space representation and $\mathcal{T}$-space. In our future research, we will attempt to integrate the costs and heuristics based on kinetic and potential energy in sampling-based planning algorithm using KG.

Author Contributions: B.C.: supervision, project administration, and funding acquisition. A.S.: conceptualisation, methodology, investigation, visualisation, validation, and writing—original draft preparation; B.C. and A.S.: writing—review and editing. All authors have read and agreed to the published version of the manuscript.

Funding: Funded by the Deutsche Forschungsgemeinschaft (DFG, German Research Foundation) under Germany's Excellence Strategy—EXC-2023 Internet of Production—390621612.

Institutional Review Board Statement: Not applicable.

Informed Consent Statement: Not applicable.

Data Availability Statement: Not applicable.

Acknowledgments: The authors would like to thank Vincent Brünjes and Thomas Kinzig for the constructive comments, discussions, and kind support during the software development.

Conflicts of Interest: The authors declare no conflict of interest. The funders had no role in the design of the study; in the collection, analyses, or interpretation of data; in the writing of the manuscript; or in the decision to publish the results.

Nomenclature

The following list of symbols is used in this manuscript:

Symbol	Description
$\mathcal{C}$	Configuration space
$\mathcal{C}_{\mathrm{free}}$	Free $\mathcal{C}$, $\mathcal{C}_{\mathrm{free}} \triangleq \mathcal{C} \setminus \mathcal{C}_{\mathrm{obs}}$
$\mathcal{C}_{\mathrm{res}}$	Sampling resolution of $\mathcal{C}$
cent	Centroid of the voxel in $\mathcal{O}$
C	Cluster of nodes in voxel
$\mathrm{c}_{\mathrm{idx}}$	Index of a cluster
c	A cluster in the set of the cluster objects
$\mathscr{C}$	The amount of the cost function
$\mathfrak{e}$	An edge of a graph
$\mathfrak{E}$	The edges of a graph
$\mathfrak{E}_{\mathrm{k}}$	The edges of the Kinematic Graph
$\mathfrak{G}$	A graph
$\mathfrak{G}_{\mathrm{k}}$	The Kinematic Graph
$\mathscr{H}$	The amount of the heuristic function
$\mathscr{K}$	Forward kinematics ($\mathscr{K} : \mathcal{C} \rightarrow \mathcal{T}$)
μ	Linear manipulability
N	Set of node objects
n	A node $\in$ N
$\mathrm{n}_{\mathrm{idx}}$	Index of a node
N_{C}	Set of nodes in a cluster C
O	Open set of nodes for clustering
P_{w}	The centre point of the wrist
pos	CARTESIAN coordinate of the node in $\mathcal{O}$-space
q	Generalized coordinates of $\mathcal{C}$
r	Reach of the point P_{w}
$\mathbb{R}$	Field of real numbers
$\mathcal{T}$	Task space
$\mathcal{T}_{\mathrm{res}}$	Sampling resolution of $\mathcal{T}$
$\mathfrak{v}$	A vertex of a graph
$\mathfrak{V}$	The vertices of a graph
$\mathfrak{V}_{\mathrm{k}}$	The vertices of the Kinematic Graph
V	Set of voxel objects
v	A voxel $\in$ V
$\mathrm{v.N_v}$	Set of nodes in a voxel v
$\mathcal{O}$	Environment (World)

References

1. Biagiotti, L.; Melchiorri, C. *Trajectory Planning for Automatic Machines and Robots*; Springer: Berlin/Heidelberg, Germany, 2008.
2. Lozano-Pérez, T.; Wesley, M.A. An algorithm for planning collision-free paths among polyhedral obstacles. *Commun. ACM* **1979**, *22*, 560–570. [CrossRef]
3. Lozano-Pérez, T. A simple motion-planning algorithm for general robot manipulators. *IEEE J. Robot. Autom.* **1987**, *3*, 224–238. [CrossRef]
4. Khatib, O. Real-time obstacle avoidance for manipulators and mobile robots. In Proceedings of the 1985 IEEE International Conference on Robotics and Automation, St. Louis, MO, USA, 25–28 March 1985.
5. LaValle, S.M. *Planning Algorithms*; Cambridge University Press: Cambridge, UK, 2006.
6. Brooks, R.A.; Lozano-Pérez, T. A subdivision algorithm in configuration space for find path with rotation. *IEEE Trans. Syst. Man, Cybern.* **1985**, *SMC-15*, 224–233. [CrossRef]
7. Siciliano, B.; Khatib, O. (Eds.) *Springer Handbook of Robotics*; Springer International Publishing: Berlin/Heidelberg, Germany, 2016. [CrossRef]
8. Koren, Y.; Borenstein, J. Potential field methods and their inherent limitations for mobile robot navigation. In Proceedings of the IEEE Conference on Robotics and Automation, Sacramento, CA, USA, 7–12 April 1991; Volume 2, pp. 1398–1404.
9. Lindemann, S.R.; LaValle, S.M. Current issues in sampling-based motion planning. In *Robotics Research, Proceedings of the Eleventh International Symposium, Siena, Italy, 19–22 October 2003*; Springer: Berlin/Heidelberg, Germany, 2005; pp. 36–54.
10. Koenig, S.; Likhachev, M. Fast replanning for navigation in unknown terrain. *IEEE Trans. Robot.* **2005**, *21*, 354–363. [CrossRef]

11. Stentz, A. The focussed D* algorithm for real-time replanning. In Proceedings of the International Joint Conference on Artificial Intelligence, Montreal, QC, Canada 20–25 August 1995; Volume 95, pp. 1652–1659.
12. Holladay, R.; Salzman, O.; Srinivasa, S. Minimizing task-space Frechet error via efficient incremental graph search. *IEEE Robot. Autom. Lett.* **2019**, *4*, 1999–2006. [CrossRef]
13. Husty, M.L.; Pfurner, M.; Schröcker, H.P. A new and efficient algorithm for the inverse kinematics of a general serial 6R manipulator. *Mech. Mach. Theory* **2007**, *42*, 66–81. [CrossRef]
14. Lynch, K.M.; Park, F.C. *Modern Robotics, Mechanics Planning, and Control*; Cambridge University Press: Cambridge, UK, 2017.
15. Shahidi, A.; Hüsing, M.; Corves, B. Kinematic Control of Serial Manipulators Using Clifford Algebra. *IFAC-PapersOnLine* **2020**, *53*, 9992–9999. [CrossRef]
16. Hauser, K.; Emmons, S. Global redundancy resolution via continuous pseudoinversion of the forward kinematic map. *IEEE Trans. Autom. Sci. Eng.* **2018**, *15*, 932–944. [CrossRef]
17. Berenson, D.; Srinivasa, S.S.; Ferguson, D.; Collet, A.; Kuffner, J.J. Manipulation planning with workspace goal regions. In Proceedings of the 2009 IEEE International Conference on Robotics and Automation, Kobe, Japan, 12–17 May 2009; pp. 618–624.
18. LaValle, S.M. Rapidly-Exploring Random Trees: A New Tool for Path Planning. 1998. Available online: http://msl.cs.illinois.edu/~lavalle/papers/Lav98c.pdf (accessed on 27 February 2022).
19. Cohen, B.; Chitta, S.; Likhachev, M. Single-and dual-arm motion planning with heuristic search. *Int. J. Robot. Res.* **2014**, *33*, 305–320. [CrossRef]
20. Rickert, M.; Sieverling, A.; Brock, O. Balancing exploration and exploitation in sampling-based motion planning. *IEEE Trans. Robot.* **2014**, *30*, 1305–1317. [CrossRef]
21. Scheurer, C.; Zimmermann, U.E. Path planning method for palletizing tasks using workspace cell decomposition. In Proceedings of the 2011 IEEE International Conference on Robotics and Automation, Shanghai, China, 9–13 May 2011. [CrossRef]
22. Mesesan, G.; Roa, M.A.; Icer, E.; Althoff, M. Hierarchical path planner using workspace decomposition and parallel task-space rrts. In Proceedings of the 2018 IEEE/RSJ International Conference on Intelligent Robots and Systems (IROS), Madrid, Spain, 1–5 October 2018; pp. 1–9.
23. Yang, Y.; Merkt, W.; Ivan, V.; Li, Z.; Vijayakumar, V. HDRM: A Resolution Complete Dynamic Roadmap for Real-Time Motion Planning in Complex Scenes. *IEEE Robot. Autom. Lett.* **2017**, *3*, 551–558. [CrossRef]
24. Denavit, J.; Hartenberg, R.S. A kinematic notation for lower-pair mechanisms based on matrices. *J. Appl. Mech.* **1955**, *22*, 215–221. [CrossRef]
25. Khalil, W.; Kleinfinger, J. A new geometric notation for open and closed-loop robots. In Proceedings of the 1986 IEEE International Conference on Robotics and Automation, San Francisco, CA, USA, 7–10 April 1986; Volume 3, pp. 1174–1179.
26. Angeles, J. *Rational Kinematics*; Springer: New York, NY, USA, 2013; Volume 34. [CrossRef]
27. Khalil, W.; Dombre, E. *Modeling Identification and Control of Robots*; CRC Press: Boca Raton, FL, USA, 2002.
28. Uicker, J.J.; Ravani, B.; Sheth, P.N. *Matrix Methods in the Design Analysis of Mechanisms and Multibody Systems*; Cambridge University Press: Cambridge, UK, 2013.
29. Müller, A. Screw and Lie group theory in multibody kinematics. *Multibody Syst. Dyn.* **2018**, *43*, 37–70. [CrossRef]
30. Angeles, J. *Fundamentals of Robotic Mechanical Systems*; Springer: Berlin/Heidelberg, Germany, 2007.
31. Elbanhawi, M.; Simic, M. Sampling-based robot motion planning: A review. *IEEE Access* **2014**, *2*, 56–77. [CrossRef]
32. Dijkstra, E.W. A note on two problems in connexion with graphs. *Numer. Math.* **1959**, *1*, 269–271. [CrossRef]
33. Hart, P.E.; Nilsson, N.J.; Raphael, B. A formal basis for the heuristic determination of minimum cost paths. *IEEE Trans. Syst. Sci. Cybern.* **1968**, *4*, 100–107. [CrossRef]
34. Likhachev, M.; Gordon, G.J.; Thrun, S. ARA*: Anytime A* with provable bounds on sub-optimality. In Proceedings of the Advances in Neural Information Processing Systems, Vancouver, BC, Canada, 8–13 December 2003; pp. 767–774.
35. Koenig, S.; Likhachev, M.; Furcy, D. Lifelong Planning A*. *Artif. Intell.* **2004**, *155*, 93–146. [CrossRef]
36. Koenig, S.; Likhachev, M. D* Lite. *Aaai/iaai* **2002**, *15*, 476–483.
37. Kavraki, L.E.; Svestka, P.; Latombe, J.C.; Overmars, M.H. Probabilistic roadmaps for path planning in high-dimensional configuration spaces. *IEEE Trans. Robot. Autom.* **1996**, *12*, 566–580. [CrossRef]
38. Pearl, J. *Heuristics: Intelligent Search Strategies for Computer Problem Solving*; Addison-Wesley Longman Publishing Co., Inc.: San Francisco, CA, USA, 1984.
39. Shahidi, A.; Kinzig, T.; Hüsing, M.; Corves, B. Kinematically Adapted Sampling-Based Motion Planning Algorithm for Robotic Manipulators. In *Advances in Robot Kinematics*; Springer International Publishing: Cham, Switzerland, 2022; Volume 24, pp. 453–461. [CrossRef]
40. Peixoto, T.P. The Graph-Tool Python Library. 2017. Available online: https://doi.org/10.6084/M9.FIGSHARE.1164194.V14 (accessed on 27 February 2022).
41. Kinzig, T. Search-Based Path Planning for Positioning of Robot Manipulators. Master's Thesis, RWTH Aachen University, Aachen, Germany, 2021.

robotics

MDPI

Article

Partial Lagrangian for Efficient Extension and Reconstruction of Multi-DoF Systems and Efficient Analysis Using Automatic Differentiation

Takashi Kusaka [1,*,†] and Takayuki Tanaka [2,†]

[1] Independent Researcher, Sapporo 063-0867, Japan
[2] Graduate School of Information Science and Technology, Hokkaido University, Sapporo 060-0814, Japan
* Correspondence: kusaka@frontier.hokudai.ac.jp
† These authors contributed equally to this work.

Abstract: In the fields of control engineering and robotics, either the Lagrange or Newton–Euler method is generally used to analyze and design systems using equations of motion. Although the Lagrange method can obtain analytical solutions, it is difficult to handle in multi-degree-of-freedom systems because the computational complexity increases explosively as the number of degrees of freedom increases. Conversely, the Newton–Euler method requires less computation even for multi-degree-of-freedom systems, but it cannot obtain an analytical solution. Therefore, we propose a partial Lagrange method that can handle the Lagrange equation efficiently even for multi-degree-of-freedom systems by using a divide-and-conquer approach. The proposed method can easily handle system extensions and system reconstructions, such as changes to intermediate links, for multi-degree-of-freedom serial link manipulators. In addition, the proposed method facilitates the derivation of the equations of motion-by-hand calculations, and when combined with an analysis algorithm using automatic differentiation, it can easily realize motion analysis and control the simulation of multi-degree-of-freedom models. Using multiple pendulums as examples, we confirm the effectiveness of system expansion and system reconstruction with the partial Lagrangians. The derivation of their equations of motion and the results of motion analysis by simulation and motion control experiments are presented. The system extensions and reconstructions proposed herein can be used simultaneously with conventional analytical methods, allowing manual derivations of equations of motion and numerical computer simulations to be performed more efficiently.

Keywords: Lagrange equation; automatic differentiation; divide-and-conquer approach

Citation: Kusaka, T.; Tanaka, T. Partial Lagrangian for Efficient Extension and Reconstruction of Multi-DoF Systems and Efficient Analysis Using Automatic Differentiation. *Robotics* **2022**, *11*, 149. https://doi.org/10.3390/robotics11060149

Academic Editor: Raffaele Di Gregorio

Received: 14 October 2022
Accepted: 6 December 2022
Published: 9 December 2022

1. Introduction

In dynamics, the Newton–Euler and Lagrange methods are well known for formulating and calculating the equations of motion and used for different purposes [1–4]. The Newton–Euler method is a computational procedure that can be used in multi-degree-of-freedom (DoF) models with the systematic algorithm for one DoF at a time [5–7]. Therefore, it is a suitable method for obtaining numerical solutions, but not for analytical solutions. Conversely, the Lagrange method is a formulation procedure that yields analytical solutions that can be understood as physically meaningful terms. However, the amount of computation is enormous when analyzing multi-DoF systems. Therefore, it is generally used for the analysis of systems with fewer DoF.

Numerical solutions that can be computed with the Newton–Euler method are useful for implementing equipment that uses equations of motion, such as real-time feedback control of power assistance in human–machine coordination systems in the field of robotics [8,9]. However, when designing algorithms or conducting theoretical analysis of systems, the meaning of each physical term in the analytical solution obtained by the Lagrange method is extremely important. For example, it is used in the behavior analysis and control of walking robots and drones [10–13]. We use these methods for different applications. In our past research, we used the Lagrange method for the analysis of human body motion modeled in four DoF and the design of control systems [14–16]. We also used the Newton–Euler method to analyze the dynamics of a 7-link (21-DoF) spinal column model for a wearable system [17,18].

A major problem with the Lagrange method is that the amount of computation increases explosively when the number of DoF increases, and the results become complicated. While computer algebra systems can be used to perform large-scale system analysis, manual calculations are extremely difficult to handle even with four DoF, and calculation errors are likely to occur. Because the equations of motion with multi-DoF are extremely complex, many techniques to generate them automatically by computer have been studied. In fact, the equations of motion for the robot arm can be obtained with mathematical processing software, such as Mathematica's Robotica [19] or MatLab's TMTDyn [20], and calculations such as the TMT method are used to optimize the computer calculations [21]. However, the results of large systems collectively are very complex and not reusable.

Therefore, we propose to solve this problem by introducing a partial Lagrangian and postural operator. The partial Lagrangian uses the divide-and-conquer approach [22,23] to divide the equations of motion, which become complex when the number of DoF increases, into the smallest units. The use of the partial Lagrangian reduces the computational load because the terms in the equations of motion for increasing DoF can be treated independently. In other words, a similar process of division can be performed for each DoF, and the results integrated to obtain an exact analytical solution. This has two advantages. First, when the system is extended or reconstructed, only the affected part of the system needs to be calculated, making this analysis method flexible in terms of system configuration. The second advantage is that the modularization of the calculation unit minimizes the burden of obtaining the equations of motion by manual calculation. The modularized calculation is also compatible with computer processing because it involves iterative calculations and the final integration of similar processes.

Studies analyzing multibody dynamics using the divide-and-conquer algorithm (DCA) include the generalized DCA [24], DCAe [25], and a study of sensitivity analysis using DCA [26]. The generalized DCA is an extension method of the DCA for modeling constrained multibody systems, the DCAe is a reconstruction of DCA for efficient handling of multibody dynamics, and the sensitivity analysis using DCA treats the DCA as a critical tool for efficient analysis of multibody dynamics. All of these are based on dividing the generalized force in a binary tree and applying the divide-and-conquer method. Therefore, mechanical constraints are important in all of these studies. In other past studies, such as those mentioned above, no method has focused on Lagrangian dividing. Since the proposed method applies to the Lagrangian (energy), which is abstract, it can be applied to any system that can be described by a Lagrangian that is linearly independent.

Recursive algorithms, such as an algorithm applying the Gibbs–Appel equation [27] and the harmony search algorithm [28], are also known to be very effective for complex systems, such as parallel robots. Our approach differs from recursive algorithms in that it exploits the linear independence of the Lagrangian for partitioning and reuse of computed results. As a benefit of partitioning, we can derive analytical solutions and gain computational efficiencies in computerized numerical solutions. Therefore, the final result of the proposed method is completely equal to the usual Lagrangian method and can be used without the need for a recursive algorithm.

We propose a numerical analysis method based on automatic differentiation as the optimal analysis method for the partial Lagrangian. Automatic differentiation is a method used for training neural networks [29–34] We also confirm that the proposed partial Lagrangian and automatic differentiation can be used to simulate multi-linked manipulators easily and that the system can be easily extended and reconstructed.

2. Methods

2.1. Partial Lagrangian

This section provides an overview of the partial Lagrangian. Specific examples are given in the next section. The usual Lagrangian requires the total kinetic and potential energy of the system, so the term explodes with increasing DoF. There is a known efficient method of recursive computation using the linear independence of the Lagrangian [35,36]. We extend this idea and consider how to design a more efficient system using the divide-and-conquer approach by organizing it in units of the partial Lagrangian.

First, the Lagrangian $\mathcal{L}$ of the n-DoF system is the difference between the total kinetic energy K and the total potential energy P, as follows

$$\mathcal{L} = K - P \tag{1}$$

Here, if we decompose each type of energy into DoF using the distributive property, we can transform it as follows.

$$\mathcal{L} = \sum_{i=1}^{n} K_i - \sum_{i=1}^{n} P_i = \sum_{i=1}^{n}(K_i - P_i) = \sum_{i=1}^{n} \mathcal{L}_i \tag{2}$$

This $\mathcal{L}_i$ is defined as a partial Lagrangian. The subscript i denotes the division into partial Lagrangians. An n-DOF system will have n partial Lagrangians, corresponding to each DoF as $i = 1, 2, \cdots, n$. Since the Lagrangian $\mathcal{L}$ is linearly independent, the partial Lagrangian $\mathcal{L}_i$ for the i-th DoF consists of the partial kinetic energy K_i and the partial potential energy P_i.

Considering the generalized coordinate q_k, the equation of motion with the Lagrangian is as follows:

$$\frac{d}{dt}\left(\frac{\partial}{\partial \dot{q}_k}\mathcal{L}\right) - \frac{\partial}{\partial q_k}\mathcal{L} = \tau_k \tag{3}$$

where the subscript k represents the k-th equation of motion and $k \leq n$ for the n-DoF system. τ_k represents the generalized force of k-th DoF since the above equation is the usual Lagrangian equation of motion. For simplicity, let the differential operator on the left-hand side be defined formally as $D_k = \frac{d}{dt}\frac{\partial}{\partial \dot{q}_k} - \frac{\partial}{\partial q_k}$. In other words, the equation of motion for the k-th DoF is $D_k\mathcal{L} = \tau_k$.

Here, considering the partial Lagrangian $\mathcal{L}_i$,

$$D_k\mathcal{L} = D_k \sum_{i=1}^{n} \mathcal{L}_i = \sum_{i=1}^{n}(D_k\mathcal{L}_i) = \sum_{i=1}^{n} \tau_{ki} \tag{4}$$

because the order of the sum and derivative can be exchanged using term-wise differentiation. This τ_{ki} is defined as the partial generalized force.

Now consider the components of $\mathcal{L}$. If $i \geq k$, $\mathcal{L}_k$ contains q_i, but if $i < k$, $\mathcal{L}_k$ does not contain q_i. Therefore, the partial generalized force $D_k\mathcal{L}_i$ is as follows.

$$D_k\mathcal{L}_i = \begin{cases} \tau_{ki} & (i \geq k) \\ 0 & (i < k) \end{cases} \tag{5}$$

This can be summarized as shown in Table 1. Looking at this table, the equation of motion $D_k\mathcal{L} = \tau_k$ using the Lagrangian method corresponds to calculating all of the entries in row k simultaneously. Therefore, the calculation explodes as the number of DoF increases. However, the partial Lagrangian $D_k\mathcal{L}_i = \tau_{ki}$ is equivalent to splitting this calculation and performing the calculation with respect to $\mathcal{L}_i$. In other words, the analysis is column-wise. Therefore, the final equivalent analytical solution is obtained by summing, but it can be computed by dividing the solution in order, starting from $i = 1$.

In addition, the calculation results of $\mathcal{L}_{(i-1)}$ can be diverted for the calculation of $\mathcal{L}_i$, thus reducing the amount of the calculation. Furthermore, in the case of the Lagrangian equation $D_k\mathcal{L} = \tau_k$, if the robot's link is extended or the number of DoF is changed after the analytical solution is obtained, all calculations must be redone. Conversely, when the partial Lagrangian is used, the calculation is independent for each DoF, so the results of the root side calculation can be reused. Therefore, by managing the system in units of partial Lagrangian modules, system extensions, and changes can be handled in a prepared manner.

For example, to extend the system, simply add a new term regarding $\mathcal{L}_{(n+1)}$ as follows, and the summing part can be reused.

$$\text{Current system: } \tau_k = \sum_{i=1}^{n} D_k\mathcal{L}_i \tag{6}$$

$$\downarrow$$

$$\text{Extended system: } \tau_{k,\,\text{new}} = \underbrace{\sum_{i=1}^{n} D_k\mathcal{L}_i}_{\text{Reusable term}} + \underbrace{D_k\mathcal{L}_{(n+1)}}_{\text{New term by the extension}} \tag{7}$$

In addition, system changes can be realized in the same way.

$$\text{Current system: } \tau_k = \sum_{i=1}^{n} D_k\mathcal{L}_{i,\,\text{old}} = \sum_{i=1}^{n-1} D_k\mathcal{L}_i + D_k\mathcal{L}_{n,\,\text{old}} \tag{8}$$

$$\downarrow$$

$$\text{Changed system: } \tau_{k,\,\text{new}} = \sum_{i=1}^{n} D_k\mathcal{L}_{i,\,\text{new}} = \underbrace{\sum_{i=1}^{n-1} D_k\mathcal{L}_i}_{\text{Reusable term}} + \underbrace{D_k\mathcal{L}_{n,\,\text{new}}}_{\text{Replaced term}} \tag{9}$$

Table 1. Partial Lagrangian vs. Lagrange method.

	Partial Lagrangian								Lagrange Method
	$\mathcal{L}_1$	$\mathcal{L}_2$	$\mathcal{L}_3$	$\cdots$	$\mathcal{L}_i$	$\cdots$	$\mathcal{L}_n$	$\xrightarrow{\Sigma}$	$\mathcal{L}$
D_1	τ_{11}	τ_{12}	τ_{13}	$\cdots$	τ_{1i}	$\cdots$	τ_{1n}	$\rightarrow$	τ_1
D_2	0	τ_{22}	τ_{23}	$\cdots$	τ_{2i}	$\cdots$	τ_{2n}	$\rightarrow$	τ_2
D_3	0	0	τ_{33}	$\cdots$	τ_{3i}	$\cdots$	τ_{3n}	$\rightarrow$	τ_3
$\vdots$	$\vdots$	$\vdots$	$\vdots$	$\ddots$	$\vdots$		$\vdots$	$\vdots$	$\vdots$
D_k	0	0	0	$\cdots$	τ_{ki}	$\cdots$	τ_{kn}	$\rightarrow$	τ_k
$\vdots$	$\vdots$	$\vdots$	$\vdots$		$\vdots$	$\ddots$	$\vdots$	$\vdots$	$\vdots$
D_n	0	0	0	$\cdots$	0	$\cdots$	τ_{nn}	$\rightarrow$	τ_n

In summary, when the partial Lagrangian is generalized to n DoF, the following procedure can be used to calculate the partition.

1. Dynamics: Differentiation of state variables as physical constraints.
2. Kinematics: Position $\boldsymbol{p}_i$ and velocity $\dot{\boldsymbol{p}}_i$ as geometric constraints.
3. Partial Lagrangian.
   (a) Quadratic form: calculate $\boldsymbol{p}_i^T \boldsymbol{p}_i$ and $\dot{\boldsymbol{p}}_i^T \dot{\boldsymbol{p}}_i$ to find the energies. Calculate the partial energies: K_i and P_i.
   (b) Compute the partial Lagrangian: $\mathcal{L}_i$.
4. Find the partial generalized force: $D_k\mathcal{L}_i = \tau_{ki}$. If it is a multi-degree-of-freedom system, find the sum $\tau_k = \sum_{i=1}^{n} \tau_{ki}$.

This calculation procedure is performed in order from the smallest to the largest value of i, and finally, the equation of motion can be derived by computing the sum. Therefore, the complex multi-DoF Lagrangian equations can be obtained with relative ease using a divide-and-conquer approach. Censoring in the middle is equivalent to the equation of motion for a short robot arm, and it is also easy to add $i = n + 1$ later.

2.2. Postural Operator for Hand Calculations

By using the partial Lagrangian, the Lagrangian is divided and treated independently with respect to each DoF. In this subsection, we define a posture operator for more efficient computations when dealing with partial Lagrangians. Let the attitude operator in the planar coordinates be a vector of length 1, such that

$$\boldsymbol{e}_\theta = \begin{bmatrix} \cos\theta \\ \sin\theta \end{bmatrix} \tag{10}$$

This has the following trivial properties:

- Cancellation: $\boldsymbol{e}_\theta^T \boldsymbol{e}_\theta = 1$
- Interference: $\boldsymbol{e}_{\theta_{12}}^T \boldsymbol{e}_{\theta_1} = \boldsymbol{e}_{12}^T \boldsymbol{e}_1 = \cos\theta_2$
- Derivative: $\frac{d}{dt}\boldsymbol{e}_\theta = \begin{cases} \dot{\theta}\boldsymbol{e}_{\theta+\frac{\pi}{2}} = \dot{\theta}\boldsymbol{e}'_\theta & \text{if } \theta = \theta(t) \\ 0 & \text{if } \theta = \text{const.} \end{cases}$

where $\theta_{12} = \theta_1 + \theta_2$ and $\boldsymbol{e}_{\theta_i} = \boldsymbol{e}_i$ for notational simplicity. The phase difference of $+\frac{\pi}{2}$ is defined as $\boldsymbol{e}_{\theta+\frac{\pi}{2}} = \boldsymbol{e}'_\theta$.

This is similar to the stationary phasor [37–39] in electrical engineering, which uses a polar form to improve the perspective of the formula expansion. Here, we treat it as a vector rather than a complex form. This also makes it intuitively consistent with the orthogonal form expression expansion. The description of motion in a two-dimensional plane is simplified by using the above properties. In the three-dimensional case, a similar argument can be made using a versor [40–43] with a magnitude of 1 as a rotation by quaternions or using a rotation matrix [44]. We deal with the two-dimensional case because time is used as the third axis for the visualization of the simulation results.

2.3. System Analysis Using Automatic Differentiation for Numerical Calculation

Next, we consider the numerical analysis using the partial Lagrangian. The advantage of the partial Lagrangian is that extensions and reconstructions of the system can easily be realized by splitting the Lagrangian. If the equations are transformed to the form of linear differential equations for each DoF for dynamics analysis, this is equivalent to the usual Lagrangian and loses the advantage of the split calculation.

An analytical method that can directly handle $D_k\mathcal{L}_i = \tau_i k$, which is a formal description using the partial Lagrangian, is desirable. Therefore, we use an analysis method that applies automatic differentiation. The solution of the equation of motion by the partial Lagrangian is obtained by generating a computational graph by automatic differentiation and performing a gradient calculation using the balance of the partial differential equation as a constraint. Applications to the analysis of partial differential equations other than neural networks are used in the fields of rigorous simulations, such as the finite element method analysis [45,46], dynamics calculations [47,48], and electronic circuit analysis [49–51].

An example of dynamics computed by automatic differentiation is shown in Appendix A. As indicated in the function in the Appendix, the program for generating a graph of differential equation calculations for automatic differentiation is written in the following flow.

1. Dynamics: The time evolution of the state variable is registered in the calculation graph as a constraint.
2. Kinematics: Register a geometric constraint on a calculation graph.
3. Register a partial Lagrangian in the computed graph
4. Register the partial generalized force, and if it is a multi-DoF, find the sum for each DoF.

5. Backpropagate and find the coefficients as the gradient of the state variable.

This can be described in exactly the same flow as the calculation algorithm for the partial Lagrangian method presented in Section 2.1. Therefore, by using automatic differentiation and calculating dynamics by the partial Lagrangian method, the calculation of the equations of motion can be realized using a divide-and-conquer approach. In addition, each calculation is divided and modularized, allowing the system to be reconstructed instantly.

3. Results

We present some examples and simulations to confirm the effect of the partial Lagrangian. For simplicity, we assume an n-link manipulator moving on a plane and describe it using the postural operator. Although the pendulum model is assumed for the simplicity of explanation, rigid body links can be treated in the same way as long as the Lagrangian is a linear sum. The link parameter for link i has a link length of l_i, mass m_i, and stiffness $k_i = 0$. The link length is read as the total length when dealing with connections between links or the length to the center of gravity when dealing with the center of gravity, as appropriate.

3.1. Example of 1-DoF: Effect of Postural Operator

In the case of one DoF, $\tau_{11} = D_1\mathcal{L}_1 = D_1\mathcal{L} = \tau_1$; thus, the partial Lagrangian result is exactly equal to the usual Lagrangian method. Here, we use the single pendulum [52] as an example of a rotational joint. This calculation is very simple, but to confirm the effect of the postural operator and the flow of the partial Lagrangian processing, we show it in detail as an example.

First, as a geometric constraint, the link end position and its velocity are as follows from the kinematics.

$$\boldsymbol{p}_1 = l_1\boldsymbol{e}_1 \tag{11}$$

$$\dot{\boldsymbol{p}}_1 = l_1\frac{d}{dt}\boldsymbol{e}_1 = l_1\dot{\theta}_1\boldsymbol{e}_1' \tag{12}$$

These quadratic forms are then obtained as inner products.

$$\boldsymbol{p}_1^T\boldsymbol{p}_1 = l_1^2 \underbrace{\boldsymbol{e}_1^T\boldsymbol{e}_1}_{\text{cancellation}} = l_1^2 \tag{13}$$

$$\dot{\boldsymbol{p}}_1^T\dot{\boldsymbol{p}}_1 = l_1^2\dot{\theta}_1^2 \underbrace{\boldsymbol{e}_1'^T\boldsymbol{e}_1'}_{\text{cancellation}} = l_1^2\dot{\theta}_1^2 \tag{14}$$

From these, the kinetic energy K and potential energy P can be calculated as follows.

$$K_1 = \frac{1}{2}m_1\dot{\boldsymbol{p}}_1^T\dot{\boldsymbol{p}}_1 = \frac{1}{2}m_1l_1^2\dot{\theta}_1^2 \tag{15}$$

$$P_1 = \frac{1}{2}k_1\boldsymbol{p}_1^T\boldsymbol{p}_1 + m_1gy_1 = m_1g\sin\theta_1. \tag{16}$$

The partial Lagrangian calculated using these is $\mathcal{L}_1 = K_1 - P_1 = \mathcal{L}$. Therefore, its equation of motion can be obtained as $\tau_{11} = D_1\mathcal{L}_1 = D1\mathcal{L} = \tau_1$. This should be equivalent to the equation of motion for an inverted pendulum in a general textbook.

To confirm this, we analyze the behavior of the equations of motion using automatic differentiation. For one link, the analysis can be performed using the sample of automatic differentiation shown in Appendix A. The link parameters are $m_1 = 3$, $k_1 = 0$, $l_1 = 2$, with an appropriate damping term $d_1 = 1$ to make the behavior easier to understand. The results are shown in Figure 1a. Posture control can also be simulated simply by writing a control input to the generalized forces of the equations of motion in the calculation graph. The results of the proportional–derivative (PD) control with a target angle of 110° are shown in

Figure 1b. The PD gains were set to low values ($K_{p1} = 200$ and $K_{d1} = 50$, respectively) to make the behavior easy to understand.

The results show how to calculate the partial Lagrangian by hand for the simplest case and analyze it by automatic differentiation. This is the minimum unit of the divide-and-conquer approach for the Lagrangian, and from the next section, we will confirm that this procedure can be repeated according to Table 1 to obtain the desired equations of motion.

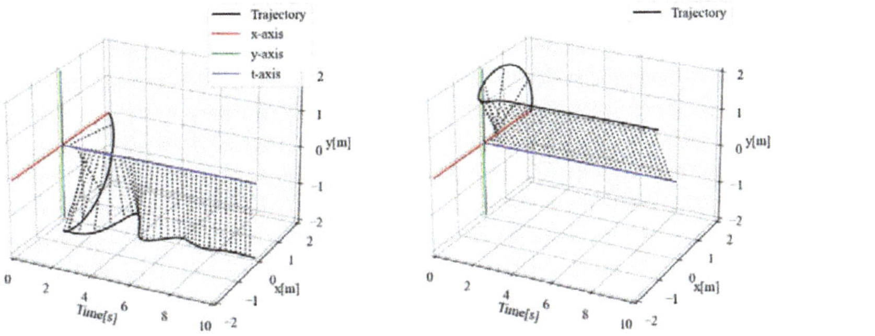

(a) (b)

Figure 1. Motion analysis of the 1-link system using automatic differentiation. (**a**) Damping oscillation; (**b**) PD control.

3.2. Example of 2-DoF: Effect of Divide-and-Conquer by Partial Lagrangian

In the previous chapter, we confirmed the reduction in computational complexity due to the postural operator e_θ on a 1-DoF example. Next, we will use a 2-DoF example to confirm this effect. According to Table 1, only τ_{12} and τ_{22} need to be added. As a simple example, we consider a double pendulum [53,54], which extends the 1-DoF example.

The same procedure as for the 1-DoF is used to obtain the partial Lagrangian $\mathcal{L}_2$. First, the geometric constraints are as follows.

$$p_2 = p_1 + l_2 e_{12} = l_1 e_1 + \underbrace{l_2 e_{12}}_{\text{new information}} \tag{17}$$

$$\dot{p}_2 = \dot{p}_1 + l_2 \frac{d}{dt} e_{12} = l_1 \dot{\theta}_1 e'_1 + \underbrace{l_2 \dot{\theta}_{12} e'_{12}}_{\text{new information}} \tag{18}$$

Then, their quadratic forms are

$$\begin{aligned} p_2^T p_2 &= l_1^2 \underbrace{e_1^T e_1}_{\text{cancellation}} + l_2^2 \underbrace{e_{12}^T e_{12}}_{\text{cancellation}} + l_1 l_2 \underbrace{e_1^T e_{12}}_{\text{interference}} \\ &= l_1^2 + l_2^2 + l_1 l_2 \cos\theta_2 \end{aligned} \tag{19}$$

$$\begin{aligned} \dot{p}_2^T \dot{p}_2 &= l_1^2 \dot{\theta}_1^2 \underbrace{e_1'^T e_1'}_{\text{cancellation}} + l_2^2 \dot{\theta}_{12}^2 \underbrace{e_{12}'^T e_{12}'}_{\text{cancellation}} + l_1 l_2 \dot{\theta}_1 \dot{\theta}_{12} \underbrace{e_1'^T e_{12}'}_{\text{interference}} \\ &= l_1^2 \dot{\theta}_1^2 + l_2^2 \dot{\theta}_{12}^2 + l_1 l_2 \dot{\theta}_1 \dot{\theta}_{12} \cos\theta_2. \end{aligned} \tag{20}$$

The kinetic and potential energies consisting of the link parameters for $i = 2$ are

$$K_2 = \frac{1}{2} m_2 \dot{p}_2^T \dot{p}_2 \tag{21}$$

$$P_2 = \frac{1}{2} k_2 p_2^T p_2 + m_2 g y_2 \tag{22}$$

Since this partial Lagrangian is $\mathcal{L}_2 = K_2 - P_2$, the equations of motion are

$$\tau_1 = \tau_{11} + \tau_{12} = \tau_{11} + D_1\mathcal{L}_2 \tag{23}$$
$$\tau_2 = \quad \tau_{22} = \quad D_2\mathcal{L}_2, \tag{24}$$

also using the result of τ_{11} with one DoF. In other words, the calculation was reduced by one previous result, τ_{11}, in finding the equations of motion for two DoF. To obtain the equation of motion for n DoF, only the n columns of Table 1 must be calculated, and the calculations from columns 1 to $n-1$ are unnecessary because they can be reused. References [53,54] show that the equation of motion of a double pendulum is complicated to be solved during the process if it is obtained by the usual Lagrangian method; however, it can be described in a simple manner by the partial Lagrangian method. This simplifies the calculation even for an increased DoF and prevents calculation mistakes in manual calculations.

We confirm that this result is correct by simulation with automatic differentiation. The simulation can take full advantage of the divide-and-conquer effect of the partial Lagrangian. In the example program shown in Appendix A (Listing A1), to obtain $\mathcal{L}_2$, only two lines of the geometric constraint need to be rewritten according to the system. Therefore, the idea of division of the process in the partial Lagrangian corresponds to the program, and the same calculation results obtained using the complicated equations of motion can also be obtained as an iteration of this module.

The results of the 2-DoF example are shown in Figure 2a as damped oscillations and that of adding the PD control to the joints are shown in Figure 2b. The link parameters are $m_1 = 3,\ k_1 = 0,\ l_1 = 2,\ d_1 = 1$ for the first joint and $m_2 = 1,\ k_2 = 0,\ l_2 = 1,\ d_2 = 1$ for the second. Target values of $\theta_{1ref} = 80°$ and $\theta_{2ref} = 30°$ were used for PD control, and its gains are $K_{p1} = 100$, $K_{p2} = 50$, $K_{d1} = 50$, and $K_{d2} = 20$. It can be shown that the desired behavior can be analyzed even in the case of multi-DoF.

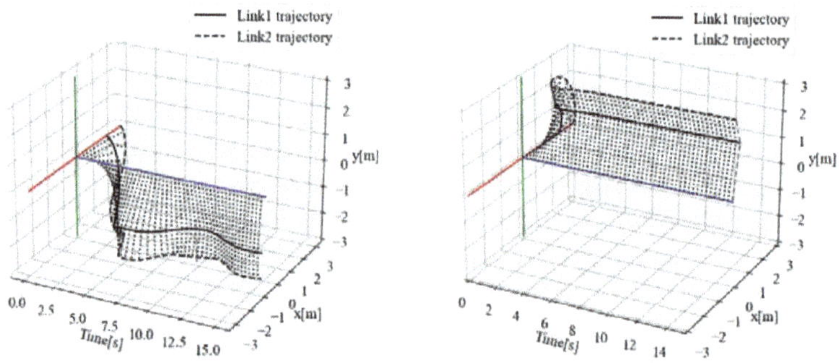

(**a**) (**b**)

Figure 2. Motion analysis of a two-link system using automatic differentiation. (**a**) Damping oscillation; (**b**) PD control.

3.3. *Example of 3-DoF and Changing System Construction*

First, we show an example of a triple pendulum [55,56] that extends the 2-DOF example. Then, we look at the system reconstruction with partial Lagrangian when the second link is changed from a rotational joint to a linear motion joint. The calculation of $\mathcal{L}_3$ is an iteration of Equations (8)–(17) and is abbreviated here. The required equations of motion are as follows.

$$\tau_1 = \tau_{11} + \tau_{12} + \tau_{13} = \tau_{11} + \tau_{12} + D_1\mathcal{L}_3 \tag{25}$$
$$\tau_2 = \quad \tau_{22} + \tau_{23} = \quad \tau_{22} + D_2\mathcal{L}_3 \tag{26}$$
$$\tau_3 = \quad \tau_{33} = \quad D_3\mathcal{L}_3 \tag{27}$$

Since the only new term that needs to be calculated is the term related to $\mathcal{L}_3$, we can extend the link easily even by hand calculation, by using the postural operator as in the case up to the 2-DoF example. References [55,56] show that the Lagrangian becomes very complex with three DoF; thus, this approach has the advantage of performing the partitioning calculation with the partial Lagrangian.

This behavior is confirmed by simulation with automatic differentiation. The results of the damped oscillation are shown in Figure 3a, and the results with the PD control are shown in Figure 3b. The target angles for PD control are $\theta_{1ref} = 125^\circ$, $\theta_{2ref} = -30^\circ$, and $\theta_{3ref} = -30^\circ$, and its gains are $K_{p1} = 1000$, $K_{p2} = 1000$, $K_{p3} = 500$, $K_{d1} = 200$, $K_{d2} = 200$, and $K_{d3} = 100$. As with the 2-DoF case, this simulation can be handled simply by modifying two lines of the geometric constraint for $\mathcal{L}_3$; thus, the implementation cost is very small.

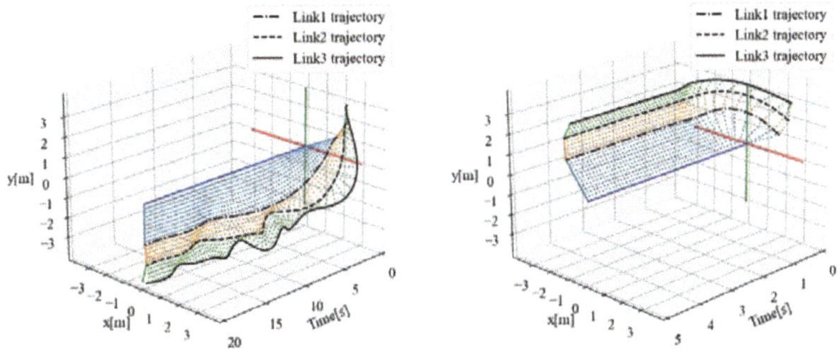

(a) (b)

Figure 3. Motion analysis of the 3-link system using automatic differentiation. (**a**) Damping oscillation; (**b**) PD control.

Next, as an example of system reconstruction, a simulation of a system in which the second link is changed to a linear joint is shown in Figure 4a,b. The second link is controlled to shorten its length by the PID control, and the other joints are the damped vibrations. The initial length of the second link is $l_2 = 1$ and its target length is $l_{2ref} = 0.1$. The results show that the desired behavior can be achieved even when the intermediate link is changed as an example of the system reconstruction. As mentioned above, in the automatic differentiation program, the simulation can be realized merely by rewriting the time-varying parameters of the geometric constraint.

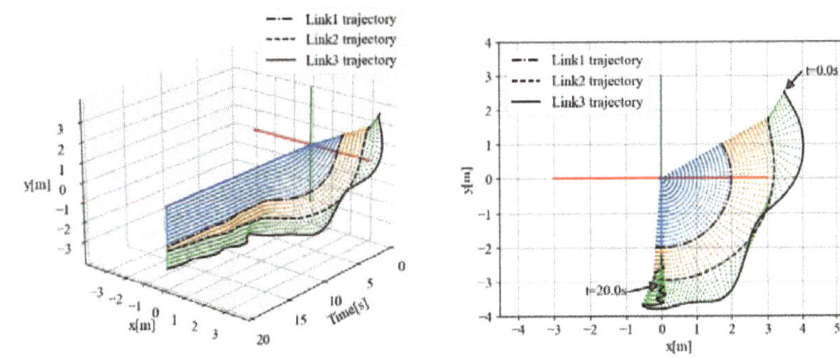

(a) (b)

Figure 4. System reconstruction example (link 2 is changed from rotational joint to linear slider). (**a**) Side view; (**b**) Front view.

4. Discussion

In this section, we discuss the advantages of the proposed partial Lagrangian. As mentioned above, the purpose of the usual Lagrangian is to derive an analytical solution; thus, its application to multiple degrees of freedom is difficult. However, by dividing it into partial Lagrangians, it can be handled relatively easily even with multiple degrees of freedom, and can be used to compute numerical solutions. In this section, we discuss computational complexity and scalability by considering extensions up to 10 links, as shown in Figure 5.

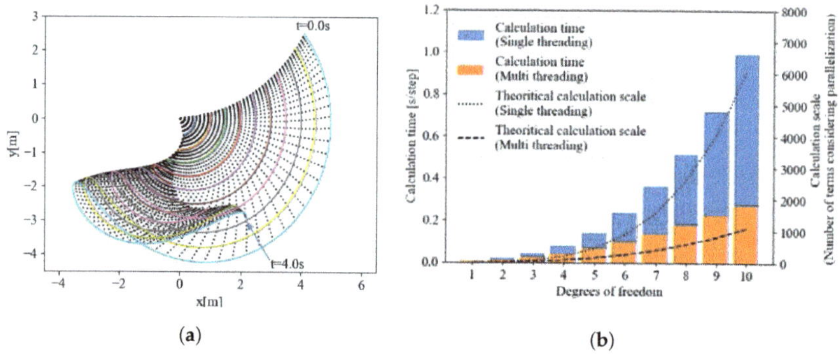

Figure 5. Experimental results of a large-scale model using the partial Lagrangian . (**a**) Damping oscillation by the 10-DoF pendulum; (**b**) computational cost of automatic differentiation (all simulations were performed by Intel Core i5-10400 CPU @2.90 GHz).

4.1. Computational Advantages of Partial Lagrangian

The usual Lagrangian is generally difficult to use with the multi-DoF system because the number of terms explodes with increasing degrees of freedom. The partial Lagrangian uses the divide-and-conquer approach to divide the computation of the Lagrange equation into its smallest module, $D_k\mathcal{L}_i = \tau_{ik}$. This makes it relatively easy to obtain an exact analytical solution eventually by repeating the same simple procedure, even if the number of DoF is increased.

We consider the advantages of the partial Lagrangian from two perspectives: the derivation of the equations of motion by hand calculations and computer simulations.

First, for manual calculations, the introduction of the postural operator together with the partial Lagrangian simplifies the calculation with respect to rotational joints and significantly reduces the number of calculations compared to those that would otherwise be required. In addition, divide-and-conquer by the partial Lagrangian prevents calculation errors due to manual calculations because the number of calculations is small for each module.

Second, for computer simulations, the gradient calculation of the calculation graph by automatic differentiation was used as an efficient method to process the formal description of the divide-and-conquer approach by the partial Lagrangian, $D_k\mathcal{L}_i = \tau_{ik}$. Automatic differentiation is a different method of analysis from mathematical and numerical differentiation. Automatic differentiation can solve the inefficiency of mathematical differentiation and the accuracy problem of numerical differentiation. It is compatible with partial Lagrangians because it can describe partial differential equations directly and perform the calculation of the balance. The partial Lagrangian allows for efficient simulation by iterating the modularized process through automatic differentiation and integrating it at the necessary stages.

As an example of this, a simulation in which the model is extended to 10 links is shown in Figure 5a. Here, we consider Figure 5b as an efficiency improvement by dividing it into partial Lagrangians. By dividing, computers can compute independent components using multi-threading. The order of computational amount of n-DoF without dividing can be estimated as the number of terms as follows. The $\mathcal{O}$ means Landau symbol.

$$\underbrace{\frac{1}{2}(n^2+n)}_{\text{Number of all } \tau_{ki}} \times \underbrace{(n^2+2n)}_{\text{Number of terms in } \tau_{ki}} = \mathcal{O}(n^4) \tag{28}$$

The fact that the computational cost is $\mathcal{O}(n^4)$ in the analysis of general dynamics is stated in the reference [26] and supports this result.

On the contrary, using partial Lagrangian, the generation of the computational graph and the gradient calculation (i.e., torque calculation) with multi-threading is shown below.

$$\underbrace{n}_{\substack{\text{Max number of dividing} \\ \text{into partial Lagrangian}}} \times \underbrace{(n^2+2n)}_{\text{Number of terms in } \tau_{ki}} = \mathcal{O}(n^3) \tag{29}$$

As shown above, the division into partial Lagrangians allows the linearly independent parts to be computed in parallel, lowering the order of computation from $\mathcal{O}(n^4)$ to $\mathcal{O}(n^3)$. These computational amounts are consistent with the results of the actual simulation shown in Figure 5b.

4.2. Application of Partial Lagrangian as an Extension or Restructuring of the System

Another advantage of the partial Lagrangian is that it can handle system extensions and reconstructions in units of divided modules. This is due to the linear independence of each column of Table 1. The pendulum examples shown in Section 3 show the flow of the system extension from one to three DoF. The system extended to 10-DoF to evaluate the computational amount in the previous section is an extension of the 3DoF model with additional link parameters, all computed with the same partial Lagrangian module.

With the partial Lagrangian, the calculations required for energy and partial derivatives are about half of those of the usual Lagrangian. This makes it easier to handle the multi-DoF system than the usual Lagrangian, but the difficulty increases with the DoF number. Moreover, if the middle link is changed, it is necessary to go back that far in Table 1. However, in the case of simulation with automatic differentiation, energy calculation and partial differentiation are performed automatically, so system reconstruction, such as the intermediate link changes, can be performed merely by rewriting the geometric constraint part. Therefore, similar to the conclusion of the computational advantage, the divide-and-conquer approach using the partial Lagrangian and module-by-module processing using an automatic differentiation algorithm allows for the easy dynamics analysis of complicated systems and their configuration changes with only a formal description of the partial Lagrangian.

5. Conclusions

We proposed the partial Lagrangian to handle the Lagrange equation efficiently using a divide-and-conquer approach. The partial Lagrangian makes it possible to handle extensions and reconstructions of the system easily to obtain analytical solutions, which was not previously possible. In addition, the introduction of the postural operator together with the partial Lagrangian facilitates the derivation of the equations of motion by hand calculation. The division of computations by the partial Lagrangian and reduction of computational complexity by the postural operator reduce the computational cost of manual calculations of energy and partial derivatives, even with multi-DoF.

Furthermore, we proposed a numerical method of computing the partial Lagrangian using automatic differentiation for simulation. Since automatic differentiation calculates the energy and partial derivatives, it is compatible with the partial Lagrangian formal description, and the system designer needs only to describe the geometric constraints to analyze the motion using the partial Lagrangian's divide-and-conquer approach.

We confirmed that the division into partial Lagrangians allows us to take advantage of multi-threading, reducing the computational complexity that normally requires $\mathcal{O}(n^4)$ to $\mathcal{O}(n^3)$. Even when extended to the multi-DoF system, the numerical solution can be obtained efficiently by repeatedly calling the partial Lagrangian module. As an example of a multi-DoF, we confirmed its effectiveness with a 10-DoF pendulum model.

By replacing or combining the proposed method with conventional systems, the system extensions and simulations can be realized more efficiently than previously possible in the field of design and control using analytical solutions of the conventional equations of motion. However, because the partial Lagrangian uses the linear independence of each DoF to split its computation, the split may not work well if its assumptions change. Several recent studies have shown that recursive algorithmic solutions are also effective in parallel-linked systems [27,28]. Therefore, our future work will include applications to systems other than the serial link manipulators treated in this study, such as closed-link systems and parallel-link systems.

Author Contributions: Conceptualization, T.K.; investigation, T.K.; methodology, T.K.; supervision, T.T.; validation, T.K. and T.T.; writing—original draft, T.K.; writing—review and editing, T.T. All authors have read and agreed to the published version of the manuscript.

Funding: This research received no external funding

Institutional Review Board Statement: Not applicable.

Informed Consent Statement: Not applicable.

Data Availability Statement: Not applicable.

Conflicts of Interest: The authors declare no conflict of interest.

Appendix A. A Sample Analysis of Partial Lagrangian Using Automatic Differentiation

A sample analysis of an edited differential equation with automatic differentiation is shown below. We begin with a sample program for the case of one degree of freedom, implemented using PyTorch [57]. In item 0, we first define the constraints for each state variable and time evolution on the computational graph. In item 1, define the geometric constraints for each system. This is the only part that needs to be rewritten when the system changes. In item 2, the partial Lagrangian is defined on the computational graph. The equations of motion with friction are used here to make the behavior easier to understand in the sample graph. In item 3, the constraints of the equations of motion are defined on the computational graph as edited differential equations using partial Lagrangians. Item 4 performs the backpropagation of the computed graph. In item 5, finally, the dynamics-aware inertia term I is obtained as a coefficient of $\ddot{\theta}$: acc, and the state variables updated by backpropagation are stored as internal variables.

As a precaution, since the Lagrangian is an edited differential equation, the Lagrangian must be registered in the computational graph as functions of θ, ω, and t. Therefore, the time evolution constraint is written in the program so that the variable theta is a function of omega and dt.

The computational graph generated by this sample is shown in Figure A1. In practice, programmers can analyze partial differential equations automatically without dealing directly with this complex graph, but only by writing algorithms similar to the sample code.

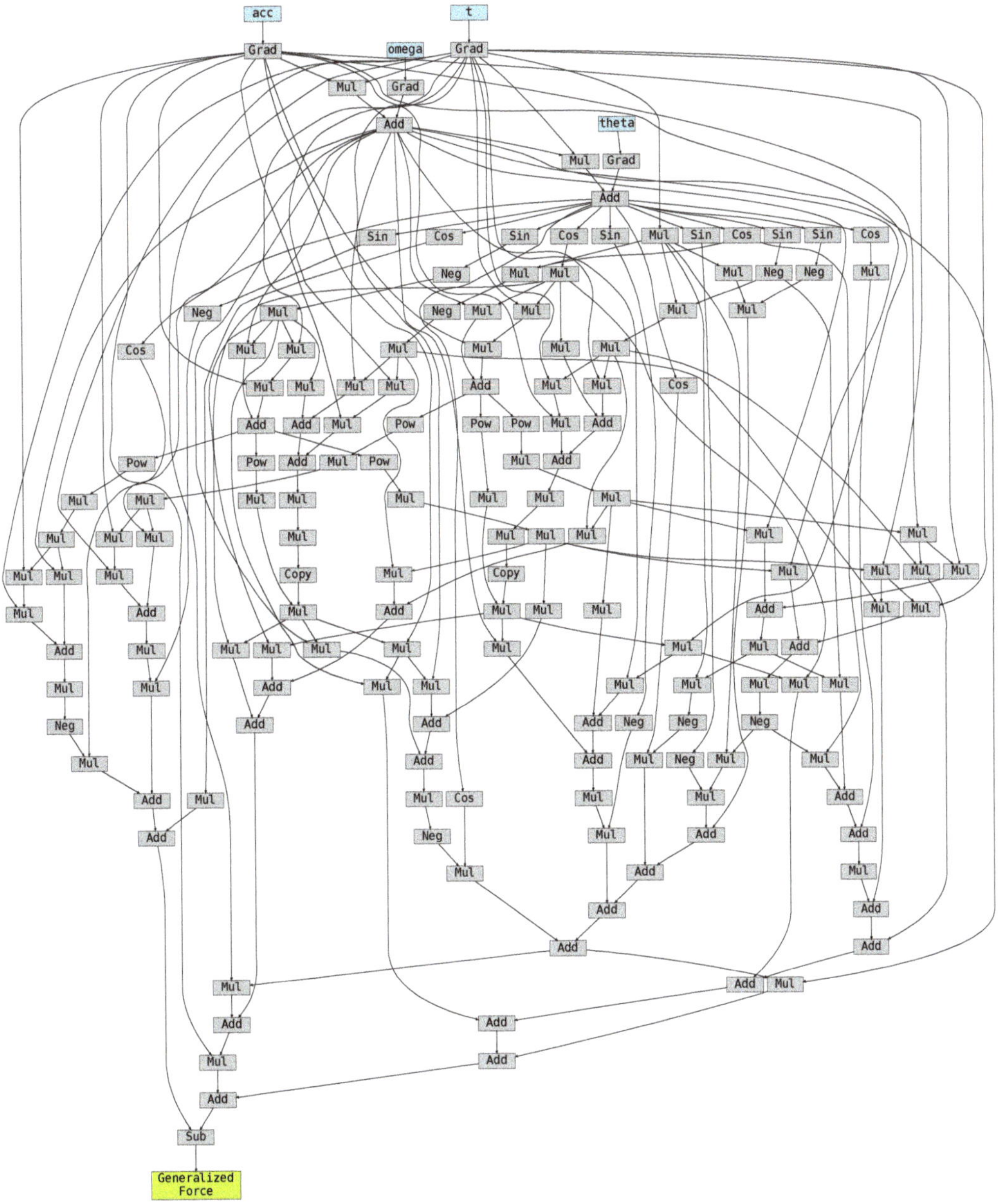

Figure A1. Example of a generated calculation graph of a partial Lagrangian (reading downward: forward calculations as inverse dynamics; reading upward: backward calculations as dynamics).

Listing A1. Example of the partial Lagrangian module.

```
# 0. State variables and their constraints of time evolution for AD
(_theta, _omega, _acc) = state
acc = torch.tensor(_acc, requires_grad=True)
omega = torch.tensor(_omega, requires_grad=True) + acc * self.dt
theta = torch.tensor(_theta, requires_grad=True) + omega *self.dt
# 1. Geometric constraints and their derivatives
x = self.l * torch.cos(theta)
y = self.l * torch.sin(theta)
dx = torch.autograd.grad(x, self.dt, create_graph=True)
dy = torch.autograd.grad(y, self.dt, create_graph=True)
# 2. (Partial) Lagrangian with friction
K = 0.5*self.m * (dx[0]**2+dy[0]**2)
P = 0.5*self.k * (x**2+y**2) + self.m*9.8*y
R = 0.5*self.c * (dx[0]**2+dy[0]**2)
L = K-P
# 3. Lagrange Equation of Motion
dLdth = torch.autograd.grad(L, theta, create_graph=True)
dLdomega = torch.autograd.grad(L, omega, create_graph=True)
dLdt = torch.autograd.grad(dLdomega, self.dt, create_graph=True)
dRddth = torch.autograd.grad(R, omega, create_graph=True)
Iacc = dLdth[0] - dRddth[0]
T = dLdt[0] - Iacc # balance
# 4. Backpropagation
T.backward(retain_graph=True)
# 5. Convert force to acc
I = acc.grad
acc = float(Iacc/I)
omega += acc * self.dt
state = (theta, omega, acc)
return state
```

References

1. de Jesus Rubio, J.; Humberto Perez Cruz, J.; Zamudio, Z.; Salinas, A.J. Comparison of two quadrotor dynamic models. *IEEE Lat. Am. Trans.* **2014**, *12*, 531–537.
2. Ali, S. Newton-Euler Approach for Bio-Robotics Locomotion Dynamics: From Discrete to Continuous Systems. Ph.D. Thesis, Ecole des Mines de Nantes, Nantes, France, 2011; p. 251.
3. Sciavicco, L.; Siciliano, B.; Luigi, V. Lagrange and Newton-Euler dynamic modeling of a gear-driven robot manipulator with inclusion of motor inertia effects. *Adv. Robot.* **1995**, *10*, 317–334. [CrossRef]
4. Robotic Systems Lab, ETH Zurich Robot Dynamics Lecture Notes. 2017. Available online: https://ethz.ch/content/dam/ethz/special-interest/mavt/robotics-n-intelligent-systems/rsl-dam/documents/RobotDynamics2017/RD_HS2017script.pdf (accessed on 20 September 2022).
5. De Luca, A.; Ferrajoli, L. A modified newton-euler method for dynamic computations in robot fault detection and control. In Proceedings of the 2009 IEEE International Conference on Robotics and Automation, Kobe, Japan, 12–17 May 2009; pp. 3359–3364. ISSN 1050-4729. [CrossRef]
6. Buondonno, G.; De Luca, A. A recursive Newton-Euler algorithm for robots with elastic joints and its application to control. In Proceedings of the 2015 IEEE/RSJ International Conference on Intelligent Robots and Systems (IROS), Hamburg, Germany, 28 September–2 October 2015; pp. 5526–5532. [CrossRef]
7. Li, X.; Nishiguchi, J.; Minami, M.; Matsuno, T.; Yanou, A. Iterative calculation method for constraint motion by extended newton-euler method and application for forward dynamics. In Proceedings of the 2015 IEEE/SICE International Symposium on System Integration (SII), Nagoya, Japan, 11–13 December 2015; pp. 313–319. [CrossRef]
8. Hirata, Y.; Iwano, T.; Tajika, M.; Kosuge, K. *Motion Control of Wearable Walking Support System with Accelerometer Based on Human Model*; Advances in Human-Robot Interaction; IntechOpen: London, UK, 2009.
9. Benaddy, A.; Labbadi, M.; Bouzi, M. Adaptive Nonlinear Controller for the Trajectory Tracking of the Quadrotor with Uncertainties. In Proceedings of the 2020 2nd Global Power, Energy and Communication Conference (GPECOM), Izmir, Turkey, 20–23 October 2020; pp. 137–142. [CrossRef]

10. Rameez, M.; Khan, L.A. Modeling and dynamic analysis of the biped robot. In Proceedings of the 2015 15th International Conference on Control, Automation and Systems (ICCAS), Busan, South Korea, 13–16 October 2015; pp. 1149–1153. ISSN 2093-7121. [CrossRef]
11. Nicotra, M.M.; Garone, E. Control of Euler-Lagrange systems subject to constraints: An Explicit Reference Governor approach. In Proceedings of the 2015 54th IEEE Conference on Decision and Control (CDC), Osaka, Japan, 15–18 December 2015; pp. 1154–1159. [CrossRef]
12. Su, B.; Gong, Y. Euler-Lagrangian modeling and exact trajectory following controlling of Ballbot-like robot. In Proceedings of the 2017 IEEE International Conference on Robotics and Biomimetics (ROBIO), Macau, Macao, 5–8 December 2017; pp. 2325–2330. [CrossRef]
13. Al-Shuka, H.F.N.; Corves, B.; Zhu, W.H. *Dynamics of Biped Robots during a Complete Gait Cycle: Euler-Lagrange vs. Newton-Euler Formulations*; Research Report; School of Control Science and Engineering, Shandong University: Jinan, China, 2019. Available online: https://hal.archives-ouvertes.fr/hal-01926090 (accessed on 19 September 2022).
14. Kusaka, T.; Tanaka, T.; Kaneko, S.; Suzuki, Y.; Saito, M.; Kajiwara, H. Assist Force Control of Smart Suit for Horse Trainers Considering Motion Synchronization. *Int. J. Autom. Technol.* **2009**, *3*, 723–730.
15. Kusaka, T.; Tanaka, T.; Kaneko, S.; Suzuki, Y.; Saito, M.; Kajiwara, H. Smart suit for horse trainers-power and skill assist based on semi-active assist and energy control. In Proceedings of the 2010 IEEE/ASME International Conference on Advanced Intelligent Mechatronics, Montreal, QC, Canada, 6–9 July 2010; pp. 509–514. ISSN 2159-6255. [CrossRef]
16. Kusaka, T.; Tanaka, T.; Kaneko, S.; Suzuki, Y.; Saito, M.; Seki, S.; Sakamoto, N.; Kajiwara, H. Assist force control of Smart Suit for horse trainer considering motion synchronization and postural stabilization. In Proceedings of the 2009 ICCAS-SICE, Fukuoka, Japan, 18–21 August 2009; pp. 770–775.
17. Tsuchiya, Y.; Imamura, Y.; Tanaka, T.; Kusaka, T. Estimating Lumbar Load During Motion with an Unknown External Load Based on Back Muscle Activity Measured with a Muscle Stiffness Sensor. *J. Robot. Mechatron.* **2018**, *30*, 696–705. [CrossRef]
18. Tsuchiya, Y.; Kusaka, T.; Tanaka, T.; Matsuo, Y. Wearable Sensor System for Lumbosacral Load Estimation by Considering the Effect of External Load. In *Proceedings of the Advances in Human Factors in Wearable Technologies and Game Design*; Advances in Intelligent Systems and Computing; Ahram, T.; Falcão, C., Eds.; Springer International Publishing: Cham, Switzerland, 2018; pp. 160–168. ._16. [CrossRef]
19. Nethery, J.; Spong, M. Robotica: A Mathematica package for robot analysis. *IEEE Robot. Autom. Mag.* **1994**, *1*, 13–20.
20. Sadati, S.H.; Naghibi, S.E.; Shiva, A.; Michael, B.; Renson, L.; Howard, M.; Rucker, C.D.; Althoefer, K.; Nanayakkara, T.; Zschaler, S.; et al. TMTDyn: A Matlab package for modeling and control of hybrid rigid–continuum robots based on discretized lumped systems and reduced-order models. *Int. J. Robot. Res.* **2021**, *40*, 296–347.
21. Sadati, S.M.H.; Naghibi, S.E.; Naraghi, M. An Automatic Algorithm to Derive Linear Vector Form of Lagrangian Equation of Motion with Collision and Constraint. *Procedia Comput. Sci.* **2015**, *76*, 217–222. [CrossRef]
22. Cormen, T.H.; Leiserson, C.E.; Rivest, R.L.; Stein, C. *Introduction to Algorithms*, 3rd ed.; The MIT Press: Cambridge, MA, USA; London, UK, 2009.
23. Knuth, D.E. *The Art of Computer Programming*; Addison-Wesley Pub. Co.: Reading, MA, USA, 1968.
24. Poursina, M.; Anderson, K.S. An extended divide-and-conquer algorithm for a generalized class of multibody constraints. *Multibody Syst. Dyn.* **2013**, *29*, 235–254. [CrossRef]
25. Critchley, J.; Binani, A.; Anderson, K. Design and Implementation of an Efficient Multibody Divide and Conquer Algorithm. In Proceedings of the International Design Engineering Technical Conferences and Computers and Information in Engineering Conference, Las Vegas, NV, USA, 4–7 September 2007. [CrossRef]
26. Mukherjee, R.M.; Bhalerao, K.D.; Anderson, K.S. A divide-and-conquer direct differentiation approach for multibody system sensitivity analysis. *Struct. Multidiscip. Optim.* **2008**, *35*, 413–429. [CrossRef]
27. Zahedi, A.; Shafei, A.M.; Shamsi, M. On the dynamics of multi-closed-chain robotic mechanisms. *Int. J. Non-Linear Mech.* **2022**, *147*, 104241. [CrossRef]
28. Mazare, M.; Taghizadeh, M. Geometric Optimization of a Delta Type Parallel Robot Using Harmony Search Algorithm. *Robotica* **2019**, *37*, 1494–1512.
29. Baydin, A.G.; Pearlmutter, B.A.; Radul, A.A.; Siskind, J.M. Automatic Differentiation in Machine Learning: A Survey. *J. Mach. Learn. Res.* **2018**, *18*, 43.
30. Neidinger, R.D. Introduction to Automatic Differentiation and MATLAB Object-Oriented Programming. *SIAM Rev.* **2010**, *52*, 545–563. [CrossRef]
31. Naumann, U. *The Art of Differentiating Computer Programs*; Software, Environments, and Tools, Society for Industrial and Applied Mathematics: Philadelphia, PA, USA, 2011. [CrossRef]
32. Harrison, D. A Brief Introduction to Automatic Differentiation for Machine Learning. *arXiv* **2021**, arXiv:2110.06209.
33. Al Seyab, R.; Cao, Y. Nonlinear system identification for predictive control using continuous time recurrent neural networks and automatic differentiation. *J. Process Control* **2008**, *18*, 568–581. [CrossRef]
34. Griewank, A.; Walther, A. *Evaluating Derivatives: Principles and Techniques of Algorithmic Differentiation*, 2nd ed.; Society for Industrial and Applied Mathematics: Philadelphia, PA, USA, 2008. [CrossRef]
35. Hollerbach, J.M. A Recursive Lagrangian Formulation of Maniputator Dynamics and a Comparative Study of Dynamics Formulation Complexity. *IEEE Trans. Syst. Man Cybern.* **1980**, *10*, 730–736.

36. Luh, J.Y.; Walker, M.W.; Paul, R.P. On-line computational scheme for mechanical manipulators. *Trans. ASME J. Dyn. Syst. Meas. Control* **1980**, *102*, 69–76. [CrossRef]
37. Fox, H.; Bolton, W. *Mathematics for Engineers and Technologists*; Butterworth-Heinemann: Oxford, UK, 2002.
38. Rawlins, J.C.; Fulton, S.R.B.A.C. *Basic AC Circuits*; Newnes: Boston, MA, USA, 2000.
39. Robbins, A.; Miller, W.C. *Circuit Analysis: Theory and Practice*; DELMAR Cengage Learning: Clifton Park, NY, USA, 2013.
40. McCarthy, J.M. *An Introduction to Theoretical Kinematics*; MIT Press: Cambridge, MA, USA, 1990.
41. Vicci, L. *Quaternions and Rotations in 3-Space: The Algebra and its Geometric Interpretation*; Technical Report; UNC: Chapel Hill, NC, USA, 2001.
42. Shoemake, K. Quaternions. 1994. Available online: https://web.archive.org/web/20200503045740/http://www.cs.ucr.edu/~vbz/resources/quatut.pdf (accessed on 11 September 2022).
43. Dam, E.B.; Koch, M.; Lillholm, M. *Quaternions, Interpolation and Animation*; Technical Report; Datalogisk Institut, Kobenhavns Universitet: Copenhagen, Denmark, 1998.
44. Bar-Itzhack, I. New Method for Extracting the Quaternion from a Rotation Matrix. *J. Guid. Control. Dyn.* **2000**, *23*, 1085–1087. [CrossRef]
45. Ma, F.; Liu, W.; Ma, T. Automatic differentiation application on stochastic finite element. In Proceedings of the 2010 International Conference on Computer, Mechatronics, Control and Electronic Engineering, Changchun, China, 24–26 August 2010; Volume 2, pp. 219–221. ISSN 2159-6034. [CrossRef]
46. Jeßberger, J.; Marquardt, J.E.; Heim, L.; Mangold, J.; Bukreev, F.; Krause, M.J. Optimization of a Micromixer with Automatic Differentiation. *Fluids* **2022**, *7*, 144.
47. Enciu, P.; Gerbaud, L.; Wurtz, F. Automatic Differentiation Applied for Optimization of Dynamical Systems. *IEEE Trans. Magn.* **2010**, *46*, 2943–2946.
48. Kaheman, K.; Brunton, S.L.; Kutz, J.N. Automatic Differentiation to Simultaneously Identify Nonlinear Dynamics and Extract Noise Probability Distributions from Data. *arXiv* **2020**, arXiv:2009.08810.
49. Feldmann, P.; Melville, R.; Moinian, S. Automatic differentiation in circuit simulation and device modeling. In Proceedings of the 1992 IEEE/ACM International Conference on Computer-Aided Design, Santa Clara, CA, USA, 8–12 November 1992; pp. 248–253. [CrossRef]
50. Toshiji, K. Generalization of Circuit Simulator by Automatic Differentiation. *IEEJ Trans. Electron. Inf. Syst.* **2004**, *124*, 404–410. [CrossRef]
51. Christoffersen, C. Implementation Of Exact Sensitivities in a Circuit Simulator Using Automatic Differentiation. In Proceedings of the 20th European Conference on Modelling and Simulation, Bonn, Germany, 28–31 May 2006. [CrossRef]
52. Ueding, M. Lagrange Examples, 2013. Available online: https://martin-ueding.de/posts/lagrange-examples/ (accessed on 19 September 2022).
53. Assencio, D. The Double Pendulum: Lagrangian Formulation—Diego Assencio. Available online: https://diego.assencio.com/?index=1500c66ae7ab27bb0106467c68feebc6 (accessed on 19 September 2022).
54. Weisstein, E.W. *Double Pendulum—From Eric Weisstein's World of Physics*; Wolfram Research, Inc.: Champaign, IL, USA, 2018. Available online: https://scienceworld.wolfram.com/physics/DoublePendulum.html (accessed on 19 September 2022).
55. Harman, Nick N. Motion of a Triple Rod Pendulum. Available online: https://www.authorea.com/users/259349/articles/412491-motion-of-a-triple-rod-pendulum (accessed on 19 September 2022).
56. Jake VanderPlas, Triple Pendulum CHAOS! | Pythonic Perambulations. Available online: https://jakevdp.github.io/blog/2017/03/08/triple-pendulum-chaos/ (accessed on 19 September 2022).
57. Automatic Differentiation Package—torch.autograd—PyTorch 1.12 Documentation. Available online: https://pytorch.org/docs/stable/autograd.html (accessed on 21 September 2022).

robotics

MDPI

Review

Current Designs of Robotic Arm Grippers: A Comprehensive Systematic Review

Jaime Hernandez [1,†,‡], Md Samiul Haque Sunny [2,*,‡], Javier Sanjuan [1,‡], Ivan Rulik [2,‡], Md Ishrak Islam Zarif [3,‡], Sheikh Iqbal Ahamed [3], Helal Uddin Ahmed [4] and Mohammad H Rahman [1,2,4]

1 Department of Mechanical Engineering, University of Wisconsin-Milwaukee, Milwaukee, WI 53212, USA
2 Department of Computer Science, University of Wisconsin-Milwaukee, Milwaukee, WI 53212, USA
3 Department of Computer Science, Marquette University, Milwaukee, WI 53233, USA
4 Biorobotics Laboratory, University of Wisconsin Milwaukee, Milwaukee, WI 53212, USA
* Correspondence: msunny@uwm.edu
† Current address: Biorobotics Lab, University of Wisconsin Milwaukee, 115 East Reindl Way, USR 281, Milwaukee, WI 53212, USA.
‡ These authors contributed equally to this work.

Abstract: Recent technological advances enable gripper-equipped robots to perform many tasks traditionally associated with the human hand, allowing the use of grippers in a wide range of applications. Depending on the application, an ideal gripper design should be affordable, energy-efficient, and adaptable to many situations. However, regardless of the number of grippers available on the market, there are still many tasks that are difficult for grippers to perform, which indicates the demand and room for new designs to compete with the human hand. Thus, this paper provides a comprehensive review of robotic arm grippers to identify the benefits and drawbacks of various gripper designs. The research compares gripper designs by considering the actuation mechanism, degrees of freedom, grasping capabilities with multiple objects, and applications, concluding which should be the gripper design with the broader set of capabilities.

Keywords: robotic arm; gripper design; actuation mechanism; grasping capabilities; object manipulation

Citation: Hernandez, J.; Sunny, M.S.H.; Sanjuan, J.; Rulik, I.; Zarif, M.I.I.; Ahamed, S.I.; Ahmed, H.U.; Rahman, M.H. Current Designs of Robotic Arm Grippers: A Comprehensive Systematic Review. *Robotics* **2023**, *12*, 5. https://doi.org/10.3390/robotics12010005

Academic Editor: Raffaele Di Gregorio

Received: 7 November 2022
Revised: 8 December 2022
Accepted: 28 December 2022
Published: 2 January 2023

1. Introduction

The ability to grip and manipulate objects has been central to the advancement of robots [1–10]. Manufacturers can use end-effector tooling for picking, placing, and packing objects using advances in gripper technology to reap the benefits of precision, performance, and productivity [11]. Grippers are classified depending on their design, how they are powered, and their application. For example, when considering industrial grippers, one of the simplest designs is the parallel motion two-jaw gripper, commonly used to lift objects [12–15]. Several other design types include the O-ring gripper [16], and the needle gripper [17]. Industrial grippers can be hydraulic, pneumatic, or electric, depending on the application requirements [2,18,19]. However, although the number of grippers currently available on the market has been increasing over the years, this does not change the fact that there are still many complex tasks that robots cannot accomplish.

A limitation of robotic grippers occurs when holding fragile objects with the correct force [3,6,8–10]. For example, a gripper handling fruit or food must grasp the fruit firmly enough so it will not slip out of their grasp but be gentle so the fruit will not get damaged, while human fingers are soft and can conform to objects, this is not inherent in a robotic gripper, typically made of metal or other materials with a hard surface. To mitigate this issue, designers developed grippers with softer materials, allowing robotic grippers to handle fragile objects, creating the subject of soft robotics. Soft robotics is a sub-field of robotics that features robots made with soft materials similar to living organisms, such as an octopus' tentacles or a human's fleshy finger. Recent advancements in soft robotics allow robots to overcome traditional challenges and expand into new fields [20,21].

Another challenge for gripper design is dexterity. Many traditional gripper designs have two or three fingers made of rigid material. Even though they can do pick and place tasks effectively, they are not suited to more complex manipulation activities [22,23]. For the design to be functional and successful, it needs to generate complex geometries, mechanically adapt to the shape of an object, specialize in grasping and manipulating with ultra-sensitive touch sensors, and have a low impact energy to achieve close resemblance to a human hand [24].

This article explores grippers' most recent industrial and research designs to answer the question: What gripper design can handle most objects independently of their fragility, shape, and weight? Thus, we classified the grippers considering the type of mechanical design, number of degree of freedom (DOF), the type of actuation, and the form of the grasping objects, concentrating our study on analyzing which gripper yields the best handling capabilities.

The article is organized as follows: Section 2 presents the classification of grippers based on their degrees of freedom and design, focusing on the advantages and limitations of each gripper design. Section 3 shows the organization of the grippers considering their grasping capability in terms of the size, the shape, and the material of the handled object. Section 4 presents the conclusions. Finally, the Appendix A presents the methodology for selecting and organizing the articles for review.

2. Design Configurations for Robotic Arm Grippers

To understand how grippers are designed, first understand how humans interact with, hold, and move objects during daily activities. A Max Planck Institute for Intelligent Systems study trained computers to understand, model, and synthesize human grasping [25]. The analysis of the study includes complex 3D object shapes, detailed contact information, hand pose and shape, and 3D body motion. Similar analyses were conducted in [26,27], in which the types of grips used were classified based on the type of object, its shape, and its weight. Figure 1 shows different grasping types. According to what was previously stated, gripping modes are also classified based on the object's shape, dividing them into three broad categories [28], which are shown in Figure 2: parallel or flat gripping mode, cylindrical gripping mode, and spherical gripping mode [29]. Other categories derived from these three, such as Tip mode, Hooke mode, and Lateral mode, are presented as subsets of the main categories. Lateral mode, for example, is a subset of parallel mode in which the object's thickness is hundreds of times less than its perpendicular area. Based on the mobility of the robotic grippers, exists three main categories that classify the design of the grippers: Completely constrained, underconstrained, and deformable. Inside those categories, there are various subdivisions based on the actuation mechanisms, as presented in Table 1. This study reviews each group of robot arm grippers, focusing on the advantages and disadvantages of each classification.

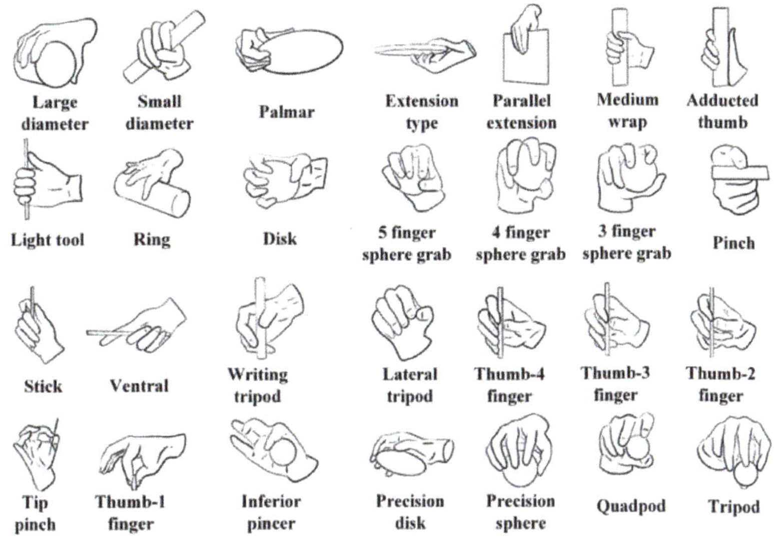

Figure 1. Categorization of grasping considering the required power and precision for different object shapes and wrap types.

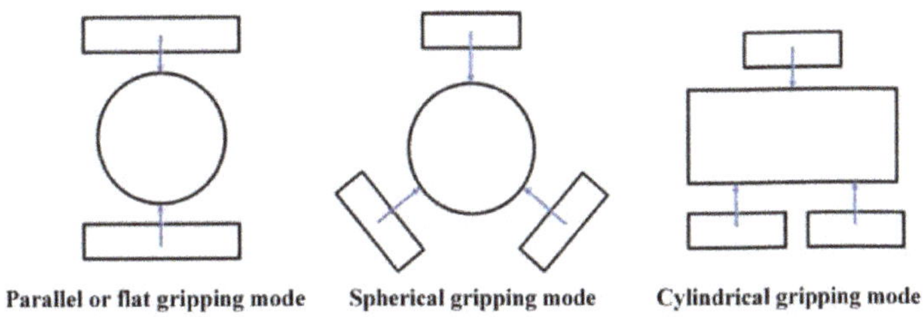

Figure 2. Generalized classification of gripping modes based on the object shapes.

Table 1. Categories of robotic grippers based on design configurations and actuation mechanism.

Completely constrained	Compliant mechanism	Cable driven	[6]
		linear actuator	[1,30–32]
	Rigid links	Linear actuator	[33–35]
		Rotary actuator	[36–38]
		Cable driven	[4,39,40]
		Electromagnet	[41,42]
Underconstrained	Compliant mechanism	unspecified	[43]
		Cable driven	[14,44]
		Rotary actuator	[45]
		Linear actuator	[46]
		Piezo actuator	[47,48]
	Rigid links	Linear actuator	[15,49,50]
		Rotary actuator	[10,51–54]
		Cable driven	[5,9,12,13,55–62]
		Pneumatic actuation	[63]
Deformable	Single mass gripper	Vacuum	[2,11,64]
		Cable driven	[8,65,66]
	single mass finger	Pneumatic/Hydraulic actuation	[7,67–73]
		Dielectric elastomer (DE) actuator	[74]
		Linear actuator	[75]
	Square continuum robot	Cable driven	[76]

2.1. Completely Constrained Gripper Mechanism

Completely constrained finger mechanisms are devices with a DOF equivalent to their number of actuators, which allows the trajectory of the tip of the finger to follow a predefined path. Note that the number of DOF is computed using the Gruebler-Kutzbach [77] equation presented below:

$$M = 3L - 2J \tag{1}$$

where M is the total DOF; L is the number of links, and J is the number of joints. Then, consider the completely constrained mechanism presented in Figure 3, with three links (L_i) and four joints(j_i); applying Equation (1), the total number of DOF is one. Thus, this robot only needs one actuator to generate motion. This property permits the device to generate high output torque, allowing the gripper to hold heavy-weight objects [30,35]. Most of these devices use 1 DOF to control the motion of the gripper, which limits the complexity of the items that the gripper can handle [33]. To solve this issue, researchers have included more DOF in the grippers to enhance their ability to handle complex objects at the expense of their output torque [37]. Based on their design, completely constrained finger mechanisms have two classifications: compliant mechanisms and rigid links.

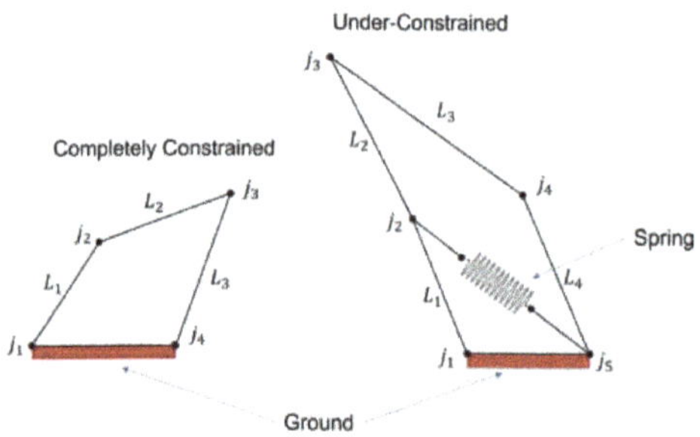

Figure 3. Comparison of a fully constrained mechanism with an underconstrained mechanism.

2.1.1. Compliant Mechanism

A compliant mechanism is a flexible mechanism that transmits force and motion through elastic deformations. Compliant mechanisms have a reduced number of moving parts which makes them light. Besides, friction impacts compliant mechanisms more than rigid links because they require fewer assembly parts. Moreover, the fewer assembly parts reduce undesirable nonlinear effects like backlash and noise on compliant mechanisms. Compliant mechanisms are usually of 3D printed materials, reducing their manufacturing cost. However, since the links are flexible, their considerably weaker than rigid links, reducing the output torque capabilities [78]. A usual design strategy for compliant mechanisms is topological optimization as proposed in [6] for a gripper of 3-flexible fingers. The principal purpose of the optimization was to facilitate the modeling of interactions between the gripper and the objects. Figure 4 presents the stages of optimization for the gripper design. The optimization model considers the loading pressure and traction frictions for this case to obtain the objective function. The obtained device uses pulleys and cables for actuation.

Another example of a compliant mechanism is used for each finger of a 3-finger flexible gripper in [1]. The finger mechanism performs linear motions as presented in Figure 5. The finger mechanism was fabricated with a thermoplastic elastomer (TPE) and was optimized for interactions with unpredictable environments and to handle delicate objects of different sizes. Additionally, the gripper mechanism has only one linear actuator to actuate the three fingers simultaneously, which generates the same displacement on each finger.

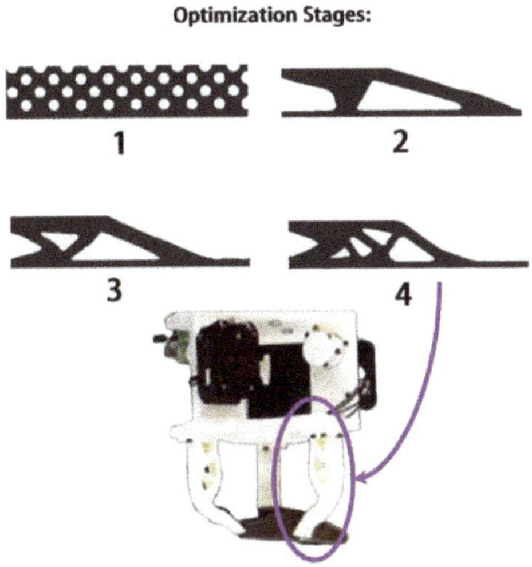

Figure 4. Topological optimization stages of the finger compliant mechanism. Reprinted with permission from ref. [6]. Copyright 2018 IEEE.

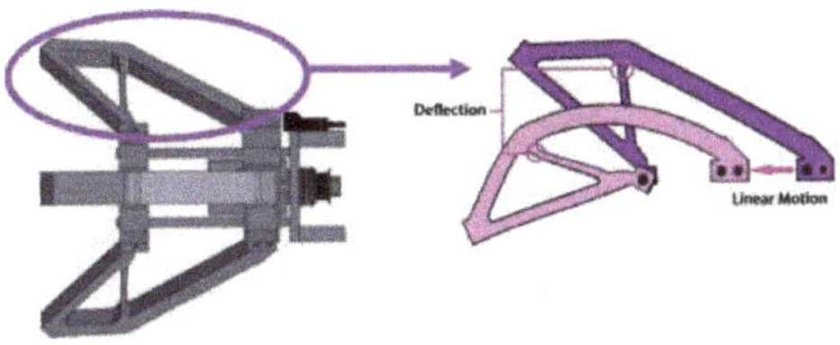

Figure 5. Flexible finger performing linear motion. Reprinted with permission from ref. [1]. Copyright 2020 IEEE.

A similar design was proposed in [30], focusing on the optimal design of a 3D-printed constant force-compliant finger. This finger mechanism uses a force regulation strategy for high-speed handling of fragile objects. Details of the regulation strategy are presented in [31]. Figure 6 shows the gripper actuation system, including the jaw and the FRM. The gripper mechanism uses pneumatic actuation and has two complaint mechanisms acting as springs.

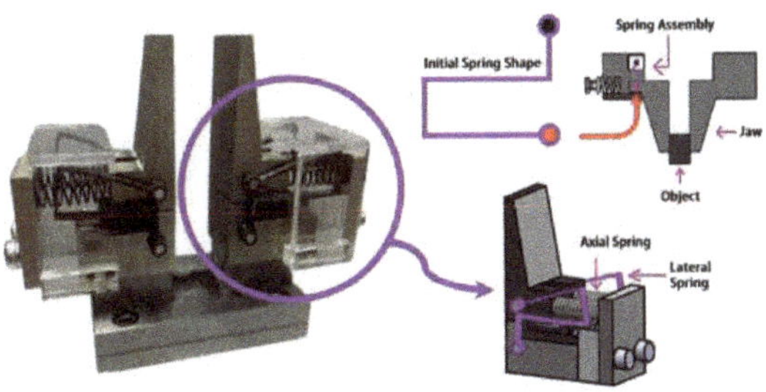

Figure 6. Actuation system of a gripper including the jaw and FRM. Reprinted with permission from ref. [31]. Copyright 2018 IEEE.

Other grippers integrate sensors in the design of the compliant gripper, as in [32]. This compliant gripper uses an integrated position and grasping/interaction force sensor for automated micro-assembly tasks. However, the integrated sensor limits the workspace to 2.2 mm with a grasping force of 16 mN. Although, this may not be an impending feature for the particular application of this gripper. Figure 7 presents the complaint mechanism with the integrated position and force sensors and a piezoresistive strain gauge for controlling the end-effector. This gripper mechanism also uses finite elements topological optimization. The author compared the results from the optimization with an experimental setup, validating the feasibility of this procedure.

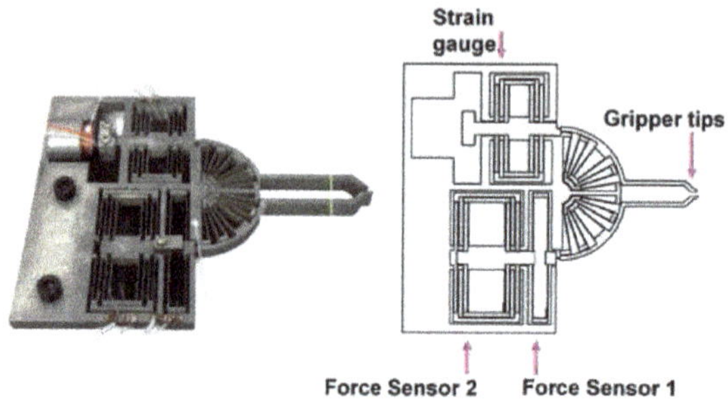

Figure 7. Structure of the complaint mechanism with integrated sensors. Reprinted with permission from ref. [32]. Copyright 2017 IEEE.

2.1.2. Rigid Links

Opposite to compliant mechanisms, rigid links can generate a high output torque while maintaining their stiffness. However, this gripper mechanism requires force sensors to avoid damaging the handling objects. This architecture is presented in [39], which proposes a cable-driven adaptive multi-DOF finger with a mechanical sensor integrated to control the position and output torque. This finger mechanism is made of Acrylonitrile Butadiene Styrene (ABS) and can generate motions in a single plane. The design of this finger mechanism maximizes the output forces in a predefined path, obtaining a gripper mechanism that can hold objects of 55mm in diameter and 800 grams in weight. Figure 8 presents the kinematical diagram of the gripper mechanism for the design trajectory. Another rigid gripper example with similar capabilities is in [4]. This rigid gripper is a 4-finger hand gripper, each finger with three DOFs and actuated by cables. The hand gripper's fingers have like dimensions and are composed of two phalanges. Other rigid grippers include variable friction surfaces which increase the manipulability dexterity of the gripper [40]. The texture of this gripper is a compound of Polylactic Acid (PLA) and TPU, and its friction changes by the actuation of two pulleys attached to DC motors as described by Figure 9.

One limitation of rigid links is the need for an accurate control strategy. This feature is complex to obtain because of the nature of the force sensor. Thus, some authors use a fuzzy logic controller to approximate the experimental results into an accurate control strategy [34]. Figure 10 presents the experimental setup used by the author to verify the behavior of this approach. Other rigid link designs use closed-loop links [79]. This finger gripper has two parallel grippers composed of symmetric parallelograms, as seen in Figure 11.

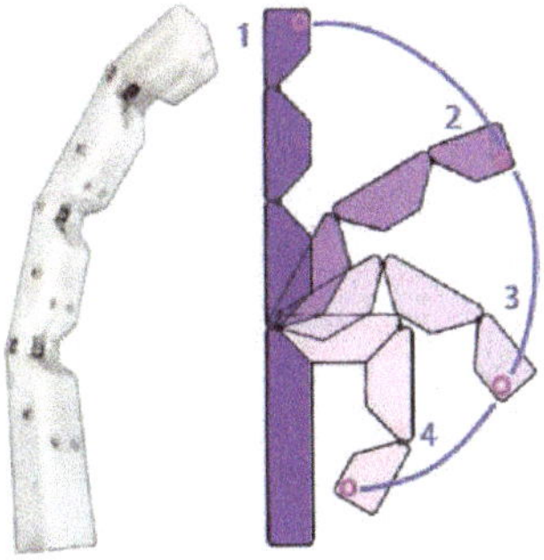

Figure 8. Kinematical diagram of the finger mechanism for the design trajectory [39].

The advantage of this kinematical design is that the model is simpler to implement, ensuring that the tool reacts well relative to the gripping forces and the spring stiffness. Moreover, to increase the accuracy, some rigid links grippers use lead screws examples of this approach are in [33,35]. The former uses a gripper inspired by a chuck clamping device, as presented in Figure 12. This gripper has a closing motion mechanism that provides the position of objects. The latter uses a slider-crank-mechanism as shown in Figure 13. This rigid link mechanism can handle items up to 5kg or fragile objects like eggs. Both gripper mechanisms have the advantage of self-locking, which reduces energy consumption because the motors do not need to be active all the time. However, this type of gripper has a slow-motion issue because of its high mechanical advantage.

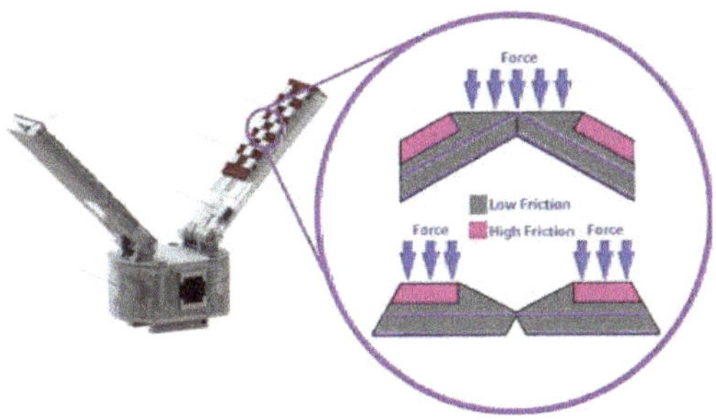

Figure 9. System of friction changing surface. Reprinted with permission from ref. [40]. Copyright 2020 IEEE.

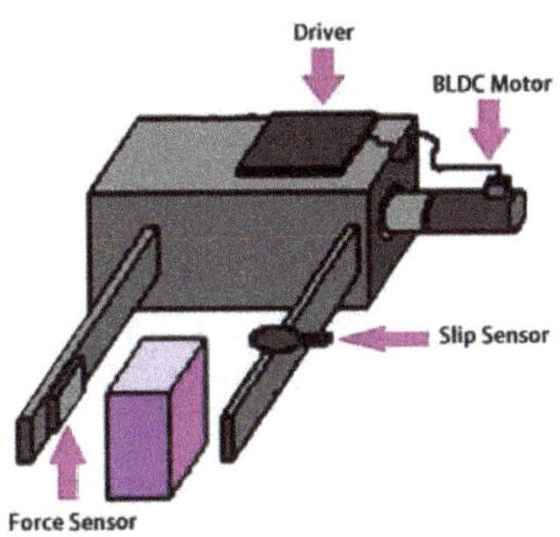

Figure 10. Gripper configuration with one movable finger (with force sensor) and one fixed finger (with slip sensor) to ease the control [34].

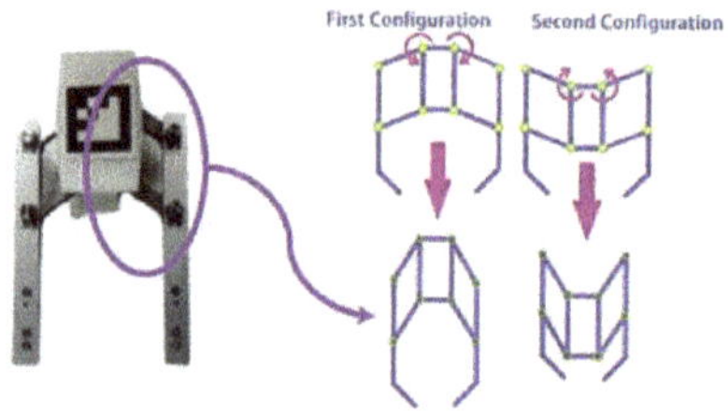

Figure 11. Finger gripper composed of two parallel grippers of symmetric parallelograms. Reprinted with permission from ref. [79]. Copyright 2019 IEEE.

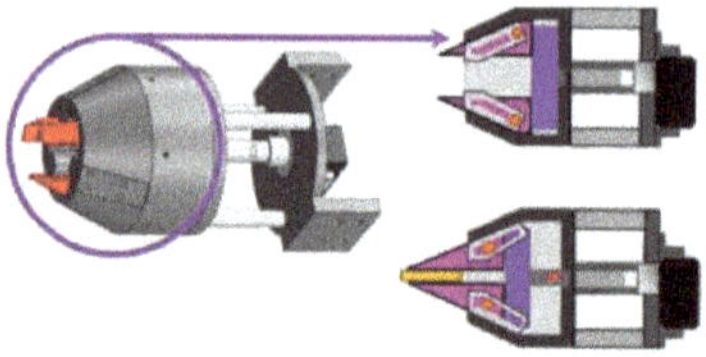

Figure 12. Assembly and operation of the chuck type system. Reprinted with permission from ref. [35]. Copyright 2018 IEEE.

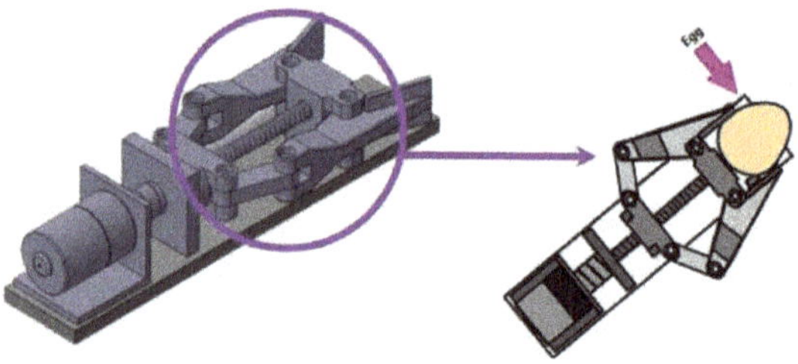

Figure 13. Representation of the gripper holding an egg without breaking it. Reprinted with permission from ref. [33]. Copyright 2018 IEEE.

Other designs use electromagnets to actuate rigid links. For example, in [41] is proposed an electromagnet actuated gripper for the manipulation of fabrics. As presented in Figure 14, the gripper uses a slider-crank mechanism. More complex electromagnet-actuated grippers use multiobjective genetic algorithms for optimal design. An example is presented in [42] of this optimization strategy. For this purpose, the authors modeled the actuator as a stack consisting of individual actuator elements arranged in series and parallel arrays in four combinations. As a result of this optimization process, the gripper has increased accuracy compared with others grippers of the same type. However, using electromagnets demands a high energy input, making them unsuitable for autonomous applications.

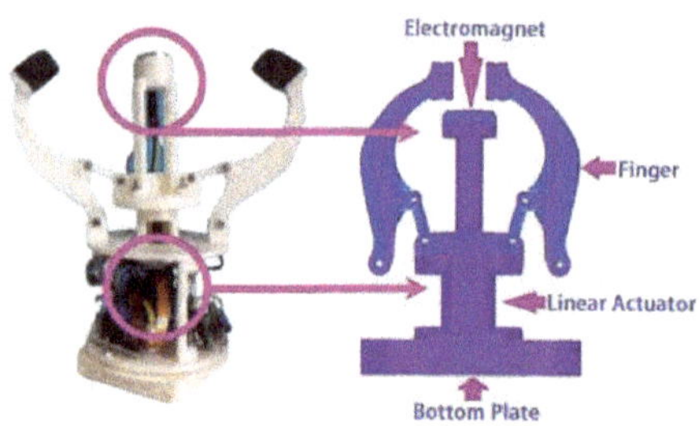

Figure 14. Electromagnet actuated gripper. Reprinted with permission from ref. [41]. Copyright 2020 IEEE.

The Salisbury hand [80] shown in Figure 15 was the first successful humanoid robot hand built as a sophisticated end-effector for grasping investigations. Each finger on this hand has three joints, allowing it to mimic the dexterous gripping of the human hand to some extent. Steel wires that pass through Teflon-coated flexible tubes activate the fingers. Each cable is tensioned by a DC brush-type motor that works through a gear reducer. The flexible conduit that allows wires to be routed around the wrist allows the actuator package to be mounted on the robot's forearm.

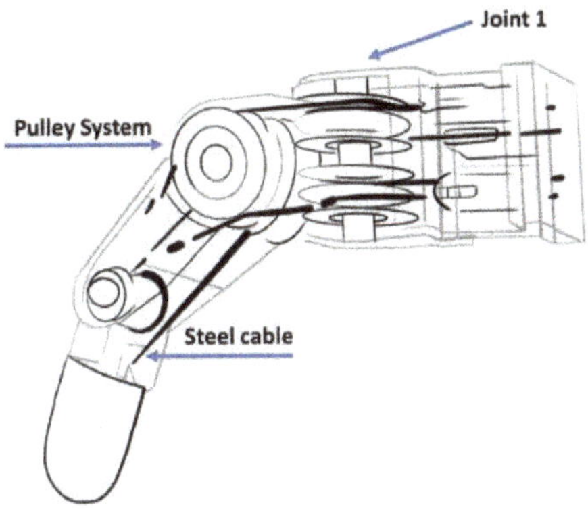

Figure 15. Finger joints of Salisbury hand.

The DHL Hand [81] is an open skeleton hand made of aluminum and steel that can manipulate a variety of objects with great dexterity and accuracy. Three separate joints in each finger are controlled by their own actuators. Brushless dc motors, tooth belts, harmonic drive gears, and bevel gears at the base joint are used in all actuation systems. As shown in Figure 16, the base joint is a differential bevel gear type, allowing for two independent motions. The two actuators can be used to their full capacity, allowing the joint to flex or extend as needed.

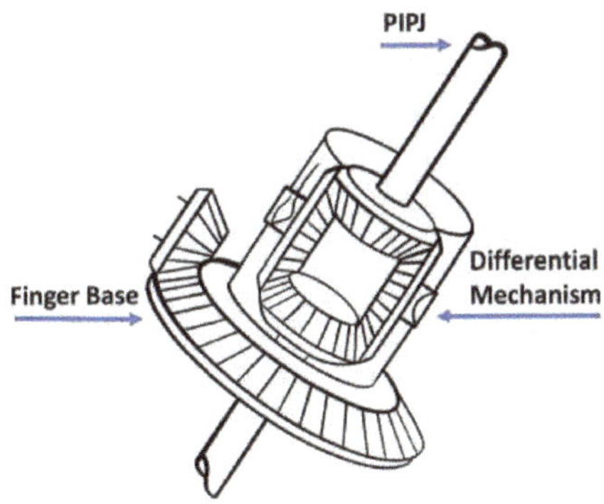

Figure 16. Actuation systems use brushless dc motors, tooth belts, harmonic drive gears, and bevel gears at the base joint to control three finger joints of DHL hand.

The Barrett hand [82] is a popular example of a hand used in industry and for grasping and manipulation research. Each Barrett Hand's finger shown in Figure 17 is powered by a motor, and each motor controls two joint axes. Torque is applied to these joints via a Torque switch mechanism. When a fingertip makes contact with an object for the first time, it locks both joints, deactivates motor currents, and waits for further instructions from the microprocessors.

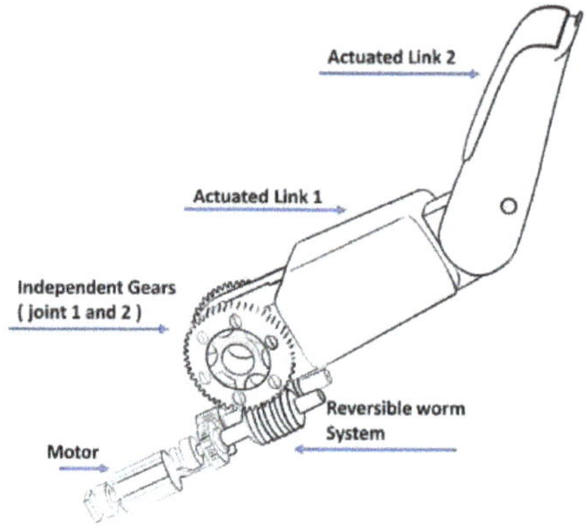

Figure 17. Mechanism of a Barrett hand's finger.

2.2. *Underconstrained Mechanism*

2.3. *Compliant Mechanism*

Underconstrained mechanisms allow a broader motion capability compared to completely constraints. The higher motion is due to the DOF, which is higher than the number of actuators, adding more flexibility to handle irregular shape objects. The extra DOF is usually passively actuated by springs for maintaining the structure, as presented in Figure 3, in which a 2-DOF under-constrained mechanism is presented. However, as is in the case of underconstrained compliant mechanisms, the shape of the gripper can include the effects of the spring without adding it. An example of this kind of mechanism is presented by the authors in [43], which developed a robotic gripper with compliant cell stacks for industrial part handling, shown in Figure 18.

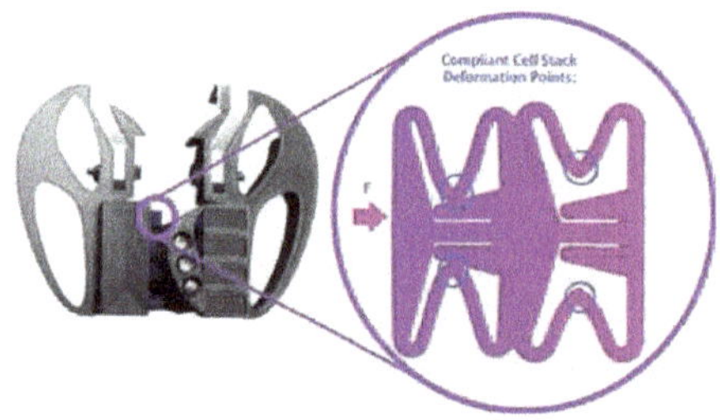

Figure 18. Robotic gripper with compliant cell stacks mechanism [43].

Another example is presented in [44], who developed an underactuated robotic gripper of three fingers inspired by an origami twisted tower shown in Figure 19. Each of the fingers of the gripper uses cable-driven actuation controlled by a central servomotor. Although the gripper mechanisms can handle objects with complex shapes, their payload capacity is limited to 1.5 N at most. Other examples with the same issue include the works of the authors of [47], who developed a passive-compliant piezo actuated microgripper (Figure 20); and the designs in [45], which presents a 3D printed Gripper for Cloth Manipulation and position control (Figure 21). However, the latter implemented a variable friction finger surface, controlled by a small motor that pushes the high friction surface at the tip of the top finger, incrementing the payload capacity.

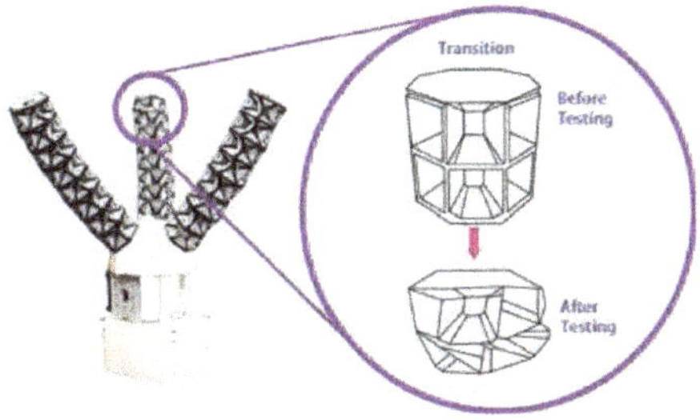

Figure 19. Underactuated robotic gripper of three fingers inspired in an origami twisted tower [44].

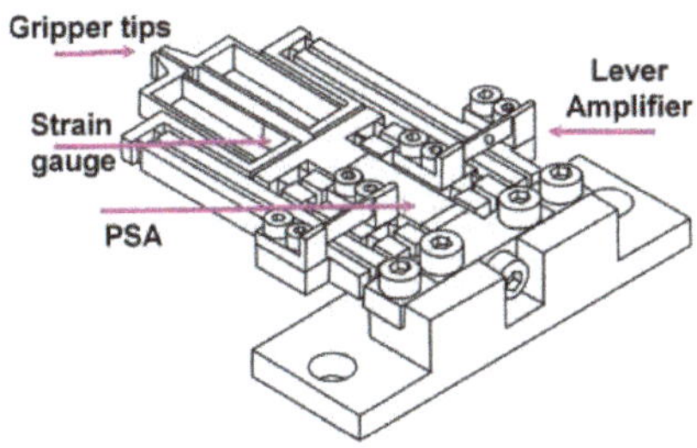

Figure 20. Passive-compliant piezo actuated micro-gripper. Reprinted with permission from ref. [47]. Copyright 2021 IEEE.

Another limitation of this type of gripper is the need for a force sensor capable of measuring the distribution of forces on a surface. To this end, the authors developed a gecko-inspired gripper that uses ABS polyimide or mylar polyester with a metalized surface for the sensors (Figure 22). The sensor is fabricated in situ with thin adhesive films on each finger and measures the change in capacitance when a region of adhesive makes contact with a surface. Another sensor is developed by the authors of [46] for a compliant adaptive gripper, which integrated an implicit force. The gripper can measure the output force by knowing the deformations on the gripper itself. To compute the deformation of the system, the authors used a general numerical network model (NTM). The NTM calculates the node coordinates of the mechanism using a hand-eye camera (Figure 23). Then, using the node information, the NTM computes the grasping force of the gripper. The proposed mechanism is the first fin-ray-based gripper that simultaneously achieves adaptive grasping and intrinsic force-sensing without any force sensor. A gripper with similar capabilities is presented in [48], who developed a novel compliant constant-force gripper based on buckled fixed-guided beams (Figure 24). The gripper has a passive type of compliant constant-force mechanism. The gripper can generate a constant-force output using a combination of positive and negative stiffness mechanisms. The negative stiffness mechanism is a bi-stable buckled fixed-guided beam.

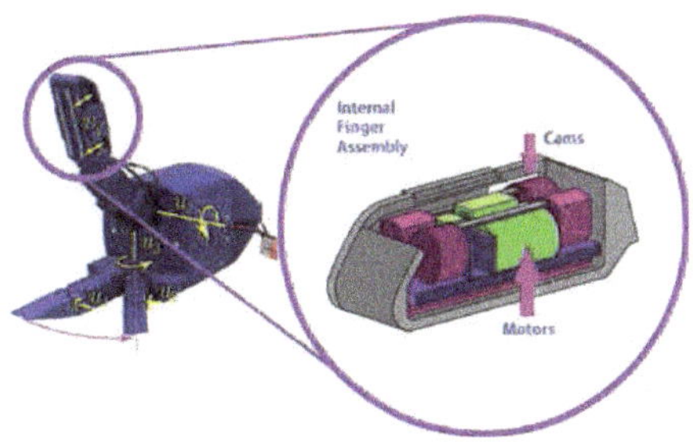

Figure 21. 3D printed Gripper for Cloth Manipulation. Reprinted with permission from ref. [45]. Copyright 2020 IEEE. The figure shows the motors used to change the friction in the gripper.

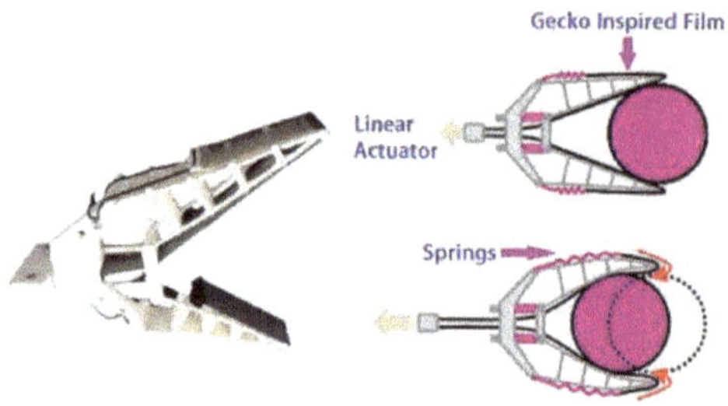

Figure 22. Movement sequence to perform a full grasping. Reprinted with permission from ref. [14]. Copyright 2019 IEEE.

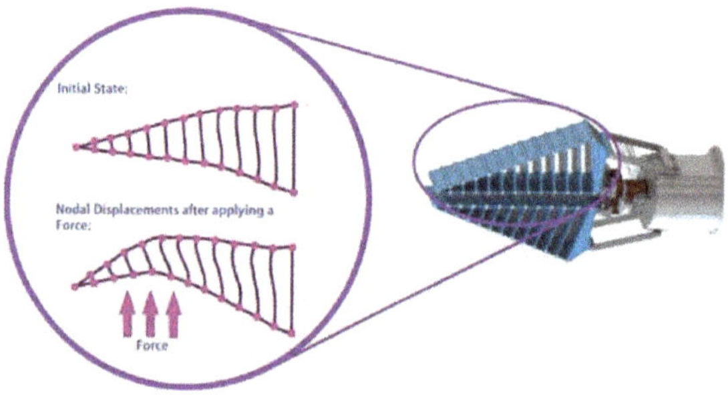

Figure 23. Representation of the theoretical deformation of the gripper given by the general numerical network model. Reprinted with permission from ref. [46]. Copyright 2021 IEEE.

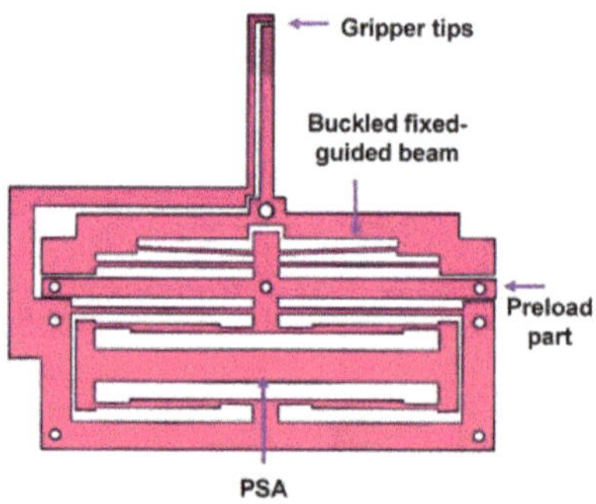

Figure 24. Compliant mechanism structure model.

Rigid Links

Underconstrained rigid gripper mechanisms have more load capacity compared to underconstrained compliant mechanisms. However, their load capacity is still low compared to totally constrained gripper mechanisms with rigid links. Thus, they are capable of mean output load capacity. Furthermore, most underconstrained grippers designs consider optimization methods to increase the kinematic capabilities. An example of an optimized gripper is presented in [55] the authors proposed a geometric design of three-phalanx underactuated fingers. In this study, the stability of two classes of three-phalanx cable-driven underactuated fingers is under analysis. Moreover, the theory for the optimal design of the gripper is presented, including an objective function that maximizes the forces normal to the contact trajectory while avoiding loss of contact and ejection (Figure 25). Another optimization example is found in [15], which presents a multi-modal adaptive gripper with the optimal design of a re-configurable finger developed for improving robotic manipulation without sacrificing grasping efficiency (Figure 26). The optimization problem maximizes the workspace volume for a wide range of objects using a parallel multi-start search algorithm. This algorithm uses all possible positions of the items during the in-hand manipulation; to compute the dexterous manipulation workspace. All the configurations are clustered in a set of points by the algorithm generating a planar point cloud. The bounding volume of the point cloud is calculated using the alpha-Shape method, which formalizes the abstract shape of the given set of points using Delaunay triangulation.

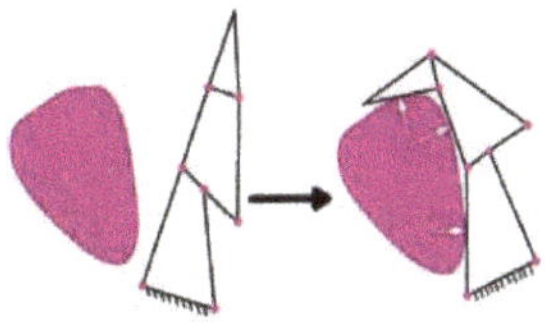

Figure 25. Clamping sequence for an object preventing ejection from the gripper design [55].

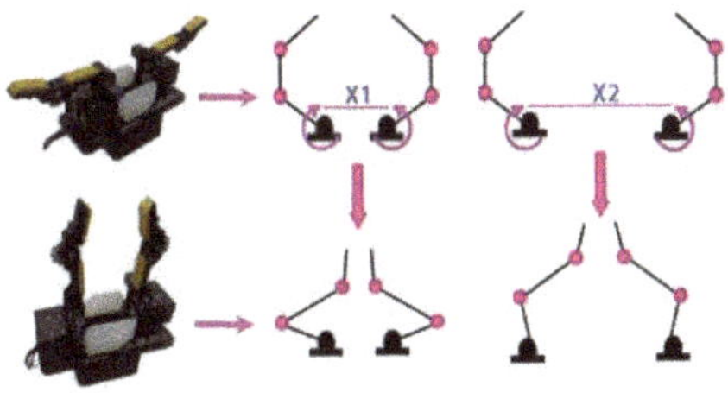

Figure 26. Multimodal adaptive gripper with the optimal design of a reconfigurable finger [15].

Other approaches add spines to increase the clamping capabilities of a gripper. For example, the authors in [5] presents a passive spine gripper that can hold rough rocky surfaces designed for a climber robot. This gripper has six fingers, making it suitable for spatial exploration in unknown environments. The mechanism (Figure 27) has dual spines that allow it to clamp to different surfaces. The finger part connects to a preload spring inside the gripper. A servomotor-pulley actuator controls the finger mechanism. Furthermore, to release or detach the gripper from the surface, a nylon gut attached to each fingertip is easily pulled by a servo motor. The gripper has a range of 120 ($\phi = 60$) degrees and can hold 4.7 N. The author identified the stiffness of the spring by energy methods.

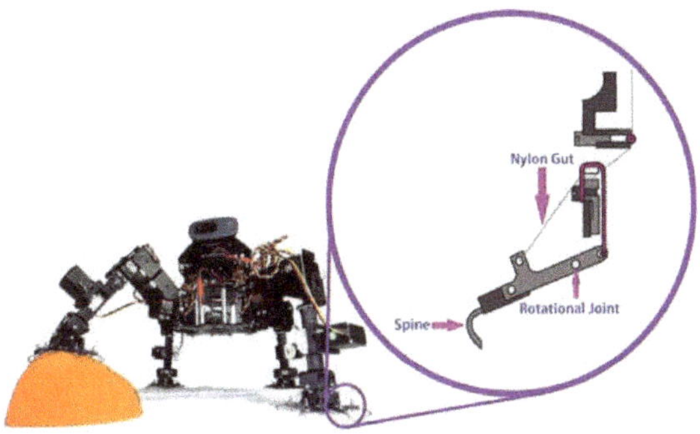

Figure 27. Dual spines that allow the mechanism to clamp to different surfaces. Reprinted with permission from ref. [5]. Copyright 2021 IEEE.

A limitation of rigid links is that they require more rigid links than completely constrained mechanisms. Thus, some gripper mechanisms may suffer from bulkiness. Some authors implemented cable-driven actuation to avoid this issue. Consider the design for an open-loop gripper shown in [12], the device is a two-fingers underactuated hand with cable-driven actuation (Figure 28). Each finger is tendon-driven with a 21-mm pulley diameter attached to an MX-28 Dynamixel servo, producing a stall torque of 2.5 Nm at 12 V. The gripper can hold square and circular objects. The authors tested the gripper using an Ascension trakSTAR sensor to measure the displaced position and orientation at the center of the gripper. The sensor can track 6-DOF with a spatial resolution of 0.5mm and 0.002 rad. Moreover, the gripper has a total stroke from 0 to 103 mm and a holding grip force at the fingertips of 8.9 ± 0.35 N. The article is limited to presenting the design, not the mathematical consideration. However, in [83], according to the authors, introduces a library for developing underactuated grippers. The Schunk SVH hand [84] shown in Figure 29 is one of the most compact designs ever created. The humanoid Schunk hand's motors are all housed in the wrist, saving a lot of space for the mechanisms. The human hand has 20 individual joints, and the majority of the SVH's joints are controlled by leadscrew mechanisms, which convert linear motion to rotational motion. There are 22 joints in total, but only 9 of them are fully actuated, turning this hand into an underactuated mechanism.

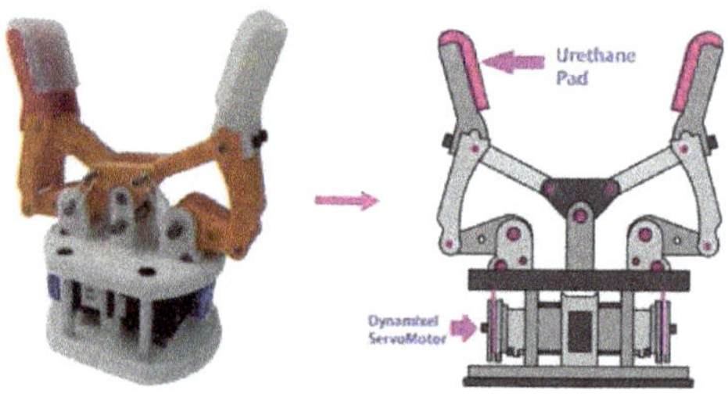

Figure 28. Internal gripper assembly. Reprinted with permission from ref. [12]. Copyright 2016 IEEE.

Another design is found in [13], which presents another similar two-fingers underactuated mechanism with cable-driven actuation and active tactile manipulation. The tip contact objects of the fingers are of a rubber-like material skin, with white pins (1 mm diameter) on its inside surface. The tip is entirely 3D-printed using a multi-material 3D-printer (Stratasys Objet 260 Connex), with the rigid parts printed in Vero White material and the compliant skin in the rubber-like TangoBlack+. An acrylic lens separates the electronic components from the tip, filled with RTV27905 silicon gel for compliance. A circuit of 6 LEDs illuminates the rubber pins, which protrude from inside the tip surface (Figure 30). The full range of object orientations depends on object size and shape, ranging from −34.4° to 32.3° for the 20 mm diameter cylinder to −21.8° to 20.6° for the 35 mm cylinder.

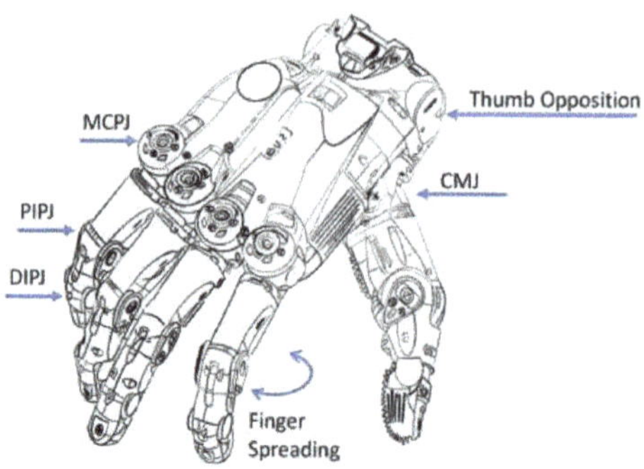

Figure 29. Schunk SVH hand.

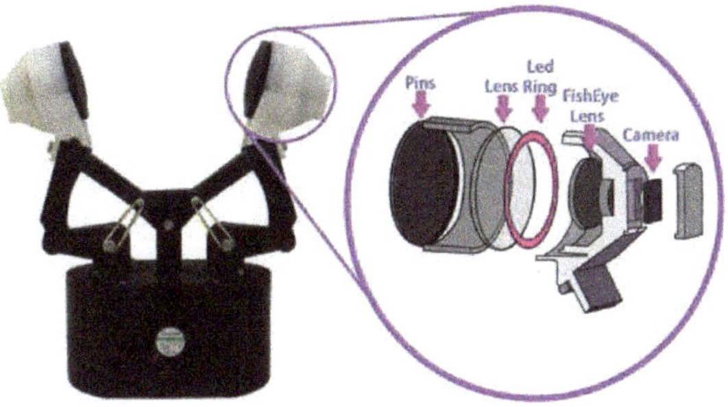

Figure 30. Layers used to use the LEDs as sensors at the tip of the gripper [13].

The authors of [58] designed an adaptive gripper with transition capabilities between a precise pinch and compliant grasp for interaction with an unexpected environment. Each finger has a minimum number of components using one rigid link, one belt, one fingertip frame, and one motor (40 Watt ECX16 motor) for flexion motion (Figure 31). The finger structure enables precise parallel pinching and highly compliant stable grasping with evenly distributed pressure. The gripper is composed of flexible belt materials with high stiffness while the fingers of ABS. The grasping force of the gripper is near 13 N, and it can hold a wide range of objects like a driller, a baseball ball, a hammer, a cup, tape, etc. The optimization of the ginger is based on the kinematic; finding the lengths for an optimal Compliant Grasping Pose.

Moreover, the authors in [59] presented an underactuated origami gripper for changing the stiffness of the gripper joints (Figure 32). This two-fingers gripper is of shape memory polymers, actuated by a tendon-driven system with adjustable stiffness joints. The controllable compliance of the fingers limits the contact forces at the desired magnitude without requiring any Feedback without a control strategy. Thus, the gripper does not need a sensor, becoming the control easier and allowing it to grasp delicate and small objects such as an egg, foam, and a coin. The minimum diameter held by the gripper was a coin of 31.5 mm with a maximum load bearing capacity of the joint of 0.97 Nm. The authors correlated the tension in the tendon with the joint angle by considering the system energy.

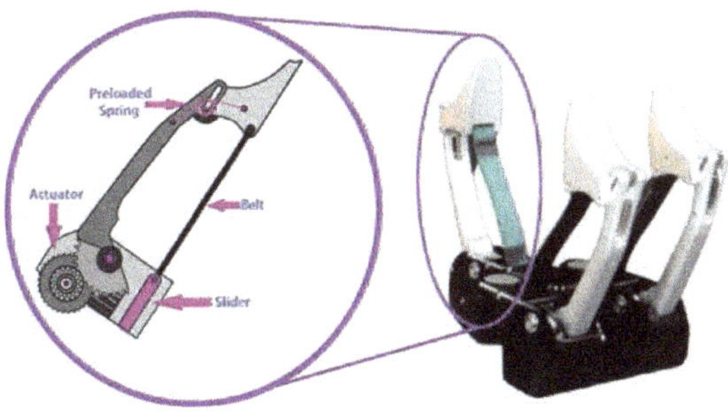

Figure 31. Double actuated gear and belt system of the gripper. Reprinted with permission from ref. [58]. Copyright 2020 IEEE.

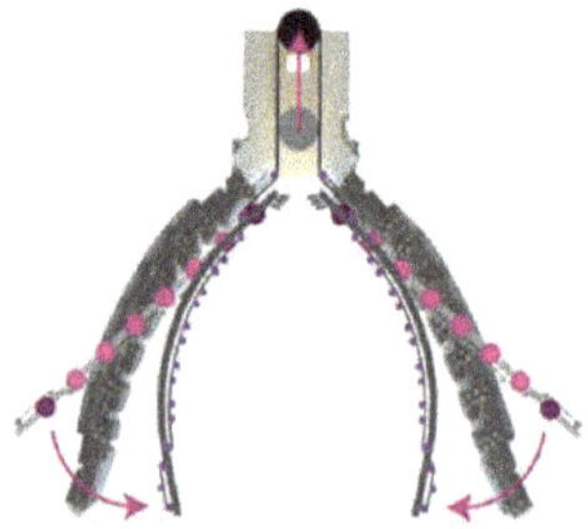

Figure 32. Underactuated origami gripper for changing the stiffness of the gripper joints. Reprinted with permission from ref. [59]. Copyright 2017 IEEE.

Another cable-driven robot gripper with a passively switchable underactuated surface is shown in [60], including a physic simulation based on parameter optimization for its design. The author proposed a gripper with an underactuated surface on the fingertip (Figure 33). With the spring-loaded passive switching mechanism, the actuation of a single motor generates three grasp modes in series: approaching the object as a standard parallel gripper, pulling items inside the hand with actuated fingertip crawler, and power grasping the object as an underactuated gripper. The authors experimentally showed that a prototyped gripper with the proposed structure successfully picked a 3-mm thin sheet and a softcover book from a flat surface. Moreover, the gripper can lift cylindrical-shaped objects from surface to end with an enveloping grasp. The workspace for this gripper is around 200 mm, with a grasping force that is flat against the object size and exceeds 20 N.

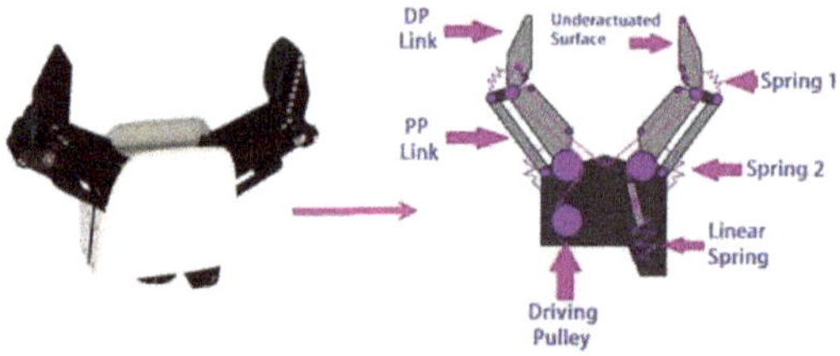

Figure 33. Cable-driven robot gripper with a passively switchable underactuated surface. Reprinted with permission from ref. [60]. Copyright 2020 IEEE.

The article presented in [9] shows an adaptive three-fingers prismatic gripper with passive rotational joints (Figure 34). The body of the hand is made with laser-cut 3 mm Delrin and 3D-printed ABS components (printed on a Fortus 250mc). The fingers are also 3D-printed, and the finger pads are from a cast using Smooth-On VytaFlex 30 urethane rubber. Each finger consists of a single joint finger connected to the prismatic joint via a perpendicular passive rotational joint to the palm. The rotational joints allow the fingers to passively switch between spherical and cylindrical grasps, while the finger joint allows the fingers to wrap about the grasped object. According to the article, the gripper can hold items from 17.4 mm to 145 mm, a set of washers ranging in size from 9.8 mm to 50.8 mm, a credit card, various tools, and other items.

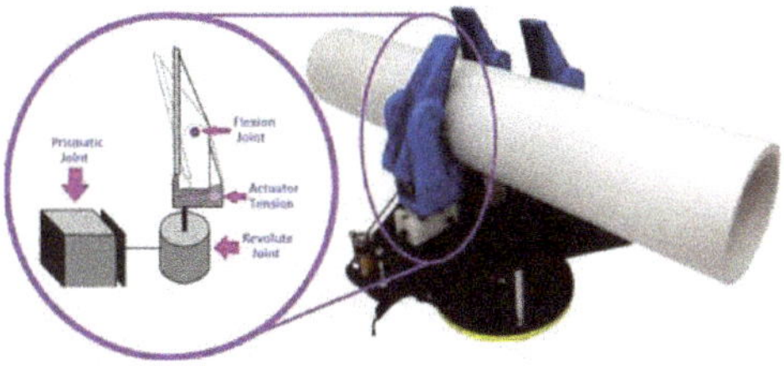

Figure 34. Adaptive three-fingers prismatic gripper with passive rotational joints. Reprinted with permission from ref. [9]. Copyright 2016 IEEE.

The authors of [61] present an underactuated gripper exploiting joint compliance with an efficient mathematical representation of soft robotic fingers based on screw theory (Figure 35). The mathematical model enables the gripper designer to analyze the influence of specific properties such as the trajectory of the fingertips, the overall stiffness, the distribution of contact force, etc. The gripper ranges 85 mm for cylindrical and spherical objects like a cup, tennis ball, small box, etc. Moreover, the material of the gripper is ABS and can hold 43 N. Lastly, the design and analysis of a novel robotic gripper integrated with a three-phalanx finger for medical applications are shown in [62]. The mechanism works like two grippers in one, a small one for grasping small objects (the embedded one) (Figure 36) and a bigger one for handling wide items. This novelty allows the gripper to reach more activities of daily living.

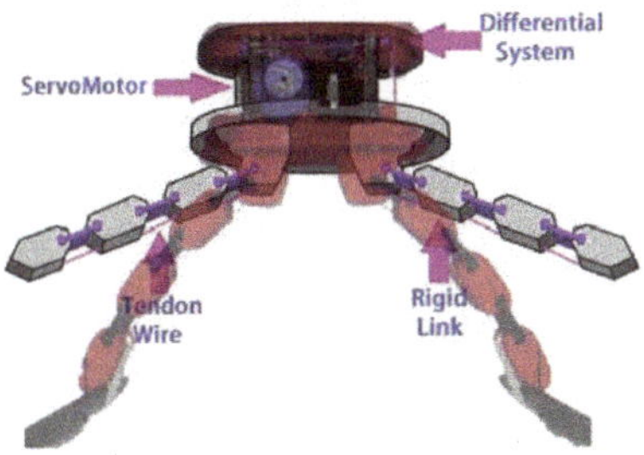

Figure 35. Underactuated gripper exploiting joint compliance. Reprinted with permission from ref. [61]. Copyright 2018 IEEE.

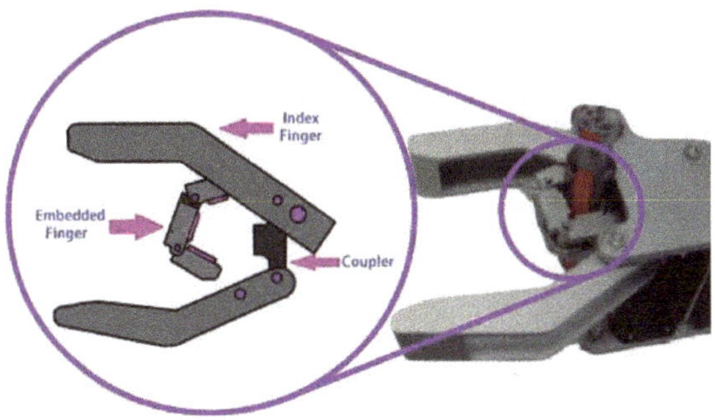

Figure 36. Novel robotic gripper integrated with a three-phalanx finger for medical applications. Reprinted with permission from ref. [62]. Copyright 2005 ASME.

Other approaches use rigid links with linear actuators in [50], where the authors presented a re-configurable gripper for robotic autonomous depalletizing for supermarket logistics. The depalletizing gripper has two extendable forks that can slide along a rail and two suction systems endowed with suction cups controlled by a closed-loop controller

(Figure 37). The gripper can hold objects of a size between 15 and 50 cm and a weight of 43 N.

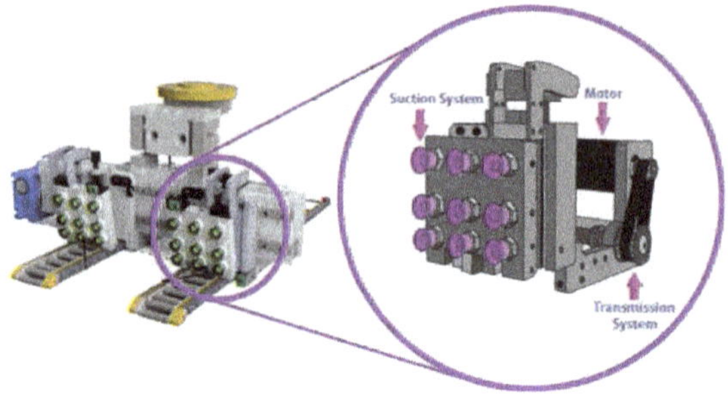

Figure 37. Depalletizing gripper with rigid links and linear actuators. Reprinted with permission from ref. [50]. Copyright 2020 IEEE.

Another type of actuation implemented in rigid links is rotary actuators. In [10], an underactuated four-bar linkage (Figure 38) is proposed. The gripper is actuated using a single actuator (Maxon EC45 70-W) that performs a robust pinch under various environmental constraints. The fingertips can slide on sloped surfaces of objects ranging from 11 mm to 85 mm in diameter. The gripper is also capable of handling lightweight objects. The research also analyzes the kinematic and static using the Plücker coordinates to determine the operation principle of the actuator. The results were used for the dimensional synthesis of the linkage according to several criteria for sliding and lifting. Furthermore, the design proposed in [52] uses rotary actuators, proposing a 3D-printed robot hand with three linkage-driven underactuated fingers with the capacity to reorient two or three of its fingers (Figure 39). The mechanism of each finger is made of a chain of rigid links, making three phalanges for each finger. The gripper design can interact with objects of different sizes and shapes, e.g., cylinders with dimensions up to 81 × 19 mm and spheres with diameters up 70 mm, while maintaining a contact force of 15 N. The gripper includes springs for recovering the initial position. The dimensions of the mechanism were obtained by implementing a grasping optimization for different objects.

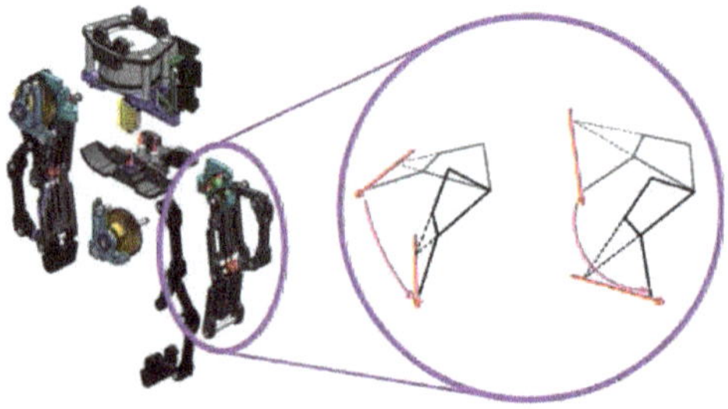

Figure 38. Underactuated four-bar linkage based gripper. Reprinted with permission from ref. [10]. Copyright 2021 IEEE.

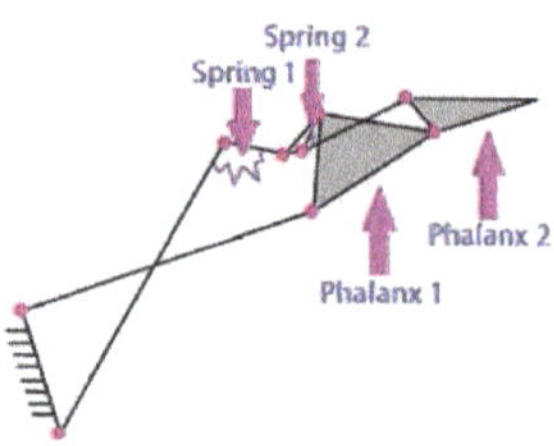

Figure 39. Kinematic representation of one of the three linkage-driven under-actuated fingers [52].

Lastly, another rotary actuator gripper is found in [51]. This research presents an underactuated adaptive 3D printed robotic gripper for interactions with unpredictable environments. The gripper has three fingers; each finger has an underactuated mechanism composed of 5 joints and one spring (Figure 40). The gripper materials are thermoplastic elastomer (TPE), PLA, and ABS. Furthermore, the gripper can hold different objects of daily living such as pencils, bottles, and whiteboard erasers, including spherical objects up to 75mm and a weight of 2.5 kg. The authors also presented a kinematic and quasi-static analysis of the finger for selecting the spring [53].

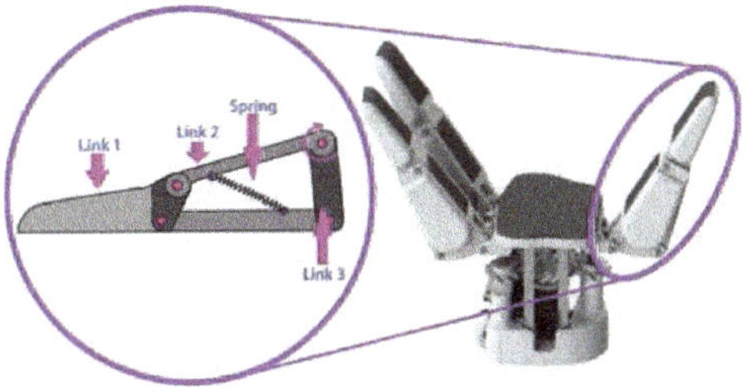

Figure 40. Underactuated adaptative 3D printed robotic gripper for interactions with unpredictable environments. Reprinted with permission from ref. [51]. Copyright 2014 IEEE.

Other designs use pneumatic actuators, as is the case of [63]. The authors presented the design of a high-payload hybrid robotic gripper with soft origamic actuators. The proposed actuator has one DOF linear translation along its axis. The repetitive trapezoid facets lead to the simple design of geometric parameters for customization and stable linear movement (Figure 41). Both the actuator body and the bottom cover are molded using polypropylene rubber. An air vent at the top of the actuator is designed to connect to pneumatic fittings. The gripper consists of two main components: a soft-actuator joint and rigid supporting structures with motion constraints. An analytical model of the actuator is derived based on the geometric parameters to capture the relations between the output force, inner pressure, and axial displacement.

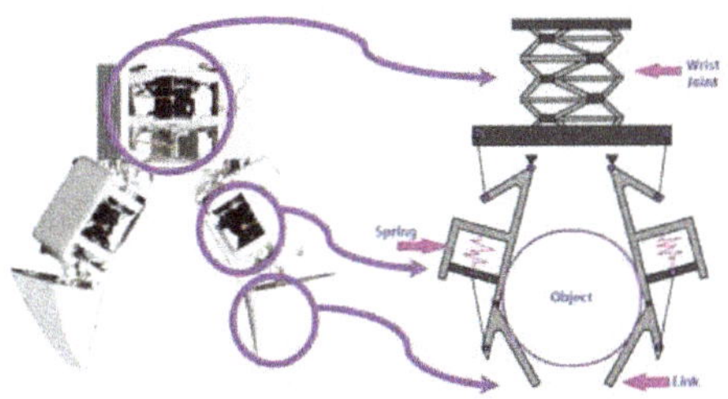

Figure 41. High-payload hybrid robotic gripper with soft origamic actuators. Reprinted with permission from ref. [63]. Copyright 2020 IEEE.

2.4. *Deformable Grippers*

2.4.1. Single Mass Gripper

Single mass grippers handle items by deforming their shape. This kind of gripper does not have a straightforward relationship between the deformation and the actuator mechanism, as with completely-constrained or underconstrained grippers. Thus, single mass grippers deform until wrapping the desired object. This type of gripper uses pneumatic or cable-driven actuation, covering the broader scope of objects, independent of their shape. However, despite their abilities to manipulate objects, the use of pneumatic actuation limits the mobility of the gripper [8], which presents a prestressed soft gripper with three fingers for food handling. The actuator has 3D printed in two parts: a soft chamber with a rigid connector and a sealed cover. The soft chamber is prestressed by stretching and gluing a non-stretched cover (Figure 42). The gripper can realize a large contact area while grasping with a wide initial opening without deflating the soft actuators. The fingers/actuators are of Rubber-like material. The authors used an air compressor (JUN-AIR 3-4) and an electro-pneumatic regulator (SMC ITV2030) to pressurize the actuator. The soft actuator has a length of 87 mm and can pick up objects of 75.2 g with an accuracy of 80%. The author did a finite element simulation to obtain the optimal dimensions.

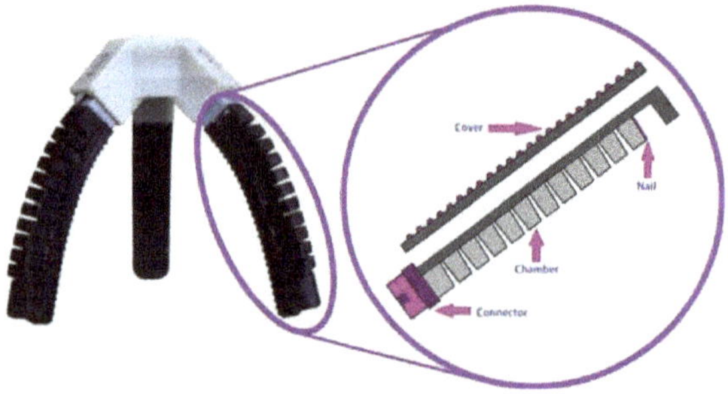

Figure 42. The prestressed soft gripper with three fingers for food handling has a soft chamber with a rigid connector and a sealed cover. Reprinted with permission from ref. [8]. Copyright 2017 IEEE.

Other authors have similar issues [64], which presents a single mass soft robotic gripper embedded with Microneedles for handling delicate fabrics. The gripper material is of the elastomer kind. The gripper hooks the delicate fabrics using four microneedles. The actuator of the gripper is a vacuum pump that deforms the elastomer once it is in position for handling the fabrics (Figure 43). Although the author does not use any sensor for controlling the pressure, some solutions include embedding soft pressure sensors at the tip of the gripper. Besides its bulkiness, the vacuum pump requires more energy than a regular air compressor. Likewise, an origami-inspired gripper controlled by an SMA actuator is designed [11] for picking objects with variable shapes and sizes. The design of the gripper takes inspiration from a reconfigurable suction gripper (Figure 44). Constructed from rigid and soft components and driven by compact shape memory alloy actuators, the gripper can effectively self-fold into three shape modes. The main objective is to pick large and small, flat, narrow, cylindrical, triangular, and spherical objects ranging from 2 mm to 43 mm in diameter and less than 5.2 N of weight.

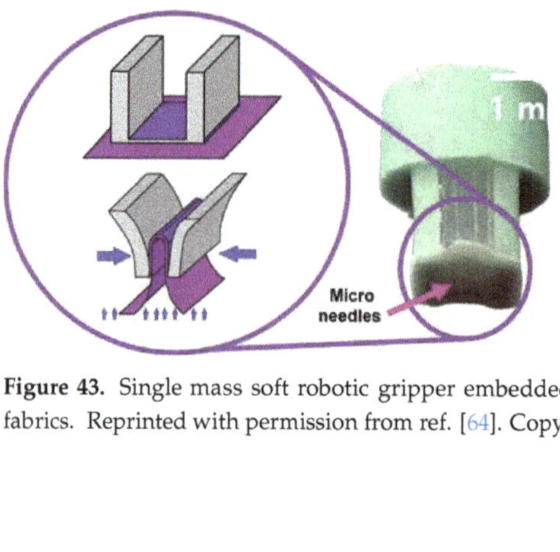

Figure 43. Single mass soft robotic gripper embedded with Microneedles for handling delicate fabrics. Reprinted with permission from ref. [64]. Copyright 2020 IEEE.

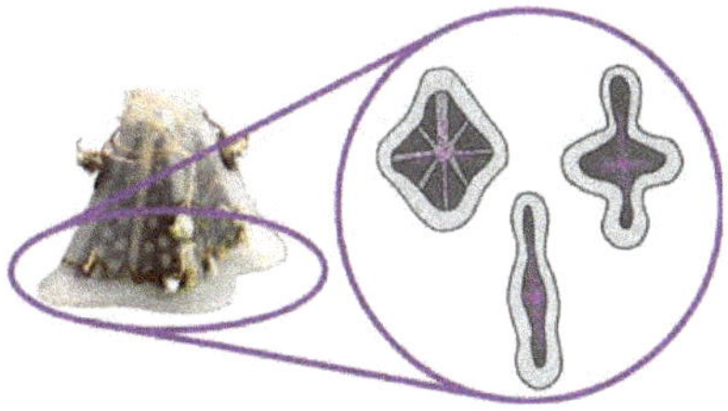

Figure 44. Origami-inspired gripper controlled by an SMA actuator and the possible forms that the gripper can take due to the SMA actuators that it has inside [11].

Other approaches use cable-driven actuation to reduce energy consumption and increase manipulability. However, controlling the gripper motion becomes an issue because of the complexity of the deformation model. The authors of [65] present an example of this kind of gripper. The authors developed a multi-legged gripper inspired by a gecko with a controllable adhesion parameter. The gripper can manipulate flat and curve objects by a self-adaptive dry adhesion system. The system consists of four symmetric adhesive units, each modulated by two adaptive-locking mechanisms for compression and rotation, respectively, and one peeling mechanism. The two adaptive-locking mechanisms can adapt to surfaces with height and curvature differences to ensure intimate contact with the objects (Figure 45). Moreover, the lock adaption configuration enables equal load sharing for a firm attachment. The peeling mechanism rapidly peels the adhesive surfaces from the substrate for easy detachment.

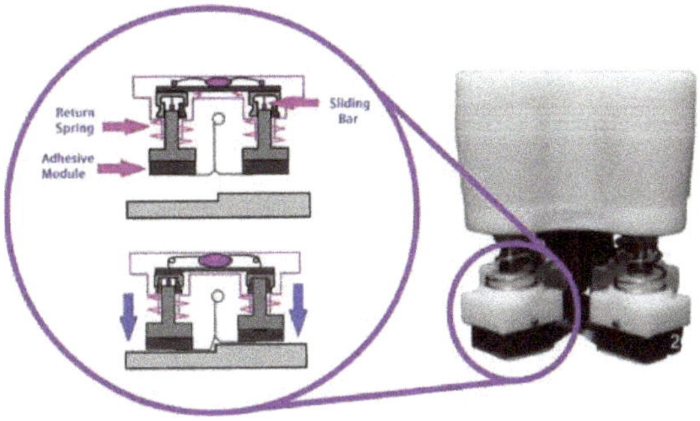

Figure 45. Multi-legged gripper inspired by a gecko with a controllable adhesion parameter generating a firm grip thanks to the adaptability of the gripper. Reprinted with permission from ref. [65]. Copyright 2021 IEEE.

Another flat dry adhesive soft gripper was presented in [66], integrating a soft actuator, micro spine, and a bioinspired design based on a gecko's toe and a cat's foot. The soft gripper has an improved design that enhances the comprehensive grasping ability of the soft gripper on smooth or rough surfaces. The design emulates two phalanges, proximal and distal, using SMA coils (Figure 46). The SMA coils are on the backside of the base layer opposite an adhesive layer. A flexible sensor measures the force inside the finger between the two layers. The fingers have micro-needles for increased grasping ability. The viscoelastic mechanics model is used to formulate the preloading process of the adhesive relating the stress encountered by the adhesive with the contact area considered during the preloading process.

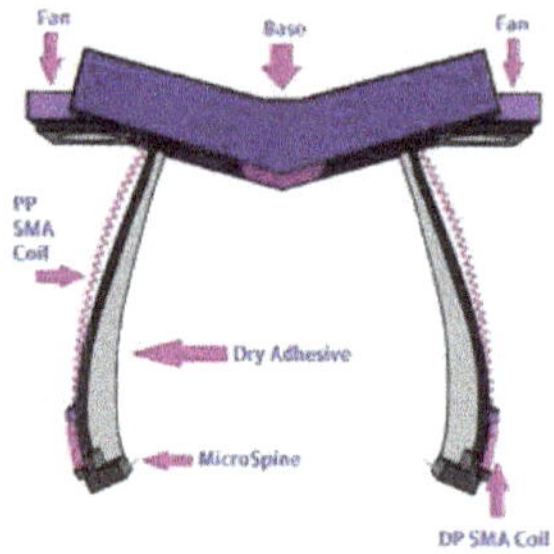

Figure 46. System composed of the SMA actuators, the structure and the cooling system for the soft gripper. Reprinted with permission from ref. [66]. Copyright 2021 IEEE.

2.4.2. Single Mass Finger

Single mass fingers deform their shape to actuate each of their fingers. This gripper mechanism can hold objects of different forms because of the actuation of the fingers, having similar advantages and disadvantages to single mass grippers. Among the benefits, using multiple grippers facilitates controlling the gripper; an increase in the number of grippers allows a better distribution of the forces generated, relaxing the dependency on an accurate control strategy. However, for this kind of grippers, the issue of bulkiness is more notorious because the actuator requires to deform multiple fingers instead of only one structure. To present some examples, The soft robotic gripper shown in [7] uses a particle transmission and has three fingers. The fingers can grasp a wide range of objects such as a nipper plier, tape, haptic device, and an electric screwdriver. Moreover, the gripper has a vertical force gauge that measures a grasping force of around 20 N at the tip. The authors modeled the system using mass conservation and the principle of incompressible homogenous neo-Hookean materials [85]. Moreover, the authors used molded silicone rubber reinforced by double-stranded woven fiberglass thread and PLA to make the actuators (Figure 47). The proposed actuator design is a slightly modified model of the widely researched fiber-reinforced soft pneumatic actuator [86].

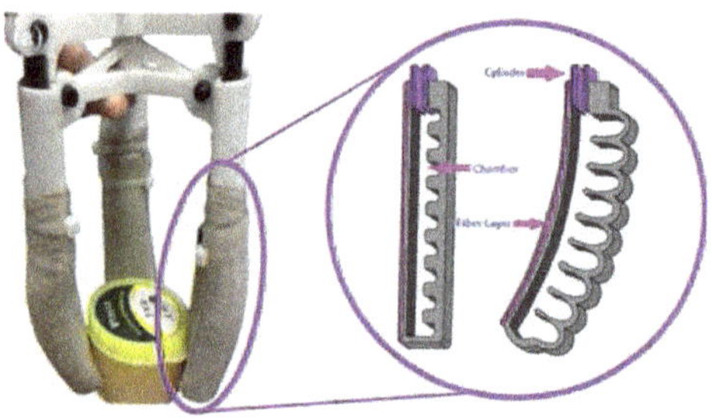

Figure 47. Soft robotic gripper actuated by particle transmission. Reprinted with permission from ref. [7]. Copyright 2019 IEEE.

Moreover, a soft robotic gripper with an active palm and reconfigurable fingers is presented in [68] performing complex motions such as rolling a pen or pouring a glass of water. The gripper performs in multiple applications, such as robotic manipulation, medical applications, mobility, rehabilitation, or assistive robotics. The fingers of the gripper are of silicone elastomers EcoFlex. Stepper motors, micropumps, and solenoids control the position of the fingers (Figure 48). Each finger has three pneumatic chambers, which are independent, giving each finger a wide range of mobility. Optimization over a previous iteration is mentioned but not explained. However, the author performs a workspace analysis to determine the total active area. Additionally, the soft gripper shown in [67] is based on pre-charged pneumatic soft actuators. The gripper has a pre-charged pneumatic

(PCP) with a silicone chamber with one air tube for pressurizing it (Figure 49). A check valve controls the pressure inside the silicone chamber. When the fingers are pressurized, their shape is corrected using tensile cables or tendons. The actuator body material is silicone rubber with an inextensible layer attached to the bottom of the actuator with a range of 150 mm and capable of holding soft objects like tomatoes or eggs.

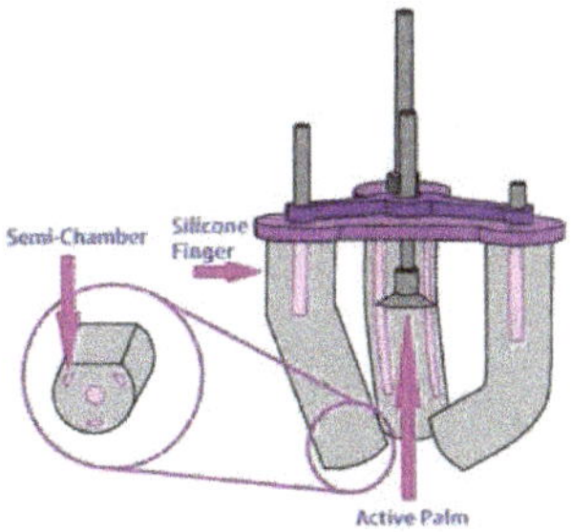

Figure 48. Soft robotic gripper with an active palm and reconfigurable fingers. Reprinted with permission from ref. [68]. Copyright 2021 IEEE.

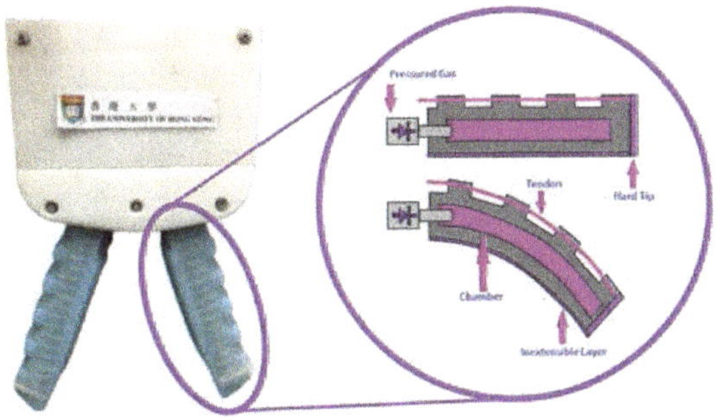

Figure 49. Soft gripper using a tendon to pre-charge the soft pneumatic actuators. Reprinted with permission from ref. [67]. Copyright 2019 IEEE.

Another gripper is in [69], including rigid and soft materials. The gripper has a pressuring single internal chamber that controls the position. The authors proposed this design to improve the fingertip force and actuation speed simultaneously, optimizing parameters like the degree of bending, the ratio of the rigid structure, the longitudinal strain by modifying the shape of the chamber, and the relation between soft and rigid materials in the same finger (Figure 50). Furthermore, two pneumatic pumps (DAO-370A) control the gripper, allowing a broad workspace suitable for teleoperation. The gripper can hold objects like a drill driver, a coffee cup, and a banana with a maximum allowable weight of 28.7 N. The design parameters were optimized by the finite elements method (FEM) and a simulation based on the hyperrealistic Mooney-Rivlin model.

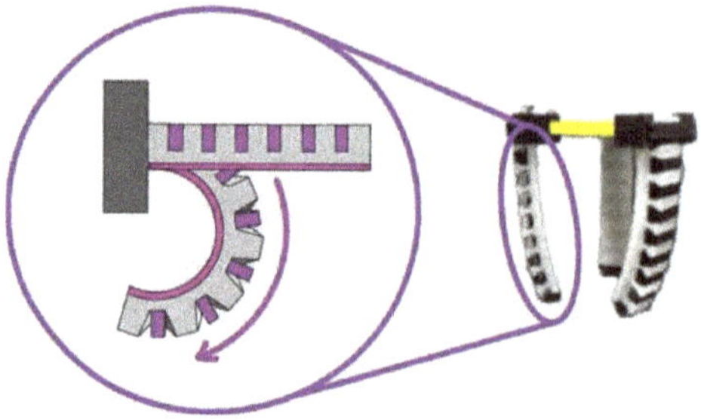

Figure 50. Representation of the circular path that the fingertip travels when pressure is applied in its chamber for the soft gripper [69].

Other designs include a pneumatically driven gripper with retractable, telescopic fingers as developed in [70] for Medical applications. The authors used two low-pressure mini-air pumps to actuate this silicone rubber gripper. Additionally, the authors implemented an optical motion capture system composed of eight cameras to track the angular displacement of the gripper (Figure 51). The gripper has retroreflective markers at the top surface of the actuator to increase the accuracy of the measure. The range of motion of the soft actuator is 0 to 105 degrees. The gripper can hold objects like a medium mustard bottle, a water bottle, an egg, and a drill driver whit a grasp up to 14.53 N. A Finite Element Analysis (FEA) model of the soft actuator deformation was developed to understand the structure's inflation behavior.

Similarly, the pneumatic two-finger soft robotic gripper shown in [71] can handle objects with enveloping and pinching grasping modes (Figure 52). This gripper consists of chambers and channels in a series arrangement. Moreover, the gripper includes a main body with a bottom both an inextensible elastomer. It combines two dual-module pneumatic actuators with a variable chamber height. The gripper has a workspace with a Bending angle of up to 250 degrees but a low payload of up to 4 N. A digital force gauge inside the finger measures the force at different chamber points. The pinching grasping mode was mainly analyzed by FE analysis and experiments.

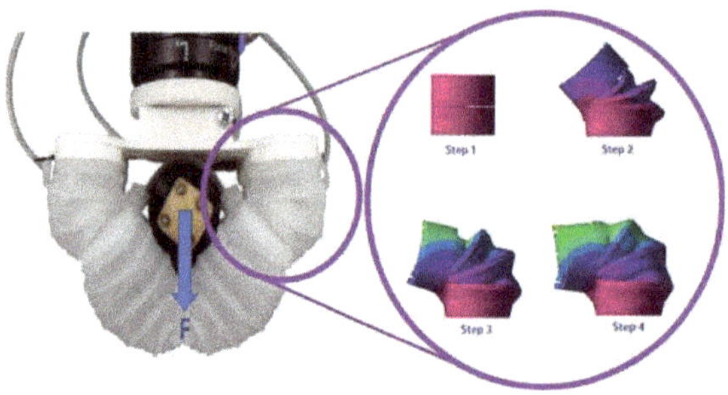

Figure 51. Deformation sequence captured by a camera to track movement at specific points on the gripper. Reprinted with permission from ref. [70]. Copyright 2021 IEEE.

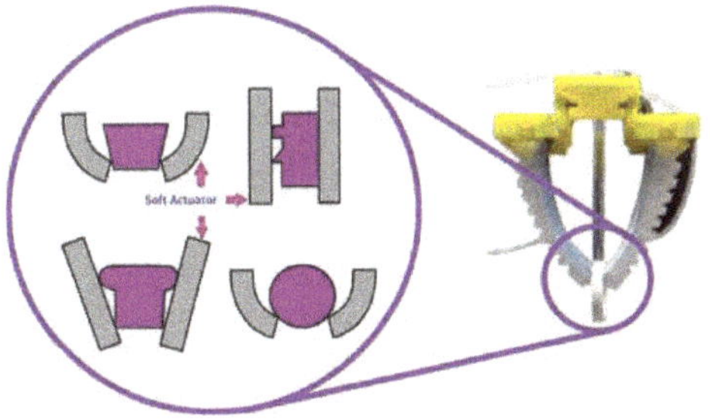

Figure 52. Different grip modes for the pneumatic two-finger soft robotic gripper and different shapes and objects. Reprinted with permission from ref. [71]. Copyright 2021 IEEE.

Other approaches use a soft robotic gripper with Gecko-inspired adhesive as proposed in [72], which can handle rocky or dirty surfaces where adhesion is limited. Gecko-inspired grippers use a combination of fluidic elastomer actuators as an actuation mechanism that goes through a circular cross-section (Figure 53). Gecko elastomer actuators provide improved control authority for manipulation tasks. The ability to achieve higher ultimate grip strengths on many objects allows manipulation of heavier objects and higher accelerations of objects during motion. This property is significant for pick and place operations, where speed is critical. The gecko elastomer actuator maintains a low energy input and a fast actuation since the actuators are optimized for these properties while using adhesion-enhanced friction for higher strength grips.

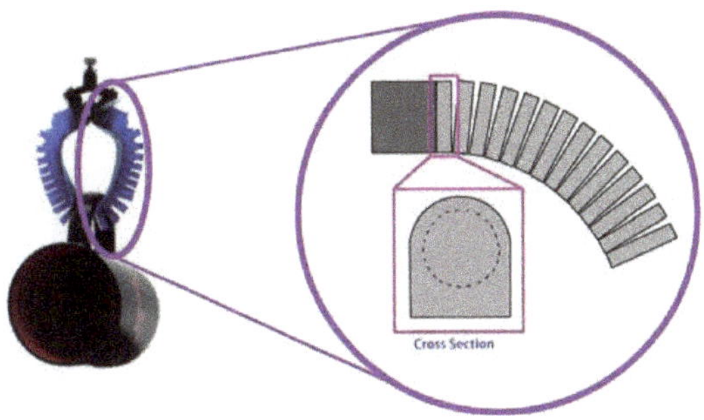

Figure 53. Cross-sectional area used for elastomer actuation of gecko-inspired gripper. Reprinted with permission from ref. [72]. Copyright 2018 IEEE.

Lastly, an underwater gripper studying the deformation characteristics of water hydraulic flexible actuators is shown in [73]. The gripper has three fingers. Each of the fingers has an inner skeleton made of 3J1, 3J21, TC4, and Carbon fiber with a wall thickness of 1 mm and 30 mm for the first knuckle and 80 mm for the second knuckle. A nonlinear equation expresses the workspace. According to the equation, when the inlet pressure is 0 or 10 MPa, the minimum and maximum deformation of the flexible actuator are 0 mm and 0.20213 mm, respectively. The authors investigated using simulations the effects of different inlet pressure, knuckle length, wall thickness, and material of the inner skeleton and external surface on the deformation characteristics of the flexible actuator. Figure 54 shows the theoretical deflection of the internal bar of the gripper and its repercussion on the external coating. According to the authors, the wall thickness and the length between the knuckles significantly affect the gripper deformation.

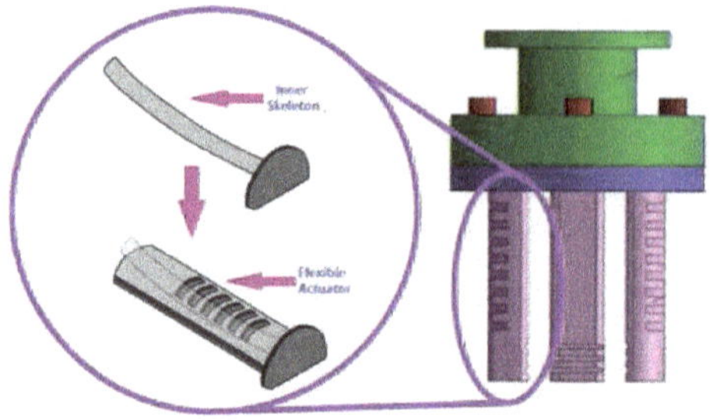

Figure 54. Theoretical deflection of the internal bar of the underwater gripper and its repercussion on the external coating [73].

2.4.3. Materials of the Deformable Grippers

Most single-mass grippers and single-mass fingers are of silicone or rubber-like materials. Rubber-like materials have elasticity, and structural compliance, which increases the safety and adaptability of the device when interacting with delicate or fragile objects [70]. However, rubber-like materials come with disadvantages; the fabrication of soft grippers often requires an iterated casting process, which is usually complex and time-consuming. Furthermore, the air bubbles within the material often result in significant individual differences, which limits the robot's repeatability [8]. An example of rubber-like material is the commercial ECOFLEX silicone 00-30, which is easy and fast to actuate thanks to its flexibility [87]. Moreover, ECOFLEX silicone 00-30 has a high-power-to-weight ratio, which allows it to have large deformations with a small input [88]. However, the analysis and design of this rubber-like material are complex due to its highly nonlinear response. Thus, when analytical solutions are required, researchers use models similar to the neo-Hookean, which are limited in representing the material behavior at large stretches [88]. Thus, an option is using finite element (FE) analysis, especially when considering the response of silicone rubber actuators [86,87,89–91].

Other rubber-like materials include the Object Full cure 930 TangoPlus, which can be 3D printed. Although this material has a higher resistance than the ECOFLEX silicone 00-30, it has the limitation of less elongation break and more cost [88]. Another example of 3D printed rubber-like material is found in [92], NinjaFlex. Although this material has a high resistance, it has a high hardness, which is unsuitable for applications requiring low pressure or force.

Some rubber-like materials can be prepared with a wide range of cured stiffnesses; such materials include the Dragonskin 20, and VTV800 [93]. However, these materials may need an external force or vibrators to recover their initial shape [7].

Other materials include the Dielectric Elastomer Actuators (DEA) or smart material. This material can be used as an actuator for the gripper. Moreover, grippers with this material have reported good performance while grasping various objects [94]. Additionally, soft grippers with this technology report fast response while consuming very-low energy [95]. However, most DEA grippers require a rigid frame to pre-stretch the high elastomer, which is a complex process. Another issue is the reliability of the flexible electrode, which deteriorates with time. Lastly, DEA grippers are limited to low-weight objects because of limitations in the voltage [73].

3. Principal Findings

Table 2 summarizes our findings of the robotic arm grippers handling different sizes, shapes, and materials of objects. In the size category, this table groups the gripper by small, medium, and large size of the handling object. Additionally, this table considers the most common shapes to classify the handling objects, i.e., circular, square, and irregular. The last criteria of comparison consider the types of objects handled by the grippers as delicate, fabric, electronics, rocks and soils, and food. According to the findings presented in the table, the best grippers to handle irregular objects are deformable single-mass grippers;

this is because of the adapting capabilities of those grippers. However, the most widely used gripper for daily objects is the underconstrained-rigid links; this is due to this type of gripper is easier to build, requiring a more available actuation system as is linear or rotary actuators. Finally, Table 3 presents the principal findings of each main design category of grippers. This table presents the principal qualities of each gripper type, the load capacity, the range capacity, and the type of objects that each category can hold.

Table 2. Summary of the findings of actuation grippers for different sizes, shapes, and materials of objects.

Type 1	Type 2	Gripper Design	Papers	Attributes
Size	Small	Deformable-Single mass	[16]	Size of 25 mm to 31.5 mm
		Completely constrained-rigid links	[36,96]	
		Underconstrained-compliant mechanism	[47]	
		Completely constrained-Clompliant mechanism	[97]	
		Underconstrained-Rigid links	[59]	
	Medium	Completely constrained-Clompliant mechanism	[1,6]	Size of 31.5 mm to 80 mm
		Completely constrained-rigid links	[57]	
		Underconstrained-Rigid links	[13]	
		Deformable-single gripper	[7,74]	
		Deformable-Single mass	[11,75]	
	Large	Underconstrained-Rigid links	[50]	Size of 10 cm to 50
Shape	Circular	Completely constrained-Rigid links	[31,33,35,39]	circular objects like eggs, fruits, tennis balls, or water bottles
		Underconstrained-Rigid links	[13,49]	
		Deformable-Single mass	[66,71]	
	Squared	Completely constrained-rigid links	[98]	square objects like cardboard boxes, cellphones, or plastic cards
		Underconstrained-Rigid links	[50]	
	irregular	Completely constrained-rigid links	[34]	Irregular objects like foam or rocks
		Underconstrained-Rigid links	[12,15,59]	
		Deformable-Single mass	[8,11,65,66,68,73,75]	
		Underconstrained-compliant mechanism	[46]	
Material	Delicate	Completely constrained-Clompliant mechanism	[1]	Delicate objects like eggs
		Completely constrained-Rigid links	[31,33]	
		Deformable-single gripper	[7]	
		Deformable-Single mass	[8,75]	
		Underconstrained-compliant mechanism	[43,44,48]	
	Fabric	Underconstrained-compliant mechanism	[44,45]	Types of fabric like cotton or linen
		Deformable-Single mass	[64]	
		Completely constrained-rigid links	[40,41]	
	Electronic	Deformable-Single mass	[16]	electrical objects like coils
		Completely constrained-rigid links	[32]	
	Rocks and Soils	Underconstrained-Rigid links	[5]	General shape rocks
	Food	Deformable-Single mass	[8]	food like spaghetti, salmon, fried chicken, among others.
Daily Objects		Underconstrained-Rigid links	[10,15,53,54,58,61,63]	objects like pencils, bottles, whiteboard erasers, or balls
		Underconstrained-compliant mechanism	[14]	
		Deformable-single mass	[69,70]	

Table 3. Principal findings of each main design category of grippers.

Type	Description	Load Capacity	Range of Motion	Type of Objects
Completely constrained grippers	This type of mechanism can exert greater forces, which is why it is especially used in applications where heavy objects must be moved. However, it cannot be attached to different shapes with ease.	These mechanisms can support very heavy objects (more than 10 kg).	The ranges depend on the application, but being rigid, they have geometric limitations due to their mechanism, so they have a range of movement between 2.2 mm and 170 mm	They are excellent at holding rigid objects. They can also hold more fragile objects if they have a force sensor. However, they are not recommended for this application.
Underconstrained grippers	These mechanisms offer a balance between flexibility and strength. Possessing rigid joints, it can support heavy weights while adapting to most objects' shapes. As a result, it is ideal for applications where the environment is uncontrolled or unpredictable.	They have a maximum descending load, up to 5 kg.	They have a descending range from a few millimeters to 120 mm. However, again, this range will depend on your design and application.	This mechanism can grab a wide range of objects such as a glass of water; pill bottle, book; smartphone; pringle; shoes; cereal boxes; apples; bread, among many others.
Deformable grippers	In contrast to the two previously mentioned mechanisms, this one cannot exert large amounts of force. This could be an advantage or a disadvantage, depending on the application. However, being flexible, they can adapt to all shapes, and their lack of strength is a positive factor when holding fragile objects.	They have little carrying capacity, ranging from grams to a few kilograms.	By being able to deform, they can twist their fingers backward, giving a much greater range than previous mechanisms. Some of these grippers can hold as much as a pill, up to a soccer ball (between 8 mm and 200 mm).	It practically conforms to the contour of the object you want to hold, no matter how irregular it is. this includes amorphous objects, such as rocks or any complicated surface.

4. Conclusions

The main contribution of this paper is the review of the majority of robotic grippers from the last four years. We classified the gripper according to the number of DOF, the actuation system, the design approach, and the shape of the grasping objects. For each classification criterion, a comparison of the advantages and disadvantages is presented, obtaining insights into which is the gripper design with the broader capabilities. Thus, the principal conclusions of our study are as follows:

- The sensing of the forces generated by the grasping is not accurate. Thus, to avoid breaking fragile objects, engineers use deformable grippers.
- Another issue is the glide of objects, which creates issues in the control strategy. Thus, a solution for this subject is variable friction in the gripper found in gecko-inspired grippers.
- The issue of the sensing forces is handled by completely constrained grippers which can exert greater forces with precision, especially in applications where heavy objects must be moved. However, it cannot be attached to different shapes with ease.
- Another option for deformable grippers is passive-compliant mechanisms that add an extra DOF to increase the manipulability. Passive-compliant mechanisms have the advantage of exerting a moderate amount of output force, adequate for handling objects with a moderate weight, especially if they are built using rigid links.
- Passive compliant mechanisms offers a balance between flexibility and strength. Possessing rigid joints, it can support heavy weights while adapting to most objects' shapes. As a result, it is ideal for applications where the environment is uncontrolled or unpredictable

Thus, based on our analysis, we conclude that the gripper design with the best capabilities for handling objects of different weights and shapes is a passive-compliant mechanism with rigid links and a gecko-inspired surface. The only disadvantage of this application depends on the exposure to environmental contaminants that may deteriorate the capabilities of the gecko-inspired surface to graduate friction.

Author Contributions: Conceptualization, J.H., M.S.H.S. and J.S.; methodology, J.H. and M.S.H.S.; formal analysis, J.H., M.S.H.S. and J.S.; investigation, J.H., M.S.H.S., I.R., M.I.I.Z. and J.S.; resources, M.H.R.; writing—original draft preparation, J.H., M.S.H.S., I.R., M.I.I.Z. and J.S.; writing—review and editing, J.H., M.S.H.S., H.U.A., J.S. and M.H.R.; visualization, J.H. and M.S.H.S.; supervision, M.H.R.; project administration, M.H.R.; funding acquisition, M.H.R. and S.I.A. All authors have read and agreed to the published version of the manuscript.

Funding: The contents of this research were supported by a grant from the National Institute on Disability, Independent Living, and Rehabilitation Research (NIDILRR grant number 90DPGE0018-01-00). NIDILRR is a Center within the Administration for Community Living (ACL), Department of Health and Human Services (HHS). The contents of this research do not necessarily represent the policy of NIDILRR, ACL, or HHS, and you should not assume endorsement by the Federal Government.

Institutional Review Board Statement: Not applicable.

Informed Consent Statement: Not applicable.

Data Availability Statement: Not applicable.

Conflicts of Interest: The authors declare no conflict of interest.

Abbreviations

The following abbreviations are used in this manuscript:

DOF	Degree of Freedom
TPE	Thermoplastic Elastomer
FRM	Flexible Redundant Robot Manipulators
ABS	Acrylonitrile Butadiene Styrene
PLA	Polylactic Acid
TPU	Thermoplastic polyurethane
BLDC	Brushless DC Motor
PID	Proportional-Integral-Derivative feedback control
NTM	Numerical Network Model
RTV	soft, adherent, clear silicone elastomer gel
3J1	Nickel-based high elastic alloy
3J21	Cobalt based high elastic alloy
TC4	Titanium Alloy

Appendix A. Journal Selection

Studies focused on current design and control approaches for the robotic arm gripper were chosen by performing a systematic electronic search in a handful of databases in January 2022. The timeline of the studies was limited to the last five years to focus on the recent advancement in this field to focus on the recent advances in this field, the study was restricted to the last five years. Keywords used to search include "robotic gripper", "robotic hand", "gripper design", "robotic manipulation", and "robotic grasping". Several databases were searched for this research: Google Scholar, IEEE Xplore, ScienceDirect, Engineering Village, Microsoft academic search, Google Patent Search, Scopus, Springer, PubMed, MDPI, IOS Press, Hindawi, SAGE, PLOS, Frontiers in Robotics and AI, etc. Around 235 publications are identified for consideration in this search. However, after some initial screening, 190 studies were shortlisted for review. A few publications were excluded based on eligibility criteria (specific goals, duplication, review, etc.), which resulted in 64 studies being selected for an in-depth review. All the selected papers (n = 64) were reviewed, including abstracts, introductions, design approach, experiments, conclusions, and future work sections to identify any other noteworthy information, such as the addressed problem, contribution, control theory, applications, experiments, used material, and sensors. Figure A1 shows the systematic approach with inclusion and exclusion criteria for the selected studies.

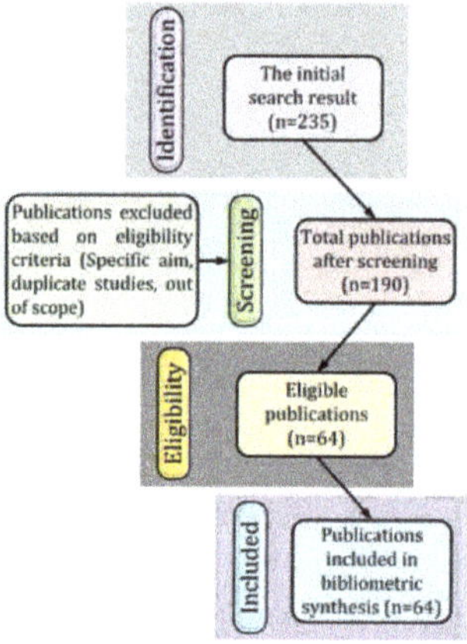

Figure A1. Inclusion and exclusion criteria of the selected studies.

References

1. Liu, C.H.; Chung, F.M.; Chen, Y.; Chiu, C.H.; Chen, T.L. Optimal Design of a Motor-Driven Three-Finger Soft Robotic Gripper. *IEEE/ASME Trans. Mechatron.* **2020**, *25*, 1830–1840. [CrossRef]
2. Birglen, L.; Schlicht, T. A statistical review of industrial robotic grippers. *Robot. Comput.-Integr. Manuf.* **2018**, *49*, 88–97. [CrossRef]
3. Choi, B.; Choi, H.R.; Kang, S. Development of tactile sensor for detecting contact force and slip. In Proceedings of the 2005 IEEE/RSJ International Conference on Intelligent Robots and Systems, Edmonton, AB, Canada, 2–6 August 2005; pp. 2638–2643. [CrossRef]
4. Cannata, G.; Maggiali, M. An embedded tactile and force sensor for robotic manipulation and grasping. In Proceedings of the 5th IEEE-RAS International Conference on Humanoid Robots, Tsukuba, Japan, 5–7 December 2005; pp. 80–85. [CrossRef]
5. Nagaoka, K.; Minote, H.; Maruya, K.; Shirai, Y.; Yoshida, K.; Hakamada, T.; Sawada, H.; Kubota, T. Passive Spine Gripper for Free-Climbing Robot in Extreme Terrain. *IEEE Robot. Autom. Lett.* **2018**, *3*, 1765–1770. [CrossRef]
6. Chen, F.; Xu, W.; Zhang, H.; Wang, Y.; Cao, J.; Wang, M.Y.; Ren, H.; Zhu, J.; Zhang, Y.F. Topology Optimized Design, Fabrication, and Characterization of a Soft Cable-Driven Gripper. *IEEE Robot. Autom. Lett.* **2018**, *3*, 2463–2470. [CrossRef]
7. Li, Y.; Chen, Y.; Yang, Y.; Li, Y. Soft Robotic Grippers Based on Particle Transmission. *IEEE/ASME Trans. Mechatron.* **2019**, *24*, 969–978. [CrossRef]
8. Wang, Z.; Torigoe, Y.; Hirai, S. A Prestressed Soft Gripper: Design, Modeling, Fabrication, and Tests for Food Handling. *IEEE Robot. Autom. Lett.* **2017**, *2*, 1909–1916. [CrossRef]
9. Backus, S.B.; Dollar, A.M. An Adaptive Three-Fingered Prismatic Gripper With Passive Rotational Joints. *IEEE Robot. Autom. Lett.* **2016**, *1*, 668–675. [CrossRef]
10. Yoon, D.; Choi, Y. Analysis of Fingertip Force Vector for Pinch-Lifting Gripper With Robust Adaptation to Environments. *IEEE Trans. Robot.* **2021**, *37*, 1127–1143. [CrossRef]
11. Zhakypov, Z.; Heremans, F.; Billard, A.; Paik, J. An Origami-Inspired Reconfigurable Suction Gripper for Picking Objects With Variable Shape and Size. *IEEE Robot. Autom. Lett.* **2018**, *3*, 2894–2901. [CrossRef]
12. Rojas, N.; Ma, R.R.; Dollar, A.M. The GR2 Gripper: An Underactuated Hand for Open-Loop In-Hand Planar Manipulation. *IEEE Trans. Robot.* **2016**, *32*, 763–770. [CrossRef]
13. Ward-Cherrier, B.; Rojas, N.; Lepora, N.F. Model-Free Precise in-Hand Manipulation with a 3D-Printed Tactile Gripper. *IEEE Robot. Autom. Lett.* **2017**, *2*, 2056–2063. [CrossRef]
14. Hashizume, J.; Huh, T.M.; Suresh, S.A.; Cutkosky, M.R. Capacitive Sensing for a Gripper With Gecko-Inspired Adhesive Film. *IEEE Robot. Autom. Lett.* **2019**, *4*, 677–683. [CrossRef]
15. Elangovan, N.; Gerez, L.; Gao, G.; Liarokapis, M. Improving Robotic Manipulation Without Sacrificing Grasping Efficiency: A Multi-Modal, Adaptive Gripper With Reconfigurable Finger Bases. *IEEE Access* **2021**, *9*, 83298–83308. [CrossRef]
16. Brown, E.; Rodenberg, N.; Amend, J.; Mozeika, A.; Steltz, E.; Zakin, M.R.; Lipson, H.; Jaeger, H.M. Universal robotic gripper based on the jamming of granular material. *Proc. Natl. Acad. Sci. USA* **2010**, *107*, 18809–18814. [CrossRef]
17. Makiyama, Y.; Wang, Z.; Hirai, S. A Pneumatic Needle Gripper for Handling Shredded Food Products. In Proceedings of the 2020 IEEE International Conference on Real-Time Computing and Robotics (RCAR), Hokkaido, Japan, 28–29 September 2020; IEEE: Piscataway, NJ, USA, 2020; pp. 183–187.

18. Zhang, B.; Xie, Y.; Zhou, J.; Wang, K.; Zhang, Z. State-of-the-art robotic grippers, grasping and control strategies, as well as their applications in agricultural robots: A review. *Comput. Electron. Agric.* **2020**, *177*, 105694. [CrossRef]
19. Hughes, J.; Culha, U.; Giardina, F.; Guenther, F.; Rosendo, A.; Iida, F. Soft manipulators and grippers: A review. *Front. Robot. AI* **2016**, *3*, 69. [CrossRef]
20. Wang, J.; Gao, D.; Lee, P.S. Recent Progress in Artificial Muscles for Interactive Soft Robotics. *Adv. Mater.* **2021**, *33*, 2003088. [CrossRef]
21. Laschi, C.; Mazzolai, B.; Cianchetti, M. Soft robotics: Technologies and systems pushing the boundaries of robot abilities. *Sci. Robot.* **2016**, *1*, eaah3690. [CrossRef]
22. Bicchi, A.; Kumar, V. Robotic grasping and contact: A review. In Proceedings of the 2000 ICRA. Millennium Conference. IEEE International Conference on Robotics and Automation. Symposia Proceedings (Cat. No. 00CH37065), San Francisco, CA, USA, 24–28 April 2000; IEEE: Piscataway, NJ, USA, 2000; Volume 1, pp. 348–353.
23. Samadikhoshkho, Z.; Zareinia, K.; Janabi-Sharifi, F. A brief review on robotic grippers classifications. In Proceedings of the 2019 IEEE Canadian Conference of Electrical and Computer Engineering (CCECE), Edmonton, AB, Canada, 5–8 May 2019; IEEE: Piscataway, NJ, USA, 2019; pp. 1–4.
24. Mukhtar, M. Design, Modelling, and Control of an Ambidextrous Robot Arm. Ph.D. Thesis, Brunel University London, London, UK, 2020.
25. Taheri, O.; Ghorbani, N.; Black, M.J.; Tzionas, D. GRAB: A dataset of whole-body human grasping of objects. In Proceedings of the European Conference on Computer Vision, Glasgow, UK, 23–28 August 2020; Springer: Cham, Switzerland, 2020; pp. 581–600.
26. Cini, F.; Ortenzi, V.; Corke, P.; Controzzi, M. On the choice of grasp type and location when handing over an object. *Sci. Robot.* **2019**, *4*, eaau9757. [CrossRef]
27. Feix, T.; Bullock, I.M.; Dollar, A.M. Analysis of human grasping behavior: Object characteristics and grasp type. *IEEE Trans. Haptics* **2014**, *7*, 311–323. [CrossRef]
28. Prakash, B.; Veeregowda, B.; Krishnappa, G. Biofilms: A survival strategy of bacteria. *Curr. Sci.* **2003**, *85*, 1299–1307.
29. Song, E.J.; Lee, J.S.; Moon, H.; Choi, H.R.; Koo, J.C. A Multi-Curvature, Variable Stiffness Soft Gripper for Enhanced Grasping Operations. *Actuators* **2021**, *10*, 316. [CrossRef]
30. Liu, C.H.; Chung, F.M.; Ho, Y.P. Topology Optimization for Design of a 3D-Printed Constant-Force Compliant Finger. *IEEE/ASME Trans. Mechatron.* **2021**, *26*, 1828–1836. [CrossRef]
31. Chen, C.C.; Lan, C.C. An Accurate Force Regulation Mechanism for High-Speed Handling of Fragile Objects Using Pneumatic Grippers. *IEEE Trans. Autom. Sci. Eng.* **2018**, *15*, 1600–1608. [CrossRef]
32. Xu, Q. Design and Development of a Novel Compliant Gripper With Integrated Position and Grasping/Interaction Force Sensing. *IEEE Trans. Autom. Sci. Eng.* **2017**, *14*, 1415–1428. [CrossRef]
33. Suebsomran, A. Development of Robot Gripper and Force Control. In Proceedings of the 2018 13th World Congress on Intelligent Control and Automation (WCICA), Changsha, China, 4–8 July 2018; pp. 433–437. [CrossRef]
34. Zaki, A.M.; Soliman, A.M.; Mahgoub, O.A.; El-Shafei, A. Design and implementation of efficient intelligent robotic gripper. In Proceedings of the 2010 International Conference on Modelling, Identification and Control, Okayama City, Japan, 17–19 July 2010; pp. 710–716.
35. Nishimura, T.; Tennomi, M.; Suzuki, Y.; Tsuji, T.; Watanabe, T. Lightweight, High-Force Gripper Inspired by Chuck Clamping Devices. *IEEE Robot. Autom. Lett.* **2018**, *3*, 1354–1361. [CrossRef]
36. Nie, K.; Wan, W.; Harada, K. A Hand Combining Two Simple Grippers to Pick Up and Arrange Objects for Assembly. *IEEE Robot. Autom. Lett.* **2019**, *4*, 958–965. [CrossRef]
37. Mahmoud, R.; Ueno, A.; Tatsumi, S. Dexterous mechanism design for an anthropomorphic artificial hand: Osaka City University Hand I. In Proceedings of the 2010 10th IEEE-RAS International Conference on Humanoid Robots, Nashville, TN, USA, 6–8 December 2010; pp. 180–185. [CrossRef]
38. Wattanasiri, P.; Tangpornprasert, P.; Virulsri, C. Design of Multi-Grip Patterns Prosthetic Hand With Single Actuator. *IEEE Trans. Neural Syst. Rehabil. Eng.* **2018**, *26*, 1188–1198. [CrossRef]
39. Wu, C.; Song, T.; Wu, Z.; Cao, Q.; Fei, F.; Yang, D.; Xu, B.; Song, A. Development and Evaluation of an Adaptive Multi-DOF Finger with Mechanical-Sensor Integrated for Prosthetic Hand. *Micromachines* **2021**, *12*, 33. [CrossRef]
40. Lu, Q.; Clark, A.B.; Shen, M.; Rojas, N. An Origami-Inspired Variable Friction Surface for Increasing the Dexterity of Robotic Grippers. *IEEE Robot. Autom. Lett.* **2020**, *5*, 2538–2545. [CrossRef]
41. Marullo, S.; Bartoccini, S.; Salvietti, G.; Iqbal, M.Z.; Prattichizzo, D. The Mag-Gripper: A Soft-Rigid Gripper Augmented With an Electromagnet to Precisely Handle Clothes. *IEEE Robot. Autom. Lett.* **2020**, *5*, 6591–6598. [CrossRef]
42. Datta, R.; Pradhan, S.; Bhattacharya, B. Analysis and Design Optimization of a Robotic Gripper Using Multiobjective Genetic Algorithm. *IEEE Trans. Syst. Man Cybern. Syst.* **2016**, *46*, 16–26. [CrossRef]
43. Netzev, M.; Angleraud, A.; Pieters, R. Soft Robotic Gripper With Compliant Cell Stacks for Industrial Part Handling. *IEEE Robot. Autom. Lett.* **2020**, *5*, 6821–6828. [CrossRef]
44. Lee, K.; Wang, Y.; Zheng, C. TWISTER Hand: Underactuated Robotic Gripper Inspired by Origami Twisted Tower. *IEEE Trans. Robot.* **2020**, *36*, 488–500. [CrossRef]
45. Donaire, S.; Borràs, J.; Alenyà, G.; Torras, C. A Versatile Gripper for Cloth Manipulation. *IEEE Robot. Autom. Lett.* **2020**, *5*, 6520–6527. [CrossRef]

46. Xu, W.; Zhang, H.; Yuan, H.; Liang, B. A Compliant Adaptive Gripper and Its Intrinsic Force Sensing Method. *IEEE Trans. Robot.* **2021**, *37*, 1584–1603. [CrossRef]
47. Chen, F.; Gao, Y.; Dong, W.; Du, Z. Design and Control of a Passive Compliant Piezo-Actuated Micro-Gripper With Hybrid Flexure Hinges. *IEEE Trans. Ind. Electron.* **2021**, *68*, 11168–11177. [CrossRef]
48. Liu, Y.; Zhang, Y.; Xu, Q. Design and Control of a Novel Compliant Constant-Force Gripper Based on Buckled Fixed-Guided Beams. *IEEE/ASME Trans. Mechatron.* **2017**, *22*, 476–486. [CrossRef]
49. Birglen, L.; Gosselin, C. On the force capability of underactuated fingers. In Proceedings of the 2003 IEEE International Conference on Robotics and Automation (Cat. No.03CH37422), Taipei, Taiwan, 4–19 September 2003; Volume 1, pp. 1139–1145. [CrossRef]
50. Fontanelli, G.A.; Paduano, G.; Caccavale, R.; Arpenti, P.; Lippiello, V.; Villani, L.; Siciliano, B. A Reconfigurable Gripper for Robotic Autonomous Depalletizing in Supermarket Logistics. *IEEE Robot. Autom. Lett.* **2020**, *5*, 4612–4617. [CrossRef]
51. Telegenov, K.; Tlegenov, Y.; Shintemirov, A. An underactuated adaptive 3D printed robotic gripper. In Proceedings of the 2014 10th France-Japan/8th Europe-Asia Congress on Mecatronics (MECATRONICS2014-Tokyo), Tokyo, Japan, 27–29 November 2014; pp. 110–115. [CrossRef]
52. Li, X.L.; Wu, L.C.; Lan, T.Y. A 3D-Printed Robot Hand with Three Linkage-Driven Underactuated Fingers. *Int. J. Autom. Comput.* **2018**, *15*, 593–602. [CrossRef]
53. Tlegenov, Y.; Telegenov, K.; Shintemirov, A. An open-source 3D printed underactuated robotic gripper. In Proceedings of the 2014 IEEE/ASME 10th International Conference on Mechatronic and Embedded Systems and Applications (MESA), Senigallia, Italy, 10–12 September 2014; pp. 1–6. [CrossRef]
54. Telegenov, K.; Tlegenov, Y.; Shintemirov, A. A Low-Cost Open-Source 3-D-Printed Three-Finger Gripper Platform for Research and Educational Purposes. *IEEE Access* **2015**, *3*, 638–647. [CrossRef]
55. Birglen, L.; Gosselin, C.M. Geometric Design of Three-Phalanx Underactuated Fingers. *J. Mech. Des.* **2005**, *128*, 356–364. Available online: http://xxx.lanl.gov/abs/https://asmedigitalcollection.asme.org/mechanicaldesign/article-pdf/128/2/356/5688285/356_1.pdf (accessed on 31 December 2022). [CrossRef]
56. Birglen, L.; Gosselin, C. Kinetostatic analysis of underactuated fingers. *IEEE Trans. Robot. Autom.* **2004**, *20*, 211–221. [CrossRef]
57. Melchiorri, C.; Vassura, G. Design of a Three-Finger Gripper for Intra-Vehicular Robotic Manipulation. In Proceedings of the IFAC Workshop on Space Robotics (SPRO'98), St-Hubert, QC, Canada, 19–22 October 1998; IFAC Proceedings Volumes; Volume 31, pp. 7–12. [CrossRef]
58. Kim, Y.J.; Song, H.; Maeng, C.Y. BLT Gripper: An Adaptive Gripper With Active Transition Capability Between Precise Pinch and Compliant Grasp. *IEEE Robot. Autom. Lett.* **2020**, *5*, 5518–5525. [CrossRef]
59. Firouzeh, A.; Paik, J. Grasp Mode and Compliance Control of an Underactuated Origami Gripper Using Adjustable Stiffness Joints. *IEEE/ASME Trans. Mechatron.* **2017**, *22*, 2165–2173. [CrossRef]
60. Ko, T. A Tendon-Driven Robot Gripper With Passively Switchable Underactuated Surface and its Physics Simulation Based Parameter Optimization. *IEEE Robot. Autom. Lett.* **2020**, *5*, 5002–5009. [CrossRef]
61. Hussain, I.; Renda, F.; Iqbal, Z.; Malvezzi, M.; Salvietti, G.; Seneviratne, L.; Gan, D.; Prattichizzo, D. Modeling and Prototyping of an Underactuated Gripper Exploiting Joint Compliance and Modularity. *IEEE Robot. Autom. Lett.* **2018**, *3*, 2854–2861. [CrossRef]
62. Lee, W.C.; Wu, C.W. Design and analysis of a novel robotic gripper integrated with a three-phalanx finger. *Proc. Inst. Mech. Eng. Part J. Mech. Eng. Sci.* **2014**, *228*, 1786–1796. [CrossRef]
63. Su, Y.; Fang, Z.; Zhu, W.; Sun, X.; Zhu, Y.; Wang, H.; Tang, K.; Huang, H.; Liu, S.; Wang, Z. A High-Payload Proprioceptive Hybrid Robotic Gripper With Soft Origamic Actuators. *IEEE Robot. Autom. Lett.* **2020**, *5*, 3003–3010. [CrossRef]
64. Ku, S.; Myeong, J.; Kim, H.Y.; Park, Y.L. Delicate Fabric Handling Using a Soft Robotic Gripper With Embedded Microneedles. *IEEE Robot. Autom. Lett.* **2020**, *5*, 4852–4858. [CrossRef]
65. Li, X.; Li, X.; Li, L.; Meng, Y.; Tian, Y. Load Sharing Design of a Multi-legged Adaptable Gripper With Gecko-Inspired Controllable Adhesion. *IEEE Robot. Autom. Lett.* **2021**, *6*, 8482–8489. [CrossRef]
66. Hu, Q.; Dong, E.; Sun, D. Soft Gripper Design Based on the Integration of Flat Dry Adhesive, Soft Actuator, and Microspine. *IEEE Trans. Robot.* **2021**, *37*, 1065–1080. [CrossRef]
67. Li, Y.; Chen, Y.; Li, Y. Pre-Charged Pneumatic Soft Gripper With Closed-Loop Control. *IEEE Robot. Autom. Lett.* **2019**, *4*, 1402–1408. [CrossRef]
68. Pagoli, A.; Chapelle, F.; Corrales, J.A.; Mezouar, Y.; Lapusta, Y. A Soft Robotic Gripper With an Active Palm and Reconfigurable Fingers for Fully Dexterous In-Hand Manipulation. *IEEE Robot. Autom. Lett.* **2021**, *6*, 7706–7713. [CrossRef]
69. Park, W.; Seo, S.; Bae, J. A Hybrid Gripper With Soft Material and Rigid Structures. *IEEE Robot. Autom. Lett.* **2019**, *4*, 65–72. [CrossRef]
70. Gao, G.; Chang, C.M.; Gerez, L.; Liarokapis, M. A Pneumatically Driven, Disposable, Soft Robotic Gripper Equipped With Multi-Stage, Retractable, Telescopic Fingers. *IEEE Trans. Med. Robot. Bionics* **2021**, *3*, 573–582. [CrossRef]
71. Liu, S.; Wang, F.; Liu, Z.; Zhang, W.; Tian, Y.; Zhang, D. A Two-Finger Soft-Robotic Gripper With Enveloping and Pinching Grasping Modes. *IEEE/ASME Trans. Mechatron.* **2021**, *26*, 146–155. [CrossRef]
72. Glick, P.; Suresh, S.A.; Ruffatto, D.; Cutkosky, M.; Tolley, M.T.; Parness, A. A Soft Robotic Gripper With Gecko-Inspired Adhesive. *IEEE Robot. Autom. Lett.* **2018**, *3*, 903–910. [CrossRef]
73. Nie, S.; Liu, X.; Ji, H.; Ma, Z.; Yin, F. Simulation and Experiment Study on Deformation Characteristics of the Water Hydraulic Flexible Actuator Used for the Underwater Gripper. *IEEE Access* **2020**, *8*, 191447–191459. [CrossRef]

74. Hwang, G.; Park, J.; Cortes, D.S.D.; Hyeon, K.; Kyung, K.U. Electroadhesion-Based High-Payload Soft Gripper With Mechanically Strengthened Structure. *IEEE Trans. Ind. Electron.* **2022**, *69*, 642–651. [CrossRef]
75. Krahn, J.M.; Fabbro, F.; Menon, C. A Soft-Touch Gripper for Grasping Delicate Objects. *IEEE/ASME Trans. Mechatron.* **2017**, *22*, 1276–1286. [CrossRef]
76. Li, L.; Jin, T.; Tian, Y.; Yang, F.; Xi, F. Design and Analysis of a Square-Shaped Continuum Robot With Better Grasping Ability. *IEEE Access* **2019**, *7*, 57151–57162. [CrossRef]
77. Sanjuan, J.; Serje, D.; Pacheco, J. Closed form solution for direct and inverse kinematics of a US-RS-RPS 2-DOF parallel robot. *Sci. Iran.* **2018**, *25*, 2144–2154. [CrossRef]
78. Kim, S.; Laschi, C.; Trimmer, B. Soft robotics: A bioinspired evolution in robotics. *Trends Biotechnol.* **2013**, *31*, 287–294. [CrossRef] [PubMed]
79. Hu, Z.; Wan, W.; Harada, K. Designing a Mechanical Tool for Robots With Two-Finger Parallel Grippers. *IEEE Robot. Autom. Lett.* **2019**, *4*, 2981–2988. [CrossRef]
80. Pellerin, C. The salisbury hand. *Ind. Robot. Int. J.* **1991**, *18*, 25–26. [CrossRef]
81. Butterfass, J.; Grebenstein, M.; Liu, H.; Hirzinger, G. DLR-Hand II: Next generation of a dextrous robot hand. In Proceedings of the 2001 ICRA. IEEE International Conference on Robotics and Automation (Cat. No.01CH37164), Seoul, Republic of Korea, 21–26 May 2001; Volume 1, pp. 109–114. [CrossRef]
82. Townsend, W. The BarrettHand grasper–programmably flexible part handling and assembly. *Ind. Robot. Int. J.* **2000**, *27*, 181–188. [CrossRef]
83. Tai, K.; El-Sayed, A.R.; Shahriari, M.; Biglarbegian, M.; Mahmud, S. State of the Art Robotic Grippers and Applications. *Robotics* **2016**, *5*, 11. [CrossRef]
84. Ruehl, S.W.; Parlitz, C.; Heppner, G.; Hermann, A.; Roennau, A.; Dillmann, R. Experimental evaluation of the schunk 5-finger gripping hand for grasping tasks. In Proceedings of the 2014 IEEE International Conference on Robotics and Biomimetics (ROBIO 2014), Bali, Indonesia, 5–10 December 2014; IEEE: Piscataway, NJ, USA, 2014; pp. 2465–2470.
85. Ogden, R. *Non-Linear Elastic Deformations, Courier Corporation*; John Wiley & Sons Ltd.: Hoboken, NJ, USA, 1984.
86. Polygerinos, P.; Wang, Z.; Overvelde, J.T.; Galloway, K.C.; Wood, R.J.; Bertoldi, K.; Walsh, C.J. Modeling of soft fiber-reinforced bending actuators. *IEEE Trans. Robot.* **2015**, *31*, 778–789. [CrossRef]
87. Wakimoto, S.; Suzumori, K.; Ogura, K. Miniature pneumatic curling rubber actuator generating bidirectional motion with one air-supply tube. *Adv. Robot.* **2011**, *25*, 1311–1330. [CrossRef]
88. Steck, D.; Qu, J.; Kordmahale, S.B.; Tscharnuter, D.; Muliana, A.; Kameoka, J. Mechanical responses of Ecoflex silicone rubber: Compressible and incompressible behaviors. *J. Appl. Polym. Sci.* **2019**, *136*, 47025. [CrossRef]
89. Suzumori, K.; Endo, S.; Kanda, T.; Kato, N.; Suzuki, H. A bending pneumatic rubber actuator realizing soft-bodied manta swimming robot. In Proceedings of the 2007 IEEE International Conference on Robotics and Automation, Roma, Italy, 10–14 April 2007; IEEE: Piscataway, NJ, USA, 2007; pp. 4975–4980.
90. Wakimoto, S.; Ogura, K.; Suzumori, K.; Nishioka, Y. Miniature soft hand with curling rubber pneumatic actuators. In Proceedings of the 2009 IEEE International Conference on Robotics and Automation, Kobe, Japan, 12–17 May 2009; Piscataway, NJ, USA, 2009; pp. 556–561.
91. Zhang, J.; Wang, H.; Tang, J.; Guo, H.; Hong, J. Modeling and design of a soft pneumatic finger for hand rehabilitation. In Proceedings of the 2015 IEEE International Conference on Information and Automation, Lijiang, China, 8–10 August 2015; IEEE: Piscataway, NJ, USA, 2015; pp. 2460–2465.
92. Yap, H.K.; Ng, H.Y.; Yeow, C.H. High-force soft printable pneumatics for soft robotic applications. *Soft Robot.* **2016**, *3*, 144–158. [CrossRef]
93. Lin, H.T.; Leisk, G.G.; Trimmer, B. GoQBot: A caterpillar-inspired soft-bodied rolling robot. *Bioinspir. Biomim.* **2011**, *6*, 026007. [CrossRef]
94. Wang, Y.; Gupta, U.; Parulekar, N.; Zhu, J. A soft gripper of fast speed and low energy consumption. *Sci. China Technol. Sci.* **2019**, *62*, 31–38. [CrossRef]
95. Xu, L.; Gu, G. Bioinspired Venus flytrap: A dielectric elastomer actuated soft gripper. In Proceedings of the 2017 24th International Conference on Mechatron. and Machine Vision in Practice (M2VIP), Auckland, New Zealand, 21–23 November 2017; IEEE: Piscataway, NJ, USA, 2017; pp. 1–3.
96. Chan, K.C.; Cheung, N. Grasping of delicate objects by a novel two-finger variable reluctance gripper. In Proceedings of the Conference Record of the 2001 IEEE Industry Applications Conference, 36th IAS Annual Meeting (Cat. No.01CH37248), Chicago, IL, USA, 30 September–4 October 2001; Volume 3, pp. 1969–1974. [CrossRef]
97. Kuang, L.; Lou, Y.; Song, S. Design and Fabrication of a Novel Force Sensor for Robot Grippers. *IEEE Sensors J.* **2018**, *18*, 1410–1418. [CrossRef]
98. Barsky, M.; Lindner, D.; Claus, R. Robot gripper control system using PVDF piezoelectric sensors. *IEEE Trans. Ultrason. Ferroelectr. Freq. Control* **1989**, *36*, 129–134. [CrossRef]

robotics

MDPI

Article

A Novel Gripper with Integrated Rotary Unit and Force Control for Pick and Place Applications

Alexey M. Romanov *, Ntmitrii Gyrichidi and Mikhail P. Romanov

Institute of Artificial Intelligence, MIREA-Russian Technological University (RTU MIREA), 119454 Moscow, Russia
* Correspondence: romanov@mirea.ru

Abstract: Modern electrical grippers have lower life-cycle costs compared to pneumatic ones. Furthermore, they provide force control, making it possible to grasp objects with different fragility using a single device. At the same time, electrical grippers have a higher end-effector weight, installed on the robot's flange and lower closing speed, preventing them from replacing pneumatic solutions in high dynamic Pick and Place applications. This research faces both issues by synthesizing a novel gripper mechanism based on a Torque Distribution Gearbox, which makes it possible to relocate the electric motors to the static frame of a delta robot. The proposed gripper not only has a lower mass and a higher closing speed than competitive electric solutions, but it also provides unlimited rotation around the vertical axis. The performance of the gripper was tested in experimental studies, which showed that a created aluminum prototype provides a precise force control in the range from 3 N to 48 N with an accuracy not worse than 1.27 N. Moreover, its finger's speed is 3.1–56 times higher than market available electrical grippers, which makes it comparable by this parameter with pneumatic solutions used in high dynamic Pick and Place applications.

Keywords: mechanism synthesis; grasping; force control; manipulators; end effectors; robotics and automation

Citation: Romanov, A.M.; Gyrichidi, N.; Romanov, M.P. A Novel Gripper with Integrated Rotary Unit and Force Control for Pick and Place Applications. *Robotics* **2022**, *11*, 155. https://doi.org/10.3390/robotics11060155

Academic Editor: Raffaele Di Gregorio

Received: 23 November 2022
Accepted: 15 December 2022
Published: 18 December 2022

1. Introduction

Pick and Place operations are frequently used in different industrial applications [1,2], especially in the food industry [3,4]. Typically Pick and Place operations are implemented using SCARA or Delta robots with four degrees of freedom: three transnational degree and one rotation around a vertical axis [5,6]. Another feature of these operations is their high speed: for example, Delta robots are designed to perform up to 150 Pick and Place operations per minute [7]. To ensure a high dynamics, it is necessary to significantly reduce the mass of the moving parts of the Delta robot, especially the gripper mounted on its flange. At the same time, adding a rotary unit, force sensors, or attempting to enlarge grasping force leads to heavier designs and significantly limits robot's accelerations.

This paper introduces a new approach to the design of an electrical gripper, which makes it possible to reduce its mass by relocating both motors to the robot's frame. Our design provides the ability to turn the gripper on an unlimited angle around the vertical axis as well as to control the grasping force using only current feedback from the motor without the use of additional force sensors.

The main contributions of the paper are the following: (1) a novel gripper's design, which allows it to turn on an unlimited angle and control the grasping force with no motors or electronics installed on the robot's flange; (2) a controller for the proposed gripper, which can be implemented using standard industrial servo drives connected to the robot's control system with industrial fieldbus; (3) an aluminum prototype of the gripper, which provides an opening stroke up to 70 mm and a controlled grasping force range from 3 N to 48 N with the accuracy not worse than 1.27 N.

The rest of this paper is organized as follows. Section 2 provides a brief review of the related works. Section 3 describes the design of the novel gripper and introduces soft horizontal fingers. Section 4 proposes a controller for the novel gripper. Experimental results and analysis are given in Section 5. Section 6 summarizes the main results of this research. All the symbols that are used in the paper are listed in Table 1.

Table 1. Symbols description.

Symbol	Description
φ_g	Angular position of the gripper relative vertical axis, rad
ω_g	Angular velocity of the gripper relative vertical axis, rad/s
ε_g	Angular acceleration of the gripper relative vertical axis, rad/s^2
ω_f	Angular velocity of the Finger's gear, rad/s
ε_f	Angular acceleration of the Finger's gear, rad/s^2
φ_2	Angular position of M_2 motor, rad
ω_1, ω_2	Angular velocities of M_1 and M_2 motors, rad/s
$\varepsilon_1, \varepsilon_2$	Angular accelerations of M_1 and M_2 motors, rad/s^2
g_c	Gear ratio between motor M_2 axis and the gripper's angular position
g_{ps}	Gear ratio between the Cone gear and the Sector gear
g_{pf}	Gear ratio between the Sector gear and the Finger's gear
g_f	Gear ratio between the motor M_1 and the fingers
D_{pd}	Driver's gear pitch diameter, m
D_{pu}	Upper gear pitch diameter, m
D_{pf}	Pitch diameter of the Finger's gear, m
D_{ps1}	Pitch diameter of the cone part of the Sector gear, m
D_{ps2}	Pitch diameter of the cylindrical part of the Sector gear, m
h_{pc}	Distance from the the upper part to the pitch line of the Cone gear, m
l_a	Distance between M_1 and M_2 motors' axes, m
l_f	The finger's length, m
J_{M1}, J_{M2}	Moments of inertia of the gripper mechanics relative to the M_1 and M_2, kgm^2
F_f	The force applied to the grasped object, N
U_{M1}, U_{M2}	Control voltages of M_1 and M_2 motors, V
I_{M1}, I_{M2}	Quadrature currents of M_1 and M_2 motors, A
I_{eM1}, I_{eM2}	Current errors of the controllers corresponding to M_1 and M_2 motors, A
A_1, A_2	Integrator units outputs of the controllers corresponding to M_1 and M_2 motors, V
$A_{''}$	Output of the digital differentiator unit's integrator, rad/s^2
L_{M1}, L_{M2}	Inductances of M_1 and M_2 windings, H
R_{M1}, R_{M2}	Resistances of M_1 and M_2 windings, Ohm
$K_{t_{M1}}, K_{t_{M2}}$	Torque constants of M_1 and M_2 motors, Nm/A
$K_{e_{M1}}, K_{e_{M2}}$	Back-EMF constants of M_1 and M_2 motors, V/(rad/s)
φ_{set}	Gripper's angular position set point, rad
F_{set}	Grasping force set point, N
$L(F_{set})$	Lookup table used to evaluate M_1 motor's current set point, A
φ_{sr}	Immediate angular position set point, generated by Set value generator, rad
ω_{sr}	Immediate angular velocity set point, generated by Set value generator, rad/s
ε_{sr}	Immediate angular acceleration set point, evaluated by differentiator unit, rad/s^2
T	Digital differentiator unit's time constant, s
K_p	Gain of the gripper's angular position controller, 1/s
K_v	Gain of the gripper's angular velocity controller, As/rad
K_{c1}, K_{c2}	Proportional gains of M_1 and M_2 motors' current controllers, V/A
T_{c1}, T_{c2}	Integral times of M_1 and M_2 motors' current controllers, s
K_ε	Acceleration feed-forward gain, A/(rad/s^2)
K_ω	Velocity feed-forward gain, V/(rad/s)
t	Time, s

2. Related Works

The most popular gripper type used on Pick and Place robots is a vacuum gripper [8]. Its main limitation is that it is designed for lifting flat, smooth parts and materials [4,9], thus it is not suitable for curved parts of non-standard shapes or products with a high porosity [3].

Finger grippers are more suitable for picking non-uniform objects [9], making a parallel pneumatic gripper with two fingers [10] the second most popular solution for the industrial grasping device. To compare with their electric counterparts, pneumatic grippers of the same sizes have a smaller weight, a greater grasping force and a higher grasping speed [11]. Moreover, they usually have a lower price [3,11]. However, the pneumatic grippers generally have only two stable states (opened or closed), so it is difficult to control the grasping force while handling fragile objects. Control of the grasping force via air pressure is possible, but has limited accuracy of the grasping force control. Thus, to the best of our knowledge, the only robust and cost-effective way of controlling the grasping force of a fingered pneumatic gripper with a high precision is to install an additional force regulation mechanism on it [11].

Pneumatic grippers and other pneumatic systems have been widely used in different industries for decades. Meanwhile, the results of a critical review of pneumatic grippers' energy efficiency has shown that despite the lower price of pneumatic actuators, the electrical actuators with a comparable force output have lower life-cycle costs. This difference stems from the fact that for the operation of pneumatic grasping devices at the enterprise, a large amount of compressed air is needed to produce, for which a large amount of electricity is spent. Reducing the volume of air consumed and the compressor's power could solve this problem. However, the poor efficiency of end-use pneumatic devices, e.g., linear cylinders, makes this problematic [12].

Electrical grippers provide a precise control of fingers' position, speed, and acceleration. Moreover, they are often integrated with force sensors, making it possible to gently manipulate soft and fragile objects [3,13].

One of the most advanced solutions for the classical electric parallel gripper is a gripper installed on ABB YuMi Robot. It has a low mass and can precisely dose the grasping force, thus being suitable for manipulating fragile objects. Due to these advantages, it is often used as a reference design in scientific papers dedicated to the grasping topic [14–16].

Pick and Place operations in industrial application often require rotating an object around a vertical axis in addition to transnational movements. Such a motion is usually implemented with two separate motors: one to move the fingers and the other to rotate the whole gripper, including the first motor. In 2013 Schunk introduced a gripper-swivel unit [17], where the motor used to close gripper's jaws is statically mounted inside the gripper and does not move during rotation. This type of gripper requires that the torque of its rotating part generated by the second electric motor is multiple times greater than the torque of its gripper unit generated by the first electric motor upon reaching a terminal stroke position of the jaws. Thus, enlarging of the grasping force will either require a larger and a more powerful motor for the rotation part, which would increase the mass, or higher gear ratios, which would decrease angular accelerations. A market-available version of the gripper built under the patent [17] does not suit the requirements of Pick and Place operations performed by delta robots: it has a jaws closing speed of only 120 mm/s, rotation angle limit of 270 degrees, and no encoder on the rotation stage.

Another way to reduce end effector's mass and increase its acceleration is to place one of the motors (usually the one rotating gripper) on a static robot frame transferring torque through the telescopic shaft [18,19]. However, to reach the mass comparable to pneumatic actuators another motor should also be removed from the robot's flange. As one of the solutions several cardan shafts can be used to connect statically mounted motors and the end effector. Such design is used in some commercial robots by Codian [20].

To enhance the variability of the objects that can be successfully grasped, several grippers with more complex mechanical designs have recently been proposed.

In the paper [21] the authors introduce a set of grippers for a flexible industrial assembly. The authors propose three grippers for UR5 Universal Robots, working in collaboration. The first one is the Pincher Gripper—a classic electric finger gripper with one degree of freedom. The second one is Kitting Gripper with a soft adhesive pad to pick small objects. The third one is Rotary Gripper, which has two degrees of freedom which work as a collect clamp. This gripper's design is the most suitable for the main Delta robots application. It has two in-build motors, one to open and close the fingers and one to rotate the fingers. This gripper proved its effectiveness in WRS Industrial Assembly challenge 2018. It was used to grasp parts by the outer or inner diameter and put them in their places. The main disadvantages of this gripper are as follows: it is designed to grasp objects with a limited number of shapes (cylindrical, triangular, and hexagonal); a screw transmission provides a low closing speed and a complicated force sensing.

Kim et al. [22] introduced a three-fingers electric gripper with underactuated components, which makes it possible to switch between Precise Pinch and Compliant Grasp. The gripper proved the ability to grasp a variety of objects. The main drawback of the proposed approach is a complex mechanical design, which requires five motors and small-sized precise gearboxes. To achieve a high force and ensure that all the motors have a reasonable size, the authors propose to use high gear ratios. Thus, the ability of the grasping force sensing through the motor current is quite limited, while additional force sensors will complicate the design even more.

Another designs of two-finger underactuated grippers with enhanced grasping capabilities were introduced in [23,24]. Those designs are much more simpler than [22] and require fewer motors. In [24], the authors even managed to perform grasping using a single motor. However, this motor has to be mounted directly on a gripper, which results in an increased mass compared to pneumatic solutions. Generally, none of these grippers were designed to manipulate fragile objects, so they do not have precise force control. To solve this problem, an enhanced version of M^2 gripper equipped with an additional tactile force sensor was proposed in [25].

It should also be mentioned that underactuated grippers are often designed as tendon driven, thus due to tendon elasticity they have a limited dynamic performance compared to the similar geared actuators [26].

Another promising way to improve grasping capabilities of complex curved objects is to use soft fingers [4,27,28]. It allows one to increase the contact surface between an object and the gripper without the need for additional actuators.

The common drawback of all the described grippers is that they all have a limited angle on which they can be turned around the vertical axis. Such a constraint appears due to the need to connect the rotating part of a grasping device with the robot's frame by wires or pneumatic lines in order to operate the gripper.

Depending on the gripper's type, the robot's cabinet can be connected directly to the actuators with air tubes/power cables or interact with the gripper's internal control unit. The latter case becomes more and more popular as it makes it easy to support different types of end effectors developed independently from the robot. Simple grippers are usually controlled by 24V digital outputs, while more complex ones use bi-directional industrial fieldbuses [29].

3. A Novel Gripper Design

A novel gripper design is inspired by two degrees of freedom (DoF) epicyclic mechanism and consist of two motors mounted on the robot's frame, two telescopic shafts, a Torque Distribution Gearbox, and soft fingers (Figure 1). Motor M1 is in charge of opening/closing the fingers and providing desired grasping force, while M2 rotates the gripper around the vertical axis. Due to the design, these motors have a mechanical linkage that is compensated in real-time by the controller proposed in Section 4.

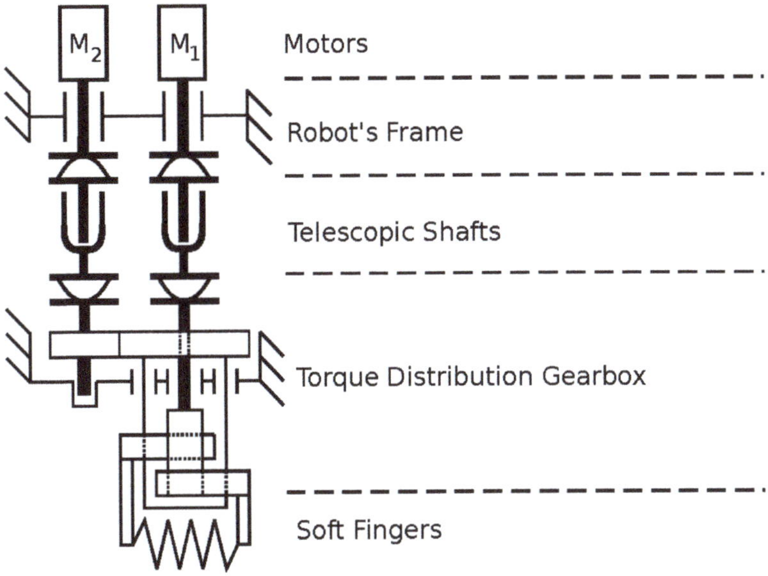

Figure 1. The novel gripper's kinematic scheme.

The key part of the gasping device is Torque Distribution Gearbox. It provides the ability to grasp and rotate handled parts without installing any DC motors on the gripper, which makes it possible to relocate them to the robot's frame. To implement the proposed solution we have developed a mechanical design shown in Figure 2. It includes the following parts: the upper cover, which connects the gripper to the robot's flange, the outboard bearing, a set of gears, implementing the mechanism, which is mathematically equivalent to 2-DoF epicyclic mechanism, and 3D-printed soft fingers, similar to the ones introduced in [30]. The set of gears under the robot's flange implements Torque Distribution Gearbox. Gearings are shown in details on Figure 3. Shafts of Driver's and Cone Gear are inputs of the Torque Distribution Gearbox. The shaft of the Cone Gear goes through the center of the Upper Gear, while the latter one is rigidly mounted to the bottom part of the gripper, including Sector and Finger's Gears. As a result the upper gear rotation angle φ_g is the same as gripper finger's orientation angle relative to vertical axis.

If during operation Driver's Gear moves simultaneously with Cone Gear in opposite directions the gripper rotates around its vertical axis. If the Driver's Gear is fixed and Cone Gear rotates, the gripper's fingers are closing or opening. The controller providing the desired rotation acceleration and grasping force, by applying specific torques to input shafts of Torque Distribution Gearbox is described in Section 4.

The purpose of the soft fingers is to dampen the impact the moment they touch an object. In the case of rigid fingers, reaching the object will result in a step change of the reaction force, which will result in the grasping force overshot due to the limited reaction speed of the force controller. The use of the soft fingers makes the reaction force rise smoothly, making it possible to avoid the above overshoot.

The gear ratios of Torque Distribution Gearbox (Figure 3) are labeled as following: g_c between Upper Gear and Driver's Gear, g_{ps} between Cone Gear and Sector Gear, and g_{pf} between Sector Gear and Finger's Gear.

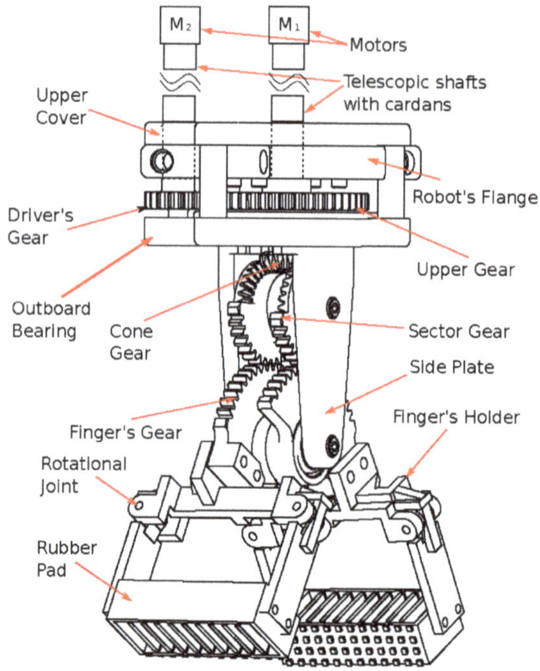

Figure 2. The novel gripper mechanical design.

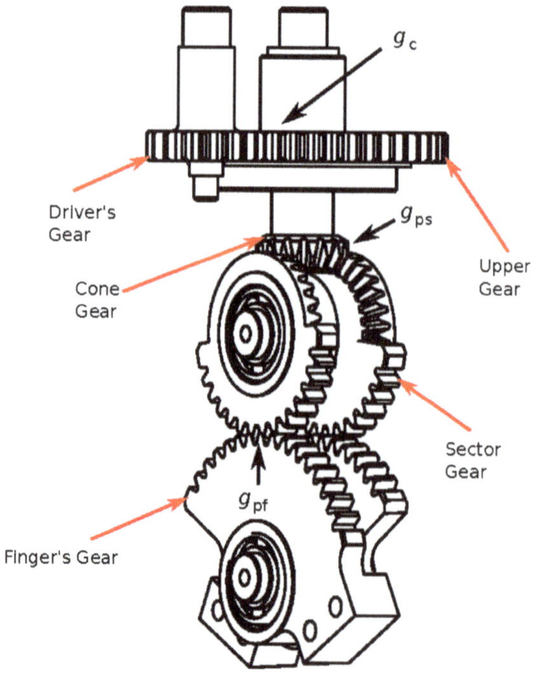

Figure 3. Torque Distribution Gearbox.

All the parts of the proposed design can be parameterized to meet specific project requirements. The upper cover size is mainly determined by the distance between motors' axes l_a and the pitch diameters of Driver's and Upper gears, defined as D_{pd} and D_{pu}, respectively, (Figure 4). Those parameters are linked through (1).

$$\begin{aligned} l_a &= \frac{D_{pd}+D_{pu}}{2}, \\ g_c &= \frac{D_{pu}}{D_{pd}}, \end{aligned} \tag{1}$$

where g_c is gear ratio between motor M_2 and the gripper.

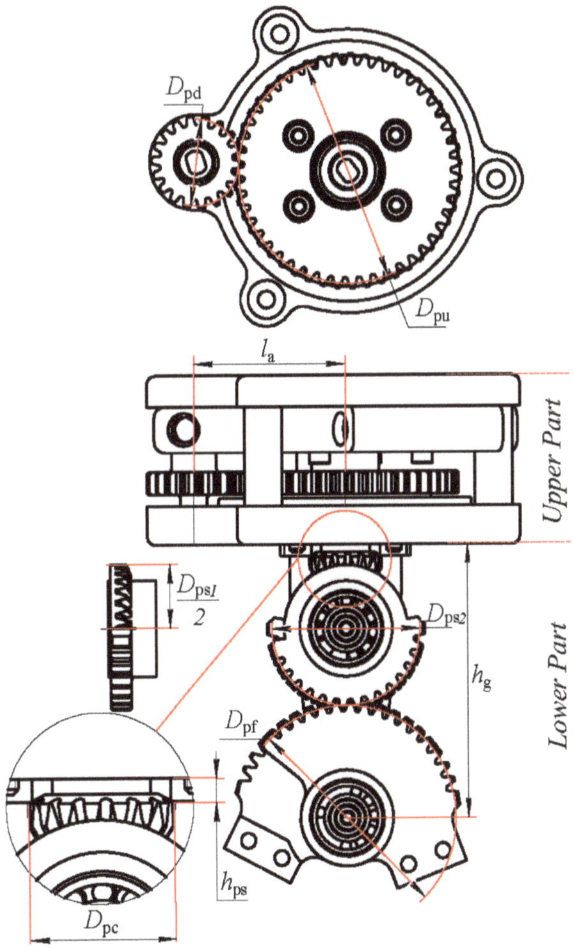

Figure 4. Key dimensions of the proposed gripper's mechanics.

The height of the lower part of the gripper mechanism h_g is defined by Equation (2).

$$h_g = \frac{D_{pf} + D_{ps1} + D_{ps2}}{2} + h_{pc}, \tag{2}$$

where D_{pf} is the pitch diameter of the Finger's gear, D_{ps1} is the pitch diameter of the cone part of the Sector gear, D_{ps2} is the pitch diameter of the cylindrical part of the Sector gear, h_{pc} is the distance from the surface of the upper part of the gripper to the pitch line of the Cone gear.

Assuming that the linked gears have the same pitch parameters, their gear ratios can be evaluated using (3).

$$g_{\mathrm{ps}} = \frac{D_{\mathrm{ps1}}}{D_{\mathrm{pc}}}, \quad g_{\mathrm{pf}} = \frac{D_{\mathrm{pf}}}{D_{\mathrm{ps2}}}, \tag{3}$$

where g_{ps} is the ratio between the Cone gear and the Sector gear, g_{pf} is the ratio between the Sector gear and the Finger's gear, and D_{pc} is the pitch diameter of the Cone gear.

The gear ratio g_{f} between the motor M_1 and the fingers can be evaluated according (4).

$$g_{\mathrm{f}} = g_{\mathrm{ps}} \cdot g_{\mathrm{pf}} = \frac{D_{\mathrm{ps1}}}{D_{\mathrm{pc}}} \cdot \frac{D_{\mathrm{pf}}}{D_{\mathrm{ps2}}} \tag{4}$$

Finally, choosing gear diameters involves a compromise between the desired gear ratios for each motor and the gripper's size, taking into account the limitations imposed by the strength of the gear's materials, considering the size of proper bearings. The choice of materials and the design of gears' pitches is not discussed in this paper as this topic is widely covered in books such as [31]. Furthermore, it is worth mentioning that the gripper's finger design is out of the scope of this research.

The proposed design ensures an unlimited gripper rotation obviating the need of passing electrical signals by slip rings and making it possible to choose motors according to the task. This can be considered as a significant benefit of the proposed solution compared to other types of grippers.

4. Rotation and Grasping Force Controller Design

To simplify the control we made a number of assumptions: (1) rotation of the gripper should start only after its fingers are closed and the desired value of F_{f} is achieved; (2) the grasping force F_{set} set point should be maintained constant between the opening and closing of the fingers; (3) the opening of the gripper should be performed only after the rotation is finished. These assumptions perfectly reflect most of Pick and Place operation performed by delta robots.

The proposed gripper design can be scaled to achieve desired size and force. Such scaling requires the controller to support a wide range of motors, in difference to ONROBOT, ABB, and SCHUNK grippers, which controllers are designed to operate with a specific motor. Thus, it was decided to use industrial servo drives from market-available families to provide desired unification. At the same time, this decision leads to limitations on controller structure that should be considered during its design.

As a reference servo drive in current research, one from the B&R ACOPOS series was used. These drives support asynchronous, synchronous, and DC-motors [32] ranging from 50 W to 120 kW. Their controller generally consists of three Proportional–Integral (PI) control loops. However, using Smart Process Technology (SPT), it can be enhanced by adding new links and filters, evaluated directly on-board of the servo drive. Even though B&R products were chosen as a reference, the controller proposed in this section can be implemented on similar industrial servo drives produced by KEBA, Beckhoff, etc.

The Torque Distribution Gearbox can be considered as a 2-DoF epicyclic mechanism. Unlike most of the grippers, it creates a mechanical linkage between the motors used to perform gripper's rotation and transnational movements of the fingers. In the mechanism the Driver's gear plays the role of a driver and the Cone gear becomes the sun, which is connected with the satellite, implemented by Sector gear and connected to the Finger's gear. Thus, the dependency between angular accelerations and velocities of the gripper's parts can be expressed by Equations (5) and (6).

$$\omega_1 - (1 + \frac{1}{g_{\mathrm{ps}}})\omega_{\mathrm{g}} = \frac{\omega_{\mathrm{f}}}{g_{\mathrm{pf}}} \tag{5}$$

$$\varepsilon_1 - (1 + \frac{1}{g_{ps}})\varepsilon_g = \frac{\varepsilon_f}{g_{pf}}, \tag{6}$$

where, ω_1 and ε_1 are angular velocity and acceleration of the motor M_1; ω_g and ε_g are angular velocity and acceleration of the gripper rotation relative to vertical axis; ω_f and ε_f are angular velocity and acceleration of the Finger's gear rotation. In the proposed gripper there is no dedicated sensor measuring its rotation angle. To measure it the encoder of the motor M_2 is used. This encoder allows to meet precision requirements in the most of real industrial applications. In this case the gripper's rotation angle φ_g, ω_g velocity and acceleration ε_g can be evaluated using (7).

$$\begin{aligned} \varphi_g &= -\varphi_2 \tfrac{1}{g_c} \\ \omega_g &= -\omega_2 \tfrac{1}{g_c} \\ \varepsilon_g &= -\varepsilon_f \tfrac{1}{g_c} \end{aligned} \tag{7}$$

The Rotation controller for the proposed gripper can be implemented using well-known three level P-P-PI structure with current, velocity, and position feedbacks (Figure 5). In this paper, we assume that a permanent magnet synchronous motors with Field-Oriented Control can be considered a DC-motor with current controller.

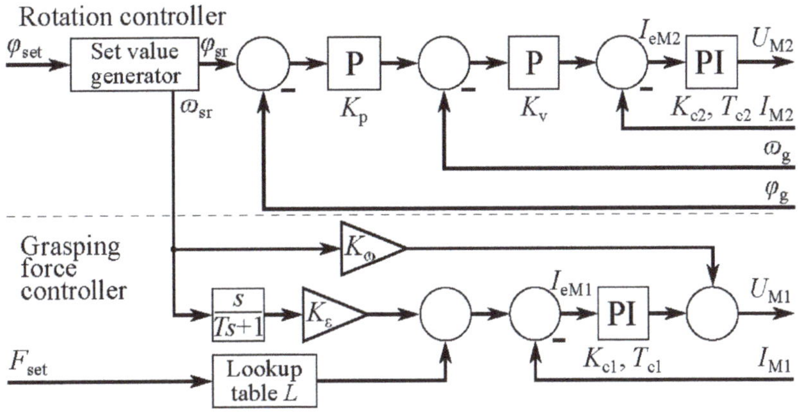

Figure 5. Rotation and Grasping force controller.

As mentioned above no rotation is performed during grasping, so epicyclic mechanism of Torque Distribution Gearbox becomes blocked and the grasping force F_f depends only of motor the M_1 current according to (8).

$$F_f = \frac{I_{M1} K_{t_{M1}} g_f}{l_f}, \tag{8}$$

where I_{M1} is motor M_1 current; $K_{t_{M1}}$ is motor M_1 torque gain; and l_f is the length of the gripper's fingers.

Thus, grasping with predefined force can be implemented using single loop current controller (Figure 5). Then, as F_{set} is assumed to be constant during rotation to maintain F_f on the desired level it is required to ensure that the fingers remain in the same position. To do so we propose to add two feed-forward channels to the Grasping force controller. The first one is connected to motor M_1 current controller input to create acceleration ε_1, required to fully compensate the rotation of the gripper and keep ε_f as close to zero as possible. The second feed-forward channel is used to compensate current I_{M1} sag caused by rising the back-EMF of M_1 during rotation. The feed-forwards were used instead of actual M_2 velocity and acceleration feedbacks to minimize their influence on thew overall stability of the controller.

The desired velocity and acceleration of M_1 can be evaluated using (5), (6), assuming that ω_f and ε_f are zero. Then the gains are found from their values following (9).

$$\begin{aligned} K_\varepsilon &= (1 + g_{ps}) \frac{J_{M1}}{K_{t_{M1}}} \\ K_\omega &= (1 + g_{ps}) K_{e_{M1}}, \end{aligned} \tag{9}$$

where J_{M1} is the moment of inertia of the gripper mechanics relative to the M_1 motor; $K_{e_{M1}}$ is motor M_1 back-EMF gain.

During the design of the controller, the telescopic shafts with the lightweight cardans, which links the gripper and the motors, were intentionally treated as inertial components whose moments of inertia are included in J_{M1} and J_{M2}. This was performed in order to simplify the controller and provide its further implementation using the market-available servo drives. Experimental studies will demonstrate the validity of this approach.

In many cases evaluation of J_{M1} is difficult and requires precise simulation in CAD software. However the initial guess regarding the moment of inertia can be made based on the gripper's size and weight. Then K_ε can be tuned using experimental data to minimize the grasping force error. Such tuning can be considered as a single-parameter optimization problem.

The input of both Rotation and Grasping force controller's feed-forward channels are generated by Set Value generator. It transforms rotation set point φ_{set} into position signal φ_{sr}, with speed ω_{sr} changing as a ramp with constant acceleration and deceleration. As such generators in market-available servo drives usually do not have acceleration output, the latter one can be obtained using a digital differentiator unit (Figure 5).

Thus, the proposed controller includes a set value generator, that interpolates position set point φ_{set} into a smooth φ_{sr} trajectory with trapezoidal ω_{sr} ramp profile, a three-level cascade position controller and a single loop current controller with two feed-forward channels, which are used to provide desired grasping force. The governing equations for the proposed controller are (10)–(18). These equations represent the general controller's behavior and for simplicity of understanding do not include current protection and PI-controllers' anti-windup units.

$$\frac{d\varphi_{sr}}{dt} = \omega_{sr} \tag{10}$$

$$\varepsilon_{sr} = \frac{\omega_{sr}}{T} - A_{''} \tag{11}$$

$$\frac{dA_{''}}{dt} = \frac{\varepsilon_{sr}}{T} \tag{12}$$

$$I_{eM1} = L(F_{set}) + \varepsilon_{sr} K_{''} - I_{M1} \tag{13}$$

$$I_{eM2} = \Big((\varphi_{sr} - \varphi_g) \cdot K_p - \omega_g \Big) \cdot K_v - I_{M2} \tag{14}$$

$$\frac{dA_1}{dt} = I_{eM1} \cdot \frac{K_{c1}}{T_{c1}} \tag{15}$$

$$\frac{dA_2}{dt} = I_{eM2} \cdot \frac{K_{c2}}{T_{c2}} \tag{16}$$

$$U_{M1} = I_{eM1} K_{c1} + A_1 + K_\omega \omega_{sr} \tag{17}$$

$$U_{M2} = I_{eM2} K_{c2} + A_2 \tag{18}$$

The grasping force set point F_{set} is transformed into the required current using a lookup table L. In case of rigid fingers this table implements a linear function with $\frac{K_{t_{M1}} g_f}{l_f}$ gain and an offset big enough to compensate friction. If the gripper is equipped with soft fingers, the use of lookup table makes it possible to implement nonlinear dependency between I_{M1} and F_f. In this case function $F_f(I_{M1})$ can be estimated from experimental data and its inverse function $I_{M1}(F_f)$ and then added into the table.

It is worth mentioning that if the rotation of the gripper is conducted with nonzero constant grasping force F_f, the current controller will automatically compensate telescopic shafts' torsion and cardans' backslashes as they will cause a sag of actual current.

Finally, the use of force control allows not to adapt of the closing angle of the grasping device for each specific object. Regardless of the object's size or texture, the grasping device will stop the closing when the desired grasping force is reached. Such behavior is an advantage of the proposed solution since it does not require a preliminary assessment of the object's size with the help of additional tools (ex., computer vision).

5. Experimental Results

Due to the fact that proposed gripper mechanics can be customized in term of size and grasping force, it is possible to adapt it for different types of robots and tasks. Before building a real prototype we decided to prove scalability of the designed concept, so we created computer models of different gripper's scales and estimated their mass using CAD software (Table 2). These models included both the gripper and the corresponding motors installed on a robot frame. As can be seen from Table 2, contrary to other industrial grippers, the increasing peak grasping force led to the rise of the overall mass, but simultaneously the mass installed on the flange staid nearly unchanged. It is explained by the fact that most of the mass was added by the motors installed on the frame. At the same time reducing the gripper's weight requires scaling down gear diameters and gear pitches, which was impossible to implement due to our manufacturing equipment limitations. Thus, the gripper part has the initially overabundant strength that insured the stability of the mass of the gripper. At the same time the stability of the mass is a significant benefit of a gripper for Pick and Place applications compared to other industrial grippers, as the rise of the mass of the flange leads to lower motion accelerations and results in a decrease in the overall robot's performance. Furthermore, it should be mentioned that our gripper includes an inbuilt rotary unit, while other grippers need additional mechanics to perform object rotation.

Table 2. Mass and Grasping Force of the Different Grippers.

Gripper Type	Peak Grasping Force, N	Mass Installed on the Robot's Flange, kg	Mass Installed on the Robot's frame, kg	Overall Mass of the Gripper and Its Motors, kg
Proposed gripper v.1	48.1	0.6	3	3.6
Proposed gripper v.2	85.3	0.6	4.9	5.5
Proposed gripper v.3	93.2	0.6	6.8	7.4
Proposed gripper v.4	119.6	0.7	8.3	9
ABB SmartGripper	19.6	0.28	0	0.28
ONROBOT RG2	39.2	0.78	0	0.78
ONROBOT 2FG7	137.3	1.1	0	1.1
ONROBOT RG6	117.7	1.25	0	1.25
SCHUNK EGS 25	30	0.45	0	0.45
GIMATIC HS-2012 (pneumatic)	110.3	0.12	0	0.12

Finally, it was decided to implement the first version (v.1) of the gripper, equipped with two B&R 8LVA23.R0015 motors. The gripper frame was made from aluminum, while the gears were created from polyacetal. All aluminum and polyacetal parts of the gripper was manufactured by using CNC milling. In the prototype, a 3D printed fingers were used. They consist of rubber pads made from REC Rubber and finger's holders printed with PLA plastic. Rubber pads are connected to finger's holders via rotational joints. These joints are designed to compensate changes in the distance between ends of rubber pads during grasping. The manufactured gripper is shown in Figure 6.

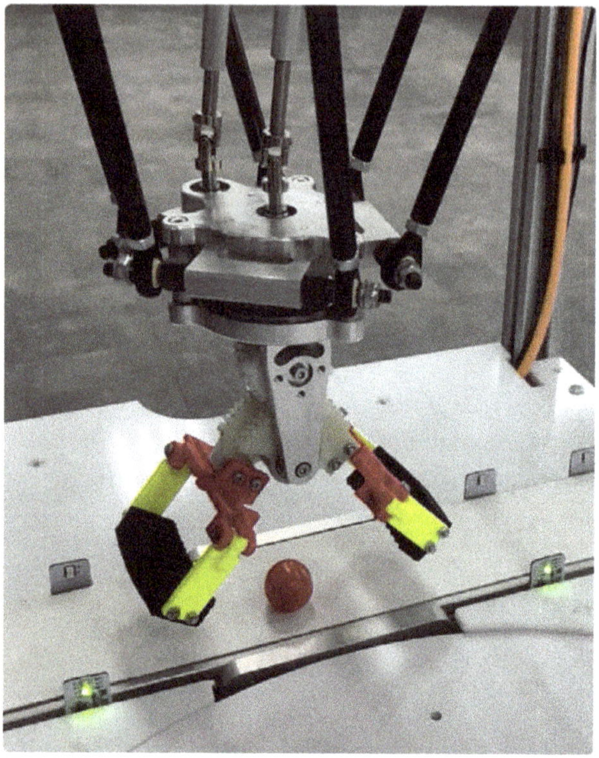

Figure 6. The gripper created for experimental studies.

The gripper has the following parameters: $D_{pd} = 0.022$ m, $D_{pu} = 0.059$ m, $D_{pf} = 0.059$ m, $D_{ps1} = 0.0335$ m, $D_{ps2} = 0.0388$ m, $D_{pc} = 0.0195$ m, $h_{pc} = 0.005$ m, $l_f = 0.08$ m, $L_{M1} = 0.0063$ H, $L_{M2} = 0.0063$ H, $R_{M1} = 2.6$ Ohm, $R_{M2} = 2.6$ Ohm, $K_{e_{M1}} = 0.24$ V/(rad/s), $K_{e_{M2}} = 0.24$ V/(rad/s), $K_{t_{M1}} = 0.42$ Nm/A, $K_{t_{M2}} = 0.42$ Nm/A, $J_{M1} = 7 \cdot 10^{-5}$ kgm^2, and $J_{M2} = 1.1884 \cdot 10^{-4}$ kgm^2.

The Rotation and Grasping Force Controller was fully implemented onboard of the B&R 80VD100PD.C022-14 ACOPOS Servo Drive using its standard cascade controller, modified with Smart Process Technology blocks. Feed-forward channels between Grasping Force and Rotation axis were created using internal cross-communication interface of the drive. Finally, current control loop cycle length of 50 μs, velocity control loop cycle length of 200 μs and positioning control loop cycle length of 400 μs were achieved. The gains of the Current controllers were chosen as: $K_{c1} = 34$ V/A, $K_{c2} = 34$ V/A, $T_{c1} = T_{Mc2} = 0.002$ s. The gains of the velocity and position controller were chosen as $K_v = 0.896$ As/rad and $K_p = 163.6$ 1/s, respectively. Feed-forward differentiator time constant was set as $T = 0.0033$ s. Feed-forward gains were evaluated according to (9). Maximal angular velocity and acceleration, which determine the speed ramp, were set as $\omega_{max} = 5.9$ rad/s and $\epsilon_{max} = 73.3$ rad/s^2, respectively. Finally, the drive was integrated into the robot's control system using Ethernet POWERLINK fieldbus.

Table 3 shows the comparison of the fingers linear speed between the proposed gripper and another industrial gripper of a comparable size. This speed is critical for the performance of Pick and Place operations as it directly determines fingers' opening/closing time. Compared to electrical grippers, the proposed solution has a much higher speed coupled with lower mass installed on the robot's flange (Table 2). The only electrical gripper involved in the comparison that is lighter than our gripper was ABB SmartGripper, designed for YuMi robot, but at the same time its grasping force is more than two times lower than that of our model. It can also be seen that a fast pneumatic gripper outperforms all electrical ones both in terms of mass to peak grasping force ratio and the finger's linear speed. Still it cannot provide any online force control, so it is not suitable for manipulating fragile objects.

Table 3. Comparison of the Industrial Gripper Fingers Linear Speed.

Gripper Type	Max. Linear Speed of the Finger's End, mm/s	Max. Speed of the Proposed Gripper Relative to the Market-Available Product
Proposed gripper v.1	1400	1x
ABB SmartGripper	25	56x
ONROBOT RG2	127	11x
ONROBOT 2FG7	450	3.1x
ONROBOT RG6	160	8.75x
SCHUNK EGS 25	120	56x
GIMATIC HS-2012 (pneumatic)	1700	0.82x

Due to the ultra-low closing time, there is no practical reason to adapt the opening/closing angle for grasping objects of various sizes. It is an additional benefit of the proposed solution, as other electrical grippers may require such adoptions to achieve timing suitable for high-performance Pick and Place operation.

To perform the rest of the experimental studies, the gripper prototype was installed on DeltaCube (https://delta-robot.ru/, accessed date: 23 November 2022) educational robot and integrated into its control system.

The accuracy analysis of the force control was verified using two different sensors. The steady state values of the grasping force were measured using B&R X20AI1744-3 module and a 10 kg range strain gauge mounted on the gripper's fingers. This sensor was used to estimate dependency $I_{M1}(F_f)$ and to inform Grasping force controller's lookup table. Then to verify repeatability we performed 50 grasps for each F_{set} value from 3 N to 48 N range. The results are shown on Figure 7a. As it can be seen, the steady state error never exceeds 0.8 N. Moreover, in most of the cases it was below 0.5 N (Figure 7b).

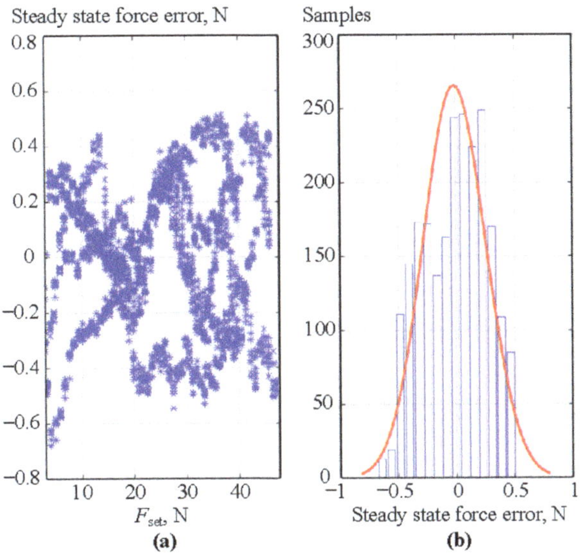

Figure 7. (**a**) Steady state grasping force error. (**b**) Histogram of steady state grasping force error.

As the X20AI1744-3 module, used in previous experiment, has 5 Hz input filter and is not suitable for dynamic measurements to estimate deviations of grasping force during highly accelerated rotations, we changed it to an additional B&R 8LVA13.R0015 motor, which was mounted on the axis of the Finger's gear. It was controlled by a separate servo drive, configured to hold its position. F_f was estimated according to (19).

$$F_f = \frac{I_{Ms} K_{t_s}}{l_f}, \tag{19}$$

where I_{Ms} is the quadrature current of the added motor and K_{t_s} is the its torque constant. Due to an additional inertia introduced by adding the motor, Rotation and Grasping force controller's gains were adjusted to provide the same timings. The measurements using these sensors were performed only in the low part of the gripper's force range, as the 8LVA13 motor has quite low a nominal torque. At the same time, more powerful motors will have a higher mass and impart more inertia that will critically limit range of possible acceleration.

Before performing the experiment on the real gripper, the built prototype was simulated using SimInTech (https://en.simintech.ru/, accessed date: 23 November 2022) software. First, we simulated ideal mechanics without friction. Then, we switched on friction simulation. Furthermore, for ease of comparison we performed additional experiment with no feed-forward ($K_\varepsilon = 0$ and $K_\omega = 0$). The grasping force set point F_{set} was constant during each rotation.

The actual grasping force F_f transient processes in all simulated cases are shown in Figure 8. As it can be seen, force sag during rotation without the use of feed-forward channels reaches up to 35% of the set point. In the case of friction simulation, this sag can be decreased down to 15%, making grasping force control precision suitable for a wide range of industrial applications. In ideal case with no friction, force error can be minimized below 5% of the set point.

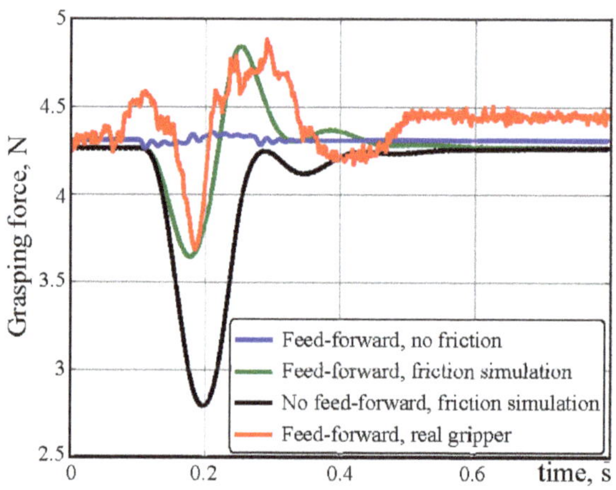

Figure 8. Grasping force F_f transient processes during rotation. Black, blue, green lines—simulated values; red line—real experiment.

The simulated behavior matches the results acquired during the experiment with the real prototype both in term of the F_f transient process form and amplitude (Figure 8, red line). The difference between the real and simulated grasping force steady state value, seen on Figure 8 can be determined by the steady state grasping force error, which was measured in the previous experiment (Figure 7). Most likely it is caused by friction in different parts of the gripper. The absence of this error in simulation can be explained by the simplicity of the used friction model. As it can be seen from the experimental data (Figure 8) the steady state error is generally smaller than the peaks caused by highly accelerated rotations.

To analyze how force control precision depends on rotation parameters, we performed a series of experiments. During each experiment the grasping force was set as a constant value from the 3–5.7 N range, and then two 40° rotations were performed, in forward and backward directions. This range of grasping force was chosen to fit the torque measurement device's capabilities. The parameters of these rotations were taken from the following ranges: velocity—from 0.9 rad/s to 5.6 rad/s; acceleration—from 17.5 rad/s^2 to 73.3 rad/s^2. All measurements were made with the position of the grasping device at the same point in the robot's working area. The position of the grasping device does not affect the measurement results in any way since the telescopic shafts transmit the moment equally efficiently throughout their entire operating range. During all the rotations, transient processes of grasping force error were recorded with 1 ms sample time. Those processes were then analyzed and a maximum grasping force error was evaluated for each pair of the rotation acceleration and velocity (Figure 9). At any acceleration and velocity, the surface of maximum force error demonstrated in Figure 9 never exceeds 1.27 N, which can be considered as the maximum error level for the developed prototype. Table 4 compares this value to the force control error of competitive grippers. As it can be seen, the force control precision of the proposed solution is 2.3–23.2 times higher compared to similar industrial grippers available on the market (Table 4). It should be mentioned that force errors for competitive grippers provided in Table 4 were taken from their datasheets and represent the worst-case scenario for each gripper. Thus, those values are achieved at different grasping speeds. Still, according to Tables 3 and 4, our design outperforms other compared electrical grippers not only in terms of force control precision but also in terms of finger speed.

If we analyze the dependency of the force error from rotation acceleration and velocity (Figure 9), it can be seen that it reaches its maximum values while moving with medium and low velocities. Moreover, one of the surface peaks is near the point corresponding to the lowest acceleration and velocity. This can be explained by friction being the main source of the grasping force error. This hypothesis is also supported by previously obtained experimental results (Figure 8). It is worth mentioning that real experimental results match simulated ones. The model used for simulation did not consider cardans' backslashes and telescopic shafts torsion. Neither of these factors had any significant impact on the behavior of the real gripper during the last experimental series, which included more than 1500 consequent grasps and rotations. This supports the above statement that the proposed controller minimizes the influence of backslashes and shafts' torsion. As a result, friction can be considered a critical factor in reducing grasping force precision, and its reduction is the primary goal of designing an industrial version of the proposed gripper.

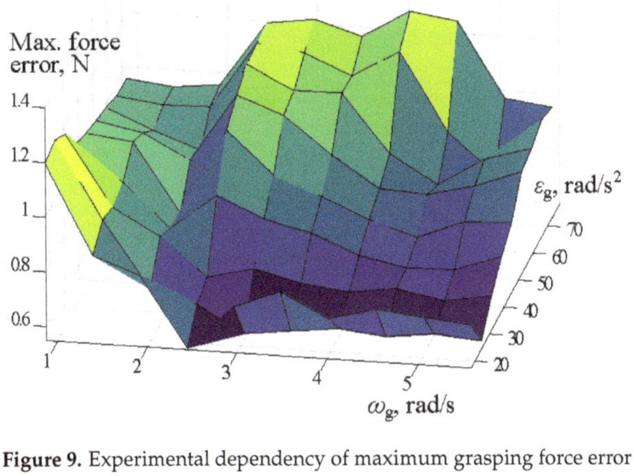

Figure 9. Experimental dependency of maximum grasping force error from rotation acceleration ε_g and velocity ω_g.

Table 4. Accuracy of the Grasping Force Control.

Gripper Type	Maximum Grasping Force Error, N	Error Relative to the Proposed Gripper
Proposed gripper v.1	1.27	×1
ABB SmartGripper	2.94	×2.3
ONROBOT RG2	9.81	×7.7
ONROBOT 2FG7	4.9	×3.9
ONROBOT RG6	29.42	×23.2

The gripper's performance in real Pick and Place operations performed on vulnerable soft objects was evaluated during the final experiment. The developed prototype, installed on the delta robot, was used to pick a cherry tomato, lift it, turn on 45° , put down, and release the tomato. This turn angle was chosen because in most of the industrial Pick and Place applications, the angle correction performed by a robot is below 45°. According to [33] cherry tomatoes are damaged if squeezed with the force 15.2–23.8 N (depending on their variety). Thus, the grasping force should be at least 1.5 times lower than this limit to prevent damage. As our gripper has a force error of 1.27 N (Table 4), the grasping force during the experiment was set to 8.2 N to guarantee the integrity of the manipulated tomato. The Pick and Place operation was recorded by a high-speed camera at 240 FPS. We performed a series of experiments on different tomatoes, and none of them were damaged. The average opening/closing time was 92 ms. The average time required to lift and rotate the tomato by 45° was 130 ms. As a result, the average overall time for Pick and Place operation was 314 ms, equaling 191 Pick and Place operation productivity. It is generally higher than the average rate of modern delta robots provided in [7]. Thus, the proposed gripper design will not be the limiting factor for the overall performance. The Pick and Place operation performed on cherry tomato is demonstrated on the video attached to the paper as media data (Video S1). For the demonstration purposes, the gripper on the attached video was rotated after the robot finished lifting and before it started moving, in deference to the described above experiment when those movements were performed simultaneously.

To the best of our knowledge, none of the currently developed grippers can provide such productivity at such low grasping force. Table 5 compares our experimental results with the competitive solutions based on data from their datasheets. The closing time of ONROBOT force-controlled grippers significantly increases when low grasping force is applied and become 1.5–2 times worse than provided by our solution, even disregarding it was measured during only 8 mm fingers translation. SCHUNK EGS 25 cannot grasp with force below 15 N, which is unsuitable for operations with such vulnerable objects as cherry tomatoes. Furthermore, it has maximal fingers travel range of only 6 mm, while the diameter of the abovementioned cherry tomatoes may vary from 25 to 35 mm (considered optimal size) and may even be greater than 35 mm [34]. Finally, compared to ABB SmartGripper, our solution has a much higher closing speed and better precision of force control (Table 4).

Table 5. Gripping time and minimal verified grasping force comparison.

Gripper Type	Gripping Time/Gripper's Fingers Travel Range That Was Used to Measure Gripping Time	Minimal Verified Grasping Force
ABB SmartGripper	320 ms/8 mm	3 N
ONROBOT RG2	60(@40N)–210(@3N) ms/8 mm	3 N
ONROBOT 2FG7	300 ms/38 mm	20 N
ONROBOT RG6	50(@120N)–150(@25N) ms/8 mm	25 N
SCHUNK EGS 25	50 ms/6 mm	15 N
Proposed gripper v.1	92 ms/98 mm	3 N

6. Conclusions

In this paper, we proposed a synthesized novel gripper mechanism for delta robots. It has a lower mass of end-effector mounted on a robot's flange and 3.1–56 times higher finger speed compared to the market-available electrical grippers with force control. At the same time, it has better 2.3–23.2 grasping force accuracy, providing an error not worse than 1.27 N in the set point range of 3 N to 48 N. Finally, it has an integrated rotary unit, providing unlimited rotation around a vertical axis. The Rotation and Grasping force controller of the novel gripper is simple enough to be implemented using a single dual-channel servo drive and can be easily integrated into the control system of modern Pick and Place robot using a real-time fieldbus. All of the above makes the designed electrical gripper a perfect fit for a high-dynamic Pick and Place applications.

Supplementary Materials: The following supporting information can be downloaded at: https://www.mdpi.com/article/10.3390/robotics11060155/s1, Video S1: Demonstration of a force-controlled Pick and Place operation on a soft, fragile object using a proposed gripper.

Author Contributions: Conceptualization, A.M.R. and M.P.R.; methodology, A.M.R.; software, N.G.; validation, N.G. and A.M.R.; formal analysis, A.M.R. and M.P.R.; investigation, A.M.R., N.G., and M.P.R.; resources, A.M.R. and M.P.R.; data curation, A.M.R.; writing—original draft preparation, A.M.R. and N.G.; writing—review and editing, A.M.R. and M.P.R.; visualization, A.M.R. and N.G.; supervision, A.M.R.; project administration, M.P.R.; funding acquisition, A.M.R. All authors have read and agreed to the published version of the manuscript.

Funding: The manufacturing of the prototype was funded by MRRobot LLC (Dimitrovgrad 433504, Russia) under the agreement on cooperation in research activities with RTU MIREA No. TC-121-232 (11.01.2022).

Institutional Review Board Statement: Not applicable.

Informed Consent Statement: Not applicable.

Data Availability Statement: Not applicable.

Conflicts of Interest: The authors declare no conflicts of interest. The funders had no role in the design of the study; in the collection, analyses, or interpretation of data; in the writing of the manuscript; or in the decision to publish the results.

Abbreviations

The following abbreviations are used in this manuscript:

SCARA	Selective Compliance Assembly Robot Arm
EMF	Electric and magnetic field
WRS	The World Robot Summit
DoF	Degrees of freedom
DC	Direct current
SPT	Smart Process Technology
CAD	Computer-aided design
CNC	Computer numerical control
PLA	Polylactic acid
FPS	Frames per Second

References

1. Najafi, E.; Ansari, M. Model-Based Design Approach for an Industry 4.0 Case Study: A Pick and Place Robot. In Proceedings of the 2019 23rd International Conference on Mechatronics Technology (ICMT), Salerno, Italy, 23–26 October 2019; pp. 1–6. [CrossRef]
2. Han, S.D.; Feng, S.W.; Yu, J. Toward Fast and Optimal Robotic Pick-and-Place on a Moving Conveyor. *IEEE Robot. Autom. Lett.* **2020**, *5*, 446–453. [CrossRef]
3. Blanes, C.; Mellado, M.; Ortiz, C.; Valera, A. Technologies for robot grippers in pick and place operations for fresh fruits and vegetables. *Span. J. Agric. Res.* **2011**, *9*, 1130–1141. [CrossRef]

4. Zhang, B.; Xie, Y.; Zhou, J.; Wang, K.; Zhang, Z. State-of-the-art robotic grippers, grasping and control strategies, as well as their applications in agricultural robots: A review. *Comput. Electron. Agric.* **2020**, *177*, 105694. [CrossRef]
5. Xie, Z.; Wu, P.; Ren, P. A comparative study on the pick-and-place trajectories for a delta robot. In Proceedings of the International Design Engineering Technical Conferences and Computers and Information in Engineering Conference. American Society of Mechanical Engineers, Charlotte, NC, USA, 21–24 August 2016; Volume 50152, p. V05AT07A040.
6. Yang, Z.; Zhao, B.; Bo, L.; Zhu, X.; Xu, K. CurviPicker: A continuum robot for pick-and-place tasks. *Assem. Autom.* **2019**, *39*, 410–421. [CrossRef]
7. Barreto, J.P.; Corves, B. Resonant delta robot for pick-and-place operations. In *Proceedings of the IFToMM World Congress on Mechanism and Machine Science*; Springer: Berlin/Heidelberg, Germany, 2019; pp. 2309–2318.
8. Björnsson, A.; Jonsson, M.; Johansen, K. Automated material handling in composite manufacturing using pick-and-place systems—A review. *Robot. Comput.-Integr. Manuf.* **2018**, *51*, 222–229. [CrossRef]
9. Littlefield, Z.; Zhu, S.; Kourtev, H.; Psarakis, Z.; Shome, R.; Kimmel, A.; Dobson, A.; De Souza, A.F.; Bekris, K.E. Evaluating end-effector modalities for warehouse picking: A vacuum gripper vs a 3-finger underactuated hand. In Proceedings of the 2016 IEEE International Conference on Automation Science and Engineering (CASE), Fort Worth, TX, USA, 21–25 August 2016; pp. 1190–1195.
10. Birglen, L.; Schlicht, T. A statistical review of industrial robotic grippers. *Robot. Comput.-Integr. Manuf.* **2018**, *49*, 88–97. [CrossRef]
11. Chen, C.C.; Lan, C.C. An accurate force regulation mechanism for high-speed handling of fragile objects using pneumatic grippers. *IEEE Trans. Autom. Sci. Eng.* **2017**, *15*, 1600–1608. [CrossRef]
12. Harris, P.; Nolan, S.; O'Donnell, G.E. Energy optimisation of pneumatic actuator systems in manufacturing. *J. Clean. Prod.* **2014**, *72*, 35–45. [CrossRef]
13. Chua, P.Y.; Ilschner, T.; Caldwell, D.G. Robotic manipulation of food products—A review. *Ind. Robot. Int. J.* **2003**, *30*, 345–354. [CrossRef]
14. Guo, M.; Gealy, D.V.; Liang, J.; Mahler, J.; Goncalves, A.; McKinley, S.; Ojea, J.A.; Goldberg, K. Design of parallel-jaw gripper tip surfaces for robust grasping. In Proceedings of the 2017 IEEE International Conference on Robotics and Automation (ICRA), Singapore, 29 May–3 June 2017; pp. 2831–2838. [CrossRef]
15. Alebooyeh, M.; Wang, B.; Urbanic, R.J.; Djuric, A.; Kalami, H. *Investigating Collaborative Robot Gripper Configurations for Simple Fabric Pick and Place Tasks*; Technical report, SAE Technical Paper; SAE International: Warrendale, PA, USA, 2019.
16. Xu, J.; Danielczuk, M.; Steinbach, E.; Goldberg, K. 6DFC: Efficiently Planning Soft Non-Planar Area Contact Grasps using 6D Friction Cones. In Proceedings of the 2020 IEEE International Conference on Robotics and Automation (ICRA), Paris, France, 31 May–31 August 2020; pp. 7891–7897. [CrossRef]
17. Becker, R.; Drab, M.; Quaas, M. Gripper-swivel unit. Patent WO2014170235A1. 23 October 2014.
18. Tanaka, J.; Sugahara, A. Parallel gripper with displacement-magnification mechanism and extendable finger mechanism. In Proceedings of the 2020 IEEE International Conference on Robotics and Automation (ICRA), Paris, France, 31 May–31 August 2020; pp. 9988–9993.
19. Pettersson, A.; Davis, S.; Gray, J.O.; Dodd, T.J.; Ohlsson, T. Design of a magnetorheological robot gripper for handling of delicate food products with varying shapes. *J. Food Eng.* **2010**, *98*, 332–338. [CrossRef]
20. Brinker, J.; Funk, N.; Ingenlath, P.; Takeda, Y.; Corves, B. Comparative study of serial-parallel delta robots with full orientation capabilities. *IEEE Robot. Autom. Lett.* **2017**, *2*, 920–926. [CrossRef]
21. Hughes, J.; Gilday, K.; Scimeca, L.; Garg, S.; Iida, F. Flexible, adaptive industrial assembly: Driving innovation through competition. *Intell. Serv. Robot.* **2020**, *13*, 169–178. [CrossRef]
22. Kim, Y.J.; Song, H.; Maeng, C.Y. BLT Gripper: An Adaptive Gripper With Active Transition Capability Between Precise Pinch and Compliant Grasp. *IEEE Robot. Autom. Lett.* **2020**, *5*, 5518–5525. [CrossRef]
23. Ma, R.R.; Spiers, A.; Dollar, A.M. M 2 gripper: Extending the dexterity of a simple, underactuated gripper. In *Advances in Reconfigurable Mechanisms and Robots II*; Springer: Berlin/Heidelberg, Germany, 2016; pp. 795–805.
24. Ko, T. A Tendon-driven Robot Gripper with Passively Switchable Underactuated Surface and its Physics Simulation Based Parameter Optimization. *IEEE Robot. Autom. Lett.* **2020**, *5*, 5002–5009. [CrossRef]
25. Ward-Cherrier, B.; Cramphorn, L.; Lepora, N.F. Tactile manipulation with a TacThumb integrated on the open-hand M2 gripper. *IEEE Robot. Autom. Lett.* **2016**, *1*, 169–175. [CrossRef]
26. Nazma, E.; Mohd, S. Tendon driven robotic hands: A review. *Int. J. Mech. Eng. Robot. Res.* **2012**, *1*, 350–357.
27. Wang, Z.; Torigoe, Y.; Hirai, S. A prestressed soft gripper: Design, modeling, fabrication, and tests for food handling. *IEEE Robot. Autom. Lett.* **2017**, *2*, 1909–1916. [CrossRef]
28. Gao, Y.; Huang, X.; Mann, I.S.; Su, H.J. A novel variable stiffness compliant robotic gripper based on layer jamming. *J. Mech. Robot.* **2020**, *12*, 1–10. [CrossRef]
29. Romanov, A.M. A review on control systems hardware and software for robots of various scale and purpose. Part 1. Industrial robotics. *Russ. Technol. J.* **2019**, *7*, 30–46. [CrossRef]
30. Xu, W.; Zhang, H.; Yuan, H.; Liang, B. A Compliant Adaptive Gripper and Its Intrinsic Force Sensing Method. *IEEE Trans. Robot.* **2021**, *37*, 1584–1603. [CrossRef]
31. Radzevich, S.P. *Dudley's Handbook of Practical Gear Design and Manufacture*; CRC Press: Boca Raton, FL, USA, 2016.

32. Romanov, A.; Petrovskaya, A.; Slashchev, B. Usage of modern alternating current servo drives for direct current motors control. *Mekhatronika Avtom. Upr.* **2014**, *12*, 34–40.
33. Kabas, O.; Ozmerzi, A. Determining the mechanical properties of cherry tomato varieties for handling. *J. Texture Stud.* **2008**, *39*, 199–209. [CrossRef]
34. Santamaria, P.; Campanile, G.; Parente, A.; Elia, A. Subirrigation vs. drip-irrigation: Effects on yield and quality of soilless grown cherry tomato. *J. Hortic. Sci. Biotechnol.* **2003**, *78*, 290–296. [CrossRef]

robotics

Article

Development of an End-Effector Type Therapeutic Robot with Sliding Mode Control for Upper-Limb Rehabilitation

Md Mahafuzur Rahaman Khan [1], Asif Al Zubayer Swapnil [1], Tanvir Ahmed [2], Md Mahbubur Rahman [1], Md Rasedul Islam [3], Brahim Brahmi [4,*], Raouf Fareh [5] and Mohammad Habibur Rahman [1]

1 Department of Mechanical Engineering, University of Wisconsin-Milwaukee, Milwaukee, WI 53211, USA
2 Biomedical Engineering, University of Wisconsin-Milwaukee, Milwaukee, WI 53211, USA
3 Richard J. Resch School of Engineering, University of Wisconsin—Green Bay (UWGB), Green Bay, WI 54311, USA
4 Electrical Engineering Department, College Ahuntsic, Montreal, QC H2M 1Y8, Canada
5 Electrical Engineering Department, University of Sharjah, Sharjah P.O. Box 27272, United Arab Emirates
* Correspondence: brahim.brahmi@collegeahuntsic.qc.ca

Abstract: Geriatric disorders, strokes, spinal cord injuries, trauma, and workplace injuries are all prominent causes of upper limb disability. A two-degrees-of-freedom (DoFs) end-effector type robot, iTbot (intelligent therapeutic robot) was designed to provide upper limb rehabilitation therapy. The non-linear control of iTbot utilizing modified sliding mode control (SMC) is presented in this paper. The chattering produced by a conventional SMC is undesirable for this type of robotic application because it damages the mechanical structure and causes discomfort to the robot user. In contrast to conventional SMC, our proposed method reduces chattering and provides excellent dynamic tracking performance, allowing rapid convergence of the system trajectory to its equilibrium point. The performance of the developed robot and controller was evaluated by tracking trajectories corresponding to conventional passive arm movement exercises, including several joints. According to the results of experiment, the iTbot demonstrated the ability to follow the desired trajectories effectively.

Keywords: upper-limb rehabilitation; end-effector robot (iTbot); sliding mode control; reaching law; trajectory tracking

Citation: Khan, M.M.R.; Swapnil, A.A.Z.; Ahmed, T.; Rahman, M.M.; Islam, M.R.; Brahmi, B.; Fareh, R.; Rahman, M.H. Development of an End-Effector Type Therapeutic Robot with Sliding Mode Control for Upper-Limb Rehabilitation. *Robotics* **2022**, *11*, 98. https://doi.org/10.3390/robotics11050098

Academic Editor: Raffaele Di Gregorio

Received: 27 July 2022
Accepted: 15 September 2022
Published: 21 September 2022

1. Introduction

Globally, approximately 15 million individuals suffer from neurological diseases, such as strokes. This total comprises one-third fatalities and about an equal number of patients who spend the rest of their lives with irreversible disability [1]. In the USA, approximately 750,000 persons are affected by stroke [2], leaving most survivors with varying degrees of motor dysfunction [3]. Stroke is the third most significant cause of disabilities globally [4]. Approximately 85 % of stroke survivors who suffer from hemiparesis live with acute arm impairment [5]. As a result, 60% of individuals with upper limb hemiparesis experience long-term functional limitations, which reduces their quality of life [6,7]. This poor life quality includes losing the ability to work and failing to self-care. These consequences have a significant social and economic impact on the families of those affected and society [8,9]. The recovery of function of the upper limb can play a significant role in reinstating quality of life. Robots are able to assist in recovery from upper and lower limb dysfunction arising from neurological disorders, as reported in several recent studies [10–13].

Traditional rehabilitation focuses on rehabilitative exercises, in which the patient performs a series of bodily motions under the guidance of a trained therapist. A number of researchers have set their sights on developing medical rehabilitation devices [14]. Robot-aided rehabilitation provides high-intensity therapy and decreases the workload of medical staff, unlike traditional rehabilitative treatment. Additionally, this modern therapy can facilitate the quantitative recording of data related to the recovery progress. Robots can provide passive, active, and active-assistive therapies for impaired limbs [15,16].

In general, upper limb robotic devices are subdivided into two types: end-effector type and exoskeleton type robots [17]. Exoskeleton robots are designed to be worn on human limbs [18–20]. The design of this type of robot mimics human anatomical joints and the length of limb segments. Due to the multiple "bundling" of the exoskeleton robot and the upper limbs, the patient cannot detach from the robot quickly if unexpected danger occurs [21]. Many exoskeleton robots can provide rehabilitation therapy for full arm motion and rehabilitation therapy in passive and active modes and provide endpoint-based therapy [22–25]. However, these exoskeleton robots are frequently costly, cumbersome, structurally complicated, and lack mobility. In contrast, end-effector type rehabilitation robots are designed to attach to users at a single point, usually at the wrist or forearm. The primary advantage of end-effector type devices over exoskeletons is their simplicity in design and manufacture. End-effector robots are more compact and lightweight than other robots, and they are easier to build. Furthermore, because of the single-point interaction between the two entities, end-effector type devices are the most prevalent type of assistive device [26–28]. An end-effector-type robot is easy to install in a patient's home because of its simple structure and presents a low risk.

The control techniques used for therapeutic devices directly impact the success of robotic rehabilitation training. Patients with acute hemiplegia can benefit from a patient-passive training program, which requires the afflicted limb to reach a specified trajectory while passively executing repetitive movements [29]. The rehabilitation training system is more complicated than traditional manipulators due to the robot's highly non-linear dynamics, unknown external disturbances, and the viscoelastic features of biological joints [30]. In order to improve position control during repetitive reach training, rehabilitation robots have been designed that use a variety of control methods, such as adaptive control [31], flatness-based control [32], EMG-based control [33], admittance control [34], fuzzy and backstepping control [35]. Sliding mode control (SMC) is an effective technique for controlling robotic systems with unknown dynamics and constrained disturbances [36]. Theoretically, SMC's robustness guarantees optimal tracking performance regardless of parameter or model uncertainty [37]. Furthermore, SMC has a simple construction, strong transient performance, and rapid response. As a result, we evaluate SMC as a viable option for providing reliable, high-quality tracking in dynamic environments. The control input that carries the switching function signum (sign(.)) is the primary source of issues in conventional SMC [38]. In real-time implementation, this function's switching results in undesired chattering due to the control effort. Therefore, the system's performance degrades, and unmodeled high-frequency dynamics may be triggered. Islam et al. [39] proposed a unique upper limb exoskeleton with sliding mode fractional control (FSMC), due to its excellent tracking performance and durability against external disturbances. Using this method, the suggested controller's settling time and maximum pitch angle control were enhanced. Furthermore, in comparison to traditional SMC, tracking and chatter were found to be enhanced. Babaiasl et al. [40] introduced SMC to a three DoFs exoskeleton for shoulder joint rehabilitation, using a genetic algorithm to modify the SMC parameters. They found that SMC performed well in this non-linear control process and that uncertainties and disturbances (such as patients' hand tremors) were effectively rejected. Rahman et al. [29] developed a rehabilitative exoskeleton for the lateral side of the upper limb with non-linear SMC to aid in rehabilitation and improve upper-limb movement. In the device developed, they combined the concept of a saturation (sat) function [41] with an ERL [42] to implement trajectory tracking. The authors utilized a non-linear SMC approach to move the ETS-MARSE so that it could deliver a variety of passive rehabilitation activities, including single-joint movement exercises and workouts involving the movement of multiple joints. This enabled the controller to regulate the exoskeleton's movement, allowing it to conduct passive rehabilitation treatment.

Therefore, the goal of the present investigation was to develop a novel sliding mode controller (SMC) reaching law capable of quickly converging a system trajectory to an equilibrium point. In the proposed approach, fast convergence does not provoke any chattering as occurs with conventional sliding mode controllers. The proposed approach utilizes adjustable parameters so that the control law applied significantly reduces the unwanted chattering phenomenon. Furthermore, the designed approach can provide a faster convergence time than the exponential reaching law, which is among the best solutions to avoid chattering problems and produce a fast convergence time.

The remainder of the article is organized as follows. Section 2 provides an overview of the iTbot. In Section 3, the iTbot's kinematics and dynamics are discussed. Section 4 presents the control design and stability analysis. The experimental findings and a comparative study assessment are reported in Section 5. Finally, Section 6 presents the conclusions and suggestions for future work.

2. Overview of the iTbot

The iTbot was developed based on the human reachable workspace to provide therapy covering the full range of the workspace. Figure 1, depicts the structural design of the iTbot, which was designed to be a minimally feasible solution for a functional robot-aided rehabilitation treatment system.

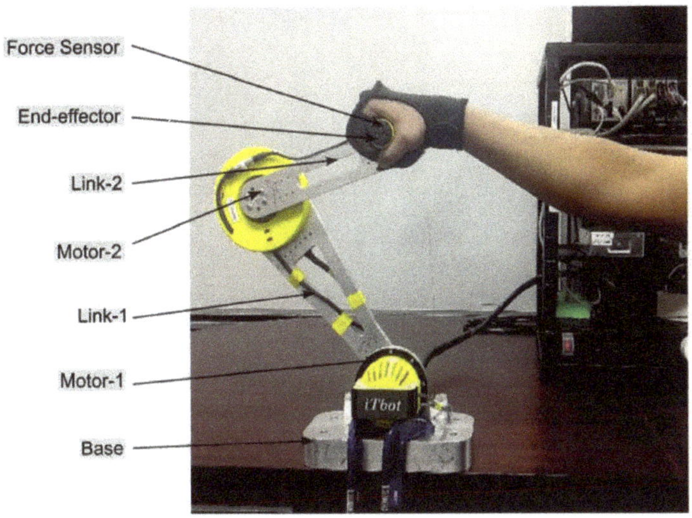

Figure 1. Mechanical design of the iTbot.

The body of the iTbot is made up of the base, as well as two linkages (Link-1 and Link-2). The robot's base contains two fabricated aluminum parts, one being the bottom base of the robot, and the other a mounting for the Motor-1 (Joint-1) hardware shown in Figure 1. The base is designed as a heavy aluminum block to provide stability during the robot's operation during experiments. It also carries a plastic bumper with rubber stoppers to stop the robot at the limit of its range of motion. The Joint-1 (Motor-1) consists of a harmonic drive gear reducer mounted directly on the top base part, with the motor mounted on its back with a custom-designed motor adapter. Link-1, which consists of an aluminum portion with a gear reducer mounted directly to it, is directly connected to the harmonic drive gear reduction unit's output. In the Link-2 assembly, the fabricated aluminum part contains the second half of the Motor-2 wire spool holder. The Joint-2 (Motor-2) is composed of two links (Link-1 and Link-2). There is an end-effector at the end of the connection, which holds the force sensor and the handle. The handle is custom-designed to match the user's hand profile; in our prototype, the average size of the two adult male laboratory members who participated in the development of the iTbot. The handle has a base part, 3D-printed in polycarbonate plastic, that mounts on the force sensor. An inner tube with mounting features for two bearings on both ends is used—this tube screws into the base part.

Specification of the iTbot

The completed CAD model was created by assigning materials to each component in SolidWorks software. The developed iTbot's parameters are listed in Table 1 below.

Table 1. Mechanical parameters of iTbot estimated from CAD model.

Joint Parameters		
Item	Joint-1	Joint-2
Joint range of motion (Degrees)	±85°	±180°
Link Parameters		
Mass (Kg)	1.79	0.65
Location of the center of gravity in link frame (m)	Center of gravity of link 1 in frame {1} $X_1 = 0.26$, $Y_1 = 0.00$, $Z_1 = 0.00$	Center of gravity of link 2 in frame {2} $X_2 = 0.15$, $Y_2 = 0.00$, $Z_2 = 0.02$
Robot Properties		
Mass (Kg)	6.67 (3.2 without base)	
Maximum horizontal reach (m)	±0.55	
Maximum vertical reach (m)	+0.1 to +0.55	

In the X-axis configuration, the iTbot can provide a 1.1 m range of motion. It can achieve the design goal of supporting human upper limb motion [43] in the desired workspace [44,45] fitting any patients with a height from 1.21 m to 1.82 m for the horizontal and vertical configurations shown in Figure 2. In the shaded region from Figure 2, a human arm is covered from 0.5 m to 0.66 m. However, the iTbot symmetric design with asymmetric joint range of motion makes it proficient for ambidextrous use. The base design allows both positionings either in horizontal or vertical orientation.

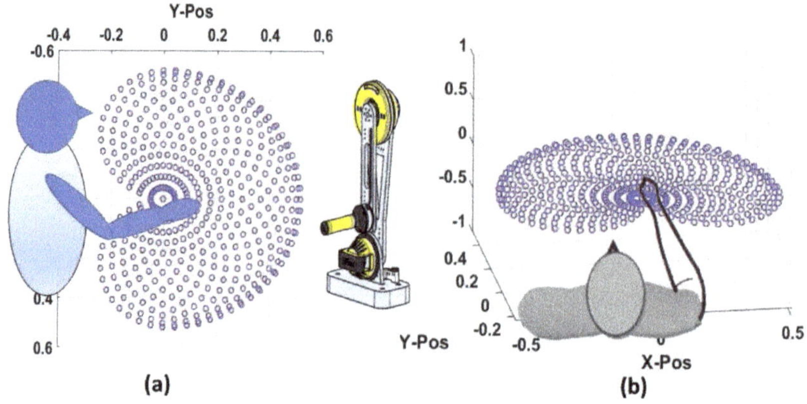

Figure 2. Workspace of the upper limb for (**a**) vertical and (**b**) horizontal configuration.

3. Description of Kinematics and Dynamics of iTbot

3.1. Kinematics of the iTbot

As shown in Figure 3, only the two joint angles move during the rehabilitation training; therefore, the iTbot's forward kinematics analysis and updated Denavit–Hartenberg (DH) [46] parameters are utilized. The iTbot's kinematic model was created using modified DH notations. Each robot connection has a coordinate frame (link frame) connected to it that describes its location in relation to its neighbors. Figure 3 depicts the system's serial link-frame end-effector-type robot attachments.

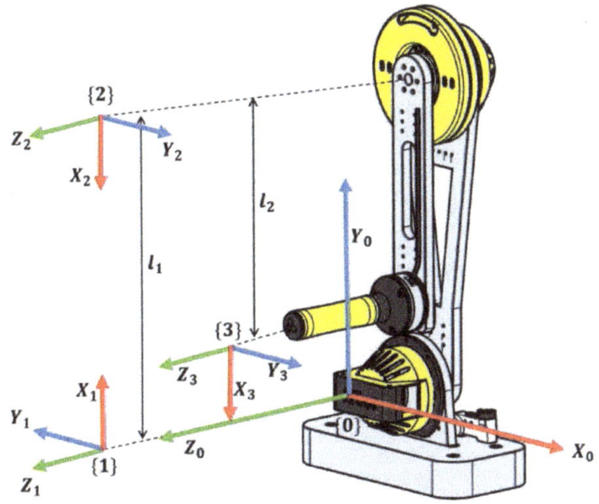

Figure 3. Coordinate frame assignment for 2DoF iTbot.

To calculate the DH parameters, we consider the co-ordinate frames (i.e., link frames that map one axis of rotation to another). This means that Frame 1 is Joint 1, Frame 2 is Joint-2, and Frame 3 is the end-effector position of the iTbot. The fixed reference frame 0 is considered to match with the initial reference frame at the base frame (world frame). Table 2 summarizes the updated DH parameters related to the location of the link frames (in Figure 3). Integrating the DH parameters yields the homogeneous transfer matrix, which defines the positions and orientations of the reference frame regarding the fixed reference frame.

Table 2. Modified Denavit–Hartenberg parameters.

Joint (i)	α_{i-1}	d_i	a_{i-1}	θ_i
1	0	0	0	$\theta_1 + \frac{\pi}{2}$
2	0	0	l_1	$\theta_2 + \pi$
3	0	0	l_2	0

Where α_{i-1} is the link twist, a_{i-1} corresponds to link length, d_i denotes the link offset, and θ_i is the joint angle of the iTbot.

3.2. Dynamics of the iTbot

The dynamics of the iTbot were analyzed (in Figure 4) in order to imitate joint motions in experiments utilizing non-linear control. The motion of bodies under the influence of external forces was calculated using dynamics. A dynamic model for the iTbot was created using the iterative Newton–Euler approach [47].

The dynamic equation for the iTbot can be expressed by Equation (1)

$$\tau = M(\theta)\ddot{\theta} + V(\theta,\dot{\theta}) + G(\theta) + F(\theta,\dot{\theta}) \tag{1}$$

where $\theta \in \Re^2$ denotes a two-vector of generalized coordinates. $M(\theta) \in \Re^{2x2}$, $V(\theta,\dot{\theta}) \in \Re^{2x1}$, $G(\theta) \in \Re^{2x1}$, are, respectively, the symmetric, bounded, inertia matrix, the Coriolis and centrifugal torques,and the gravitational torque. $\tau \in \Re^2$ is the torque input vector and $F(\theta,\dot{\theta}) \in \Re^2$ represents the external disturbances. Further details are provided in Appendix A.

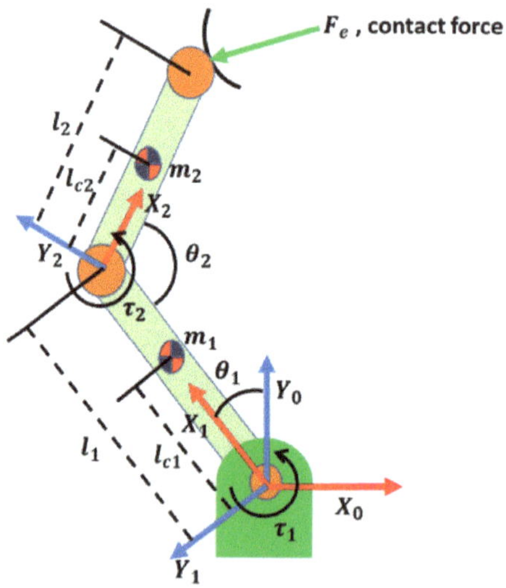

Figure 4. iTbot nomenclature for dynamic modeling with contact force at the end-effector.

4. Control Design and Stability Analysis

Introducing $x = \theta$ and $\dot{x} = \dot{\theta}$, the dynamic model expressed in Equation (1) can be rewritten in the form of Equation (2) as follows:

$$\ddot{x} = f(x,\dot{x}) + g(x)u \tag{2}$$

with:

- $u = \tau$
- $g(x) = g = M^{-1}(\theta)$
- $f(x,\dot{x}) = f = M^{-1}(\theta)\left[-C(\theta,\dot{\theta})\dot{\theta} - G(\theta) - F(\theta,\dot{\theta})\right]$

The tracking position error can be given by: $e = x - x^d$, where $x^d \in \Re^n$ is the reference trajectory and x is the actual position.

The first step in conceiving an SMC approach is to choose the switching surface ν as follows:

$$\nu = \dot{e} + \lambda e \tag{3}$$

where $\lambda \in \Re^{n\times n} > 0$. Note that λ plays a decisive role in the convergence rate of the error tracking to zero.

Consider the Lyapunov function: $V(\nu) = \dfrac{1}{2}\nu^T\nu$, with its time derivative:

$$\dot{V} = \nu^T\dot{\nu} \tag{4}$$

The criterion for stability is therefore: $\dot{V} < 0$. This requires $\dot{\nu} < 0$ for $\nu > 0$ and $\dot{\nu} > 0$ for $\nu < 0$, which gives rise to the commonly known control law switching phenomenon around $\nu = 0$. Based on (Equation (3)) and its derivative, the following control input is proposed:

$$u = g^{-1}\left[\ddot{x}^d - \lambda\dot{e} - f + \dot{\nu}\right] \tag{5}$$

Note that, from (Equation (5)), the control input is substantially based on $\dot{\nu}$, which in turn defines the rate of ν. So, if $\dot{\nu} \ll 0$ for $\nu > 0$ (with the opposite being too correct), the dynamics' pushed trajectory converges to $\nu = 0$. Therefore, $\dot{\nu}$ is usually referred to as the "reaching" law. When the trajectory of the system is in the vicinity of $\nu = 0$, with $\dot{V} < 0$, $\dot{\nu} < 0$ determines precisely how close the system is to the sliding surface $\nu = 0$. Therefore, a "switching" phenomenon arises in order to maintain the condition: $\nu\dot{\nu} < 0$.

To cope with the high gain value of SMC and to produce rapid and high tracking performance, a kind of adaptive reaching known as the exponential reaching law (ERL) was proposed. This approach is considered to be an effective approach to overcome this issue.

The ERL technique solves the problem associated with the high gain of the SMC reaching law by allowing the controller to dynamically respond to variations in the switching function. This method allows the SMC control gain to be readily varied between an appropriate amount of control gain that does not cause chattering K_{1i} and K_{1i}/ϕ_i where $\phi_i < 1$ (for definition of variables see Equation (6) below). As a result, the ERL approach can guarantee the convergence rate in the period indicated by Equation (7). The fundamental issue with this strategy is that it cannot prevent chattering at the start of motions, which reduces the accuracy of the control performance. The suggested controller was created to overcome this issue, while maintaining the high converge time lower than the ERL. The designed reaching law is defined as follows:

$$\begin{aligned}\dot{\nu}_i = &-\frac{K_{1i}}{\phi_i + (1-\phi_i)e^{-\alpha_i|\nu_i|^{p_i}}}|\nu_i|^{\varphi} sign(\nu_i)\\ &-\omega_i\frac{K_{1i}(1-\varphi)}{\phi_i} sign(\nu_i)\end{aligned} \tag{6}$$

where $\phi_i > 0$, $\alpha_i > 0$ and $p_i > 0$ with $\phi_i < 1$ and $0 < \varphi < 0.5$. ω_i is defined by $\lim_{t\to\infty}(\omega_i) = 0$ and $\int_0^t \omega_i(w)dw = Q_i < \infty$, where $\omega_i = 1/(1+t_i^2)$ and t_i is the execution time of the reference trajectory. The second term of the control law, Equation (6) is designed to preserve its robustness around the beginning of the executing trajectory. Note that, as time goes on, this term would disappear based on the definition of ω_i.

Note that, the term φ is commonly defined as a high value in the classical power rate law to guarantee fast convergence to the origin while causing unwanted chattering. In the designed law, a restriction on φ was imposed: $0 < \varphi < 0.5$. This would guarantee not only fast convergence, but also reduced chattering.

Proposition 1. *For $K_{1i} > 0$, and in conformity with the selection of φ determined in advance, the proposed reaching law (Equation (6)) consistently provides faster convergence to the origin than ERL [42] and ensures stability of the closed loop dynamical system Equation (2).*

Proof. The ERL's reaching time is provided in [42]:

$$Tr_{1i} = \frac{1}{K_i}\left(\phi_i|\nu_i(0)| + (1-\phi_i)\int_0^{|\nu_i(0)|} e^{-\alpha_i|\nu_i|^{p_i}} d\nu_i\right) \tag{7}$$

To obtain the reaching time (Tr_{2i}) of the designed reaching law (6), it is initially rewritten as:

$$dt_i = \frac{\left(\phi_i + (1-\phi_i)e^{-\alpha_i|v_i|^{p_i}}\right)dv_i}{-K_{1i}|v_i|^{\varphi}sign(v_i)} + \frac{\phi_i dv_i}{-\omega_i K_{1i}(1-\varphi)sign(v_i)}. \tag{8}$$

Integrating (Equation (8)) from zero to Tr_{2i}, with $v_i(Tr_{2i}=0)$, gives:

$$\begin{aligned} Tr_{2i} &= \int_{v_i(0)}^{0} \frac{\left(\phi_i + (1-\phi_i)e^{-\alpha_i|v_i|^{p_i}}\right)dv_i}{-K_{1i}|v_i|^{\varphi}sign(v_i)} + \int_{v_i(0)}^{0} \frac{\phi_i dv_i}{-\omega_i K_{1i}(1-\varphi)sign(v_i)}. \\ &= \int_{0}^{v_i(0)} \frac{\left(\phi_i + (1-\phi_i)e^{-\alpha_i|v_i|^{p_i}}\right)dv_i}{K_{1i}|v_i|^{\varphi}sign(v_i)} + \int_{0}^{v_i(0)} \frac{\phi_i dv_i}{\omega_i K_{1i}(1-\varphi)sign(v_i)}. \end{aligned} \tag{9}$$

if $v_i < 0$ for all $ti < Tr_{2i}$, so:

$$Tr_{2i} = \int_{0}^{-v_i(0)} \frac{\left(\phi_i + (1-\phi_i)e^{-\alpha_i|v_i|^{p_i}}\right)dv_i}{K_{1i}|v_i|^{\varphi}} + \int_{0}^{-v_i(0)} \frac{\phi_i dv_i}{\omega_i K_{1i}(1-\varphi)}. \tag{10}$$

Else, if $v_i > 0$ for all $ti < Tr_{2i}$, gives:

$$Tr_{2i} = \int_{0}^{v_i(0)} \frac{\left(\phi_i + (1-\phi_i)e^{-\alpha_i|v_i|^{p_i}}\right)dv_i}{K_{1i}|v_i|^{\varphi}} + \int_{0}^{v_i(0)} \frac{\phi_i dv_i}{\omega_i K_{1i}(1-\varphi)}. \tag{11}$$

Based on Equations (10) and (11):

$$Tr_{2i} = \int_{0}^{|v_i(0)|} \frac{\phi_i dv_i}{K_{1i}|v_i|^{\varphi}} + \int_{0}^{|v_i(0)|} \frac{(1-\phi_i)e^{-\alpha|v_i|^{p_i}}dv_i}{K_{1i}|v_i|^{\varphi}} + \int_{0}^{|v_i(0)|} \frac{\phi_i dv_i}{\omega_i K_{1i}(1-\varphi)}. \tag{12}$$

Then:

$$Tr_{2i} = \frac{1}{K_{1i}}(\phi_i \frac{|v_i(0)|^{(1-\varphi)}}{(1-\varphi)} + \frac{\phi_i |v_i(0)|}{\omega_i(1-\varphi)} + (1-\phi_i)\int_{0}^{|v_i(0)|} e^{-\alpha_i|v_i|^{p_i}}|v_i|^{-\varphi}dv_i). \tag{13}$$

In [42], the authors applied the properties of Euler's gamma function (φ) to demonstrate that the reaching time Tr_{1i} fulfils the next conditions:

$$Tr_{1i} \leq \frac{\phi_i}{K_{1i}}|v_i(0)| + \frac{(1-\phi_i)}{K_{1i}\alpha_i^{1/p_i}}. \tag{14}$$

By adopting similar properties of Euler's gamma function for the designed reaching law, the last term of (Equation (13)) can be rewritten in terms of the φ function as:

$$\int_0^{|v_i(0)|} e^{-\alpha_i |v_i|^{p_i}} |v_i|^{-\varphi} dv_i = \alpha_i^{\varphi/p_i} \frac{\left[\varphi^{-}\left(\frac{\varphi-1}{p_i}\right) - \varphi\left(-\left(\frac{\varphi-1}{p_i}\right), \alpha_i |v_i(0)|^{p_i}\right)\right]}{p_i \alpha_i^{1/p_i}}. \tag{15}$$

According to the properties of the φ function:

$$\varphi\left(-\left(\frac{\varphi-1}{p_i}\right), \alpha_i |v_i(0)|^{p_i}\right) \ll \varphi^{-}\left(\frac{\varphi-1}{p_i}\right). \tag{16}$$

Thus, it is valid to suppose that: $\varphi\left(-\left(\frac{\varphi-1}{p_i}\right), \alpha_i |v_i(0)|^{p_i}\right) \approx 0$, and, hence:

$$\int_0^{|v_i(0)|} e^{-\alpha_i |v_i|^{p_i}} |v_i|^{-\varphi} dv_i = \alpha_i^{\varphi/p_i} \frac{\varphi^{-}\left(\frac{\varphi-1}{p_i}\right)}{p_i \alpha_i^{1/p_i}}. \tag{17}$$

substituting Equation (17) into Equation (13), it is established that the reaching time achieves the following condition:

$$Tr_{2i} \leq \frac{\phi_i}{K_{1i}} \left[\frac{|v_i(0)|^{(1-\varphi)} + \omega_i |v_i(0)|}{(1-\varphi)} \right] + \left(\frac{1-\phi_i}{K_{1i}}\right) \frac{\varphi^{-}\left(\frac{\varphi-1}{p_i}\right)}{p_i \alpha_i^{\left(\frac{1-\varphi}{p_i}\right)}} \tag{18}$$

To show that the designed reaching law presents a reaching time shorter than that given by ERL [42], it is important to rewrite the reaching time of the designed law as:

$$Tr_{2di} = \frac{\phi_i}{K_{1i}} \left[\frac{|v_i(0)|^{(1-\varphi)} + \omega_i |v_i(0)|}{(1-\varphi)} \right] + \left(\frac{1-\phi_i}{K_{1i}}\right) \frac{\varphi^{-}\left(\frac{\varphi-1}{p_i}\right)}{p_i \alpha_i^{\left(\frac{1-\varphi}{p_i}\right)}}. \tag{19}$$

Thus, the reaching time Tr_{2i} should be lower than the desired reaching time Tr_{2di} for each value of α as:

$$\alpha_i \gg \left[\frac{(1-\phi_i)\varphi^{-}\left(\frac{\varphi-1}{p_i}\right)(1-\varphi)}{\phi_i\left(|v_i(0)|^{(1-\varphi)} + \omega_i |v_i(0)|\right)} \right]^{\frac{p_i}{1-\varphi}}. \tag{20}$$

Therefore, the desired reaching law can be re-estimated as:

$$Tr_{2di} \approx \frac{\phi_i}{K_{1i}}\left[\frac{|\nu_i(0)|^{(1-\varphi)} + \omega_i|\nu_i(0)|}{(1-\varphi)}\right]. \tag{21}$$

In addition, the gain K_{1i} should fulfil:

$$K_{1i} \approx \frac{\phi_i}{Tr_{2di}}\left[\frac{|\nu_i(0)|^{(1-\varphi)} + \omega_i|\nu_i(0)|}{(1-\varphi)}\right]. \tag{22}$$

If both condition Equations (20) and (22) are fulfilled, it can then be guaranteed that $Tr_{2i} < Tr_{2di}$. Since the suggested reaching law will be against the ERL [42], it would be advantageous to indicate the desired reaching law, along with the tuning gain, provided by the ERL proposition:

$$Tr_{1di} \approx \phi_i \frac{|\nu_i(0)|}{K_{1i}} \tag{23}$$

$$K_{1i} \approx \phi_i \frac{|\nu_i(0)|}{Tr_{1di}} \tag{24}$$

Subtracting (21) from (23) gives:

$$\begin{aligned} Tr_{1di} - Tr_{2di} &\approx \phi_i \frac{|\nu_i(0)|}{K_{1i}} - \frac{\phi_i}{K_{1i}}\left[\frac{|\nu_i(0)|^{(1-\varphi)} + \omega_i|\nu_i(0)|}{(1-\varphi)}\right] \\ &\approx \frac{\phi_i}{K_{1i}}|\nu_i(0)|\left[1 - \left(\frac{|\nu_i(0)|^{-\varphi} + \omega_i}{(1-\varphi)}\right)\right] \end{aligned} \tag{25}$$

Since $\phi_i > 0$ and $K_{1i} > 0$, it is then obvious that $\frac{\phi_i}{K_{1i}}|\nu_i(0)|$ is always positive.

It is important to demonstrate that the second term of (Equation (25)) is always positive. According to the definition of ω_i in (Equation (6)), as $t \longrightarrow \infty$, the term $\omega_i \longrightarrow 0$. In this case, to prove that the second term of (Equation (25)) is always positive, the next should hold:

$$\frac{1}{|\nu_i(0)|^{\varphi}(1-\varphi)} < 1 \tag{26}$$

which means that the next must hold:

$$|\nu_i(0)| > (1-\varphi)^{-1/\varphi} \tag{27}$$

Therefore,

$$\left(1 - \frac{1}{|\nu_i(0)|^{\varphi}(1-\varphi)}\right) > 0, \forall|\nu_i(0)| > (1-\varphi)^{-1/\varphi} \tag{28}$$

Then, (Equation (25)) can be rewritten as:

$$\begin{aligned} Tr_{1di} - Tr_{2di} \approx \frac{\phi_i}{K_{1i}}|\nu_i(0)|\left[1 - \left(\frac{|\nu_i(0)|^{-\varphi}}{(1-\varphi)}\right)\right] > 0, \\ \forall|\nu_i(0)| > (1-\varphi)^{-1/\varphi} \end{aligned} \tag{29}$$

Note that, according to Equations (18) and (19), $Tr_{2i} \leq Tr_{2di}$. Moreover, based on [42], $Tr_{1i} \leq Tr_{1di}$. Therefore, according to the condition provided by (Equation (29)), the next can be rewritten:

$$Tr_{1i} - Tr_{2i} > 0, \forall |\nu_i(0)| > (1-\varphi)^{-1/\varphi} \tag{30}$$

Thus, based on the value of φ, the reaching time of the designed law is shorter than that given by the ERL. So, the proof is finished. □

5. Experimental and Comparative Study Evaluation

5.1. Real-Time System Setup

Figure 5 depicts the experimental configuration of the iTbot architecture. Three levels of computing units comprise the rehabilitation robot instrumentation system. To communicate with the rehabilitation robot, a user interface (UI) runs on a personal computer (PC). The UI is used to send control and operation commands to the rehabilitation robot, as well as to visualize the live data for joint positioning and force the sensor inputs. The second level of computing is performed in a National Instruments CompactRIO, running as a RealTime target. LabVIEW RealTime applications running in the RealTime target perform the computational tasks of trajectory generation, position control, and feedback processing using the programmed algorithms. The RealTime applications also collect and store data during robot operation and send the data to the UI application on the PC to be saved for analysis. The third level of computing is performed in the field programmable gate array (FPGA) built into the CompactRIO, which runs at 50 µs to run a PI control algorithm to control the motor current and to process hall-sensor signals from the motors to calculate the joint position. The joints of the iTbot rehabilitation robot are powered by Brushless DC (BLDC) motors (Maxon EC-45 Flat, 70 and 30-watt variants) with Harmonic Drive strain wave reducers that provide a 100:1 gear ratio. The motors are driven by ZB12A brushless servo amplifiers. The motor's built-in hall-sensors are used for both commutations by the servo amplifiers and for position feedback by the FPGA system.

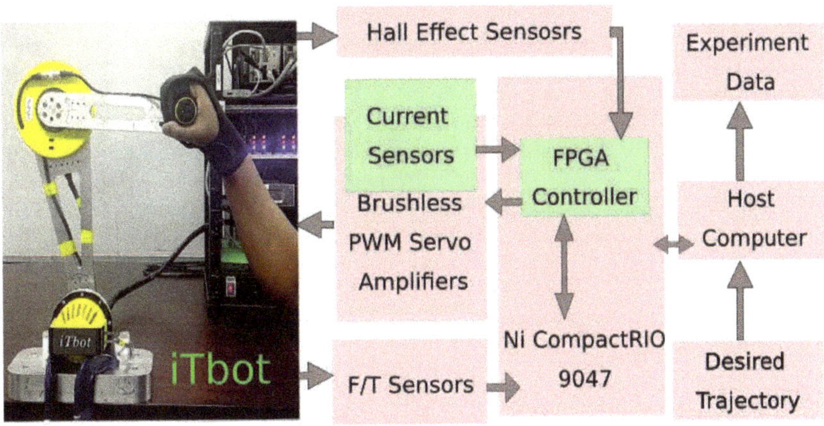

Figure 5. Experimental setup with iTbot architecture.

5.2. Experimental Results

To assess the proposed new sliding mode exponential reaching law (nSMERL), our developed iTbot was deployed to provide passive arm movement exercises to a healthy human subject. Passive arm movement therapy is the first form of physiotherapy treatment that patients receive, aiming to improve their passive range of motion. The proposed nSMERL controller was compared to a SMERL [48] and conventional SMC [38]. Multi-joint movement (desired trajectory) was performed in this experiment, using all of the robot's joints. The purpose of this test was to demonstrate the proposed nSMERL controller's performance during passive rehabilitation exercises. The proposed nSMERL technique was used to monitor the required motion in the first scenario. In the second scenario, which employed the same individual (age: 28 years; height: 5ft 4 in; Weight: 125 lbs.) and same exercise, the precise target trajectory was tracked using SMERL and a conventional sliding mode controller, which was used to monitor the first. The goal was to demonstrate how the nSMERL controller differed from existing controllers. The individual was seated on a chair that had been adjusted to their comfort level. The controller gain parameters were manually selected by trial and error, as shown in Table 3.

Table 3. Gain parameters.

λ_i	K_i	Φ_i	α_i	P_i	Φ
40	2	0.1	2	0.5	0.25
50	0.8	0.1	2	0.5	0.25

The elimination of chattering is the major improvement achieved by the proposed nSMERL. Chattering causes high-frequency vibration to the mechanical structure, which might cause harm to the participants involved; therefore, it has to be mitigated or eradicated. We carried out a multi-joint rehabilitation exercise to evaluate the efficacy of the proposed nSMERL in reducing chattering. The desired trajectories (reference trajectories, dotted line) were compared to the measured trajectories (solid line) and the tracking error, as shown in Figures 6 and 7. Figure 8 show the joint angles for three controllers. Error, or the difference between the desired and actual trajectories, is shown as a function of time in Figure 9 for both controllers. Figure 10 displays a plot of the joint torques that were created to follow the trajectory; the result demonstrates that the method provided more stable tracking and less chattering than the conventional SMC method. The error plots also show that the proposed SMC outperformed the conventional SMC in terms of tracking accuracy. For example, we sought to keep the higher tracking errors below two degrees for the three controllers.

Figure 11 displays force graphs which show that the additional weight (1lb) applied to participants' hands caused maximum resistive forces of approximately Fx = 14 N and Fy = 10 N. The extra weight cannot precisely simulate spasticity, but it was sufficient to demonstrate that the controller was capable of withstanding patients' spasticity. As a result, despite the fact that the robot's dynamic model and external disturbances were linearized to a simple linear system, the end-effector robot (iTbot) performed satisfactorily under the proposed SMC controller, in comparison to conventional SMC, which retains the non-linear dynamic method in its control strategy. The proposed nSMERL controller continuously provided appropriate tracking with modest control input and less chattering. While the conventional SMC controller produced satisfactory results, its control input was significantly greater than that of the proposed controller (high chattering). These findings demonstrate that the proposed nSMERL controller improved the other sliding mode approach.

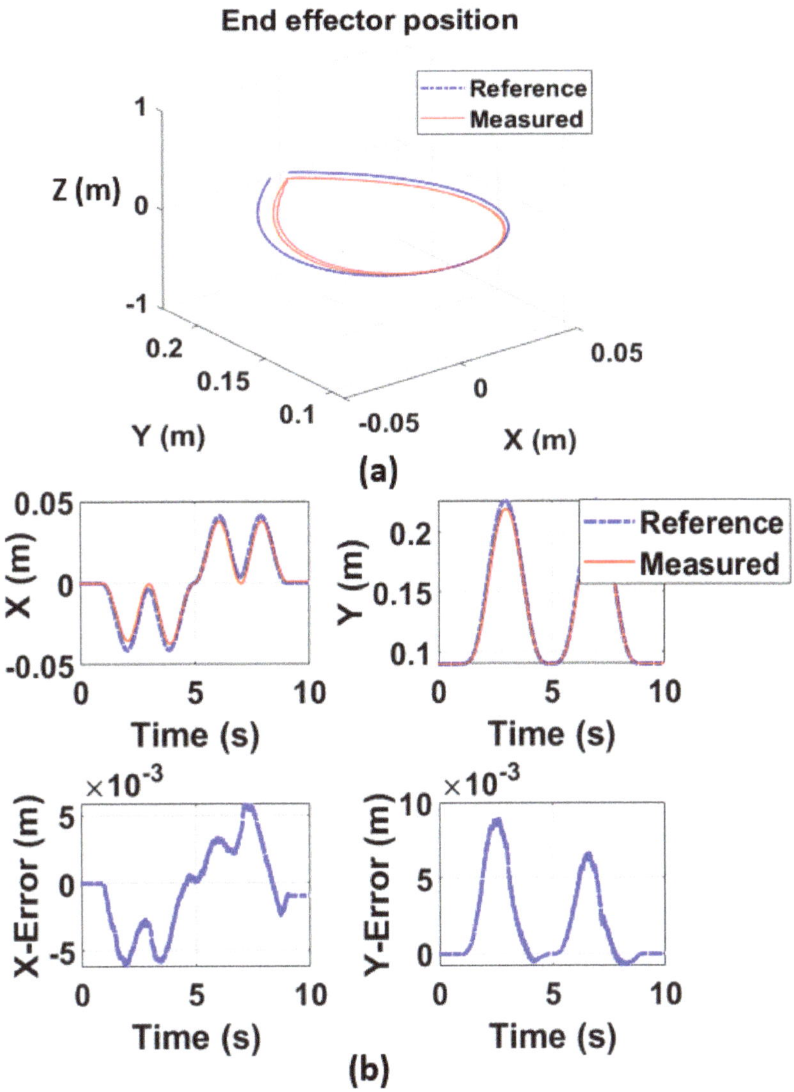

Figure 6. (**a**) End-effector position under the proposed nSMERL, (**b**) End-effector tracking error under the proposed nSMERL.

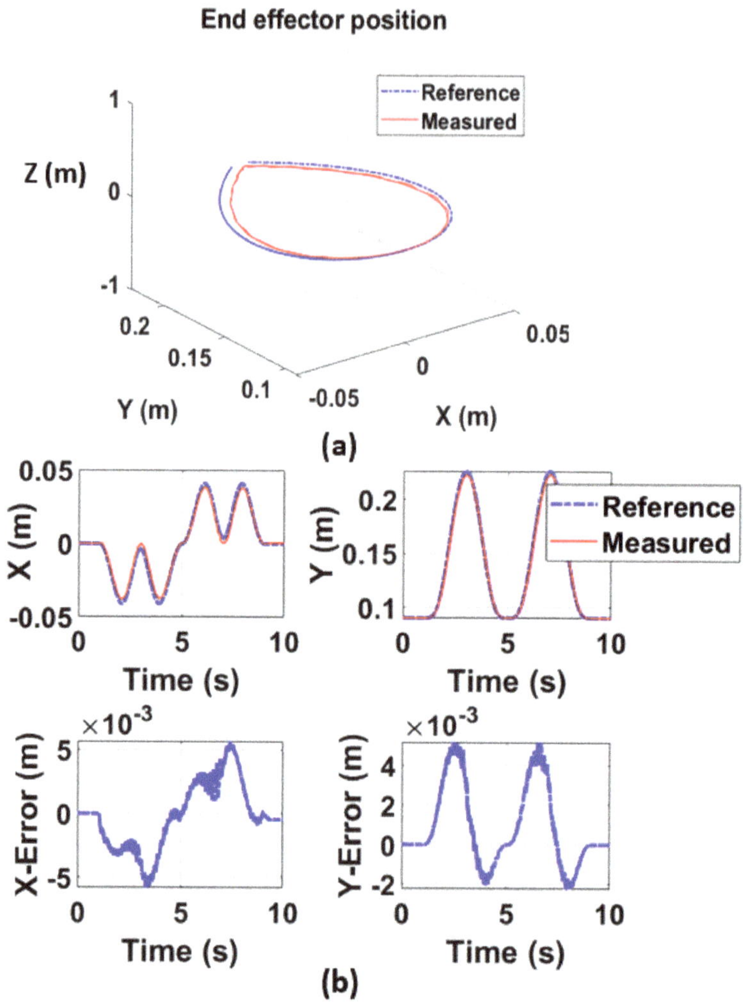

Figure 7. (**a**) End-effector position under the SMERL and conventional SMC, (**b**) End-effector tracking error under the SMERL and conventional SMC.

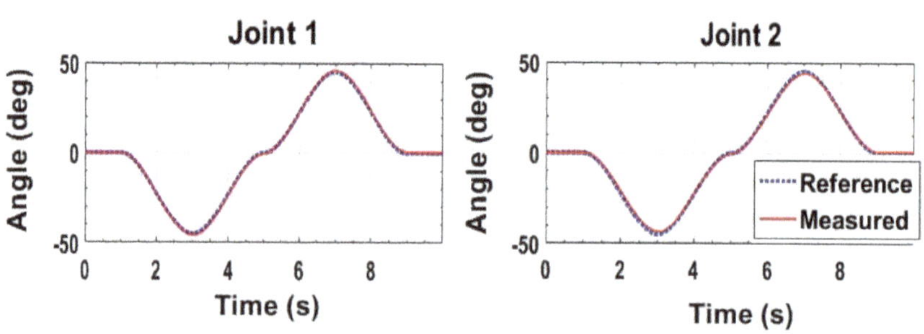

Figure 8. Joint angle for the proposed nSMERL, SMERL, and conventional SMC.

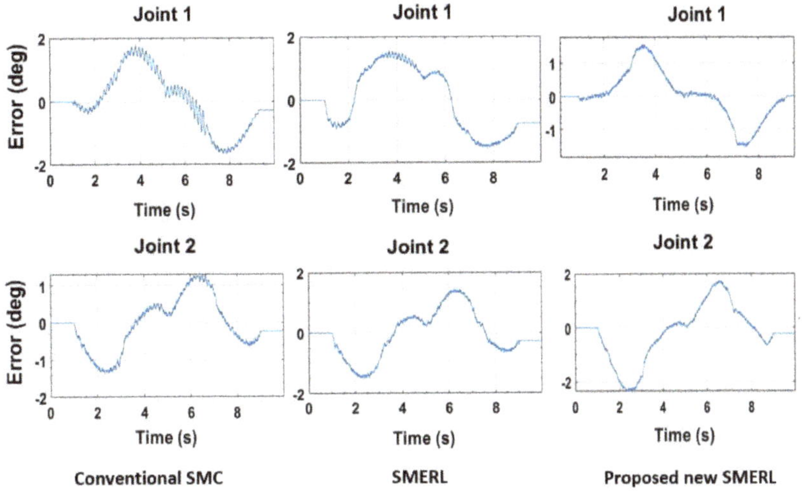

Figure 9. Tracking of joint errors under the proposed nSMERL, SMERL, and conventional SMC.

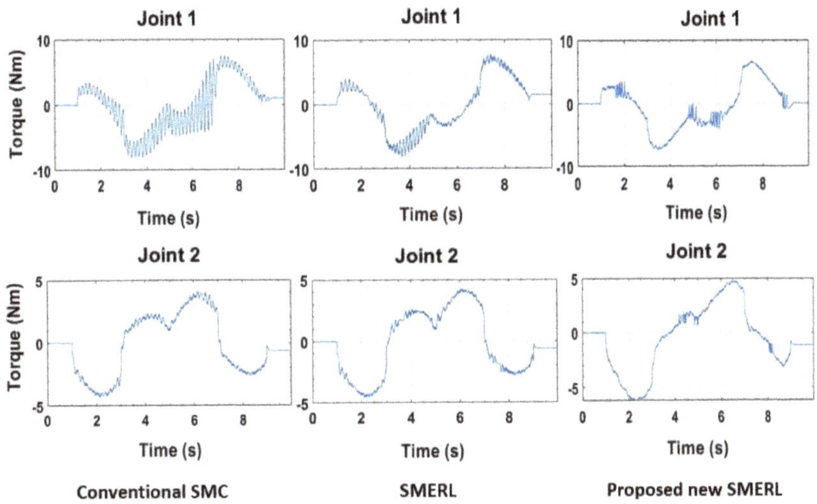

Figure 10. Joint torque for the proposed nSMERL, SMERL, and conventional SMC.

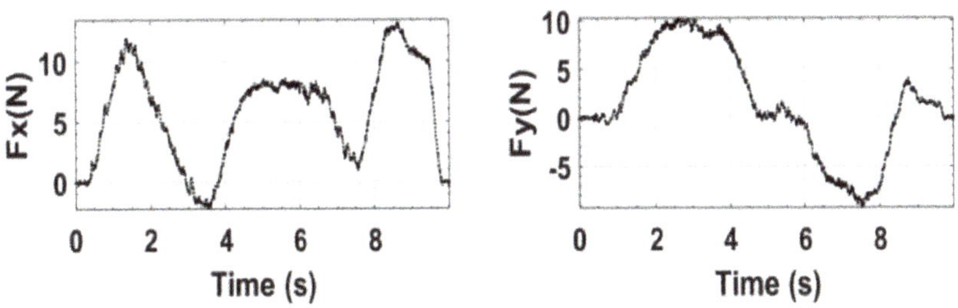

Figure 11. Force (end-effector) data during experiments.

6. Conclusions and Future Work

Robot-assisted treatment has emerged as a valuable technique in the rehabilitation and reinforcement of motor skills in individuals suffering from neuromuscular ailments. An end-effector-type robot was presented to assist in the rehabilitation training of elbow flexion and extension motions. To evaluate the efficacy of the proposed method, multiple trajectory tracking experiments were carried out using a real-time control system. The proposed control law demonstrated its potential to both overcome and improve the performance of SMERL. To evaluate the efficacy of the proposed nSMERL, multiple trajectory tracking experiments were carried out using a real-time control system. The proposed nSMERL, a novel non-linear control approach, enhanced transient tracking performance and decreased chattering in SMERL and conventional SMC. Experiments were conducted to evaluate the dynamic tracking performance of SMERL and conventional SMC with the proposed nSMERL. It was shown that the proposed nSMERL could reduce chattering and provide superior tracking performance. A Kinect sensor will be used in future research to detect the motions of the human arm. A further investigation into EMG signals is planned to achieve more advanced rehabilitation and assistive control.

Author Contributions: Conceptualization, B.B. and M.H.R.; Data curation, M.M.R.K., T.A.; Formal analysis, R.F., M.R.I. and M.H.R.; Investigation, M.R.I. and M.H.R.; Methodology, M.M.R.K., A.A.Z.S.; Software, M.M.R.K., A.A.Z.S., T.A., M.R.I.; Supervision, M.H.R.; Validation, M.M.R.K., B.B. and M.H.R.; Visualization, M.M.R.K., A.A.Z.S.; Writing—original draft, M.M.R.K., B.B., and A.A.Z.S.; Writing—review & editing, M.M.R.K., B.B., M.R.I. and M.M.R. All authors have read and agreed to the published version of the manuscript.

Funding: This research received no external funding.

Data Availability Statement: The data that support the findings of this study are available from the corresponding author, [Md Mahafuzur Rahaman Khan], upon reasonable request.

Acknowledgments: The authors would like to thank Helal Uddin Ahmed for his participation and assistance in finalizing the reviews of this work.

Conflicts of Interest: The authors declare no conflict of interest.

Appendix A

Appendix A.1. Kinematics of the iTbot

The general form of a link transformation that relates the frame i relative to the frame $i-1$ [49] is:

$$ {}_{i}^{i-1}T = \begin{bmatrix} {}_{i}^{i-1}R^{3\times3} & {}_{i}^{i-1}P^{3\times1} \\ 0^{1\times3} & 1 \end{bmatrix} \tag{A1} $$

where, ${}_{i}^{i-1}\mathbf{R}$ is the rotation matrix that represents the frame i relative to frame **i-1** and can be articulated as follows:

$$ {}_{i}^{i-1}R = \begin{bmatrix} \cos\theta_i & -\sin\theta_i & 0 \\ \sin\theta_i \cos\alpha_{i-1} & \cos\theta_i \cos\alpha_{i-1} & -\sin\alpha_{i-1} \\ \sin\theta_i \sin\alpha_{i-1} & \cos\theta_i \sin\alpha_{i-1} & \cos\alpha_{i-1} \end{bmatrix} \tag{A2} $$

and ${}_{i}^{i-1}\mathbf{P}$ is the vector that locates the origin of the frame i relative to frame $i-1$ and can be expressed as the following:

$$ {}_{i}^{i-1}P = \begin{bmatrix} \alpha_{i-1} & -\sin(\alpha_{i-1})d_i & \cos(\alpha_{i-1})d_i \end{bmatrix}^T \tag{A3} $$

The homogenous transformation matrix that relates frame 3 to frame 0 can be obtained by multiplying individual transformation matrices that result in the generic form (A4).

$$ {}^0_3T = \left[{}^0_1T \cdot {}^1_2T \cdot {}^2_3T \right] \tag{A4}$$

Appendix A.2. Dynamics of the iTbot

Then, the joint torques of the iTbot for vertical configuration with active gravity compensation, based on the nomenclature provided in Figure 4, can be found by the iterative Newton–Euler formulation as:

$$\begin{aligned} \tau_1 = {} & \left(m_2 l_1^2 - 2m_2 l_1 l_{c2} \cos(\theta_2) + m_1 l_{c1}^2 + m_2 l_{c2}^2 + I_{z1} + I_{z2}\right)\ddot{\theta}_1 \\ & + \left(m_2 l_{c2}^2 - m_2 l_1 l_{c2} \cos(\theta_2) + I_{z2}\right)\ddot{\theta}_2 + m_2 l_1 l_{c2} \sin(\theta_2)\dot{\theta}_2^2 \\ & + 2m_2 l_1 l_{c2} \sin(\theta_2)\dot{\theta}_1\dot{\theta}_2 + \left(m_2 l_{c2} \sin(\theta_1 + \theta_2) - m_1 l_{c1} \sin(\theta_1) - m_2 l_1 \sin(\theta_1)\right) \end{aligned} \tag{A5}$$

$$\begin{aligned} \tau_2 = {} & \left(I_{z2} + m_2 l_{c2}(l_{c2} - l_1 \cos(\theta_2))\right)\ddot{\theta}_1 \\ & + \left(m_2 l_{c2}^2 + I_{z2}\right)\ddot{\theta}_2 - m_2 l_1 l_{c2} \sin(\theta_2)\dot{\theta}_1^2 + \left(m_2 l_{c2} \sin(\theta_1 + \theta_2)\right) \end{aligned} \tag{A6}$$

where m_1, m_2 is the mass of Link 1 and Link 2; l_{c1} is the distance relative to 1 and center of mass (Link 1), l_{c1} is the distance relative to 2 and center of mass (Link 2); l_1, l_2 is the length of Link 1 and Link 2; I_{z1}, I_{z2} is the inertia tensor ; τ_1, τ_2 is the inertia of Joint 1 and Joint 2.

Equations (A5) and (A6) give expressions for the torque at the actuators as a function of joint position, velocity, and acceleration. The dynamic Equation of iTbot can be written in the form given by Equation (A7):

$$\tau = M(\theta)\ddot{\theta} + V(\theta, \dot{\theta}) + G(\theta) \tag{A7}$$

where and are the 2 × 1 torque and acceleration vector. $M(\theta)$ is the 2 × 2 mass matrix given as:

$$M(\theta) = \begin{bmatrix} \kappa_1 & \kappa_2 \\ I_{z2} + m_2 l_{c2}(l_{c2} - l_1 \cos(\theta_2)) & m_2 l_{c2}^2 + I_{z2} \end{bmatrix} \tag{A8}$$

where κ_1 and κ_2 are as follows

$\kappa_1 = m_2 l_1^2 - 2m_2 l_1 l_{c2} \cos(\theta_2) + m_1 l_{c1}^2 + m_2 l_{c2}^2 + I_{z1} + I_{z2}$

$\kappa_2 = m_2 l_{c2}^2 - m_2 l_1 l_{c2} \cos(\theta_2) + I_{z2}$

$V(\theta, \dot{\theta})$ is a 2 × 1 vector of centrifugal and Coriolis terms given as:

$$V(\theta, \dot{\theta}) = \begin{bmatrix} m_2 l_1 l_{c2} \sin(\theta_2)\dot{\theta}_2^2 \cdots \\ \cdots + 2m_2 l_1 l_{c2} \sin(\theta_2)\dot{\theta}_1\dot{\theta}_2 \\ -m_2 l_1 l_{c2} \sin(\theta_2)\dot{\theta}_1^2 \end{bmatrix} \tag{A9}$$

$G(\theta)$ is a 2 × 1 vector of gravity terms given as:

$$G(\theta) = \begin{bmatrix} (m_2 l_{c2} \sin(\theta_1 + \theta_2) - m_1 l_{c1} \sin(\theta_1) \cdots \\ \cdots - m_2 l_1 \sin(\theta_1))g \\ (m_2 l_{c2} \sin(\theta_1 + \theta_2))g \end{bmatrix} \tag{A10}$$

If $F(\theta, \dot{\theta}) \in \Re^2$ is the vector of non-linear Coulomb friction and expressed by Equation (A12).

$$F(\theta, \dot{\theta}) = c.sgn(\dot{\theta}). \tag{A11}$$

Then, when friction is added to the model, Equation (A7) becomes Equation (A12):

$$\tau = M(\theta)\ddot{\theta} + V(\theta, \dot{\theta}) + G(\theta) + F(\theta, \dot{\theta}) \tag{A12}$$

References

1. National Spinal Cord Injury Statistical Center. *Facts and Figures at a Glance*; University of Alabama at Birmingham: Birmingham, AL, USA, 2016; Volume 10.
2. Burns, M.; Zavoda, Z.; Nataraj, R.; Pochiraju, K.; Vinjamuri, R. HERCULES: A Three Degree-of-Freedom Pneumatic Upper Limb Exoskeleton for Stroke Rehabilitation. In Proceedings of the 2020 42nd Annual International Conference of the IEEE Engineering in Medicine & Biology Society (EMBC), Montreal, QC, Canada, 20–24 July 2020; pp. 4959–4962.
3. Mackay, J.; Mensah, G.A.; Greenlund, K. *The Atlas of Heart Disease and Stroke*; World Health Organization: Geneva, Switzerland, 2004.
4. Johnson, W.; Onuma, O.; Owolabi, M.; Sachdev, S. Stroke: A global response is needed. *Bull. World Health Organ.* **2016**, *94*, 634. [CrossRef] [PubMed]
5. Hatem, S.M.; Saussez, G.; Della Faille, M.; Prist, V.; Zhang, X.; Dispa, D.; Bleyenheuft, Y. Rehabilitation of motor function after stroke: A multiple systematic review focused on techniques to stimulate upper extremity recovery. *Front. Hum. Neurosci.* **2016**, *10*, 442. [CrossRef] [PubMed]
6. Lawrence, E.S.; Coshall, C.; Dundas, R.; Stewart, J.; Rudd, A.G.; Howard, R.; Wolfe, C.D. Estimates of the prevalence of acute stroke impairments and disability in a multiethnic population. *Stroke* **2001**, *32*, 1279–1284. [CrossRef] [PubMed]
7. Nichols-Larsen, D.S.; Clark, P.; Zeringue, A.; Greenspan, A.; Blanton, S. Factors influencing stroke survivors' quality of life during subacute recovery. *Stroke* **2005**, *36*, 1480–1484. [CrossRef] [PubMed]
8. Liu, F.; Han, X.; Lin, M.; Wu, X.; Sun, Q.; Song, A. Remote Upper Limb Exoskeleton Rehabilitation Training System Based on Virtual Reality. In Proceedings of the 2019 16th International Conference on Ubiquitous Robots (UR), Jeju, Korea, 24–27 June 2019; pp. 323–327.
9. Bai, J.; Song, A.; Xu, B.; Nie, J.; Li, H. A novel human-robot cooperative method for upper extremity rehabilitation. *Int. J. Soc. Robot.* **2017**, *9*, 265–275. [CrossRef]
10. Khan, M.M.R.; Ahmed, T.; Pallares, J.R.H.; Islam, M.R.; Brahmi, B.; Rahman, M.H. Development of A Desktop-mounted Rehabilitation Robot For Upper Extremities. In Proceedings of the International Conference on Industrial & Mechanical Engineering and Operations Management, Dhaka, Bangladesh, 26–27 December 2021.
11. Masiero, S.; Armani, M.; Rosati, G. Upper-limb robot-assisted therapy in rehabilitation of acute stroke patients: Focused review and results of new randomized controlled trial. *J. Rehabil. Res. Dev.* **2011**, *48*, 355–366. [CrossRef]
12. Duret, C.; Courtial, O.; Grosmaire, A.G.; Hutin, E. Use of a robotic device for the rehabilitation of severe upper limb paresis in subacute stroke: Exploration of patient/robot interactions and the motor recovery process. *BioMed Res. Int.* **2015**, *2015*, 482389. [CrossRef]
13. Cao, W.; Chen, C.; Hu, H.; Fang, K.; Wu, X. Effect of hip assistance modes on metabolic cost of walking with a soft exoskeleton. *IEEE Trans. Autom. Sci. Eng.* **2020**, *18*, 426–436. [CrossRef]
14. Lim, G.H.; Suh, I.H.; Suh, H. Ontology-based unified robot knowledge for service robots in indoor environments. *IEEE Trans. Syst. Man Cybern.-Part A Syst. Humans* **2010**, *41*, 492–509. [CrossRef]
15. Proietti, T.; Crocher, V.; Roby-Brami, A.; Jarrasse, N. Upper-limb robotic exoskeletons for neurorehabilitation: A review on control strategies. *IEEE Rev. Biomed. Eng.* **2016**, *9*, 4–14. [CrossRef]
16. Gassert, R.; Dietz, V. Rehabilitation robots for the treatment of sensorimotor deficits: A neurophysiological perspective. *J. Neuroeng. Rehabil.* **2018**, *15*, 1–15. [CrossRef] [PubMed]
17. Loureiro, R.C.; Harwin, W.S.; Nagai, K.; Johnson, M. Advances in upper limb stroke rehabilitation: A technology push. *Med. Biol. Eng. Comput.* **2011**, *49*, 1103–1118. [CrossRef] [PubMed]
18. Nef, T.; Guidali, M.; Riener, R. ARMin III–arm therapy exoskeleton with an ergonomic shoulder actuation. *Appl. Bionics Biomech.* **2009**, *6*, 127–142. [CrossRef]
19. Rahman, M.H.; Rahman, M.J.; Cristobal, O.; Saad, M.; Kenné, J.P.; Archambault, P.S. Development of a whole arm wearable robotic exoskeleton for rehabilitation and to assist upper limb movements. *Robotica* **2015**, *33*, 19–39. [CrossRef]
20. Kim, B.; Deshpande, A.D. An upper-body rehabilitation exoskeleton Harmony with an anatomical shoulder mechanism: Design, modeling, control, and performance evaluation. *Int. J. Robot. Res.* **2017**, *36*, 414–435. [CrossRef]
21. Zhang, L.; Guo, S.; Sun, Q. Development and assist-as-needed control of an end-effector upper limb rehabilitation robot. *Appl. Sci.* **2020**, *10*, 6684. [CrossRef]
22. Nef, T.; Mihelj, M.; Kiefer, G.; Perndl, C.; Muller, R.; Riener, R. ARMin-Exoskeleton for arm therapy in stroke patients. In Proceedings of the 2007 IEEE 10th International Conference on Rehabilitation Robotics, Noordwijk, The Netherlands, 13–15 June 2007; pp. 68–74.
23. Chang, J.J.; Tung, W.L.; Wu, W.L.; Huang, M.H.; Su, F.C. Effects of robot-aided bilateral force-induced isokinetic arm training combined with conventional rehabilitation on arm motor function in patients with chronic stroke. *Arch. Phys. Med. Rehabil.* **2007**, *88*, 1332–1338. [CrossRef]
24. Liu, L.; Shi, Y.Y.; Xie, L. A novel multi-dof exoskeleton robot for upper limb rehabilitation. *J. Mech. Med. Biol.* **2016**, *16*, 1640023. [CrossRef]
25. Pignolo, L.; Dolce, G.; Basta, G.; Lucca, L.; Serra, S.; Sannita, W. Upper limb rehabilitation after stroke: ARAMIS a "robo-mechatronic" innovative approach and prototype. In Proceedings of the 2012 4th IEEE RAS & EMBS International Conference on Biomedical Robotics and Biomechatronics (BioRob), Rome, Italy, 24–27 June 2012; pp. 1410–1414.

26. Pan, H.; Chen, G.; Kang, Y.; Wang, H. Design and Kinematic Analysis of a Flexible-Link Parallel Mechanism With a Spatially Quasi-Translational End Effector. *J. Mech. Robot.* **2021**, *13*, 011022. [CrossRef]
27. Zhao, P.; Zhang, Y.; Guan, H.; Deng, X.; Chen, H. Design of a Single-Degree-of-Freedom Immersive Rehabilitation Device for Clustered Upper-Limb Motion. *J. Mech. Robot.* **2021**, *13*, 031006. [CrossRef]
28. Burgar, C.G.; Lum, P.S.; Scremin, A.; Garber, S.L.; Van der Loos, H.; Kenney, D.; Shor, P. Robot-assisted upper-limb therapy in acute rehabilitation setting following stroke: Department of Veterans Affairs multisite clinical trial. *J. Rehabil. Res. Dev.* **2011**, *48*, 445–458. [CrossRef] [PubMed]
29. Rahman, M.H.; Saad, M.; Kenné, J.P.; Archambault, P.S. Nonlinear sliding mode control implementation of an upper limb exoskeleton robot to provide passive rehabilitation therapy. In Proceedings of the International Conference on Intelligent Robotics and Applications, Montreal, QC, Canada, 3–5 October 2012; pp. 52–62.
30. Wu, Q.; Chen, B.; Wu, H. Rbfn-based adaptive backstepping sliding mode control of an upper-limb exoskeleton with dynamic uncertainties. *IEEE Access* **2019**, *7*, 134635–134646. [CrossRef]
31. Feng, Y.; Zhou, M.; Zheng, X.; Han, F. Continuous adaptive terminal sliding-mode control. In Proceedings of the 2016 IEEE 11th Conference on Industrial Electronics and Applications (ICIEA), Hefei, China, 5–7 June 2016; pp. 184–188.
32. Brahmi, B.; Ahmed, T.; Elbojairami, I.; Swapnil, A.A.Z.; Assaduzzaman, M.; Schultz, K.; Mcgonigle, E.; Rahman, M.H. Flatness Based Control of a Novel Smart Exoskeleton Robot. *IEEE/ASME Trans. Mechatron.* **2021**, *27*, 974–984. [CrossRef]
33. Rahman, M.H.; Ochoa-Luna, C.; Saad, M. EMG based control of a robotic exoskeleton for shoulder and elbow motion assist. *J. Autom. Control Eng.* **2015**, *3*, 270–276. [CrossRef]
34. Ayas, M.S.; Altas, I.H. Fuzzy logic based adaptive admittance control of a redundantly actuated ankle rehabilitation robot. *Control Eng. Pract.* **2017**, *59*, 44–54. [CrossRef]
35. Chen, Z.; Li, Z.; Chen, C.P. Disturbance observer-based fuzzy control of uncertain MIMO mechanical systems with input nonlinearities and its application to robotic exoskeleton. *IEEE Trans. Cybern.* **2016**, *47*, 984–994. [CrossRef] [PubMed]
36. Fallaha, C.; Saad, M.; Ghommam, J.; Kali, Y. Sliding mode control with model-based switching functions applied on a 7-dof exoskeleton arm. *IEEE/ASME Trans. Mechatron.* **2020**, *26*, 539–550. [CrossRef]
37. Brahmi, B.; Laraki, M.H.; Brahmi, A.; Saad, M.; Rahman, M.H. Improvement of sliding mode controller by using a new adaptive reaching law: Theory and experiment. *ISA Trans.* **2020**, *97*, 261–268. [CrossRef] [PubMed]
38. Utkin, V.; Poznyak, A.; Orlov, Y.; Polyakov, A. Conventional and high order sliding mode control. *J. Frankl. Inst.* **2020**, *357*, 10244–10261. [CrossRef]
39. Islam, M.R.; Rahmani, M.; Rahman, M.H. A novel exoskeleton with fractional sliding mode control for upper limb rehabilitation. *Robotica* **2020**, *38*, 2099–2120. [CrossRef]
40. Babaiasl, M.; Goldar, S.N.; Barhaghtalab, M.H.; Meigoli, V. Sliding mode control of an exoskeleton robot for use in upper-limb rehabilitation. In Proceedings of the 2015 3rd RSI International Conference on Robotics and Mechatronics (ICROM), Tehran, Iran, 7–9 October 2015; pp. 694–701.
41. Khalil, H. Chapter 13: State Feedback Stabilization. In *Nonlinear Systems*, 3rd ed.; Prentice Hall: Upper Saddle River, NJ, USA, 2002; pp. 197–227.
42. Fallaha, C.J.; Saad, M.; Kanaan, H.Y.; Al-Haddad, K. Sliding-mode robot control with exponential reaching law. *IEEE Trans. Ind. Electron.* **2010**, *58*, 600–610. [CrossRef]
43. Chen, Y.; Fan, J.; Zhu, Y.; Zhao, J.; Cai, H. A passively safe cable driven upper limb rehabilitation exoskeleton. *Technol. Health Care* **2015**, *23*, S197–S202. [CrossRef] [PubMed]
44. Ahmed, T.; Islam, M.R.; Brahmi, B.; Rahman, M.H. Robustness and Tracking Performance Evaluation of PID Motion Control of 7 DoF Anthropomorphic Exoskeleton Robot Assisted Upper Limb Rehabilitation. *Sensors* **2022**, *22*, 3747. [CrossRef] [PubMed]
45. Lauretti, C.; Cordella, F.; Guglielmelli, E.; Zollo, L. Learning by demonstration for planning activities of daily living in rehabilitation and assistive robotics. *IEEE Robot. Autom. Lett.* **2017**, *2*, 1375–1382. [CrossRef]
46. Xiao, F.; Gao, Y.; Wang, Y.; Zhu, Y.; Zhao, J. Design of a wearable cable-driven upper limb exoskeleton based on epicyclic gear trains structure. *Technol. Health Care* **2017**, *25*, 3–11. [CrossRef]
47. Luh, J.Y.; Walker, M.W.; Paul, R.P. On-line computational scheme for mechanical manipulators. *J. Dyn. Syst. Meas. Control* **1980**, *102*, 69–76. [CrossRef]
48. Rahman, M.H.; Saad, M.; Kenné, J.P.; Archambault, P.S. Control of an exoskeleton robot arm with sliding mode exponential reaching law. *Int. J. Control Autom. Syst.* **2013**, *11*, 92–104. [CrossRef]
49. Perry, J.C.; Rosen, J.; Burns, S. Upper-limb powered exoskeleton design. *IEEE/ASME Trans. Mechatron.* **2007**, *12*, 408–417. [CrossRef]

robotics

MDPI

Article

Constrained-Differential-Kinematics-Decomposition-Based NMPC for Online Manipulator Control with Low Computational Costs

Jan Reinhold *, Henry Baumann and Thomas Meurer

Automation and Control Group, Faculty of Engineering, Kiel University, Kaiserstraße 2, 24143 Kiel, Germany
* Correspondence: janr@tf.uni-kiel.de; Tel.: +49-431-880-6121

Abstract: Flexibility combined with the ability to consider external constraints comprises the main advantages of nonlinear model predictive control (NMPC). Applied as a motion controller, NMPC enables applications in varying and disturbed environments, but requires time-consuming computations. Hence, given the full nonlinear multi-DOF robot model, a delay-free execution providing short control horizons at appropriate prediction horizons for accurate motions is not applicable in common use. This contribution introduces an approach that analyzes and decomposes the differential kinematics similar to the inverse kinematics method to assign Cartesian boundary conditions to specific systems of equations during the model building, reducing the online computational costs. The resulting fully constrained NMPC realizes the translational obstacle avoidance during trajectory tracking using a reduced model considering both joint and Cartesian constraints coupled with a Jacobian transposed controller performing the end-effector's orientation correction. Apart from a safe distance from the obstacles, the presented approach does not lead to any limitations of the reachable workspace, and all degrees of freedom (DOFs) of the robot are used. The simulative evaluation in GAZEBO using the Stäubli TX2-90 commanded of ROS on a standard computer emphasizes the significantly lower online computational costs, accuracy analysis, and extended adaptability in obstacle avoidance, providing additional flexibility. An interpretation of the new concept is discussed for further use and extensions.

Keywords: kinematic analysis; robotic differential model decomposition; nonlinear model predictive control (NMPC); controller couplings; joint and Cartesian space constraints; computing time reduction; accuracy analysis; trajectory tracking; obstacle avoidance

Citation: Reinhold, J.; Baumann, H.; Meurer, T. Constrained-Differential-Kinematics-Decomposition-Based NMPC for Online Manipulator Control with Low Computational Costs. *Robotics* **2023**, *12*, 7. https://doi.org/10.3390/robotics12010007

Academic Editor: Raffaele Di Gregorio

Received: 15 November 2022
Revised: 12 December 2022
Accepted: 20 December 2022
Published: 3 January 2023

1. Introduction

Modern industry is in a constant state of change driven by the contemporary labor market, the purchase demand, and the effective use of resources or machines [1]. Robots are increasingly being used in process automation to carry out monotonous and strenuous work, also reducing the operating costs [2,3]. In addition, efficient image recognition and sensor fusion enable increasingly accurate recognition of the environment in the robot's workspace [4]. Thus, using appropriate algorithms, robots can also be deployed in varying and disturbed environments to cover further fields of activity [5].

One sector undergoing a tremendous transformation is agriculture, which motivates this paper, but does not limit the scope of the presented approach. On the one hand, farmers, industry, and governments need to keep the costs moderate, even in high-wage countries, and on the other hand, consumers appreciate a sustainable and regional production [6]. These requirements are not mutually exclusive, but this is a subject area that needs to be developed, among other fields of application [7]. In particular, image recognition has been improved and adapted to specific agricultural problems in the last decade, allowing high-quality recognition with many features in widely disturbed environments [8–10]. Precision agriculture enables, e.g., mechanical weed removal without damaging the adjacent plants, so that the use of pesticides can be reduced [11,12]. This usually requires equipment that is dedicated to a specific application and is expensive to purchase and maintain. In contrast, (industrial) robots are flexible and sustainable, as they are applicable for multiple applications throughout the agricultural season, simply by using different end-effectors. However, for the application in a distributed environment with multiple obstacles, trajectory planning and control have to be accomplished almost delay-free. Achieving low computational costs in optimal control using a robot with multiple degrees of freedom (DOFs) is addressed in this work.

Motion control is used for the adaption of planned trajectories in the Cartesian or joint space [13,14], which must subsequently be adjusted to a varying environment by means of a closed control loop. In general, either a discrete or a continuous interpretation of the workspace can be chosen. When choosing the discrete approach, the detected environment is meshed [15], and the optimal path is planned along the resulting nodes and edges [16]. Here, the Dijkstra and A* algorithms [17], as well as sampling-based methods can be applied with low computational cost to solve the planning task [18]. In general, inverse kinematics or Jacobian inverse controllers with low computational costs are subsequently used for the transformation into the joint space [19]. However, setting up the mesh is computational expensive, so that a delay-free motion control is not possible in a highly varying environment [20,21]. Examples of continuous motion planning tools include CHOMP, STOMP, and TrajOpt [22–24], which are commonly used. However, even though the computation times are short, they are not optimized to be used iteratively for delay-free control [25]. In addition, learning-based methods are increasingly applied, especially to take the aging of the robotic systems into account during motion control [26]. Using iterative learning control [27], motions are repeated until the solution is within an acceptable range. However, for motions in a varying environment, it is complicated to train these systems, as individual movements have to be run several times with the same initial and terminal states [28]. If a reference trajectory is known, the repetitive control approach can be added to be periodic and address the initialization problem [29]. Furthermore, reinforcement learning is used to improve the performance of the tracking controllers [30]. However, this paper presents a model-based control scheme that adapts the motion based on the robot's kinematic specifications. To realize a closed control loop, which iteratively considers varying environmental constraints in the Cartesian space and robotic constraints in the joint space given by the multi-DOF robot, nonlinear model predictive control (NMPC) is used [31]. As the dynamic optimization has to be solved on a receding horizon, computational efficiency is an issue for real-time application.

Two different time horizons have to be considered during the implementation of NMPC [32,33]. The prediction horizon specifies how far the movements in the disturbed environment are predicted. Governed by the sample size and the DOFs, the number of decision variables is set, which determines the computational costs to solve the optimal control problem (OCP) online [34]. Secondly, the control horizon, which is shorter than the prediction horizon, describes a kind of buffer along which the robot executes the movements of the last valid OCP solution [35]. A delay-free implementation of NMPC is not possible at the sampling rates of commonly used (robot) controllers, if an OCP for the prediction horizon were to be solved in every iteration step [36]. Thus, the control is maintained for subsequent samples along the control horizon, which is as accurate as the environment has been captured. Hence, faster solving of the OCP results in a shorter control horizon, and thus, rather optimal movements will be obtained [25]. A variety of approaches exist that perform NMPC [37–39], also involving horizon adaptions [40–42] and system reformulations [43,44]. However, in order to decrease the computational costs and, thus, the number of the decision variables, either the three-dimensional (3D) Cartesian space is only considered for the implementation of NMPC in robotics [45–47] or the number of actuated robot joints is reduced and particularly powerful hardware or software is used for the computations [25,48,49]. If only the Cartesian space is considered, the OCP neglects all nonlinearities of the robot model and does not take the reachable work and joint space into account during the motion computations [50,51]. It must be ensured that the subsequent joint space transformation is reachable; otherwise, the OCP must be solved again with a different parameter set. Some approaches include the robotic constraints, but they limit the robots' DOFs to handle the computational expenses [52,53]. This complicates the general application of multi-DOF robots in disturbed environments, where all six Cartesian DOFs must be adjusted [54].

The approach introduced here analyzes the robot kinematics, thereby reducing the number of decision variables of NMPC to reduce the computational costs. It preserves the full robot workspace by adding an additional controller. Using kinematic analysis, this contribution addresses the cause of the computational costs themselves, rather than the symptoms, by means of adjustments in the implementation. Obstacle avoidance in 3D space is primarily performed by translational movements, i.e., evasion is achieved by displacement. In general, tilting the end-effector can also avoid collisions. However, the associated robotic joints provide a significantly smaller workspace, and simultaneously, the tool cannot perform the desired task in the correct orientation [55]. Referring to the agricultural context, manual weeding would have to be interrupted to avoid adjacent plants, which is less effective. The approach introduced here decomposes the differential kinematics analogously to the inverse kinematics method to partition the relevant equations and joints, respectively [56]. The procedure is applied to an industrial robot, which can be decomposed into the anthropomorphic arm and the spherical wrist, but it can be transferred to all multi-DOF robot types, which can also be separated into a translational and rotational part [57]. By splitting the problem, the constraints caused by external obstacles are assigned a priori and, therefore, do not need to be assigned during the online processing. As a result, two coupled controllers execute a constrained translational motion combined with a rotational movement for accurate trajectory tracking in a disturbed environment. The translational motion controller is realized as NMPC and considers both joint and Cartesian constraints. Compared to the consideration of the complete robot model, a significant reduction of the computational costs can be achieved due to the limited number of decision variables in the OCP. In this way, the consideration of additional boundary conditions to adjust the behavior of the robot still allows almost delay-free evaluations. Based on the joint control for the translational motion avoiding obstacles, a Jacobian transpose controller performs the rotation correction using the spherical wrist so that the end-effector maintains the correct alignment [58].

The proposed method is suitable for applications in various fields including industrial robots, where the dynamic parameters are typically unknown and can be realized even when using standard computer hardware due to the reduction of the computing times in the optimization. The approach can be used as well, e.g., for the online motion control of welding processes [59] or in the context of collaboration [60,61], which are common applications in industry requiring delay-free adaptation to a disturbed environment. For standard industrial robots, typically not only the dynamic parameters are unknown, but it is also advantageous to use only the kinematic specifications. In the case of model-based controllers, model uncertainties lead to performance losses in operation and inaccuracies due to the robot aging [62], for which the differential kinematics is taken into account in the presented approach. Here, the standard industrial robot Stäubli TX2-90 [63] is used as an example, which is simulated in GAZEBO [64]. The communication is performed by means of ROS [65]. In the evaluation, the required computation times needed by the introduced NMPC approach are compared with the consideration of the full robot model in different setups. Further, the trajectory tracking accuracy is analyzed. Motivated by manual weed removal, a scenario is set up where the robot must adjust its initially given trajectory online to avoid damaging adjacent fixed and moving obstacles. Plants are abstracted as spheres so that objects recognized by image recognition, such as cabbages, can be easily integrated into a concrete application [66]. Weed removal itself is not shown, but the collision-free movements with correct alignment demonstrate the applicability of the approach [54], which can be transferred to various industrial applications. Before the movement starts, a polynomial planned trajectory is specified, which crosses the obstacles. The end-effector must track this trajectory using NMPC and the Jacobian transpose controller. The NMPC detects the respective obstacles only within its prediction horizon, to which the movements must be adapted. The short evaluation times of the optimization allow the additional limitation of the achievable Cartesian workspace in height, which leads the translational motion to ground-level avoidance. In the context of the trajectory planning, an automatic selection of the joint configurations is presented, which replaces a manual selection, as is common for point-to-point movements in robotics, e.g., as utilized in [63,65,67,68].

The paper is organized as follows: Section 2 recalls the concept of the inverse kinematics for anthropomorphic robots and proceeds with the explanation of the differential kinematics decomposition. Based on this, the OCP for the end-effector's translational movement including joint and Cartesian constraints is described in Section 3, which is subsequently implemented as NMPC for online control. In addition, the Jacobian transpose controller for the rotation correction in 3D space is applied and coupled with the NMPC. In Section 4, the computation time savings of the approach, the trajectory tracking accuracy, and the control in a disturbed and varying environment including fixed and moving obstacles are demonstrated. A discussion of the results is provided in Section 5, and the final remarks in Section 6 conclude this contribution.

2. Modeling and Mathematical Decomposition of the Manipulator

For the decomposition of the robotic model, a standard industrial manipulator with n revolute joints $\boldsymbol{q} \in \mathbb{R}^n$ and an anthropomorphic structure consisting of an anthropomorphic arm and a spherical wrist was considered [69]; see Figure 1. A (non-)redundant robot with $n \geq 6$ was assumed, so that the workspace included all six Cartesian DOFs. Within the reachable workspace, an end-effector's pose is expressed by the homogeneous transformation matrix $H^e_w(\boldsymbol{q}) \in SE(3)$, which comprises the translation vector $\boldsymbol{p}^e_w(\boldsymbol{q}) \in \mathbb{R}^3$ and the rotation matrix $R^e_w(\boldsymbol{q}) \in SO(3)$. The subscript clarifies the reference system, while the superscript marks the body-fixed frame to be described therein. Thereby, the index w represents the fixed world frame. Additionally, the end-effector frame $\{e\}$ describes the pose of the tool's point of interest mounted on the flange $\{f\}$. Based on the Denavit–Hartenberg (DH) convention [70], the direct kinematics of the robot can be derived.

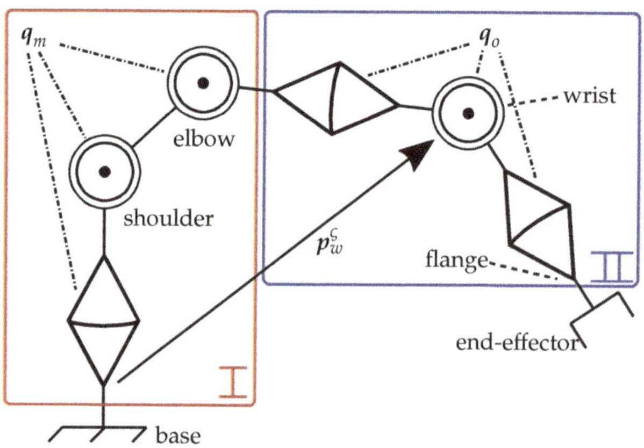

Figure 1. Manipulator with $n = 6$ revolute joints and an anthropomorphic structure consisting of an anthropomorphic arm (**I**) and a spherical wrist (**II**).

2.1. Analysis of the Inverse Kinematics

The method of inverse kinematic analysis [69] is recalled briefly in order to provide a better understanding and the motivation of the following sections. Inverse kinematics can be used to determine the associated joint angle configuration $\boldsymbol{q} = [\boldsymbol{q}_m^{\mathrm{T}}, \boldsymbol{q}_o^{\mathrm{T}}]^{\mathrm{T}}$ given a desired end-effector pose $H^e_{w,\mathrm{des}}$. For the manipulator, which is shown exemplarily in Figure 1, the first m joints $\boldsymbol{q}_m \in \mathbb{R}^m$ are assigned to the red framed anthropomorphic arm (I) and the last o joints $\boldsymbol{q}_o \in \mathbb{R}^o$ belong to the blue framed spherical wrist (II). We considered $3 \leq m < n$ and $3 \leq o < n$ so that the $n = m + o$ robot DOFs are partitioned in such a way that any pose can be achieved in the reachable workspace. To this end, the robotic model is decomposed into the $\boldsymbol{q}_m$ associated part for translational displacement and the $\boldsymbol{q}_o$ relating system for rotational alignment to reduce the number of variables in the equations describing both associated models, respectively [21,71]. Thus, instead of evaluating the entire kinematic chain, two reduced chains are considered. The connection of the two systems is defined at the wrist frame $\{\varsigma\}$, where the so-called wrist point $\boldsymbol{p}_w^\varsigma = [p_{w,x}^\varsigma, p_{w,y}^\varsigma, p_{w,z}^\varsigma]^\mathrm{T} \in \mathbb{R}^3$ is located. The wrist point can be obtained by equating the traversed kinematic chains that converge in $\{\varsigma\}$ starting in the $\{w\}$ and the $\{e\}$ frame, respectively. Starting at $\{e\}$, the desired end-effector pose can be projected onto the flange by

$$H_w^f = \begin{bmatrix} R_w^f & \boldsymbol{p}_w^f \\ \boldsymbol{0}^\mathrm{T} & 1 \end{bmatrix} = H_{w,\mathrm{des}}^e \left(H_f^e\right)^{-1} = H_{w,\mathrm{des}}^e \, H_e^f \tag{1}$$

using the tool-specific transformation matrix H_f^e.

The orientation of the z_f-axis of the flange is denoted by $\boldsymbol{r}_{w,z}^f$ and can be obtained from the third column of R_w^f. The wrist point

$$\boldsymbol{p}_w^\varsigma = \boldsymbol{p}_w^f - d_f \, \boldsymbol{r}_{w,z}^f \in \mathbb{R}^3 \tag{2}$$

can be calculated based on (1) and the length d_f of the last link ending at the flange. Using (2) as the left-hand side and the position vector of the direct kinematics $H_w^\varsigma(\boldsymbol{q}_m)$ as the right-hand side, a system of equations can be set up to determine $\boldsymbol{q}_m$. When considering a redundant robot with $m > 3$ and $n > 6$, additional constraints to the nullspace must be introduced to solve the system uniquely [72]. Apart from the possible nullspace, there are generally up to four valid solutions describing the shoulder left or right and elbow up or down configurations [73]. The so-called rotation correction can be determined by

$$R_m^f = (R_w^m)^\mathrm{T} \, R_w^f = R_m^w \, R_w^f. \tag{3}$$

Via the ZYZ-sequence [74], which is based on the joint structure of the spherical wrist, R_m^f can be implied for the joints $\boldsymbol{q}_o$. This in turn yields ambiguous solutions known as wrist top and wrist bottom [73], respectively. This doubles the maximum number of possible configurations mentioned above, so that up to eight solutions can exist for one desired pose $H_{w,\mathrm{des}}^e$. In this contribution, an automatic selection was introduced, which performs an evaluation of the most-suitable joint configuration and selects it for the movement. Jumps between the up to eight solutions are avoided, and the common boundaries are considered.

2.2. Decomposition of Differential Kinematics

Based on the method presented before, the robot model is decomposed for the following control architecture. The separation into a translational and a rotational part allows the consideration of external boundary conditions, e.g., for the avoidance of obstacles, to be directly assigned to specific joints in the robot's kinematic chain. Thus, the DOFs considered in the optimization-based control approach are reduced by kinematic analysis, reducing the computational costs. Differential kinematics rather than direct kinematics was taken into account to avoid algebraic loops in the online computations [75] and for a more straightforward restriction of the nullspace in the case of redundant robots [57,69]. The transformation of joint velocities $\dot{\boldsymbol{q}} \in \mathbb{R}^n$ into Cartesian velocities can be performed by differential kinematics using

$$\begin{bmatrix} \dot{\boldsymbol{p}}_w^e \\ \boldsymbol{\omega}_w^e \end{bmatrix} = J_w^e(\boldsymbol{q}) \, \dot{\boldsymbol{q}} \in \mathbb{R}^6. \tag{4}$$

The translational velocity of the end-effector with respect to the $\{w\}$ system is described by $\dot{\boldsymbol{p}}_w^e \in \mathbb{R}^3$, while $\boldsymbol{\omega}_w^e \in \mathbb{R}^3$ expresses the corresponding angular velocities [69]. The nonlinear geometric Jacobian matrix:

$$J_w^e(\boldsymbol{q}) = \begin{bmatrix} J_{\text{trans}} \\ J_{\text{rot}} \end{bmatrix} = \begin{bmatrix} J_{\text{trans},1} & J_{\text{trans},2} \\ J_{\text{rot},1} & J_{\text{rot},2} \end{bmatrix} \in \mathbb{R}^{6\times n} \tag{5}$$

is introduced. From (5), it can be seen that the entire Jacobian consists of a translational and a rotational part. Transferring the approach of Section 2.1 to the decomposition of the manipulator, $J_{\text{trans},1}$ and $J_{\text{rot},2}$ represent the associated terms in the differential kinematics, and the submatrices $J_{\text{trans},2}$ and $J_{\text{rot},1}$ denote cross-couplings, respectively. Instead of using the entire transformation from (4), the differential kinematics is split as well. Based on the decomposition analyzed in the inverse kinematics, translation is performed by the first m and orientation by the last o robot joints [57]. According to the general matrix computation as described in [69], the matrices $J_{w,\text{trans}}^{\varsigma}(\boldsymbol{q}_m) \in \mathbb{R}^{3\times m}$ and $J_{m,\text{rot}}^{f}(\boldsymbol{q}_o) \in \mathbb{R}^{3\times o}$ are introduced and used subsequently instead of (5). The DOFs due to the cross-couplings are eliminated as a consequence in the transformation performed in (4) for the full robot system. This means that the joints $\boldsymbol{q}_m$ no longer have an active influence on the end-effector's orientation and $\boldsymbol{q}_o$ cannot be used for the translational positioning of the $\{e\}$ frame. Further, two controllers for $\boldsymbol{q}_m$ and $\boldsymbol{q}_o$ were designed separately and then coupled.

One controller controls the positioning using $J_{w,\text{trans}}^{\varsigma}(\boldsymbol{q}_m)$, and the other controller adjusts the alignment with $J_{m,\text{rot}}^{f}(\boldsymbol{q}_o)$. Analogous to the evaluation of inverse kinematics in Section 2.1, there is no loss of DOFs in the Cartesian space, and due to controller couplings, the entire workspace is still reachable.

It should be emphasized that the translational part $J_{w,\text{trans}}^{\varsigma}(\boldsymbol{q}_m)$ refers to the wrist point $\boldsymbol{p}_w^{\varsigma}$, while the orientation of the $\{\varsigma\}$ system is irrelevant in this context. Using (1) and (2), the desired wrist point is obtained from $\boldsymbol{p}_{w,\text{des}}^e$, and an orientation error follows from the wrist positioning using the first m robot joints. In turn, the Jacobian $J_{m,\text{rot}}^{f}(\boldsymbol{q}_o)$ for the rotational part refers to the $\{m\}$ system localized in the robot's elbow, the last joint of the anthropomorphic arm, as shown in Figure 1. The link between the $\{m\}$ and the $\{\varsigma\}$ system exhibits a constant length and is aligned along the rotation axis of the first spherical wrist joint. Thus, this DOF only changes the alignment and not the displacement between both systems, and the two kinematic chains can be connected in this way.

3. Optimal Trajectory Control Using Decomposed Differential Kinematics

To implement fast online control, Section 3.1 presents a computationally effective planning scheme involving all robot joints to generate an initial trajectory that does not take external disturbances into account. It can be used when the workspace is not constrained and serves as a reference in a warm start of the following optimization. An automatic selection is introduced that identifies the most-suitable joint configuration for the desired terminal pose. The up to eight possible solutions of the inverse kinematics are checked for jumps for the planned pose transition, and the solution with the largest distance to the joint boundaries is selected. In Section 3.2, the constrained OCP is formulated, with the translational part of the decomposed system from Section 2.2 as the underlying model. The OCP is evaluated on a receding time horizon, i.e., controlling the robot using NMPC. Meanwhile, the orientation of the end-effector is considered separately using the controller presented in Section 3.3. In Table 1, the main difficulties and characteristics of the two controllers are listed as an overview. Special attention has to be paid to the wrist position, which is iteratively placed by the NMPC and determines the starting pose for the orientation controller. The combination of the two controllers provides the overall control of the robot, and both are calculated subsequently in each iteration. If the orientation controller, based on the Jacobian transpose controller here, is also implemented as a second NMPC, the controllers would have to be evaluated sequentially and, therefore, would be time consuming because of the dependency with respect to the reference pose at the $\{m\}$ frame.

Table 1. Comparison of the difficulties and characteristics of the decomposed robot model illustrated in Figure 1 for optimal trajectory control achieving low computation times in online calculations.

Properties	I: Anthropomorphic arm	II: Spherical Wrist
intended use	• translational movement • avoiding obstacles in disturbed environment	• orientation control • alignment for desired rotation
constraints	• iterative solving starting at fixed base • environmental, state and input constraints • consideration of the distance between p^C_w and end-effector	• depending on wrist movement • standard controller bounded to limits • compensation of rotation correction
control	• control of q_m using NMPC • optimization with known reference	• control of q_o with Jacobian controller • reaction based on wrist displacement (from NMPC)
singularity avoidance	• configuration bounded to objective function with regularization • preselection of the closest solution	• Jacobian transpose using no inversion in calculations • unit quaternions preventing a *Gimbal lock* [74]

Remark 1. *Here, only a multi-DOF robot with an anthropomorphic structure and revolute joints is discussed, so that an independent assignment of the joints to a translational and rotational motion in the Cartesian space can be performed. This design as an open or closed kinematic chain is the most common setup of standard industrial robots. A transfer of the approach to other manipulator types can be applied if the robotic model admits a decomposition according to the specification.*

3.1. Polynomial Trajectory Planning

An initial planning for all n robot joints is performed before the online controlled robot movements start. To generate a reference trajectory, a polynomial approach in the joint space is utilized to connect the initial end-effector pose represented by the homogeneous transformation $H^e_w(\boldsymbol{q}(t_0))$ at time $t_0 \in \mathbb{R}_{\geq 0}$ with the desired terminal pose $H^e_{w,\text{des}}$ at time $t_1 = t_0 + T$, obtaining a continuously differentiable trajectory. The transition time $T \in \mathbb{R}_{>0}$ must be chosen so that the dynamic joint limits of the robot are not violated. To check whether $H^e_{w,\text{des}}$ is an admissible pose with the mounted end-effector according to the given bounds in the robot's data sheet, (2) can be used to validate the wrist point. As mentioned in Section 2.1, up to eight possible joint configurations can be determined for the given pose at t_1. From the set of possible solutions of the inverse kinematics, the configurations that are not included in the reachable joint space:

$$\mathcal{Q} := \{\boldsymbol{q} \in \mathbb{R}^n \mid \boldsymbol{q}_{\min} \leq \boldsymbol{q} \leq \boldsymbol{q}_{\max}\} \tag{6}$$

are excluded. To connect the initial joint setup $\boldsymbol{q}(t_0)$ and the remaining $\beta \leq 8$ terminal configurations $Q_\beta(t_1) = [\boldsymbol{q}_1(t_1), \ldots, \boldsymbol{q}_\beta(t_1)] \in \mathbb{R}^{n\times\beta}$, the polynomial:

$$\gamma(t) = \sum_{j=0}^{7} \boldsymbol{\lambda}_j t^j \in \mathbb{R}^n \tag{7}$$

is introduced. The eight unknowns $\boldsymbol{\lambda}_j$, $j \in \{0,\ldots,7\}$ for each of the n joints can be determined, respectively, from the eight boundary conditions:

$$\gamma_i^{(j)}(t_0) = \boldsymbol{q}^{(j)}(t_0), \quad \gamma_i^{(j)}(t_1) = \boldsymbol{q}_i^{(j)}(t_1), \quad j \in \{0,\ldots,3\},\ i \in \{1,\ldots,\beta\} \tag{8}$$

for each configuration i. In (8), $\gamma_i(t) = \gamma_i^{(0)}(t)$ applies, and the derivatives are given by $\gamma_i^{(j)}(t)$, which describe the associated velocity, acceleration, and jerk, respectively. The velocity bounds can be measured or formed by the inverse evaluation of (4). Without loss of generality, the acceleration and jerk are chosen to be zero at the beginning and at the end of the transition. The acceleration bounds can alternatively be transformed by introducing the time derivative of the Jacobian in (5) [69]. In order to drop the solutions that contain an unnecessary change at the shoulder, elbow, or wrist of the robot, all β transitions connecting $\boldsymbol{q}(t_0)$ with the configurations in $Q_\beta(t_1)$ are sampled by $t_\gamma \in \mathbb{R}_{>0}$ and examined for jumps. From the remaining $r \leq \beta$ possibilities that do not involve jumps, the configurations that are furthest away from the joint boundaries with the corresponding joints are selected from

$$\max_{\rho}\Big\{\min_{i}\{q_{i,\rho}(t_\gamma) - q_{i,\min}, q_{i,\max} - q_{i,\rho}(t_\gamma)\}\Big\}, \quad \rho \in \{1,\ldots,r\},\ i \in \{1,\ldots,n\}. \tag{9}$$

Each joint i is evaluated individually at each sample step t_γ. If several joint configurations exhibit the same distances to the bounds, the maximum operator in (9) is used to select the configuration that maintains the greatest distance from the boundaries $\boldsymbol{q}_{\min}$ and $\boldsymbol{q}_{\max}$, considering all sampling steps t_γ. If the coupling of the two checks were reversed, a joint that is far from its bound could compensate a joint close to its respective bound in the evaluation. Since the planning is implemented in the joint space, no consideration of the *Euler angles* [74] in Cartesian space is necessary, which prevents representation singularities. Using the introduced procedure in (9), an automatic selection method of the most appropriate kinematic configuration is introduced, eliminating the need for a manual selection, required by most of the inverse kinematics tools [63,65,67,68].

3.2. Optimization-Based Translational Trajectory Control

For the translational motion in the robot's workspace considering obstacles, a constrained optimization problem with a fixed end time $\tau_1 = \tau_0 + N \in \mathbb{R}_{>0}$ is introduced. The prediction horizon of the OCP is defined by $N \in \mathbb{R}_{>0}$ and is starting at $\tau_0 = t_\delta$. Thereby, $t_\delta \in \mathbb{R}_{\geq 0}$ describes the current sampling step. With the underlying model of the decomposed differential kinematics from Section 2.2, the joint velocities:

$$\mathcal{U} := \{\boldsymbol{u} \in \mathbb{R}^m \mid -\dot{\boldsymbol{q}}_{m,\max} \leq \boldsymbol{u}(\tau) \leq \dot{\boldsymbol{q}}_{m,\max}\} \tag{10a}$$

are chosen as fictitious inputs of the OCP. As can be seen from (10a), only the first m joints of the robot are taken into account for the displacement of the wrist point $\boldsymbol{p}_w^\varsigma$. The end-effector orientation is adjusted subsequently by means of $\boldsymbol{q}_o$. Standard industrial robots are usually controlled using joint position controllers [63,65] so that the joint angles $\boldsymbol{q}_m$ to command the translational motion of the robot can be obtained by solving $\dot{\boldsymbol{q}}_m = \boldsymbol{u}$. Furthermore, the OCP:

$$\min_{\boldsymbol{u}\in\mathcal{U}} F(\boldsymbol{u}) = \int_{\tau_0}^{\tau_1} l(\boldsymbol{u}, \boldsymbol{p}_w^\varsigma(\boldsymbol{q}_m), \ddot{\boldsymbol{q}}_m, \boldsymbol{\mu})\, \mathrm{d}\tau \tag{10b}$$

is considered by minimizing the running cost:

$$l(\boldsymbol{u}, \boldsymbol{p}_w^\varsigma(\boldsymbol{q}_m), \ddot{\boldsymbol{q}}_m, \boldsymbol{\mu}) = \mu_u\, \boldsymbol{u}^{\mathrm{T}}\boldsymbol{u} + \mu_{\ddot{q}}\, \ddot{\boldsymbol{q}}_m^{\mathrm{T}}\, \ddot{\boldsymbol{q}}_m + \mu_p \big(\boldsymbol{p}_{w,\mathrm{des}}^\varsigma - \boldsymbol{p}_w^\varsigma(\boldsymbol{q}_m)\big)^{\mathrm{T}} \big(\boldsymbol{p}_{w,\mathrm{des}}^\varsigma - \boldsymbol{p}_w^\varsigma(\boldsymbol{q}_m)\big) \tag{10c}$$

over the time interval $\tau \in [\tau_0, \tau_1]$. The elements from $\boldsymbol{\mu} = [\mu_u, \mu_{\ddot{q}}, \mu_p]^{\mathrm{T}} \in \mathbb{R}^3$ can be used to weight the individual terms in (10c). The desired wrist position $\boldsymbol{p}_{w,\mathrm{des}}^\varsigma$ is derived from the desired end-effector pose using (2). If a reference trajectory is specified, e.g., with the procedure introduced in Section 3.1, using MoveIt for task-level motion planning [76] or based on the techniques summarized in [14], the Lagrange function shown in (10c) aims for trajectory tracking. Alternatively, only the terminal position could be passed to (10c) as a reference, which is called a local motion planning problem [35]. By considering the input $\boldsymbol{u}$ in (10c), the agility can be influenced, and the relating part also represents a regularization term, to prevent singular arcs [43]. In order to further prevent singularities, it is possible to include the manipulability measure into the running cost as well [77]. In practice, the integral listed in (10b) is discretized by a sum over $k \in \mathbb{N}_{>0}$ temporal grid points for the numerical implementation. Enabling an influence on the rate change, $\ddot{\boldsymbol{q}}_m$ is also included in (10c). The acceleration $\ddot{\boldsymbol{q}}_m$ and the jerk $\dddot{\boldsymbol{q}}_m$ result from the discrete differentiation of the input $\boldsymbol{u}$, respectively. Based on the system formulation:

$$\frac{\mathrm{d}}{\mathrm{d}\tau}\begin{bmatrix}\boldsymbol{p}_w^{\varsigma}\\ \boldsymbol{q}_m\end{bmatrix} = \begin{bmatrix}\boldsymbol{J}_{w,\mathrm{trans}}^{\varsigma}(\boldsymbol{q}_m)\,\boldsymbol{u}\\ \boldsymbol{u}\end{bmatrix}, \ \tau > \tau_0 \quad \text{with} \begin{bmatrix}\boldsymbol{p}_w^{\varsigma}(\tau_0)\\ \boldsymbol{q}_m(\tau_0)\end{bmatrix} = \begin{bmatrix}\boldsymbol{p}_{w,0}^{\varsigma}\\ \boldsymbol{q}_{m,0}\end{bmatrix}, \tag{10d}$$

the variables in (10b) can be determined. Here, $\boldsymbol{p}_{w,0}^{\varsigma}$ describes the initial wrist position and $\boldsymbol{q}_{m,0}$ the initial joint angles at time τ_0. Using the inequality constraints:

$$\begin{aligned} \boldsymbol{q}_{m,\min} \leq \boldsymbol{q}_m \leq \boldsymbol{q}_{m,\max} \\ -\boldsymbol{q}_{m,\max}^{(j)} \leq \boldsymbol{q}_m^{(j)} \leq \boldsymbol{q}_{m,\max}^{(j)}, \quad j \in \{2,3\}, \end{aligned} \tag{10e}$$

the system (10d) is constrained to the reachable joint space, since the selected joint angles $\boldsymbol{q}_m$ must be inside the valid bounds of (6). Applying the constraints in (10e) to the acceleration $\ddot{\boldsymbol{q}}_m$ and jerk $\dddot{\boldsymbol{q}}_m$, non-adjustable changes can be avoided. The respective bounds are usually known for standard industrial robots and can be taken from the appropriate data sheet, e.g., given by [63].

Obstacles are subsequently modeled as spheres to illustrate the approach [78], but can also be described by using sophisticated techniques as, e.g., by the evaluation of tetrahedral meshes or polyhedra [21,79,80]. Let ν denote the number of obstacles in the Cartesian space. These are assumed moveable and centered at $\boldsymbol{p}_{w,i}^{\nu}(\tau) \in \mathbb{R}^3$, $i \in \{1,\ldots,\nu\}$, imposing the inequality constraints:

$$r_i + d_f + \sqrt{(a_e)^2 + (d_e)^2} < |\boldsymbol{p}_w^{\varsigma} - \boldsymbol{p}_{w,i}^{\nu}(\tau)|, \quad \forall i = 1,\ldots,\nu. \tag{10f}$$

Due to the decomposition of the robot model, the resulting orientation of the end-effector during motion is not known in the optimization. Therefore, the length of the spherical wrist plus the DH parameters $a_e \in \mathbb{R}$ and $d_e \in \mathbb{R}$ of the end-effector are also defined as a sphere around the wrist point. This is added to the radius $r_i \in \mathbb{R}_{>0}$ of each obstacle to obtain a safe distance.

For example, to perform horizontal motion only or to prevent touching the ground, the height:

$$p_{w,z,\min}^{\varsigma} \leq p_{w,z}^{\varsigma} \leq p_{w,z,\max}^{\varsigma} \tag{10g}$$

of the robot's workspace can also be optionally bounded. Constraining of the OCP (10) by adding (10g) usually increases the computational times significantly, which will be demonstrated in Section 4.1. However, by reducing the robot model in (10d), it is possible to include further constraints influencing the robot's behavior.

NMPC can be applied by solving the presented OCP on a receding horizon with a suitable prediction length N. Both the initial wrist position and the joint angles can be obtained from the measured joint angles of the robot. Direct multiple shooting is used for the numerical evaluations, which considers an initial value problem in each time interval $[\tau_{\kappa-1}, \tau_\kappa]$, $\kappa \in \{1,\ldots,k\}$ [81]. Hence, k initial value problems have to be solved in total. Since the subintervals can be solved independently, parallelization can be used. To ensure a continuous transition between intervals, boundary conditions must be imposed so that the boundary values of the adjacent intervals are identical [82]. For a warm start, the initial estimates of the optimization variables and the input can be set for each step τ_κ by the procedure presented in Section 3.1. According to [81], an approach is used here that first discretizes and then optimizes, converging to local or global minima depending on the solver settings and the weightings chosen in the quadratic objective function (10b).

3.3. *Jacobian Transpose Controller Achieving Desired Orientation*

To achieve the desired orientation, a controller is presented to track the last joints $\boldsymbol{q}_o$ of the robot's kinematic chain accordingly. For this purpose, the Jacobian transpose controller is used and applied to the problem formulation [57]. This implies less computational effort compared to the Jacobian inverse controller commonly used in robotics and can be utilized to cross kinematic singularities [69]. When solving the NMPC formulated in Section 3.2, a joint configuration $\boldsymbol{q}_m(t_\delta)$ for the first m robot joints of the kinematic chain is obtained for each iteration step. These define the orientation of the $\{m\}$ frame at the robot's elbow, which results in the rotation matrix $R_w^m(\boldsymbol{q}_m)$. This matrix can be used in (3) to determine the deviation matrix $R_m^f(\boldsymbol{q}_m)$ between the desired end-effector orientation, transformed to the flange $\{f\}$ and the current wrist orientation governed by $\boldsymbol{q}_m$. Therefore, the corresponding unit quaternions $[\eta_m^f(\boldsymbol{q}_m), (\zeta_m^f(\boldsymbol{q}_m))^{\mathrm{T}}]^{\mathrm{T}}$ can be derived [83,84]. They specify the desired unit quaternions with respect to the $\{m\}$ frame depending on the displacement of the $\{\varsigma\}$ system at the wrist point performed by the NMPC. From the joint angles $\boldsymbol{q}_o(t_\delta)$ at sampling step t_δ, the current $[\eta_m^f(\boldsymbol{q}_o), (\zeta_m^f(\boldsymbol{q}_o))^{\mathrm{T}}]^{\mathrm{T}}$ unit quaternions can be calculated. The orientation error:

$$\tilde{\boldsymbol{e}}_m^f = \eta_m^f(\boldsymbol{q}_o)\,\zeta_m^f(\boldsymbol{q}_m) - \eta_m^f(\boldsymbol{q}_m)\,\zeta_m^f(\boldsymbol{q}_o) - S(\zeta_m^f(\boldsymbol{q}_m))\,\zeta_m^f(\boldsymbol{q}_o) \in \mathbb{R}^3 \tag{11}$$

between these unit quaternions can be inferred, where the skew symmetric operator [85] reads

$$S(s_1, s_2, s_3) = \begin{bmatrix} 0 & -s_3 & s_2 \\ s_3 & 0 & -s_1 \\ -s_2 & s_1 & 0 \end{bmatrix} \in \mathbb{R}^{3\times 3}. \tag{12}$$

It should be emphasized that $\eta = 1$ holds true for the real part of the unit quaternions when the orientation is aligned, and thus, the orientation error in (11) can be expressed as a 3D quantity [86]. Using

$$\dot{\boldsymbol{q}}_o = \left(J_{m,\mathrm{rot}}^f\right)^{\mathrm{T}} K\,\tilde{\boldsymbol{e}}_m^f, \tag{13}$$

the feedback is imposed, including the positive definite matrix $K \in \mathbb{R}^{o\times 3}$ and the Jacobian determined in Section 2.2. The weighting matrix K is bounded to the sample time and influences the speed of convergence. The required joint angles $\boldsymbol{q}_o$ to control the robot are obtained by the integration of (13). In order to analyze the stability of the orientation controller, the Lyapunov function:

$$V = \left(\eta_m^f(\boldsymbol{q}_m) - \eta_m^f(\boldsymbol{q}_o)\right)^2 + \left(\zeta_m^f(\boldsymbol{q}_m) - \zeta_m^f(\boldsymbol{q}_o)\right)^{\mathrm{T}}\left(\zeta_m^f(\boldsymbol{q}_m) - \zeta_m^f(\boldsymbol{q}_o)\right) \tag{14a}$$

is considered. After substituting the propagation equations for quaternions [86] into the rate of change:

$$\dot{V} = -\left(\tilde{\boldsymbol{e}}_m^f\right)^{\mathrm{T}} K\,\tilde{\boldsymbol{e}}_m^f \tag{14b}$$

of (14a), the asymptotic stability of the orientation controller can be concluded. Thus, the controller converges to the desired orientation and is able to cross singularities, whereas, in contrast to the Jacobian inverse, it may deviate during the transition phase [58,69].

4. Simulation Results and Evaluation

To show that standard computer hardware is sufficient for the online calculations of the introduced NMPC for the translational motion including Cartesian and joint constraints coupled with the Jacobian transpose controller for the orientation correction, a standard computer with 16GB RAM and an Intel Core i7-8550U processor running Linux Ubuntu 18.04 was utilized. The iterative solution of the OCP is calculated using the MATLAB-interface [87] from CASADI [88] with the interior-point (IPOPT) algorithm [89]. As can be seen in Figure 2, the standard industrial 6-DOF robot Stäubli TX2-90 with a black rod as the end-effector and the DH parameters from Table 2 was used to present the introduced approach. Based on the decomposition performed in Section 2.2, both controllers consider different kinematic chains, respectively. Thus, the associated DH parameters to the $\{m\}$ system and the wrist point p_w^ς are also listed in the table. The $n = 6$ revolute joints are equally partitioned with $m = 3$ and $o = 3$ for the translational and rotational controllers.

Table 2. Denavit–Hartenberg (DH) parameters of the considered 6-DOF industrial Stäubli TX2-90 manipulator.

i	a_i (mm)	α_i (rad)	d_i (mm)	θ_i (rad)
w	0	0	−478	0
1	50	$-\pi/2$	478	q_1
2	425	0	0	$q_2 - \pi/2$
$3/m$	0	$\pi/2$	50	$q_3 + \pi/2$
ς	425	0	50	q_3
4	0	$-\pi/2$	425	q_4
5	0	$\pi/2$	0	q_5
f	0	0	100	q_6
e	0	0	150	0

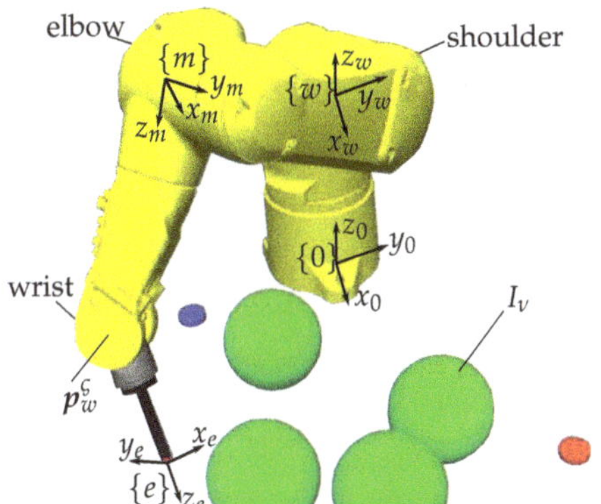

Figure 2. Standard industrial 6-DOF manipulator Stäubli TX2-90 [63] with an end-effector, visualized in GAZEBO [64,90]. The exemplary movement starts at the blue marker and ends at the red one. The green obstacles are only considered in Sections 4.3 and 4.4, but are not present in Section 4.2.

The evaluation consisted of four different demonstrations to highlight the performance of the NMPC approach based on the decomposed robot model. In Section 4.1, the computation times of the introduced approach are evaluated and compared to an NMPC considering the full robot system. This highlights the significant difference in the computational costs between those approaches. Despite the decomposition of the model and the controller couplings, no losses in the applicability by the approach occur, which is shown in the following three evaluations. The trajectory tracking accuracy in the undisturbed case is discussed in Section 4.2 to show that the method can be used as an online motion controller. Subsequently, in Section 4.3, obstacles are placed in the environment. In the presented scenario, the end-effector must be guided between them without causing collisions. The obstacles are not taken into account for the trajectory planning described in Section 3.1, but will be avoided by the online controller introduced in Section 3.2, which perceives them only within its predictive horizon. Finally, in Section 4.4, moving obstacles are considered and the collision-free guidance of the end-effector in this setup is investigated.

4.1. Quantitative Analysis of the Computation Times

In the analysis of the reduction in the computation times, the presented approach was compared to an NMPC that considers the full robotic model. Using the Stäubli TX2-90, the NMPC based on the full system utilizes $n = 6$ joints as decision variables in each optimization step and incorporates both translation and orientation by $J_w^e(\boldsymbol{q})$ from (5). On the contrary, the decomposed system requires only $m = 3$ decision variables in its OCP and controls the orientation in parallel with the remaining three DOFs using the coupled Jacobian transpose controller. It should be emphasized that (10c) must be extended in the NMPC of the full system to include orientation as well. The objective functions of the two systems differ, but each was designed for the quantitative comparison. As listed in Table 3, three different scenarios consisting of no obstacle, one obstacle, and one obstacle including height constraints were compared to analyze the computation times. Additionally, two different prediction horizons $N_1 = 100\,\text{ms}$ and $N_2 = 200\,\text{ms}$ partitioned with $k_1 = 10$ and $k_2 = 20$ grid points were considered, achieving discrete intervals with a length of 10 ms, respectively. The discretization of the inputs to be determined corresponds to the update rate of the robot controller. Various converging point-to-point (PTP) movements covering the entire workspace of the robot were run multiple times, and the average computation time $\bar{t}$ per optimization step was recorded. This time and the standard deviation σ, which expresses the fluctuation of $\bar{t}$ required for one OCP, denotes the online capability of the NMPC. Note that the optimization was carried out until an optimal solution was found, but could be further shortened by limiting the maximum iterations, as done in [91]. Here, the stop condition for the objective function (10b) was set to a tolerance of 10^{-8}, and if below this limit, the value did not change more than 10^{-6}, indicating a minimum.

Table 3. Comparison of the averaged computation times $\bar{t}$ with the standard deviation σ per optimization step of the respective NMPC.

Point-to-Point Movement		Decomp. System $\bar{t} \pm \sigma$ (ms)	Full System $\bar{t} \pm \sigma$ (ms)
without obstacles	N_1	26 ± 1	130 ± 12
	N_2	30 ± 2	216 ± 21
with obstacle	N_1	27 ± 2	153 ± 53
	N_2	33 ± 2	209 ± 9
with obstacle and height constraints	N_1	29 ± 2	230 ± 136
	N_2	36 ± 6	284 ± 112

From Table 3, the comparison between the decomposed and the full system shows that the decomposed system requires only 10 % to 20 % of $\bar{t}$ to achieve an optimal solution and possesses lower deviations σ, independent of the scenario or the prediction horizon. Both $\bar{t}$ and σ are important factors to be considered using NMPC in varying environments. The Jacobian transpose controller evaluates the orientation error in each iteration and is only constrained to the gain matrix K.

It can be seen that the average computation times $\bar{t}$ for the full system increased significantly with the complexity of the scenario. This effect did not occur with the decomposed model, as the full system involves more nonlinearities, which must be taken into account to solve the OCP. On the one hand, higher computational costs yield lower possible update rates of the NMPC, which restricts the ability to act in rapidly varying environments. On the other hand, it is evident from Table 3 that N_2 increased the computation times for the full system by up to 67 %, in contrast to $\bar{t}$, when N_1 was chosen. This means that the choice of the prediction horizon limits the online capabilities. Using the decomposed robot model approach, the evaluation times increased by a maximum of 24 % using N_2 instead of N_1. Thus, the comparison of both the absolute and the relative computation times revealed a significantly higher performance of the presented decomposition-based method.

A second important factor is σ, which is a measure of the reliability to achieve $\bar{t}$. Smaller standard deviations, even between different scenarios, indicate that $\bar{t}$ is more likely to be achieved. In contrast to the NMPC considering the full system, the lower σ of the decomposed system also allows for easier applicability to different tasks, since the NMPC does not converge for an unexpectedly long time in a more complex scenario. The choice of the control horizon is determined by the length of expected calculation times and should be kept as short as possible. Referring to Table 3, the control horizon can be set more reliably using the decomposed approach. The NMPC controller remains capable in online operation without delaying the robot's motion, resulting in $\bar{t}$ being larger than the set control horizon in the implementation.

4.2. *Trajectory Tracking Accuracy of the Controller*

In Section 4.1, the significant reduction of the computational costs is presented. Furthermore, it is shown that this did not lead to any restrictions in the motion behavior of the robot. The simulative setup for evaluating the introduced approach involving the NMPC and the orientation controller was built in GAZEBO [64]. The joint position controlled robot shown in Figure 2 is commanded by means of ROS [65,90]. The prediction horizon was chosen as $N = 100\,\text{ms}$ with $k = 10$ grid points per iteration, while the control horizon involved four discretization steps of 10 ms each. This means that, for all four consecutive updates of the robot commands, the solution from the buffer was used before being updated. Considering the average computation times from the previous subsection, this allows for online calculations without delaying the robot's motion due to the too long computations solving the OCP. Finally, the gain matrix for the orientation control was set to $K = \text{diag}(20, 20, 20)$, and the weights of the NMPC's running costs in (10c) were chosen to be $\boldsymbol{\mu} = [10^4, 10^2, 10^4]^\text{T}$.

We omit the comparison with the full system in the following evaluations, on the one hand, for the sake of readability and, on the other hand, to avoid having an unfair comparison realized. As shown in Section 4.1, no delay-free execution can be realized for the chosen N using the full system, which distorts the comparison. Depending on the controller settings, we observed only minor to no deviations between the results in internal comparisons, depending on the scenario.

As shown in Figure 2, the end-effector has to move from the blue marker with $\boldsymbol{p}^e_{w,0} = [110, -350, -405]^\text{T}\text{mm}$ at t_0 to the red marker with $\boldsymbol{p}^e_{w,\text{des}} = [780, 390, -405]^\text{T}\text{mm}$ at t_1. None of the green obstacles are considered in this subsection when performing the trajectory tracking analysis, and they are only drawn in preparation for the next scenario. The desired orientation was set to $R^e_{w,\text{des}} = \text{diag}(-1, 1, -1)$, meaning that the end-effector has to point vertically downwards. However, all other orientations reachable in the manipulator's workspace can also be realized. As explained in Section 3.3, the Jacobian transpose controller is asymptotically stable and does not induce singularities in individual joint configurations, e.g., compared to the Jacobian inverse controller. As illustrated in Figure 3b with dashed lines, the set point change of the desired position using the polynomial approach from Section 3.1 starts at $t_0 = 0.5\,\text{s}$ and ends at $t_1 = 2\,\text{s}$. The demonstration scenario involving a short transition time $T = 1.5\,\text{s}$ and a long path is representative for movements between all reachable poses in the workspace of the manipulator. If T is not set, the desired terminal pose $H^e_{w,\text{des}}$, will be approached by minimizing the objective function within the NMPC, just bounded to the given OCP constraints.

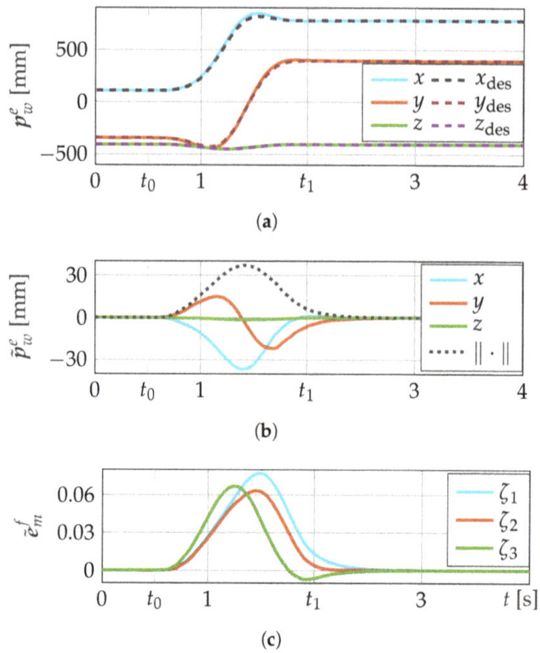

Figure 3. Trajectory of the end-effector starting at $t_0 = 0.5$ s and ending at $t_1 = 2$ s in an undisturbed environment for tracking accuracy analysis in 3D space. (**a**) Trajectory tracking of the pre-planned trajectory marked by the subscript "des", which is planned in the joint space and transformed into the Cartesian space. (**b**) Absolute displacement of the end-effector $\tilde{\boldsymbol{p}}_w^e = \boldsymbol{p}_{w,\text{des}}^e - \boldsymbol{p}_w^e$ to the reference trajectory and the individual translational parts depending on the length $d_e = 150$ mm of the end-effector. (**c**) Orientation error (11) in quaternion representation.

The pre-planned trajectory is generated by using the polynomial depicted in (7) in the joint space. The computation time of approximately 1 ms required for this involving the automatic joint configuration selection ensures an almost instantaneous start. Subsequently, this is transformed to the Cartesian space. As can be seen in Figure 3a, the reference trajectory exhibits rounded deviations, for example at $t = 1.2$ s in $p_{w,z,\text{des}}^e$, compared to a trajectory that would be directly planned in the Cartesian space, because the joints are actuated uniformly over T here. In turn, the evaluation of the orientation by, e.g., roll-pitch-yaw [74] is omitted by using the joint space, which could be singular in the representation.

In the evaluation of the end-effector's translational deviation, both controllers must be taken into account. It should be noted that the end-effector position is composed of the positioning of the wrist point by the NMPC and the alignment by the orientation controller. Both a too slow control of the wrist point and an incorrect orientation of the end-effector would lead to a deviation from the end-effector's reference trajectory. In Figure 3b, the individual error components of $\tilde{\boldsymbol{p}}_w^e = \boldsymbol{p}_{w,\text{des}}^e - \boldsymbol{p}_w^e$ and the absolute distance $\|\tilde{\boldsymbol{p}}_w^e\|$ to the reference trajectory at each time step are shown. Despite the short transition time T and the long displacement along the trajectory, resulting in a rapid change of poses, only small deviations can be detected. Compared to a common path tracking task, it must be taken into account that, in the analysis of the trajectory tracking accuracy, a slight lag also leads to notable deviations. As can be seen in Figure 3b, especially the errors of $p_{w,x}^e$ and $p_{w,y}^e$ exhibit small deviations, which converge to zero in the end, so that no stationary error remains. The small lag in the $x_w y_w$-plane during the motion results from the parameterized smoothness of the orientation controller, since it must perform the rotation correction in each iteration step due to its constantly shifting reference $\{m\}$ system. The NMPC places the wrist point $\boldsymbol{p}_w^\zeta$ very accurately so that the reference system of the upper kinematic chain, used for the orientation control, is moved further and further by the NMPC. Therefore, a permanent adaptation in (11) governed by K is necessary. The rotation error is shown in Figure 3c, where each of the imaginary unit quaternion error components can take a maximum value of one. Thus, it can be seen that the orientation error was very small in this case. Even though, the end-effector is chosen to be relatively long with $d_e = 150\,\text{mm}$. As a result, a larger deviation was enforced for a better illustration here. If d_e is chosen shorter, the amplitudes in Figure 3b decrease. In total, just small deviations from the pre-planned trajectory and, thus, accurate trajectory tracking can be observed when using the presented approach.

4.3. Trajectory Control in Disturbed Environment With Fixed Obstacles

In the evaluation involving obstacles, the same control setup as in Section 4.2 was utilized, but as illustrated in Figure 2, the scenario now included $\nu = 4$ obstacles and $\boldsymbol{p}^e_{w,0} = [110, -350, -445]^\mathrm{T}\text{mm}$ and $\boldsymbol{p}^e_{w,\text{des}} = [780, 390, -445]^\mathrm{T}\text{mm}$ were set 40 mm lower in the z_w-direction. This small lowering of the reference trajectory would cause ground contact, which should be prevented by the controller. Starting from the blue marker in Figure 4, the first obstacle was placed close to the reference trajectory so that the boundary condition (10f) had to consider the mentioned safety distance, since the NMPC has no knowledge about the orientation controller, which adjusts the desired orientation. In the extreme case, when the end-effector would be vertical, the NMPC should directly leave the reference trajectory to avoid collisions. As shown in Figure 4, the two consecutive obstacles on the left-hand side are crossed by the blue reference and disturb the tracking of the pre-planned trajectory in $x_w y_w$-plane. Additionally, the central obstacle (I_ν) presents a difficulty in conjunction with the height constraint (10g), since the NMPC must deviate significantly from the reference and take a remarkable detour to reach $H^e_{w,\text{des}}$. The NMPC only considers the obstacles within the prediction horizon and has no information about them before. In the accompanying video [92], the orientation error and the wrist point tracking are also shown, in addition to the executed robot movements. For the sake of readability, the evaluation is omitted in this section and reference is made to Section 4.2.

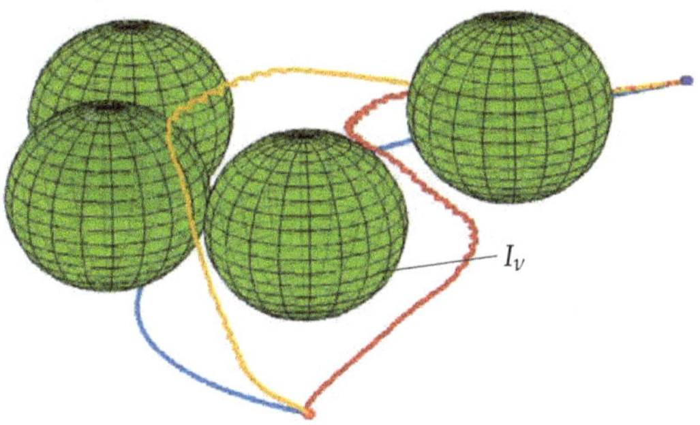

Figure 4. Resulting trajectories of the end-effector $\boldsymbol{p}^e_w(t)$ governed by the NMPC and orientation controller in the scenario from Figure 2. The motion starts at the blue marker and ends at the red marker. The reference trajectory (blue) according to (7) crosses the obstacles (green). Without a height constraint for $p^{\varsigma}_{w,z}$, the motion results in an upward swerve (yellow). Activating (10g), the spheres are avoided in the $x_w y_w$-plane (red).

From Figure 5a, the trajectory of the end-effector can be taken in the case where the wrist point $\boldsymbol{p}^{\varsigma}_w$ is only constrained by $p^{\varsigma}_{w,z,\text{min}} = -228\,\text{mm}$ in (10g) involving no upper height limit, so that the ground will not be touched. A deviation from the dashed reference trajectory due to the obstacle avoidance can be seen. Especially with respect to $p^e_{w,x}$ and $p^e_{w,y}$, it is obvious that the trajectory controller tries to follow the reference trajectory under consideration of the given constraints, but a delay is recognizable. From Figure 4, it becomes even clearer that the yellow trajectory in the $x_w y_w$-plane follows the arc of the blue reference quiet accurately. The obstacles are avoided by swerving in the z_w-direction, which is confirmed by the green line in Figure 5a. Due to the chosen IPOPT algorithm and depending on the length of the control horizon, which has to be chosen according to the computation times of the controller's online calculations, small repeated repulsions of the end-effector can be detected in Figure 4 while avoiding the obstacles tightly.

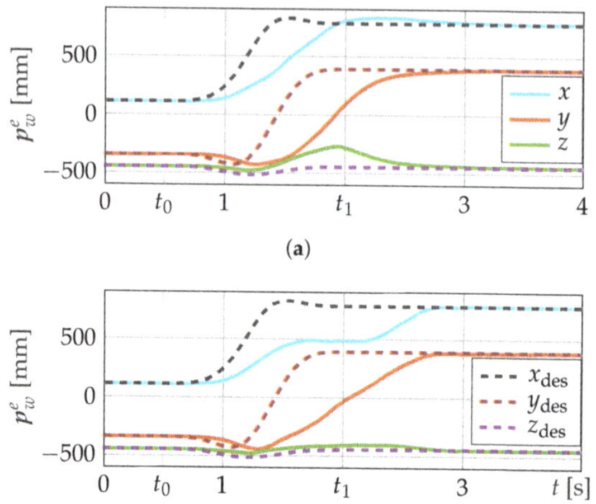

(**a**)

(**b**)

Figure 5. Trajectory tracking of the blue curve in Figure 4 starting at $t_0 = 0.5\,\mathrm{s}$ and ending at $t_1 = 2\,\mathrm{s}$ in a disturbed environment involving $\nu = 4$ fixed obstacles. (**a**) No upper height limitation of the Cartesian space. Analogous to the yellow trajectory in Figure 4, the manipulator moves over the obstacles. (**b**) Constraining the height by $-228\,\mathrm{mm} \leq p^{\varsigma}_{w,z} \leq -145\,\mathrm{mm}$ in (10g) for obstacle avoidance in the $x_w y_w$-plane, as the red curve in Figure 4.

By reducing the computational costs, short evaluation times of the NMPC can be achieved, even if additional constraints are inserted, which further influence the robot's behavior. This means that, by decomposing the differential kinematics, not only an accurate controller can be designed, but also, it can be used more flexibly. In order to demonstrate this, the maximum height in the Cartesian space was constrained in the further analysis. The height constraint $-228\,\mathrm{mm} \leq p^{\varsigma}_{w,z} \leq -145\,\mathrm{mm}$ of the wrist's workspace forces the controller to avoid the obstacles by a planar motion. The lifting of the end-effector is thus suppressed. Therefore, $H^e_{w,\mathrm{des}}$ can just be approached by a significant deviation from the reference trajectory, mainly disturbed by the central obstacle (I_ν). A noticeable change in the movement compared to the dashed lines can be noticed at Figure 5b. Even though, the motion has to be adapted and is thus slightly delayed. The online applicability of the approach is still valid. For a better interpretation, the corresponding course is illustrated as the red path in Figure 4. This shows that the introduced approach is able to control the standard industrial robot in disturbed environments.

4.4. Trajectory Control in a Varying Environment with Moving Obstacles

Based on the evaluation of the controller in a disturbed environment, the same setup as shown in Figures 2 and 4 with $p^e_{w,0} = [110, -350, -445]^{\mathrm{T}}\mathrm{mm}$ and $p^e_{w,\mathrm{des}} = [780, 390, -445]^{\mathrm{T}}\mathrm{mm}$ was utilized subsequently. The Jacobian transpose controller continued to align the end-effector downward. However, the $\nu = 4$ obstacles were in motion here, and thus, they represent a varying environment. Again, the wrist point was constrained in height using $-228\,\mathrm{mm} \leq p^{\varsigma}_{w,z} \leq -145\,\mathrm{mm}$ to avoid an upward swerve. Two different scenarios were examined to demonstrate the resulting behavior of the controller. First, all obstacles moved uniformly in one direction, continuously blocking the corridor realized Figure 5b after the end-effector deviated from the reference trajectory. Subsequently, only the central obstacle I_ν moved, which caused a dead end for a short time.

Both scenarios were chosen such that the obstacles force the NMPC to adjust the movement and the end-effector must depart from the desired trajectory analogous to Section 4.3. The reference trajectory shown in Figure 5 and, thus, the motion of the robot starts at $t_0 = 0.5\,\mathrm{s}$ and ends at $t_1 = 2\,\mathrm{s}$.

A uniform movement of the obstacles simulated the driving of the robot, as is common, e.g., in agricultural or industrial applications. Starting at $t = 0\,\mathrm{s}$, the four obstacles moved uniformly with $30\,\mathrm{mm\,s^{-1}}$ in the negative x_w-direction and with $45\,\mathrm{mm\,s^{-1}}$ in the y_w-direction, so that they moved diagonally towards the robot. The $\{w\}$ system can be taken from Figure 2, which is aligned with the axes of the $\{0\}$ frame. The change in position of the moving obstacles relative to the yellow robot can be seen in Figure 6a–g. The safe distance introduced in (10f) is illustrated in orange. The trajectory of the online controlled movement can be taken from Figure 7a. For comparison, the resulting trajectories from the previous subsection are also plotted in Figure 7 and marked with the subscript "fix". Due to the height constraint of the wrist in the NMPC, any adjustment of the movement in the z_w-direction was excluded. However, compared to the static scenario, further adjustments were conducted in the $x_w y_w$-plane. The end-effector was also guided into the corridor between the obstacles after passing the first one from the robot's point of view, as the rear ones prevent the direct tracking of the trajectory.

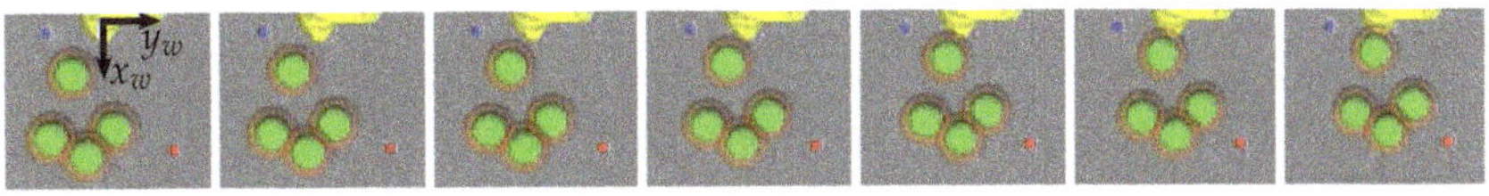

(a) $t = 0\,\mathrm{s}$ **(b)** $t = 0.5\,\mathrm{s}$ **(c)** $t = 1\,\mathrm{s}$ **(d)** $t = 1.5\,\mathrm{s}$ **(e)** $t = 2\,\mathrm{s}$ **(f)** $t = 2.5\,\mathrm{s}$ **(g)** $t = 3\,\mathrm{s}$

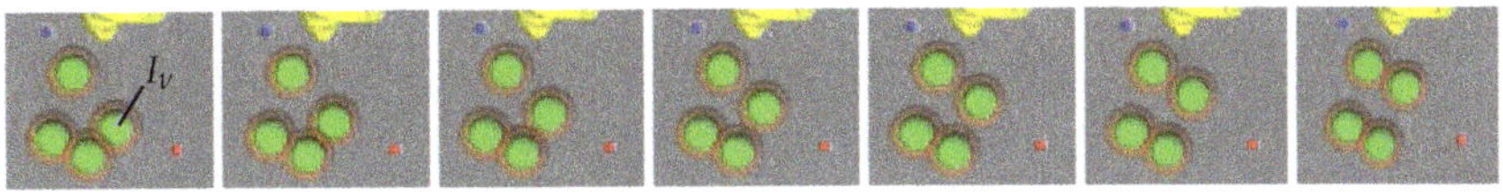

(h) $t = 0\,\mathrm{s}$ **(i)** $t = 0.5\,\mathrm{s}$ **(j)** $t = 1\,\mathrm{s}$ **(k)** $t = 1.5\,\mathrm{s}$ **(l)** $t = 2\,\mathrm{s}$ **(m)** $t = 2.5\,\mathrm{s}$ **(n)** $t = 3\,\mathrm{s}$

Figure 6. The positions of the moving obstacles (green) at different time steps. The safe distance (orange) from (10f) and the robot (yellow) are highlighted for clarification. Referring to the $\{w\}$ system in Figure 2, **(a–g)** illustrate the uniform movement of all obstacles with $30\,\mathrm{mm\,s^{-1}}$ in the negative x_w-direction and with $45\,\mathrm{mm\,s^{-1}}$ in the y_w-direction. In **(h–n)**, the displacement of the central obstacle I_v with $60\,\mathrm{mm\,s^{-1}}$ in the negative x_w-direction is visualized.

Due to the displacement of the obstacles, the end-effector can be shifted earlier in the y_w-direction, which can be observed from the comparison of the curves visualizing $p^e_{w,y}$. At about $t = 1.7\,\mathrm{s}$, the NMPC must further adjust the motion to avoid a collision with the central obstacle marked by I_v in Figure 2 and Figure 4, respectively. Therefore, the end-effector is initially pulled closer to the base, which is apparent from $p^e_{w,x}$, and then, a fast movement is performed to pass I_v. The increase in velocity can be seen in Figure 7a at about $t = 2.4\,\mathrm{s}$ by the larger slope of $p^e_{w,x}$ and $p^e_{w,y}$. The desired terminal pose is reached insignificantly later without collision.

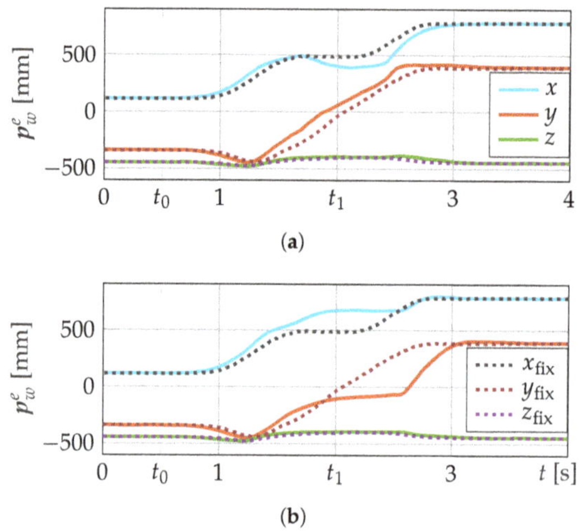

Figure 7. Trajectory tracking of the blue curve in Figure 4 starting at $t_0 = 0.5$ s and ending at $t_1 = 2$ s in a disturbed environment involving moving obstacles and an active height constraint. The directions of the obstacles' movement can be taken from Figure 2, and for comparison, the subscript "fix" indicates the end-effector's motion in the statically disturbed environment from Figure 5b. (**a**) All obstacles move uniformly with 30 mm s^{-1} in the negative x_w-direction and with 45 mm s^{-1} in the y_w-direction. (**b**) The central obstacle (I_v) moves with 60 mm s^{-1} in the negative x_w-direction, so that a dead end is created briefly.

In the second scenario evaluating the decomposition-based controller in a varying environment, only the central obstacle (I_v) moved with 60 mm s^{-1} in the negative x_w-direction, and the others were fixed. Figure 6h–n demonstrate the movement of I_v. The corresponding trajectory of the end-effector is shown in Figure 7b. It can be seen that the beginning of the movement was similar to the motion performed in the static disturbed environment. The NMPC departed from the blue reference trajectory in Figure 4 to move through the corridor between the obstacles. However, the movement of I_v was chosen in such a way that this briefly formed a dead end in combination with the other obstacles, while the end-effector tried to pass it. This led to a deceleration of the movement between approximately $t = 1.4$ s and $t = 1.9$ s, since $\boldsymbol{p}^e_{w,\text{des}}$ cannot be reached at that moment. Due to the round shape of the obstacles, a small adjustment of the motion can be detected from that time in $p^e_{w,y}$ as I_v continued to move on. From about $t = 2.6$ s, the corridor between I_v and the rear obstacles opened enough to continue the motion without any collisions, so that the desired terminal pose was reached. This demonstrates that the presented optimal control approach can be used in disturbed and varying environments.

5. Discussion

In this paper an approach that reduces the computational costs of NMPC was introduced and applied for online trajectory control in disturbed and varying environments. For this purpose, the differential kinematics was decomposed and partitioned into a translational and rotational part related to the Cartesian space containing the corresponding robot joints. The differential kinematics was considered since it can be modeled based on the DH parameters from the robot's data sheets, and thus, the approach is also applicable to position controlled industrial robots. The decomposition-based approach can be applied to all robot types that can be partitioned in this way. The translational motion control was used for obstacle avoidance. For this purpose, an OCP was introduced and implemented as the NMPC, which considered both Cartesian and joint constraints. Due to the reduced number of decision variables in the OCP, additional constraints to adjust the robot's behavior, such as the height of the workspace, can be included without significantly increasing the computation times. The NMPC moved the wrist point of the anthropomorphic robot, which led to a change in the orientation of the end-effector. For the correction of the orientation, the Jacobian transpose controller was introduced and applied to the problem using unit quaternions, avoiding singularities. In addition to the online motion control, trajectory planning with an automatic selection of the best joint configuration was introduced, eliminating the need for manual input, as usually required for PTP motions.

In the evaluation, the approach was compared in terms of the computational costs with an NMPC that considered the full robot model involving all DOFs. The comparison demonstrated that the required times in the computation of the OCP were significantly reduced by the introduced method. For the analysis, different scenarios and parameters were considered. A comparison with other approaches has been omitted here, since higher computation times resulted from the references presented in the introduction, e.g., in [23–25,48], and the consideration of the full system is most common.

The analysis of the tracking accuracy and the delay-free online control in a disturbed or varying environment with (moving) obstacles was performed in simulations. The implementation was carried out using a standard computer, MATLAB, ROS, and CasADi, although the performance in hardware and software can be further increased by replacing these tools. Larger prediction horizons and differently shaped obstacles can also be considered. However, this work serves as a proof of concept and was intended to show the possibilities of this decomposition-based approach. The extensive evaluation revealed that the coupled controllers precisely followed a trajectory and adapted the motion to the environment. This resulted in an optimal controller setup that considered external constraints with high precision and without limiting the workspace. The focus was placed on the NMPC, although the Jacobian transpose controller was also considered in the evaluation, but this has been extensively analyzed in other publications [58,69]. Further, it was examined whether the orientation was also implemented by a second NMPC whose initial conditions were given by the translational NMPC.

This contribution introduced the concept of the control architecture and evaluated it by simulations. This demonstrated the precision of the method and serves as a basis for further developments. Besides the replacement of MATLAB by another programming language, improvements will be made to the individual components for the transfer to a real experiment. ROS is suitable for the communication with Gazebo, but for a better performance, this will be replaced by EtherCAT [93], as, e.g., implemented in [94] for the control of a Stäubli TX2-90. The obstacles are detected by motion tracking, and their positions can be updated in (10f) at a high rate. Alternative solvers such as ACADOS [95,96] or GraMPC [97] are evaluated either on a Linux server or on a PLC to further reduce the computation times.

6. Conclusions and Future Work

This contribution exhibited that, by decomposing the differential kinematics of an anthropomorphic robot, the computational costs of NMPC can be significantly reduced with basically no effect on the solution's accuracy and reliability. This seemingly small adjustment has a huge impact on the computational effort and demonstrates an approach that addresses the cause, not the symptoms, of long NMPC computation times. By reducing the decision variables in the OCP, optimal online trajectory control in disturbed and varying environments for (standard industrial) robots is possible. The NMPC for the translational motion of the end-effector was coupled with a Jacobian transpose controller for the orientation correction, so that all DOFs of the robot were used. There were no special requirements for the control hardware, and a standard computer was sufficient for the NMPC evaluations. The simulation results showed that an online implementation for NMPC in the field of robotics has been elaborated without limiting the workspace due to the model's decomposition. This opens up the possibility of using standard industrial robots in various areas and applications, where many sensor data have to be processed or the interaction with a dynamically varying environment is required. The evaluation of the computation times, the tracking accuracy, the control in a disturbed environment with additional height constraints, and the trajectory adaptation in a varying environment demonstrated the performance of the approach.

The method offers many possibilities in terms of extension and transferability. In addition to self-collision avoidance, the concept can also be used to interact with objects due to the short evaluation times of the NMPC. Further on, the approach will be brought to a real experiment and coupled with a force control, so that, e.g., haptic grasping can be implemented in a disturbed environment.

Author Contributions: Conceptualization, J.R., H.B. and T.M.; methodology, J.R.; software, J.R. (decomposed system and evaluation) and H.B. (full system for comparison); validation, J.R. and H.B.; formal analysis, J.R. and H.B.; investigation, J.R. and H.B.; resources, T.M.; data curation, J.R.; writing—original draft preparation, J.R.; writing—review and editing, J.R., H.B. and T.M.; visualization, J.R.; supervision, J.R. and T.M.; project administration, J.R. and T.M.; funding acquisition, T.M. All authors have read and agreed to the published version of the manuscript.

Funding: The authors gratefully acknowledge the financial support by the federal state of Schleswig-Holstein within the funding programme Open Access Publikationsfonds. Furthermore, this research received no external funding.

Data Availability Statement: The data presented in this study are available upon request from the corresponding author.

Conflicts of Interest: The authors declare no conflict of interest.

Abbreviations

The following abbreviations are used in this manuscript:

3D	three-dimensional
des	desired
DH	Denavit–Hartenberg
DOF	degree of freedom
e.g.,	for example
Fig.	Figure
fix	fixed
IPOPT	interior-point
NMPC	nonlinear model predictive control
OCP	optimal control problem
Tab.	Table

References

1. Bootz, J.P.; Michel, S.; Pallud, J.; Monti, R. Possible changes of Industry 4.0 in 2030 in the face of uberization: Results of a participatory and systemic foresight study. *Technol. Forecast. Soc. Chang.* **2022**, *184*, 121962. [CrossRef]
2. Vaisi, B. A review of optimization models and applications in robotic manufacturing systems: Industry 4.0 and beyond. *Decis. Anal. J.* **2022**, *2*, 100031. [CrossRef]
3. Edwards, M. Robots in industry: An overview. *Appl. Ergon.* **1984**, *15*, 45–53. [CrossRef] [PubMed]
4. Singh, S.; Singh, J.; Shah, B.; Sehra, S.S.; Ali, F. Augmented Reality and GPS-Based Resource Efficient Navigation System for Outdoor Environments: Integrating Device Camera, Sensors, and Storage. *Sustainability* **2022**, *14*, 12720. [CrossRef]
5. Grau, A.; Indri, M.; Lo Bello, L.; Sauter, T. Robots in Industry: The Past, Present, and Future of a Growing Collaboration with Humans. *IEEE Ind. Electron. Mag.* **2021**, *15*, 50–61. [CrossRef]
6. Liu, F.; Wang, C.; Luo, M.; Zhou, S.; Liu, C. An investigation of the coupling coordination of a regional agricultural economics-ecology-society composite based on a data-driven approach. *Ecol. Indic.* **2022**, *143*, 109363. [CrossRef]
7. Spykman, O.; Emberger-Klein, A.; Gabriel, A.; Gandorfer, M. Autonomous agriculture in public perception-German consumer segments' view of crop robots. *Comput. Electron. Agric.* **2022**, *202*, 107385. [CrossRef]
8. Nguyen, T.H.; Nguyen, T.N.; Ngo, B.V. A VGG-19 Model with Transfer Learning and Image Segmentation for Classification of Tomato Leaf Disease. *AgriEngineering* **2022**, *4*, 871–887. [CrossRef]
9. Junior, L.C.M.; Alfredo C. Ulson, J. Real Time Weed Detection using Computer Vision and Deep Learning. In Proceedings of the 2021 14th IEEE International Conference on Industry Applications (INDUSCON), São Paulo, Brazil, 15–18 August 2021; pp. 1131–1137.

10. Wendel, A.; Underwood, J. Self-supervised weed detection in vegetable crops using ground based hyperspectral imaging. In Proceedings of the 2016 IEEE International Conference on Robotics and Automation (ICRA), Stockholm, Sweden, 16–21 May 2016; pp. 5128–5135.
11. Bručienė, I.; Buragienė, S.; Šarauskis, E. Weeding Effectiveness and Changes in Soil Physical Properties Using Inter-Row Hoeing and a Robot. *Agronomy* **2022**, *12*, 1514. [CrossRef]
12. Abdulsalam, M.; Aouf, N. Deep Weed Detector/Classifier Network for Precision Agriculture. In Proceedings of the 2020 28th Mediterranean Conference on Control and Automation (MED), Saint-Raphaël, France, 16–19 June 2020; pp. 1087–1092.
13. Gasparetto, A.; Boscariol, P.; Lanzutti, A.; Vidoni, R. Path Planning and Trajectory Planning Algorithms: A General Overview. *Mech. Mach. Sci.* **2015**, *29*, 3–27.
14. Gasparetto, A.; Boscariol, P.; Lanzutti, A.; Vidoni, R. Trajectory Planning in Robotics. *Math. Comput. Sci.* **2012**, *6*, 269–279. [CrossRef]
15. Meysami, A.; Cuillière, J.C.; François, V.; Kelouwani, S. Investigating the Impact of Triangle and Quadrangle Mesh Representations on AGV Path Planning for Various Indoor Environments: With or Without Inflation. *Robotics* **2022**, *11*, 50. [CrossRef]
16. Yanbin, L.; Jiang, Y. Robotic Path Planning Based on a Triangular Mesh Map. *Int. J. Control. Autom. Syst.* **2020**, *18*, 2658–2666.
17. Hart, P.E.; Nilsson, N.J.; Raphael, B. A Formal Basis for the Heuristic Determination of Minimum Cost Paths. *IEEE Trans. Syst. Sci. Cybern.* **1968**, *4*, 100–107. [CrossRef]
18. Elbanhawi, M.; Simic, M. Sampling-Based Robot Motion Planning: A Review. *IEEE Access* **2014**, *2*, 56–77. [CrossRef]
19. Hayat, A.A.; Sadanand, R.O.M.; Saha, S.K. Robot Manipulation through Inverse Kinematics. In Proceedings of the Proceedings of the 2015 Conference on Advances In Robotics, Goa, India, 2–4 July 2015; Association for Computing Machinery: New York, NY, USA, 2015; AIR '15.
20. Djojo, M.A.; Karyono, K. Computational load analysis of Dijkstra, A*, and Floyd-Warshall algorithms in mesh network. In Proceedings of the 2013 International Conference on Robotics, Biomimetics, Intelligent Computational Systems, Jogjakarta, Indonesia, 25–27 November 2013; pp. 104–108.
21. Reinhold, J.; Olschewski, J.; Lippross, S.; Meurer, T. Image-Based Online Command Adaptation and Guidance to Arbitrarily Shaped Objects for Robot-Assisted Medical Procedures. In Proceedings of the 2021 IEEE/RSJ International Conference on Intelligent Robots and Systems (IROS), Prague, Czech Republic, 27 September–1 October 2021; pp. 9241–9246.
22. Ratliff, N.; Zucker, M.; Bagnell, J.A.D.; Srinivasa, S. CHOMP: Gradient Optimization Techniques for Efficient Motion Planning. In Proceedings of the Proceedings of (ICRA) International Conference on Robotics and Automation, Kobe, Japan, 12–17 May 2009; pp. 489–494.
23. Kalakrishnan, M.; Chitta, S.; Theodorou, E.; Pastor, P.; Schaal, S. STOMP: Stochastic trajectory optimization for motion planning. In Proceedings of the 2011 IEEE International Conference on Robotics and Automation, Shanghai, China, 9–13 May 2011; pp. 4569–4574.
24. Schulman, J.; Duan, Y.; Ho, J.; Lee, A.; Awwal, I.; Bradlow, H.; Pan, J.; Patil, S.; Goldberg, K.; Abbeel, P. Motion planning with sequential convex optimization and convex collision checking. *Int. J. Robot. Res.* **2014**, *33*, 1251–1270. [CrossRef]
25. Hu, S.; Babaians, E.; Karimi, M.; Steinbach, E. NMPC-MP: Real-time Nonlinear Model Predictive Control for Safe Motion Planning in Manipulator Teleoperation. In Proceedings of the 2021 IEEE/RSJ International Conference on Intelligent Robots and Systems (IROS), Prague, Czech Republic, 27 September–1 October 2021; pp. 8309–8316.
26. Mnih, V.; Kavukcuoglu, K.; Silver, D.; Rusu, A.A.; Veness, J.; Bellemare, M.G.; Graves, A.; Riedmiller, M.; Fidjeland, A.K.; Ostrovski, G.; et al. Human-level control through deep reinforcement learning. *Nature* **2015**, *518*, 529–533. [CrossRef]
27. Bristow, D.; Tharayil, M.; Alleyne, A. A survey of iterative learning control. *IEEE Control Syst. Mag.* **2006**, *26*, 96–114.
28. Longman, R.W. Iterative learning control and repetitive control for engineering practice. *Int. J. Control* **2000**, *73*, 930–954. [CrossRef]
29. Cuiyan, L.; Dongchun, Z.; Xianyi, Z. A survey of repetitive control. In Proceedings of the 2004 IEEE/RSJ International Conference on Intelligent Robots and Systems (IROS) (IEEE Cat. No.04CH37566), Sendai, Japan, 28 September–2 October 2004; Volume 2, pp. 1160–1166.
30. Pane, Y.; Nageshrao, S.; Kober, J.; Babuska, R. Reinforcement learning based compensation methods for robot manipulators. *Eng. Appl. Artif. Intell.* **2019**, *78*, 236–247. [CrossRef]
31. Grüne, L. NMPC without terminal constraints. *IFAC Proc. Vol.* **2012**, *45*, 1–13. [CrossRef]
32. Grüne, L.; Pannek, J.; Seehafer, M.; Worthmann, K. Analysis of Unconstrained Nonlinear MPC Schemes with Time Varying Control Horizon. *SIAM J. Control. Optim.* **2010**, *48*, 4938–4962. [CrossRef]
33. Schwenzer, M.; Ay, M.; Bergs, T.; Abel, D. Review on model predictive control: An engineering perspective. *Int. J. Adv. Manuf. Technol.* **2021**, *117*, 1327–1349. [CrossRef]
34. Drgoňa, J.; Arroyo, J.; Cupeiro Figueroa, I.; Blum, D.; Arendt, K.; Kim, D.; Ollé, E.P.; Oravec, J.; Wetter, M.; Vrabie, D.L.; et al. All you need to know about model predictive control for buildings. *Annu. Rev. Control* **2020**, *50*, 190–232. [CrossRef]
35. Findeisen, R.; Allgöwer, F. An Introduction to Nonlinear Model Predictive Control. In Proceedings of the 21st Benelux Meeting on Systems and Control, Veldhoven, The Netherlands, 19–21 March 2002.
36. Qin, S.J.; Badgwell, T.A. An Overview of Nonlinear Model Predictive Control Applications. In Proceedings of the Nonlinear Model Predictive Control; Allgöwer, F., Zheng, A., Eds.; Birkhäuser Basel: Basel, Switzerland, 2000; pp. 369–392.

37. Al Younes, Y.; Barczyk, M. Nonlinear Model Predictive Horizon for Optimal Trajectory Generation. *Robotics* **2021**, *10*, 90. [CrossRef]
38. Arbo, M.H.; Grøtli, E.I.; Gravdahl, J.T. Mid-Level MPC and 6 DOF output path following for robotic manipulators. In Proceedings of the 2017 IEEE Conference on Control Technology and Applications (CCTA), Maui, HI, USA, 27–30 August 2017; pp. 450–456.
39. Bai, G.; Meng, Y.; Liu, L.; Luo, W.; Gu, Q.; Liu, L. Review and Comparison of Path Tracking Based on Model Predictive Control. *Electronics* **2019**, *8*, 1077. [CrossRef]
40. Ozana, S.; Krupa, F.; Nemcik, J.; Takacs, G.; Mozaryn, J.; Beremlijski, P.; Singh, A.; Kawala-Sterniuk, A. Design and Implementation of Adaptive NMPC in industrial applications: Case study. In Proceedings of the 2022 23rd International Carpathian Control Conference (ICCC), Sinaia, Romania, 29 May–1 June 2022; pp. 12–18.
41. Leva, A.; Formentin, S.; Seva, S. Overlapping-Horizon MPC: A Novel Approach to Computational Constraints in Real-Time Predictive Control. In Proceedings of the Third Workshop on Next Generation Real-Time Embedded Systems (NG-RES 2022), Budapest, Hungary, 22 June 2022; Bertogna, M., Terraneo, F., Reghenzani, F., Eds.; Schloss Dagstuhl–Leibniz-Zentrum für Informatik: Dagstuhl, Germany, 2022; Volume 98.
42. Gil, P.; Henriques, J.; Dourado, A.; Duarte-Ramos, H. Non-Linear Predictive Control Based On A Recurrent Neural Network. In Proceedings of the ESIT'99, European Symposium on Intelligent Techniques, Platanias, Greece, 3–4 June 1999; p. 6.
43. van Duijkeren, G.; Verschueren, R.; Pipeleers, G.; Diehl, M.; Swevers, J. Path-following NMPC for serial-link robot manipulators using a path-parametric system reformulation. In Proceedings of the 2016 European Control Conference (ECC), Aalborg, Denmark, 29 June–1 July 2016; pp. 477–482.
44. Cisneros, P.G.; Werner, H. Fast Nonlinear MPC for Reference Tracking Subject to Nonlinear Constraints via Quasi-LPV Representations. *IFAC-PapersOnLine* **2017**, *50*, 11601–11606. [CrossRef]
45. Erez, T.; Lowrey, K.; Tassa, Y.; Kumar, V.; Kolev, S.; Todorov, E. An integrated system for real-time model predictive control of humanoid robots. In Proceedings of the 2013 13th IEEE-RAS International Conference on Humanoid Robots (Humanoids), Atlanta, GA, USA, 15–17 October 2013; pp. 292–299.
46. Nakanishi, J.; Cory, R.; Mistry, M.; Peters, J.; Schaal, S. Operational Space Control: A Theoretical and Empirical Comparison. *Int. J. Robot. Res.* **2008**, *27*, 737–757. [CrossRef]
47. Faulwasser, T.; Kern, B.; Findeisen, R. Model predictive path-following for constrained nonlinear systems. In Proceedings of the Proceedings of the 48h IEEE Conference on Decision and Control (CDC) Held Jointly with 2009 28th Chinese Control Conference, Shanghai, China, 15–18 December 2009; pp. 8642–8647.
48. Faulwasser, T.; Weber, T.; Zometa, P.; Findeisen, R. Implementation of Nonlinear Model Predictive Path-Following Control for an Industrial Robot. *IEEE Trans. Control. Syst. Technol.* **2017**, *25*, 1505–1511. [CrossRef]
49. Mazdarani, H.; Farrokhi, M. Adaptive neuro-predictive control of robot manipulators in work space. In Proceedings of the 2012 17th International Conference on Methods & Models in Automation & Robotics (MMAR), Miedzyzdroje, Poland, 27–30 August 2012; pp. 349–354.
50. Zhou, B.; Yang, L.; Wang, C.; Chen, Y.; Chen, K. Inverse Jacobian Adaptive Tracking Control of Robot Manipulators with Kinematic, Dynamic, and Actuator Uncertainties. *Complexity* **2020**, *2020*, 5070354. [CrossRef]
51. Dulęba, I.; Opalka, M. A comparison of Jacobian-based methods of inverse kinematics for serial robot manipulators. *Int. J. Appl. Math. Comput. Sci.* **2013**, *23*, 373–382. [CrossRef]
52. Hu, Y.; Su, H.; Zhang, L.; Miao, S.; Chen, G.; Knoll, A. Nonlinear Model Predictive Control for Mobile Robot Using Varying-Parameter Convergent Differential Neural Network. *Robotics* **2019**, *8*, 64. [CrossRef]
53. Faulwasser, T.; Findeisen, R. Nonlinear Model Predictive Control for Constrained Output Path Following. *IEEE Trans. Autom. Control* **2016**, *61*, 1026–1039. [CrossRef]
54. Ding, Y.; Wang, L.; Li, Y.; Li, D. Model predictive control and its application in agriculture: A review. *Comput. Electron. Agric.* **2018**, *151*, 104–117. [CrossRef]
55. Mavrogiannis, C.I.; Liarokapis, M.V.; Kyriakopoulos, K.J. Quantifying anthropomorphism of robot arms. In Proceedings of the 2015 IEEE/RSJ International Conference on Intelligent Robots and Systems (IROS), Hamburg, Germany, 28 September–2 October 2015; pp. 4084–4089.
56. Jazar, R.N. Inverse Kinematics. In *Theory of Applied Robotics: Kinematics, Dynamics, and Control*; Springer: Boston, MA, USA, 2007; pp. 263–296.
57. Taki, S.; Nenchev, D. A novel singularity-consistent inverse kinematics decomposition for S-R-S type manipulators. In Proceedings of the 2014 IEEE International Conference on Robotics and Automation (ICRA), Hong Kong, China, 31 May–7 June 2014; pp. 5070–5075.
58. Moosavian, S.A.A.; Papadopoulos, E. Modified transpose Jacobian control of robotic systems. *Automatica* **2007**, *43*, 1226–1233. [CrossRef]
59. Pires, J.N.; Loureiro, A.; Bölmsjo, G. *Welding Robots*; Springer: London, UK, 2006.
60. Maurtua, I.; Ibarguren, A.; Kildal, J.; Susperregi, L.; Sierra, B. Human–robot collaboration in industrial applications: Safety, interaction and trust. *Int. J. Adv. Robot. Syst.* **2017**, *14*, 1729881417716010. [CrossRef]
61. Schmitz, A. Human-Robot Collaboration in Industrial Automation: Sensors and Algorithms. *Sensors* **2022**, *22*, *5848*. [CrossRef] [PubMed]
62. Murray, R.M.; Zexiang, L.; Shankar, S. *A Mathematical Introduction to Robotic Manipulation*; CRC Press: Boca Raton, FL, USA, 1994.

63. Stäubli International AG. TX2-90 Industrial Robot Range. Available online: https://www.staubli.com/en/robotics/product-range/industrial-robots/6-axis-robots/tx2-90/ (accessed on 1 January 2021).
64. Koenig, N.; Howard, A. Design and use paradigms for Gazebo, an open-source multi-robot simulator. In Proceedings of the 2004 IEEE/RSJ International Conference on Intelligent Robots and Systems (IROS) (IEEE Cat. No.04CH37566), Sendai, Japan, 28 September–2 October 2004; Volume 3, pp. 2149–2154.
65. Stanford Artificial Intelligence Laboratory. Robotic Operating System (ROS) Melodic Morenia. Available online: https://www.ros.org (accessed on 1 February 2021).
66. Lüling, N.; Reiser, D.; Stana, A.; Griepentrog, H. Using depth information and colour space variations for improving outdoor robustness for instance segmentation of cabbage. In Proceedings of the 2021 IEEE International Conference on Robotics and Automation (ICRA), Xi'an, China, 30 May–5 June 2021; pp. 2331–2336.
67. Diankov, R. Automated Construction of Robotic Manipulation Programs. Ph.D. Thesis, Carnegie Mellon University, Pittsburgh, PA, USA, 2010. AAI3448143.
68. Xie, S.; Sun, L.; Wang, Z.; Chen, G. A speedup method for solving the inverse kinematics problem of robotic manipulators. *Int. J. Adv. Robot. Syst.* **2022**, *19*, *17298806221104602*. [CrossRef]
69. Siciliano, B.; Sciavicco, L.; Villani, L.; Oriolo, G. *Robotics-Modelling, Planning and Control*; Springer: Berlin/Heidelberg, Germany, 2009.
70. Alavala, C. Difference between Denavit-Hartenberg (D-H) Classical and Modified Conventions for Forward Kinematics of Robots with Case Study. In Proceedings of the International Conference on Advanced Materials and manufacturing Technologies, Solo, Indonesia, 16–17 September 2014.
71. McCarthy, J. *Introduction to Theoretical Kinematics*; MIT Press: Cambridge, MA, USA, 1990.
72. Chevallereau, C.; Khalil, W. A new method for the solution of the inverse kinematics of redundant robots. In Proceedings of the Proceedings. 1988 IEEE International Conference on Robotics and Automation, Philadelphia, PA, USA, 24–29 April 1988; Volume 1, pp. 37–42.
73. Wang, X.; Zhang, D.; Zhao, C. The inverse kinematics of a 7R 6-degree-of-freedom robot with non-spherical wrist. *Adv. Mech. Eng.* **2017**, *9*, 1687814017714985. [CrossRef]
74. Corke, P. *Robotics, Vision and Control-Fundamental Algorithms in MATLAB*; Springer International Publishing: Berlin/Heidelberg, Germany, 2017.
75. Rülling, K.; Schröer, S. Algebraic loop groups. *arXiv* **2021**, arXiv:2109.11364.
76. Görner, M.; Haschke, R.; Ritter, H.; Zhang, J. MoveIt! Task Constructor for Task-Level Motion Planning. In Proceedings of the 2019 International Conference on Robotics and Automation (ICRA), Montreal, QC, Canada, 20–24 May 2019; pp. 190–196.
77. Ott, C. *Cartesian Impedance Control of Redundant and Flexible-Joint Robots*, 1st ed.; Springer Publishing Company, Incorporated: Berlin/Heidelberg, Germany, 2008.
78. Zube, A. Cartesian nonlinear model predictive control of redundant manipulators considering obstacles. In Proceedings of the 2015 IEEE International Conference on Industrial Technology (ICIT), Seville, Spain, 17–19 March 2015; pp. 137–142.
79. Zhang, X.; Liniger, A.; Borrelli, F. Optimization-Based Collision Avoidance. *IEEE Trans. Control. Syst. Technol.* **2020**, *29*, 972–983. [CrossRef]
80. Jiménez, P.; Thomas, F.; Torras, C. 3D collision detection: A survey. *COmputers Graph.* **2001**, *25*, 269–285. [CrossRef]
81. Grüne, L.; Pannek, J. *Nonlinear Model Predictive Control. Theory and Algorithms*; Springer: London, UK, 2011.
82. Diehl, M.; Bock, H.G.; Diedam, H.; Wieber, P.B. Fast Direct Multiple Shooting Algorithms for Optimal Robot Control. In *Proceedings of the Fast Motions in Biomechanics and Robotics*; Springer: Berlin/Heidelberg, Germany, 2005.
83. Sarabandi, S.; Thomas, F. Accurate Computation of Quaternions from Rotation Matrices. In *Proceedings of the Advances in Robot Kinematics 2018*; University of Bologna, Italy; Lenarcic, J., Parenti-Castelli, V., Eds.; Springer International Publishing: Cham, Switzerland, 2019; pp. 39–46.
84. Shuster, M.D. The nature of the quaternion. *J. Astronaut. Sci.* **2008**, *56*, 359–373. [CrossRef]
85. Andrle, M.S.; Crassidis, J.L. Geometric Integration of Quaternions. *J. Guid. Control Dyn.* **2013**, *36*, 1762–1767. [CrossRef]
86. Chiaverini, S.; Siciliano, B. The Unit Quaternion: A Useful Tool for Inverse Kinematics of Robot Manipulators. *Syst. Anal. Model. Simul.* **1999**, *35*, 45–60.
87. MATLAB. *Version 9.8.0.1323502 (R2020a)*; The MathWorks Inc.: Natick, MA, USA, 2020.
88. Andersson, J.A.E.; Gillis, J.; Horn, G.; Rawlings, J.B.; Diehl, M. CasADi—A software framework for nonlinear optimization and optimal control. *Math. Program. Comput.* **2019**, *11*, 1–36. [CrossRef]
89. Nocedal, J.; Wright, S.J. *Numerical Optimization*, 2nd ed.; Springer: New York, NY, USA, 2006.
90. vd. Hoorn, G. (TU Delft Robotics Institute). Staubli_Experimental. Available online: https://wiki.ros.org/staubli_experimental (accessed on 10 September 2020).
91. Mavrommati, A.; Osorio, C.; Valenti, R.G.; Rajhans, A.; Mosterman, P.J. An Application of Model Predictive Control to Reactive Motion Planning of Robot Manipulators. In Proceedings of the 2021 IEEE 17th International Conference on Automation Science and Engineering (CASE), Lyon, France, 23–27 August 2021; pp. 915–920.
92. Reinhold, J.; Baumann, H.; Meurer, T. mdpi-robotics-kard-2022_acon-kiel-janr_nmpc-decomp-robot.mp4. Available online: https://cloud.rz.uni-kiel.de/index.php/s/xJLRyg8WFYTTTDM (accessed on 15 November 2022).
93. EtherCAT Technology Group. EtherCAT—The Ethernet Fieldbus. Available online: https://www.ethercat.org/ (accessed on 1 January 2022).

94. Reinhold, J.; Elsner, L.; Meurer, T. Force-Based Pick-and-Place Along Curved Surfaces Exploiting Robot Repeatability. In Proceedings of the ISR Europe 2022; 54th International Symposium on Robotics, Munich, Germany, 20–21 June 2022; pp. 1–8.
95. Verschueren, R.; Frison, G.; Kouzoupis, D.; Frey, J.; van Duijkeren, N.; Zanelli, A.; Novoselnik, B.; Albin, T.; Quirynen, R.; Diehl, M. acados—A modular open-source framework for fast embedded optimal control. *Math. Program. Comput.* **2021**, *14*, 147–183. [CrossRef]
96. Reinhold, J.; Amersdorfer, M.; Meurer, T. A Dynamic Optimization Approach for Sloshing Free Transport of Liquid Filled Containers using an Industrial Robot. In Proceedings of the 2019 IEEE/RSJ International Conference on Intelligent Robots and Systems (IROS), Macau, China, 3–8 November 2019; pp. 2336–2341.
97. Käpernick, B.; Graichen, K. The gradient based nonlinear model predictive control software GRAMPC. In Proceedings of the 2014 European Control Conference (ECC), Strasbourg, France, 24–27 June 2014; pp. 1170–1175.

robotics

MDPI

Article

Application of Half-Derivative Damping to Cartesian Space Position Control of a SCARA-like Manipulator

Luca Bruzzone * and Shahab Edin Nodehi

DIME—Department of Mechanical, Energy, Management and Transportation Engineering, 16145 Genoa, Italy
* Correspondence: luca.bruzzone@unige.it; Tel.: +39-010-3352967

Abstract: In classical Cartesian space position control, KD, the end-effector follows the set-point trajectory with a stiffness expressed in the directions of the external coordinates through the stiffness matrix, K, and with a damping proportional to the first-order derivatives of errors of the external coordinates through the damping matrix, D. This work deals with a fractional-order extension of the Cartesian space position control, KDHD, which is characterized by an additional damping term, proportional to the half-order derivatives of the errors of the external coordinates through a second damping matrix, HD. The proposed Cartesian position control scheme is applied to a SCARA-like serial manipulator with elastic compensation of gravity. Multibody simulation results show that the proposed scheme was able to reduce the tracking error, in terms of mean absolute value of the end-effector position error and Integral Square Error, with the same amount of Integral Control Effort and comparable maximum actuation torques.

Keywords: Cartesian space control; fractional calculus; half-order derivative; KDHD; SCARA

Citation: Bruzzone, L.; Nodehi, S.E. Application of Half-Derivative Damping to Cartesian Space Position Control of a SCARA-like Manipulator. *Robotics* **2022**, *11*, 152. https://doi.org/10.3390/robotics11060152

Academic Editor: Raffaele Di Gregorio

Received: 7 November 2022
Accepted: 14 December 2022
Published: 16 December 2022

1. Introduction

In robotics and mechatronics, the mathematical concepts of derivatives and integrals are used both to express the dynamics of continuous-time mechanical systems, and to define their control strategies. While dynamic modelling usually exploits only integer-order (IO) derivatives and integrals, in the conception of a control algorithm it is also possible to adopt fractional-order (FO) derivatives and integrals to improve the closed-loop dynamic performance.

Fractional calculus (FC) deals with the concepts of non-integer-order derivatives and integrals. Although its origin dates back to the seventeenth century, only in the last decades has there been a renewed and growing interest in FC, also motivated by research on chaos theory [1]. FC is now used in physics [2], biology [3] and medicine [4], and is particularly effective in modeling multi-scale phenomena [5].

In addition to its application in the basic sciences, FC is applied in technological applications, principally for control system design. The PID scheme is undoubtedly the most widespread control approach for any kind of closed-loop system, and it is based on the evaluation of the first-order time derivative and integral of the error. Consequently, the most intuitive way to apply FC to the PID scheme is to generalize the orders of integration and derivation, giving rise to the well-known $PI^{\lambda}D^{\mu}$, also known as FOPID (fractional-order PID). FOPID was first proposed by Podlubny [6] for single-input single-output (SISO) systems, and its features, benefits and possible tuning criteria are extensively discussed in the scientific literature [7,8].

Instead of replacing first-order terms with FO ones, as in FOPID, an alternative way to apply FC to control synthesis is to add FO terms with intermediate orders to the first-order ones, as in the $PDD^{1/2}$ [9] and $PII^{1/2}DD^{1/2}$ [10] proposed by Bruzzone et al., which can be included in the category of distributed-order PID (DOPID) controls, recently defined by Jakovljevic et al. [11]. In the $PDD^{1/2}$ control scheme, without integral action, the classical PD control is modified by introducing a half-derivative term, proportional to the derivative of order 1/2 of the error; in the $PII^{1/2}DD^{1/2}$, with integral action, a half-integral term is also present, proportional to the integral of order 1/2 of the error. Simulations and experimental tests show that the introduction of the half-integral and half-derivative terms make it possible to reduce the error in the position control of mechatronic axes, with similar values of maximum control output [10] or settling energy [12].

Moving from single-input single-output (SISO) systems to multi-input multi-output (MIMO) robotic systems, FC can be applied to impedance control and Cartesian space position control [13,14]. In general, impedance controllers use the Jacobian matrix to transform the Cartesian generalized forces, expressed in the directions of the external coordinates of the end-effector, into the required joint generalized forces. If it is necessary to impose a compliant behavior to the end-effector in some directions in order to control the contact forces with the environment, the corresponding elements of the stiffness matrix (K) are lowered; in contrast, high stiffness is imposed for the directions which must be accurately position-controlled. For all directions, the damping matrix D adds a proper damping, proportional to the first-order derivatives of the errors of the external coordinates. If all the external coordinates must be position-controlled, impedance control becomes a PD-type Cartesian space position control, in which stiffness and damping are defined in the Cartesian space rather than in the joint space [15].

FC can be used to obtain alternative definitions of the damping term of impedance/Cartesian space position control, in order to improve the system behavior. FO extensions of impedance control were recently proposed in [16,17], in which the end-effector damping is not proportional to the first-order derivative of the position error, as usual, but to a FO derivative with generic order, thus generalizing to MIMO systems the PD^{μ} scheme for SISO systems. In contrast, the FO extension of impedance control which generalizes the $PDD^{1/2}$ scheme to MIMO systems, named KDHD, was proposed in [18] for a purely translational parallel robot and in [19] for a six-degrees-of-freedom parallel robot. In the KDHD scheme, a half-derivative damping term defined by the matrix HD is added to the stiffness and first-order damping terms, defined by the matrices K and D.

In the present paper the KDHD algorithm is applied in simulation to a serial SCARA-like RRFbR robot with elastic balancing, focusing on Cartesian space position control, with high stiffness for all the external coordinates. We propose a methodology to derive the tuning of the KDHD control starting from a given KD control. Simulation results show that the KDHD scheme can be used to reduce the trajectory tracking error with equal Integral Control Effort and comparable maximum values of the joint torques.

There are two main novelties of this paper with respect to [18,19]. The first is related to the robotic architecture: the proposed control is applied to a serial robot instead of a parallel robot; moreover, elastic balancing of gravity is present. The second is related to the aim of the KDHD algorithm: Cartesian space position control instead of impedance control, without interactions with the environment and primarily analyzing the trajectory tracking error.

In the remainder of this paper, Section 2 describes the RRFbR SCARA-like serial architecture and its main kinematic and dynamic properties; Section 3 discusses the KDHD control of the RRFbR robot, highlighting the differences with respect to the classical KD control; Section 4 discusses the multibody model used for the KD–KDHD comparison; Section 5 presents the simulation results of the comparison and proposes a method to derive the KDHD tuning from a given KD and Section 6 outlines conclusions and future developments.

2. The RRFbR SCARA-like Manipulator with Elastic Balancing

The considered RRFbR SCARA-like architecture derives from the RRPR SCARA architecture [20], replacing the third prismatic joint (P) with a four-bar mechanism (Fb) with parallelogram shape placed in a vertical plane, which provides the vertical mobility of the end-effector. The advantages of this solution are the elimination of the prismatic joint, with the consequent friction issues, and the ease of adding mass balancing or elastic balancing of the gravitational forces [21].

In this work, the RRFbR scheme with elastic balancing shown in Figure 1 is considered. Similar to the SCARA, this robot performs a 4-DOF motion with three translations and one rotation around a vertical axis (Schoenflies motion, [22]), suitable for a wide range of industrial applications.

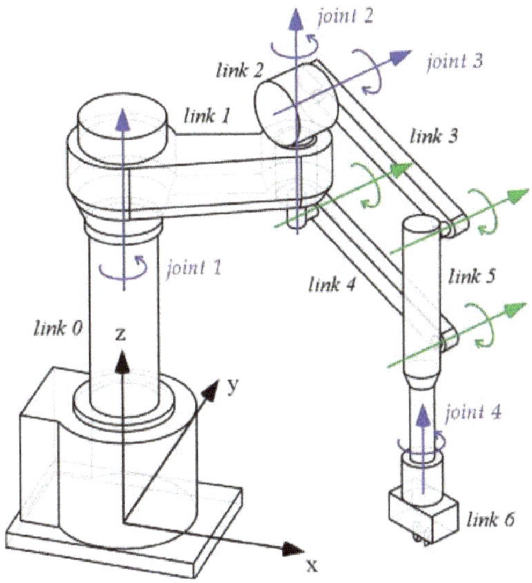

Figure 1. The RRFbR SCARA-like architecture with elastic balancing (blue: actuated revolute joints; green: passive revolute joints).

The RRFbR architecture comprises seven links, from the base (link 0) to the end-effector (link 6). In Figure 1 the actuated joints (joint 1 to joint 4) are represented in blue. Joint 3 is the only actuated revolute joint of the four-bar mechanism; the remaining three passive revolute joints of the four-bar mechanism are represented in green. The axes of the actuated revolute joints 1, 2 and 4 are vertical; consequently, the corresponding motors are not loaded by gravity. On the contrary, the actuator of joint 3 is loaded by the gravitational forces acting on links 3 to 6. Therefore, to add a partial static balancing, reducing the torque of actuator 3 in steady state, a torsional spring can be added on joint 3, acting in parallel with actuator 3.

An alternative method to obtain static balancing involves adding a counterweight placed on link 3. Mass balancing and elastic balancing of the RRFbR architecture are compared in [21]. Mass balancing has the advantage of being exact for all angular positions of link 3, while elastic balancing is exact only for one angular position of actuator 3, and only approximated in the others. Nevertheless, elastic balancing reduces the inertial forces and avoids the encumbrance of the counterweight, with possible interferences with link 1 and with the environment; therefore, in the following only elastic balancing is considered.

Figure 2 shows the simplified kinematic scheme of the RRFbR architecture with elastic balancing. The internal coordinates are collected in the vector $\boldsymbol{\theta}$:

$$\boldsymbol{\theta} = \begin{bmatrix} \theta_1 & \theta_2 & \theta_3 & \theta_4 \end{bmatrix}^T, \tag{1}$$

where θ_i is the angular position of the i-th actuated joint. The vector of the external coordinates collects the three components of the end-effector reference point E with respect to the origin of the fixed reference frame O(x,y,z) and the angle of rotation θ of the end-effector around the z axis (Figure 2):

$$\mathbf{x} = \begin{bmatrix} x & y & z & \theta \end{bmatrix}^T. \tag{2}$$

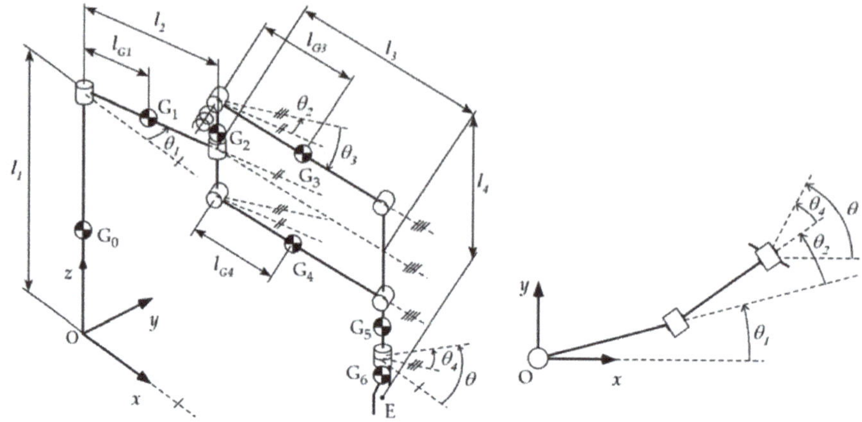

Figure 2. Kinematic scheme of the RRFbR architecture with elastic balancing.

The direct position kinematics is expressed by the following expressions, which can be obtained from the geometry represented in Figure 2:

$$x = l_2 \cos(\theta_1) + l_3 \cos(\theta_3) \cos(\theta_1 + \theta_2), \tag{3}$$

$$y = l_2 \sin(\theta_1) + l_3 \cos(\theta_3) \sin(\theta_1 + \theta_2), \tag{4}$$

$$z = l_1 - l_3 \sin(\theta_3) - l_4, \tag{5}$$

$$\theta = \theta_1 + \theta_2 + \theta_4. \tag{6}$$

Deriving Equations (3)–(6) with respect to time, it is possible to obtain the Jacobian matrix, Equation (7), which expresses the direct velocity kinematics; in Equation (7), for brevity, s_i indicates $\sin\theta_i$, c_i indicates $\cos\theta_i$ and s_{ij} indicates $\sin(\theta_i + \theta_j)$:

$$\dot{\mathbf{x}} = \mathbf{J}\dot{\boldsymbol{\theta}},$$
$$\mathbf{J} = \begin{bmatrix} -l_2 s_1 - l_3 c_3 s_{12} & -l_3 c_3 s_{12} & -l_3 s_3 c_{12} & 0 \\ l_2 c_1 + l_3 c_3 c_{12} & l_3 c_3 c_{12} & -l_3 s_3 s_{12} & 0 \\ 0 & 0 & -l_3 c_3 & 0 \\ 1 & 1 & 0 & 1 \end{bmatrix} \tag{7}$$

In [21], the detailed dynamic model of the RRFbR architecture without static balancing is obtained by means of the Lagrange equations, neglecting friction in the joints:

$$\tau_i = \frac{d}{dt}\frac{\partial L}{\partial \dot{\theta}_i} - \frac{\partial L}{\partial \theta_i} + \sum_{j=1}^{4} J_{ji} F_j, \tag{8}$$

where τ_i is the torque of the i-th actuator; L is the Lagrangian function, which is the difference of the kinetic energy E_c and of the gravitational energy E_g; J_{ji} is the element (j,i) of the Jacobian matrix $\mathbf{J}$ and F_j is the j-th element of the vector $\mathbf{F}$ of the generalized forces applied by the end-effector on the environment, in the directions of the four external coordinates:

$$\mathbf{F} = \begin{bmatrix} F_x & F_y & F_z & M_z \end{bmatrix}^T. \tag{9}$$

The expressions of the actuator torques (8), available in [21], are not reported here for the sake of brevity. The x and y components of the moment applied to the end-effector are supported by the joints and, in absence of friction, do not influence the actuation torques.

Introducing elastic balancing on joint 3, the third equation of system (8) becomes:

$$\begin{aligned}\tau_3 &= \frac{d}{dt}\frac{\partial L}{\partial \dot{\theta}_3} - \frac{\partial L}{\partial \theta_3} + \sum_{j=1}^{4} J_{j3}F_j + \tau_{el,3}(\theta_3) \\ &= \frac{d}{dt}\frac{\partial L}{\partial \dot{\theta}_3} - \frac{\partial L}{\partial \theta_3} + \sum_{j=1}^{4} J_{j3}F_j + k_3(\theta_3 - \theta_{3p}),\end{aligned} \tag{10}$$

where k_3 is the torsional spring stiffness and θ_{3p} is the angular position of joint 3 at which the elastic return force of the spring is null. A good equilibration can be obtained imposing that in static conditions, with links 3 and 4 horizontal (θ_3 = 0) and without payload, the spring moment exactly balances the gravitational effects acting on the robot arm, obtaining τ_3 = 0; this can be obtained with a negative value of θ_{3p}. This condition will be applied in the following.

Let us note that, depending on constructive requirements, the balancing spring can be placed on the actuated joint 3 or on another passive revolute joint of the four-bar mechanism without influencing the dynamic model; it is also possible to use more springs in parallel, distributing the stiffness on the four-bar joints. However, placing elastic elements only on the revolute joints connected to link 2, closer to the base, is favorable because it reduces the arm inertia.

3. KDHD Control of the RRFbR Robot

The classical KD impedance control/Cartesian space control with gravity compensation of the RRFbR arm is expressed by the following control law:

$$\boldsymbol{\tau} = \mathbf{J}^T(\boldsymbol{\theta})\left(\mathbf{K}_{KD}(\mathbf{x}_d - \mathbf{x}(\boldsymbol{\theta})) + \mathbf{D}_{KD}(\mathbf{x}_d - \mathbf{x}(\mathbf{q}))^{(1)}\right) + \boldsymbol{\tau}_g(\boldsymbol{\theta}). \tag{11}$$

In Equation (11), the vector $\boldsymbol{\tau}$ collects the four actuation torques ($\tau_1 \ldots \tau_4$); $\mathbf{x}_d$ is the time-varying vector of the set-point trajectory expressed in the external coordinates; the superscript (i) indicates the *i*-th-order time derivative; $\boldsymbol{\tau}_g(\boldsymbol{\theta})$ is the vector of the gravity compensation torques and $\mathbf{K}_{KD}$ and $\mathbf{D}_{KD}$ are the stiffness and damping matrices, which express the desired linear end-effector compliance.

In general, the matrices $\mathbf{K}_{KD}$ and $\mathbf{D}_{KD}$, on the basis of the robot mobility, define the translational impedance, the rotational impedance or both [23]. In case of robots with Schoenflies motion, such as the RRFbR arm, their size is 4 × 4, and it is reasonable that the 1-DOF rotational behavior is decoupled from the translational behavior. Consequently, $\mathbf{K}_{KD}$ and $\mathbf{D}_{KD}$ are block-diagonal, with a 3 × 3 submatrix representing the translational impedance and the fourth diagonal element representing the rotational impedance.

Since gravity acts only on joint 3, as discussed in Section 2, all the elements of $\boldsymbol{\tau}_g(\boldsymbol{\theta})$ are null except the third, which has the following expression:

$$\tau_{g,3}(\theta_3) = -(m_3 l_{G3} + m_4 l_{G4} + l_3(m_5 + m_6))g\cos\theta_3 + k_3(\theta_3 - \theta_{3p}), \tag{12}$$

where m_i is the mass of link *i*, l_{G3} and l_{G4} define the positions of the centers of mass of links 3 and 4, G_3 and G_4, and l_3 is the length of the links 3 and 4 (Figure 2). In Equation (12) there is the sum of two terms; the second term takes into account the elastic return force of the balancing spring: the first term (negative, since the torque required to compensate gravity is opposite to the positive direction of θ_3) is partially compensated by the action of the balancing spring, as desired, so the necessary gravity compensation torque $\tau_{g,3}$ is reduced by the second term, usually positive, since $\theta_{3p} < 0$, as discussed in Section 2. Imposing exact static balancing ($\tau_{g,3}$ = 0) when links 3 and 4 are horizontal (θ_3 = 0) in Equation (12) leads to the following relationship, which can be used to define the spring parameters k_3 and θ_{3p}:

$$\theta_{3p} = -(m_3 l_{G3} + m_4 l_{G4} + l_3(m_5 + m_6))g/k_3. \tag{13}$$

The KDHD control law is derived from the KD control law (11), with the addition of a damping term proportional through the half-derivative damping matrix $\mathbf{HD}_{KDHD}$ to the half-derivative (derivative of order 1/2) of the external coordinates error:

$$\boldsymbol{\tau} = \mathbf{J}^T(\boldsymbol{\theta})\left(\mathbf{K}_{KDHD}(\mathbf{x}_d - \mathbf{x}(\boldsymbol{\theta})) + \mathbf{D}_{KDHD}(\mathbf{x}_d - \mathbf{x}(\mathbf{q}))^{(1)} + \mathbf{HD}_{KDHD}(\mathbf{x}_d - \mathbf{x}(\mathbf{q}))^{(1/2)}\right) + \boldsymbol{\tau}_g(\boldsymbol{\theta}). \quad (14)$$

Accordingly, the KD and KDHD control laws, (11) and (14), respectively extend the PD and $PDD^{1/2}$ schemes from one dimension to the four-dimensional space of the Schoenflies motion. As discussed in [18,19], the half-derivative of a function of time $f(t)$ can be calculated by using the following n-th order digital filter, derived from the Grünwald–Letnikov definition [24]:

$$f(t)^{(1/2)} \cong f(kT_s)^{(1/2)} \cong \frac{1}{T_s^{1/2}}\left(\sum_{j=0}^{n} w_j f((k-j)T_s)\right), k = [t/T_s], \quad (15)$$

where $[y]$ indicates the integer part of y, and the $n + 1$ filter coefficients w_j are:

$$w_0 = 1,\ w_j = \left(1 - \frac{3}{2j}\right)w_{j-1}\,,\, j = 1\ \ldots\ n. \quad (16)$$

The digital filter (15), which is used in the control law (14) to numerically evaluate the half-derivative of the error of the external coordinates, takes into account only the recent past of the derived function, in the interval $[t–L, t]$. In contrast, the exact calculation of an FO derivative involves the complete time history of the derived function. Consequently, Equation (15) introduces an approximation, which is usually accepted according to the *short-memory principle* [24]. Nevertheless, this approximation leads to an alteration of the end-effector stiffness, which should depend only on the stiffness matrix and not on the half-derivative matrix. To solve this problem, the following modification of the stiffness term can be introduced [18]:

$$\begin{aligned}\boldsymbol{\tau} = \ & \mathbf{J}^T(\boldsymbol{\theta})\left(\left(\mathbf{K}_{KDHD} - \frac{\sum_{j=0}^{n} w_j}{T_s^{1/2}}\mathbf{HD}_{KDHD}\right)(\mathbf{x}_d - \mathbf{x}(\boldsymbol{\theta})) + \right. \\ & \left. + \mathbf{D}_{KDHD}(\mathbf{x}_d - \mathbf{x}(\mathbf{q}))^{(1)} + \mathbf{HD}_{KDHD}(\mathbf{x}_d - \mathbf{x}(\mathbf{q}))^{(1/2)}\right) + \boldsymbol{\tau}_g(\boldsymbol{\theta})\end{aligned} \quad (17)$$

The effectiveness of the stiffness compensation of the KDHD control law (17) was verified in [18,19]. The KDHD control law (17) will be compared in the following with the classical KD algorithm, expressed by Equation (11).

4. Dynamic Model of the RRFbR Arm with KD and KDHD Control

The KD and KDHD control laws were compared by means of the simulation environment Simscape Multibody™ by MathWorks. The multibody model is shown in Figure 3, comprising two equal RRFbR manipulators, one with KD control and one with KDHD control, for a quicker comparison. The robots are subject only to gravitational forces. Friction in joints is neglected, in agreement with the analytical model discussed in Section 2.

Table 1 collects the main geometrical and mass parameters of the RRFbR arm that were used in the simulations.

In this comparison, a trajectory composed of six phases is considered. In each phase, the end-effector set-point position moves from the reference position $\mathbf{x}_{ref}$ along a direction parallel to one axis of the fixed reference frame $O(x,y,z)$ for a distance d in a time t_{mov}, stops at $\mathbf{x}_{ref} + \Delta\mathbf{x}$ for a time t_{stop}, then returns to $\mathbf{x}_{ref}$ in t_{mov} and stops again for t_{stop}, always keeping constant the end-effector orientation ($\theta = 0$). The reference position $\mathbf{x}_{ref}$ is defined by the following internal coordinates: $[-45°, 90°, 0°, -45°]$; consequently, it corresponds to the robot arm semi-bent, in a central zone of the workspace. Moreover, in this position, links 3 and 4 are horizontal, and the gravity force action is compensated exactly by the balancing spring. The displacements $\mathbf{\Delta x}$ of the six phases are respectively $(d, 0, 0)$, $(0, d, 0)$, $(0, 0, d)$, $(-d, 0, 0)$, $(0, -d, 0)$, $(0, 0, -d)$. For each phase there is a gone and a return motion, with the same duration t_{mov}, while the stops have duration t_{stop}; therefore, the overall duration of the six phases is $12(t_{mov} + t_{stop})$. Moreover, each motion is divided into three parts: acceleration, with duration αt_{mov}; constant velocity, with duration $(1–2\alpha)t_{mov}$; and deceleration, with duration αt_{mov} (s-curve motion). Consequently, the set-point motion is completely defined by the parameters d, t_{mov}, t_{stop} and α.

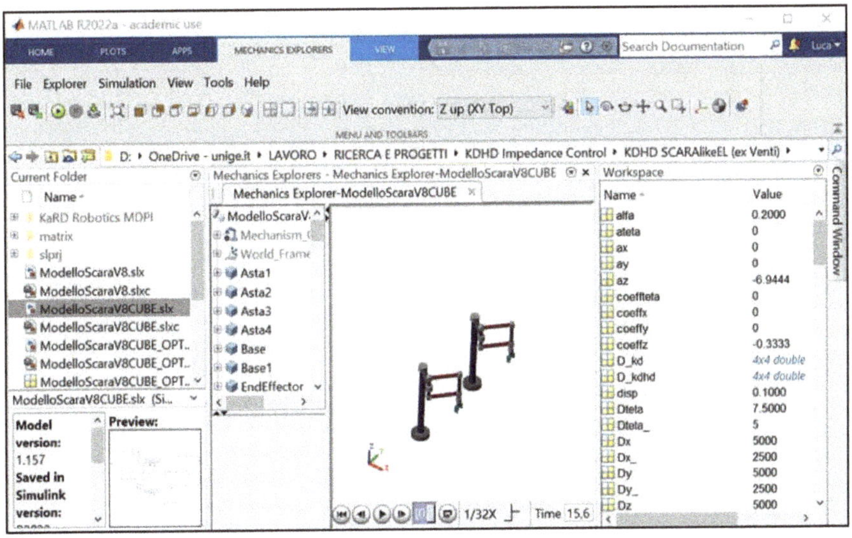

Figure 3. Multibody model of the RRFbR manipulator.

Table 1. Main geometrical parameters (Figure 2) and mass properties of the 3-PUU parallel robot.

Symbol	Parameter	Value	Unit
l_1	length of link 0	0.900	m
l_2	length of link 1	0.330	m
l_3	length of links 3 and 4	0.330	m
l_4	vertical distance between link 1 and point E	0.387	m
l_{G1}	c.o.m. position of link 1 (Figure 2)	0.165	m
l_{G3}	c.o.m. position of link 3 (Figure 2)	0.165	m
l_{G4}	c.o.m. position of link 4 (Figure 2)	0.165	m
m_1	mass of link 1	10	kg
m_2	mass of link 2	5	kg
m_3	mass of link 3	5	kg
m_4	mass of link 4	5	kg
m_5	mass of link 5	12	kg
m_6	mass of link 6	3	kg
θ_{ref}	reference central position (internal coordinates)	$[-45, 90, 0, -45]$	°
x_{ref}	reference central position (external coordinates)	$[0.467, 0, 0.513, 0]$	m, °
θ_{3p}	θ_3 position with spring in neutral position	-15	°
k_3	balancing spring stiffness	247.3	Nm/rad

Figure 4 shows the set-point motion characterized by d = 0.15 m, t_{mov} = 0.3 s, t_{stop} = 1 s and α = 0.2, in terms of position, velocity and acceleration. These parameters correspond to trapezoidal velocity profiles of the displacements with maximum velocity of ±0.625 m/s. Acceleration alternates null values and constant values of ±10.416 m/s^2 with duration

$\alpha t_{mov} = 0.06$ s. This set point motion corresponds to the time histories of the internal coordinates shown in Figure 5.

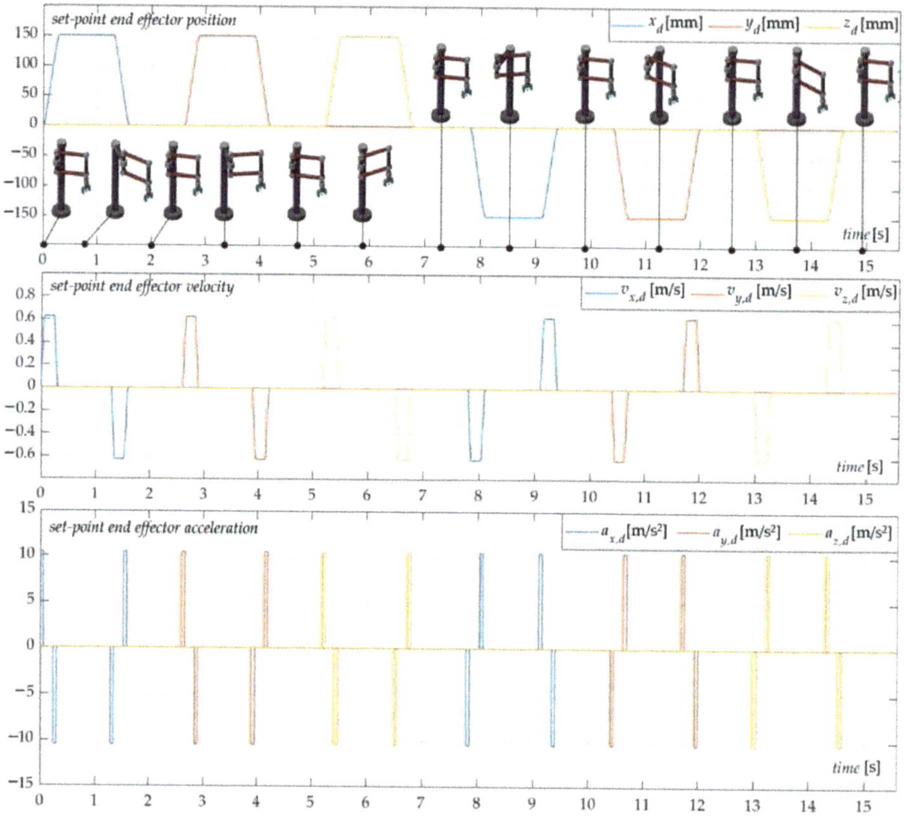

Figure 4. Position, velocity and acceleration of the end-effector during the set-point motion.

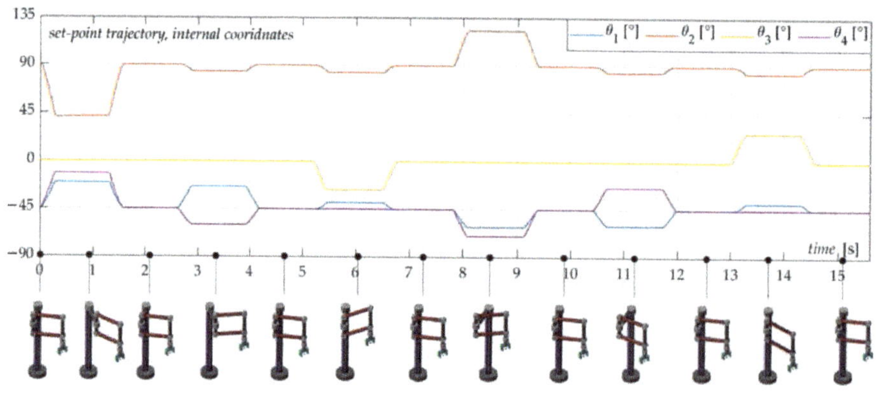

Figure 5. Time histories of the internal coordinates during the set-point motion.

The end-effector orientation is kept constant along the planned motion. Nevertheless, this assumption does not limit the generality of the analysis because, as discussed in Section 3, it is reasonable to have the 1-DOF rotational behavior decoupled from the 3-DOF translational behavior, adopting block-diagonal stiffness and damping matrices. With this hypothesis and in absence of friction in the joints, the rotational behavior related to θ_4 and τ_4 is completely decoupled, and the tuning of the corresponding stiffness and damping matrix elements can be performed by the criteria already extensively discussed for the PDD$^{1/2}$ control of SISO second-order linear systems [9,12].

5. Simulation Results

Figure 6 shows some synthetic results of the KD/KDHD comparison for the set-point motion of Figures 4 and 5 (d = 0.15 m, t_{mov} = 0.3 s, t_{stop} = 1 s and α = 0.2). The considered sampling time is T_s = 5 ms and the filter order for the calculation of the half-derivatives is n = 10; these values are suitable for a real discrete-time implementation on a commercial controller. In this comparison, the Integral Control Effort *(ICE)* and the Integral Square Error *(ISE)* of the end-effector translational coordinates are compared:

$$ICE = \sum_{i=1}^{4} \int_{0}^{T_{sim}} \tau_i^2 dt, \tag{18}$$

$$ISE = \int_{0}^{T_{sim}} \left((x_d - x)^2 + (y_d - y)^2 + (z_d - z)^2 \right) dt, \tag{19}$$

where T_{sim} is a simulation time sufficiently long to take into account the residual vibrations. Moreover, in the comparison:

- For the KD control law, the stiffness matrix $\mathbf{K}_{KD}$ is diagonal, with the first three diagonal elements $k_{KD,x} = k_{KD,y} = k_{KD,z} = 8{\cdot}10^5$ N/m, and $k_{KD,\theta} = 5{\cdot}10^2$ Nm/rad; the damping matrix $\mathbf{D}_{KD}$ is diagonal, with diagonal elements $d_{KD,x} = d_{KD,y} = d_{KD,z} = d_{xyz}$, and $d_{KD,\theta}$ = 7.5 Nms/rad; and the translational damping d_{xyz} is varied in the simulations in the range from $2{\cdot}10^3$ to $6{\cdot}10^3$ Ns/m;
- For the KDHD control law, the stiffness matrix $\mathbf{K}_{KDHD}$ is again diagonal, and the elements $k_{KDHD,x}$, $k_{KDHD,y}$, $k_{KDHD,z}$ and $k_{KDHD,\theta}$ are the same as in the $\mathbf{K}_{KD}$ matrix, but with the addition of the stiffness compensation of Equation (17); the matrix $\mathbf{D}_{KDHD}$ is diagonal, with diagonal elements $d_{KD,x} = d_{KD,y} = d_{KD,z} = d_{xyz}$, and $d_{KD,\theta}$ = 3.75 Nms/rad; the matrix $\mathbf{HD}_{KDHD}$ is also diagonal, with diagonal elements $hd_{KDHD,x} = hd_{KDHD,y} = hd_{KDHD,z} = hd_{xyz}$, and $hd_{KDHD,\theta} = 2.4{\cdot}10^4$ Nms$^{1/2}$/rad; in the simulations, the first-order translational damping d_{xyz} varies from $2{\cdot}10^3$ to $6{\cdot}10^3$ Ns/m, and the half-order translational damping hd_{xyz} varies from 0 to $1.7{\cdot}10^4$ Ns$^{1/2}$/m.

Figure 6 shows in red the 3D graphs of the Integral Control Effort (left) and of the Integral Square Error (right) as functions of the first-order translational damping d_{xyz} and of the half-order translational damping hd_{xyz}; the intersections of these surfaces with the planes at hd_{xyz} = 0 Ns$^{1/2}$/m represent the KD behavior, without half-derivative damping.

From the analysis of Figure 6, it is possible to observe that increasing d_{xyz} and hd_{xyz} (that is, increasing the damping) leads to higher control effort and lower integral error. In order to evaluate the potential benefits of the KDHD control over the classical KD, the proposed method aims to assess if it is possible to obtain lower error using the same control effort. In Figure 6 (left), the green horizontal plane at 44,200 N^{2}m^{2}s represents the ICE value of the KD controller with $d_{xyz} = 5 \cdot 10^3$ Ns/m. This plane intersects the red surface of the KDHD Integral Control Effort with good approximation along a straight, yellow dotted line in Figure 6 (left). This straight line corresponds to this linear relationship between the translational first-order and half-order coefficients:

$$hd_{xyz} = 47{,}200 - 9.44\, d_{xyz}. \tag{20}$$

This means that the KDHD controllers fulfilling this relationship have approximately the same ICE as the original KD controller with $d_{xyz} = 5 \cdot 10^3$ Ns/m. Let us note that, according to relation (20), a decrease of the first-order damping is compensated by an increase of the half-order damping.

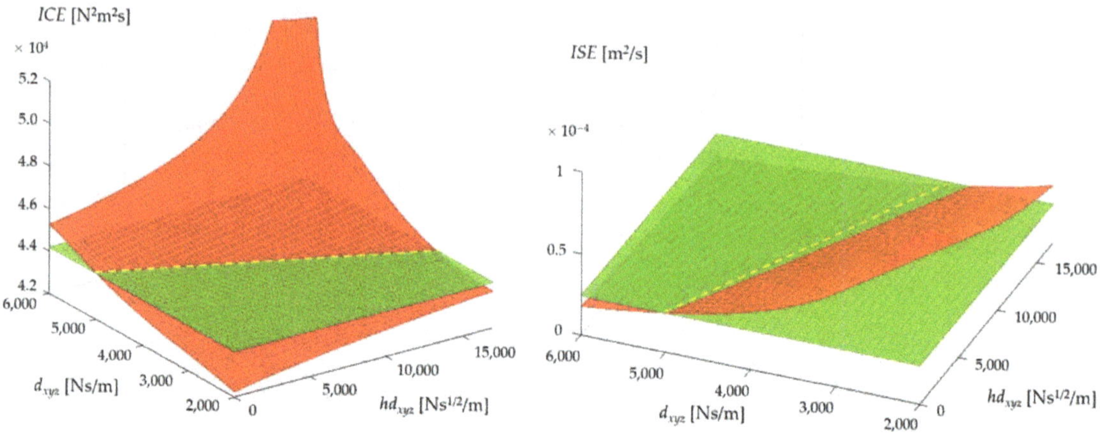

Figure 6. Integral Control Effort (ICE) and Integral Square Error (ISE) as functions of the first-order translational damping d_{xyz} and of the half-order translational damping hd_{xyz}; the red surface is the KDHD control; the KD control is represented by the curves at $hd_{xyz} = 0$ Ns$^{1/2}$/m; the green horizontal planes indicate the ICE and ISE values of the KD control with $d_{xyz} = 5 \cdot 10^3$ Ns/m.

The trace of the linear relationship (20) is also represented in Figure 6 (right), on the horizontal plane which corresponds to $2.41 \cdot 10^{-5}$ m^{2}s, which is the ISE value of the KD controller with $d_{xyz} = 5{\cdot}10^3$ Ns/m. The red surface of the ISE of the KDHD control is lower than the green plane along the linear hd_{xyz}–d_{xyz} path for any positive value of hd_{xyz}. This means that replacing an amount of first-order damping with some half-order damping reduces the Integral Square Error while keeping constant the Integral Control Effort.

This is confirmed by the plots in Figure 7, which show the ISE of the KDHD control (blue plot) and hd_{xyz} (orange plot) as a function of d_{xyz}, applying the linear relationship (20): as d_{xyz} decreases from 5000 to 0 Ns/m and hd_{xyz} increases from 0 (KD control) to 47,200 Ns$^{1/2}$/m (KDHD control without first-order damping, named KHD), the Integral Square Error decreases from $2.41 \cdot 10^{-5}$ m^{2}s to $2.02 \cdot 10^{-5}$ m^{2}s (−16%).

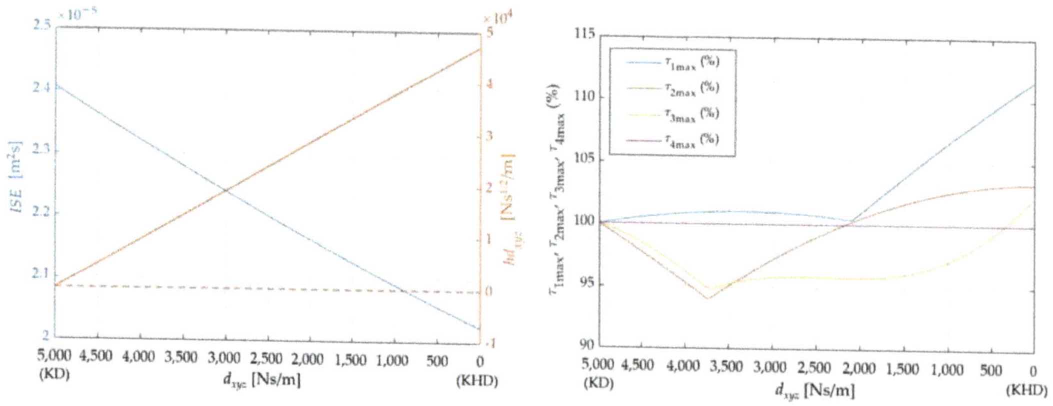

Figure 7. Integral Square Error and half-order translational damping hd_{xyz} as functions of the first-order translational damping d_{xyz} (**left**); percentage maximum values of the actuation torques as functions of d_{xyz} (**right**).

On the other hand, the complete replacement of the first-order damping with the half-order damping increases the maximum absolute values of the actuation torques, which are represented in Figure 7 (right) as percentages with respect to the KD control with $d_{xyz} = 5 \cdot 10^3$ Ns/m (τ_4 is constant, due to the planned motion, with constant orientation of the end-effector).

A possible compromise to reduce the Integral Square Error while keeping nearly constant the maximum absolute values of the actuation torques is located near the middle ($d_{xyz} = 2.5 \cdot 10^3$ Ns/m, hd_{xyz} = 23,600 Ns$^{1/2}$/m).

Figures 8–10 show the time histories of the errors of the translational external coordinates, e_x, e_y and e_z, comparing:

- The KD control with $d_{xyz} = 5 \cdot 10^3$ Ns/m;
- The KDHD control with intermediate tuning ($d_{xyz} = 2.5 \cdot 10^3$ Ns/m, hd_{xyz} = 23,600 Ns$^{1/2}$/m);
- The KHD control with $d_{xyz} = 0$ Ns/m and hd_{xyz} = 47,200 Ns$^{1/2}$/m.

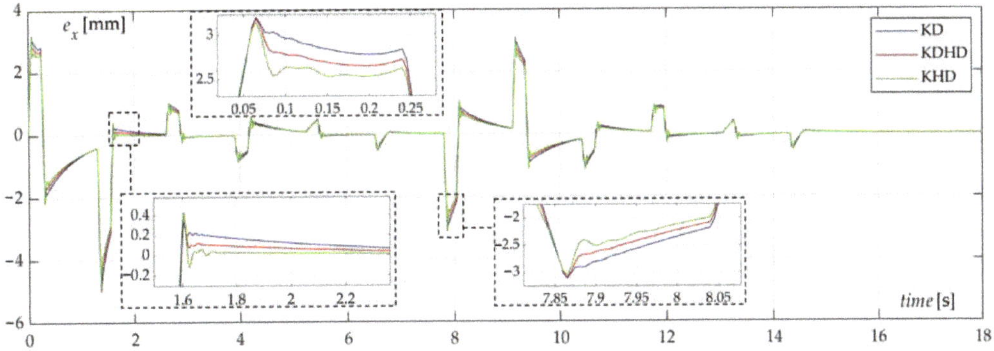

Figure 8. Comparison of the KD (blue), KDHD (red), KHD (green) controls, error e_x.

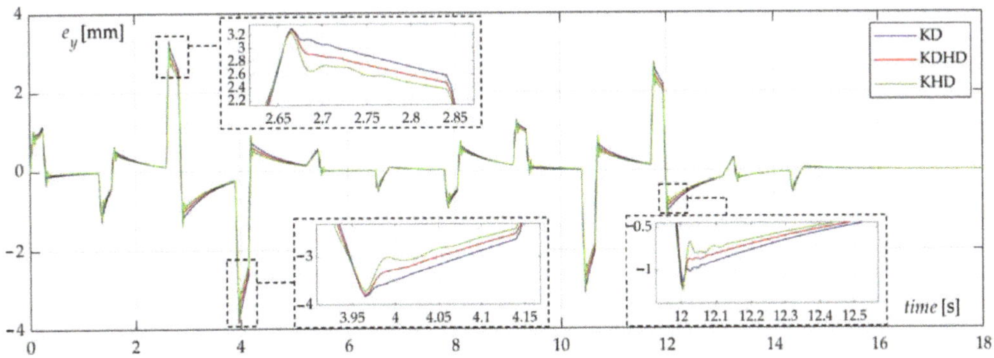

Figure 9. Comparison of the KD (blue), KDHD (red), KHD (green) controls, error e_y.

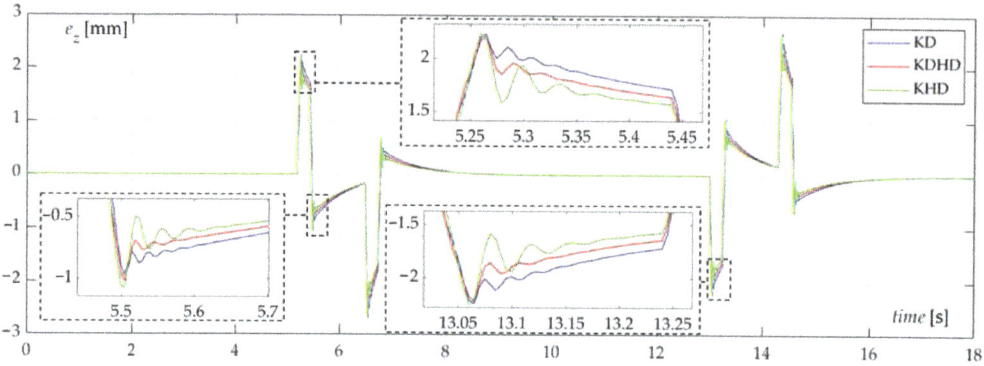

Figure 10. Comparison of the KD (blue), KDHD (red), KHD (green) controllers, error e_z.

Table 2 collects the main performance indexes for the three controllers. From the analysis of Figures 8–10, and from the data collected in Table 2, it is possible to observe that the peak errors are similar for the three controllers, with a maximum reduction of -2.45% for the KHD with respect to KD (e_y coordinate), but the mean errors decrease more remarkably moving from KD to KHD, coherently with the ISE plot in Figure 7 (left). In particular, the reductions of the mean absolute values of the translational errors with respect to KD are around $-5 \div -6\%$ for the KDHD, and $-10 \div -11\%$ for the KHD, while the Integral Square Error decreases were -8.7% for the KDHD and -16.2% for the KHD.

Table 2. Comparison of the main performance indexes of the three controllers (KD, KDHD, KHD), with percentage variations of KDHD and KHD with respect to KD.

Symbol	Parameter	KD	KDHD	KHD
$e_{x,max}$	maximum absolute value of the x error (mm)	4.979	4.970 (−0.17%)	4.882 (−1.93%)
$e_{y,max}$	maximum absolute value of the y error (mm)	3.833	3.817 (−0.41%)	3.739 (−2.45%)
$e_{z,max}$	maximum absolute value of the z error (mm)	2.676	2.652 (−0.89%)	2.642 (−1.27%)
$e_{x,mean}$	mean absolute value of the x error (mm)	0.359	0.338 (−5.65%)	0.320 (−10.7%)
$e_{y,mean}$	mean absolute value of the y error (mm)	0.352	0.334 (−5.08%)	0.318 (−9.65%)
$e_{z,mean}$	mean absolute value of the z error (mm)	0.187	0.176 (−5.51%)	0.167 (−10.7%)
τ_{1max}	maximum absolute value of the torque τ_1 (Nm)	172.8	171.6 (−0.71%)	191.6 (+10.9%)
τ_{2max}	maximum absolute value of the torque τ_2 (Nm)	90.2	94.2 (+0.98%)	157.2 (+4.31%)
τ_{3max}	maximum absolute value of the torque τ_3 (Nm)	157.2	164.1(+4.17%)	167.9 (+6.47%)
τ_{4max}	maximum absolute value of the torque τ_4 (Nm)	1.67	1.57 (−5.79%)	1.67 (0%)
ISE	Integral Square Error of the translational coordinates (m^2s)	$2.41 \cdot 10^{-5}$	$2.20 \cdot 10^{-5}$ (−8.7%)	$2.02 \cdot 10^{-5}$ (−16.2%)

On the other hand, as shown in Figure 7 (right) and Table 2, the peak values of the actuator torques are similar for the KD controller and the KDHD with intermediate tuning, while they are higher for the KD (up to +10.9% for actuator 1, which is the most loaded).

Moreover, the zooms of the peak zones in Figures 8–10 show that the behavior of the KHD is more oscillatory than those of the KD and KHD, which is generally not desirable.

6. Discussion

The procedure used here to tune the KDHD control starting from a given KD control is applicable for a generic manipulator, and can be summarized as follows:

- A set-point motion, which covers a significant part of the workspace, is selected;
- Multibody simulations are performed varying the first-order translational damping d_{xyz} and the half-order translational damping hd_{xyz}, creating the 3D map of the ICE;

- The locus of the d_{xyz} - hd_{xyz} combinations with the same ICE of the given KD controller is identified;
- Simulations are performed along this d_{xyz}–hd_{xyz} path, starting from null hd_{xyz} (KD control) and arriving at null d_{xyz} (KHD control);
- Along this path, the main performance indexes (Integral Square Error, maximum and mean absolute values of the external coordinates error, maximum absolute values of the actuation torque) are evaluated, to select the KDHD tuning on the basis of a proper compromise;
- In general, a combination of first-order damping and half-derivative damping, intermediate between pure KD and pure KHD, makes it possible to reduce the tracking error with similar values of the maximum actuation torques, while the KHD has even lower tracking error, but also higher peaks of the actuation torques.

These outcomes are coherent with the results obtained for SISO systems controlled by means of the PDD$^{1/2}$ scheme, for which a proper combination of first-order and half-order damping is preferable [9,25]. Nevertheless, the error reduction which can be achieved by introducing the half-order damping in Cartesian space control for the considered elastically balanced manipulator is much lower than the reduction achievable for a single mechatronic axis with inertial load, with equal maximum actuation torque of around −50% [26]. This is because in some phases the actuation torques must overcome not only the inertial forces, but also the elastic and gravitational forces, which are perfectly compensated only when links 3 and 4 are horizontal ($\theta_3 = 0$); these torque components are equal for the two controllers, and the half-derivative damping brings benefits only related to the transient behavior in the acceleration/deceleration phases; as a consequence, the overall performance benefit is lower with respect to an almost linear system with purely inertial load.

Some limitations of the proposed approach are as follows:

- The tuning is based on an arbitrarily selected set-point motion, which must be chosen in order to be representative of the typical working conditions;
- The first-order and half-order damping coefficients d_{xyz} and hd_{xyz} are imposed to be equal for the three translational external coordinates;
- The tuning of the rotational behavior is fixed.

In future research, these topics will be addressed in order to obtain more general results. However, the discussed simulation results confirm the potential benefits of fractional-order calculus, and in particular of the KDHD scheme, in the Cartesian space control of robotic manipulators. Moreover, an interesting research direction is the application to mechanisms with flexible links [27,28].

Author Contributions: L.B. conceived the control algorithm and designed the simulation campaign; S.E.N. developed the multibody model and performed simulations; L.B. and S.E.N. prepared the manuscript. All authors have read and agreed to the published version of the manuscript.

Funding: This research received no external funding.

Data Availability Statement: Data are contained within the article. The simulation data presented in this study are available in the figures and tables.

Conflicts of Interest: The authors declare no conflict of interest.

References

1. Miller, K.S.; Ross, B. *An Introduction to the Fractional Calculus and Fractional Differential Equations*; John Wiley & Sons: New York, NY, USA, 1993.
2. Hilfer, R. *Applications of Fractional Calculus in Physics*; World Scientific: Singapore, 2000.
3. Rihan, F.A. Numerical modeling of Fractional-Order biological systems. *Abstr. Appl. Anal.* **2013**, *2013*, 816803. [CrossRef]
4. Shaikh, A.S.; Shaikh, I.N.; Nisar, K.S. A mathematical model of COVID-19 using fractional derivative: Outbreak in India with dynamics of transmission and control. *Adv. Differ. Equ.* **2020**, *2020*, 373. [CrossRef] [PubMed]
5. Sasso, M.; Palmieri, G.; Amodio, D. Application of fractional derivative models in linear viscoelastic problems. *Mech. Time-Depend. Mater.* **2011**, *15*, 367–387. [CrossRef]
6. Podlubny, I. Fractional-Order systems and PI$^{\lambda}$D$^{\mu}$ controllers. *IEEE Trans. Autom. Control* **1999**, *44*, 208–213. [CrossRef]

7. Yeroglu, C.; Tan, N. Note on fractional-order proportional-integral-differential controller design. *IET Control Theory Appl.* **2012**, *5*, 1978–1989. [CrossRef]
8. Beschi, M.; Padula, F.; Visioli, A. The generalised isodamping approach for robust fractional PID controllers design. *Int. J. Control* **2015**, *90*, 1157–1164. [CrossRef]
9. Bruzzone, L.; Fanghella, P. Comparison of $PDD^{1/2}$ and PD^{μ} position controls of a second order linear system. In Proceedings of the IASTED International Conference on Modelling, Identification and Control, MIC 2014, Innsbruck, Austria, 17–19 February 2014; pp. 182–188. [CrossRef]
10. Bruzzone, L.; Baggetta, M.; Fanghella, P. Fractional-Order PII1/2DD1/2 control: Theoretical aspects and application to a mechatronic axis. *Appl. Sci.* **2021**, *11*, 3631. [CrossRef]
11. Jakovljevic, B.B.; Sekara, T.B.; Rapaic, M.R.; Jelicic, Z.D. On the distributed order PID controller. *Int. J. Electron. Commun.* **2017**, *79*, 94–101. [CrossRef]
12. Bruzzone, L.; Fanghella, P. Fractional-order control of a micrometric linear axis. *J. Control Sci. Eng.* **2013**, *2013*, 947428. [CrossRef]
13. Caccavale, F.; Siciliano, B.; Villani, L. Robot impedance control with Nondiagonal Stiffness. *IEEE Trans. Autom. Control* **1999**, *44*, 1943–1946. [CrossRef]
14. Seraji, K. Cartesian control of robotic manipulators. In Proceedings of the IFAC 10th Triennial World Congress, Munich, Germany, 27–31 July 1987; pp. 289–294.
15. Albu-Schäffer, A.; Hirzinger, G. Cartesian impedance control techniques for torque controlled light-weight robots. In Proceedings of the 2002 IEEE International Conference on Robotics & Automation, Washington, DC, USA, 11–15 May 2002; pp. 657–663.
16. Kizir, S.; Elşavi, A. Position-based Fractional-Order impedance control of a 2 DOF serial manipulator. *Robotica* **2021**, *39*, 1560–1574. [CrossRef]
17. Liu, X.; Wang, S.; Luo, Y. Fractional-Order impedance control design for robot manipulator. In Proceedings of the ASME 2021 International Design Engineering Technical Conferences and Computers and Information in Engineering Conference, Online, 17–19 August 2021; Volume 7, p. V007T07A028.
18. Bruzzone, L.; Fanghella, P.; Basso, D. Application of the half-order derivative to impedance control of the 3-PUU parallel robot. *Actuators* **2022**, *11*, 45. [CrossRef]
19. Bruzzone, L.; Polloni, A. Fractional Order KDHD impedance control of the Stewart Platform. *Machines* **2022**, *10*, 604. [CrossRef]
20. Makino, H.; Furuya, N. Selective compliance assembly robot arm. In Proceedings of the First International Conference on Assembly Automation, Brighton, UK, 25–27 March 1980; pp. 77–86.
21. Bruzzone, L.; Bozzini, G. A statically balanced SCARA-like industrial manipulator with high energetic efficiency. *Meccanica* **2011**, *46*, 771–784. [CrossRef]
22. Hervé, J.M. The Lie group of rigid body displacements, a fundamental tool for mechanism design. *Mech. Mach. Theory* **1999**, *34*, 719–730. [CrossRef]
23. Bruzzone, L.; Molfino, R.M. A geometric definition of rotational stiffness and damping applied to impedance control of parallel robots. *Int. J. Robot. Autom.* **2006**, *21*, 197–205. [CrossRef]
24. Das, S. *Functional Fractional Calculus*; Springer: Berlin/Heidelberg, Germany, 2011.
25. Bruzzone, L.; Bozzini, G. $PDD^{1/2}$ control of purely inertial systems: Nondimensional analysis of the ramp response. In Proceedings of the IASTED International Conference on Modelling, Identification and Control, Innsbruck, Austria, 14–16 February 2011; pp. 308–315. [CrossRef]
26. Bruzzone, L.; Fanghella, P.; Baggetta, M. Experimental assessment of fractional-order $PDD^{1/2}$ control of a brushless DC motor with inertial load. *Actuators* **2020**, *9*, 13. [CrossRef]
27. Boscariol, P.; Scalera, L.; Gasparetto, A. Nonlinear control of multibody flexible mechanisms: A model-free approach. *Appl. Sci.* **2021**, *11*, 1082. [CrossRef]
28. Gupta, S.; Singh, A.P.; Deb, D.; Ozana, S. Kalman filter and variants for estimation in 2DOF serial flexible link and joint using Fractional Order PID controller. *Appl. Sci.* **2021**, *11*, 6693. [CrossRef]

robotics

MDPI

Article

Collision Avoidance for Redundant 7-DOF Robots Using a Critically Damped Dynamic Approach

Henrique Simas [1] and Raffaele Di Gregorio [2,*]

1 Raul Guenther Lab. of Applied Robotics, Department of Mechanical Engineering, Federal University of Santa Catarina, Florianópolis 88040-900, Brazil
2 Lab. of Mechatronics and Virtual Prototyping (LaMaViP), Department of Engineering, University of Ferrara, Via Saragat, 1, 44100 Ferrara, Italy
* Correspondence: raffaele.digregorio@unife.it; Tel.: +39-0532-974828 or +39-0532-974870

Abstract: The presence of collaborative robots in industrial environments requires that their control strategies include collision avoidance in the generation of trajectories. In general, collision avoidance is performed via additional displacements of the kinematic chain that make the robot move far from the objects that are occasionally inserted into its safety workspace. The variability of the coordinates of the collision points inside the safety volume leads to abrupt movements for the robot. This paper presents a general method for smoothing abrupt movements in robots with one degree of redundancy for collision-avoidance trajectories, employing a second-order digital filter designed with adjustable critical damping. The method is illustrated by applying it to a redundant robot with a spherical–revolute–spherical type (SRS-type) kinematic chain, which is a benchmark used to test the algorithms ideated for solving this problem. This paper also presents an alternative algorithm for the inverse kinematics of the SRS-type robot and the computational experiments that show the collision avoidance proposal's performance and its properties through graphical results.

Keywords: redundant robots; collaborative robots; analytical inverse kinematics; collision avoidance; smooth transitions

Citation: Simas, H.; Di Gregorio, R. Collision Avoidance for Redundant 7-DOF Robots Using a Critically Damped Dynamic Approach. *Robotics* **2022**, *11*, 93. https://doi.org/10.3390/robotics11050093

Academic Editor: Marco Ceccarelli

Received: 23 July 2022
Accepted: 4 September 2022
Published: 8 September 2022

1. Introduction

Complex tasks require interactions between robots and humans collaboratively. This collaboration comes from sharing tools, operations, and workspaces [1–3] and requires specific control techniques to overcome possible collisions between robots and humans or devices located inside the workspace [4]. Such collaboration implies a more significant number of restrictions on the volume available to the end-effector and the respective *safety volume* (i.e., the geometric spatial volume around the links and the end-effector, where the robot is considered free of collisions) available for the displacements of the robot's kinematic chain [5]. In this sense, a new class of robots is arising, defined as *collaborative robots* [6,7], which are provided with strategies at the hardware and software levels to avoid collisions and improve the interactions between robots and humans. Collision-avoidance strategies require additional movements of the kinematic chains, which means that the robots need to have greater mobility than what is needed to perform their tasks; that is, they need to be redundant [6,8–11]

Elements inserted into a robot's safety workspace [12] usually make the robot change its coordinates randomly, causing abrupt variation in the distances to the points belonging to the robot's kinematic chain. If the values of these distances are used as references in collision-avoidance algorithms, the movements of the robot's kinematic chain are abrupt and naturally undesirable. Indeed, abrupt changes in joint variables, besides being physically difficult to obtain, generate vibrations in a wide range of frequencies that cannot be tolerated during machine motion. This is why smoothing the additional motion that avoids a possible collision is an important requirement for any collision-avoidance strategy.

Collision-avoidance strategies consist of programming a set of internal movements of a robot's kinematic chains based on sensor data to circumvent or deviate certain parts of the robot from an obstacle that has been inserted into its overall workspace, which can occur in online, offline, or autonomous tasks [3,4,12]. The internal movements are performed via changes in joint displacements, which often lead to abrupt variations in joint trajectories [13,14] and need to be treated to ensure smooth movement.

Improving the motion quality requires collision-avoidance strategies that generate smooth transients in terms of joint displacements and rates. Such strategies are, in general, based on optimization algorithms via energy minimization [15,16] or on differential geometric approaches [8,17,18], which, among other propositions, are based on the Moore–Penrose pseudoinverse approach [19,20] and are limited mainly by the problem of homogeneity and the lack of repeatability of the solutions [21,22]. A systematic proposition based on concepts of linear equations independent of numerical optimization methods [23] allows for better control over the performance and execution of the collision-avoidance strategy.

Some interesting trajectory generation proposals involve using digital filters in the definition of reference signal limits [24,25] or combined with curve formulations [26,27]. The complexity and limited adjustment possibilities make it difficult to replicate or use such strategies, limiting their application. The aim of this work is to overcome these applicative limitations.

This paper presents a new approach to collision avoidance based on the use of second-order digital filters [28], which solves the inverse kinematics through the *resolved motion control* technique (RMC) [14,29]. The proposal calculates the minimum distance straight line between the nearest collision point of a mobile obstacle and the robot's links using a geometric formulation. The length of this straight-line segment changes as the collision point coordinates change inside the robot's overall workspace. The collision-avoidance algorithm proposition calculates the internal movements (*null space* of the kinematic chain [30,31]), smoothing abrupt movements that eventually arise. The proposal uses a low-pass filter to minimize the possible peaks of the internal displacements, improving the transients of the trajectory in the kinematic chain while the robot executes a task. The proposal also considers an envelope defined by a geometric surface around the links as a safety region and obtains a smooth transition through a digital filter, which allows adjustments according to the desired performance and response-time criteria [21,22,28].

Several spherical–revolute–spherical-type (SRS-type) redundant robots have arisen as workbenches for the simulation and experimentation of such proposals [5,15,26,32–34], including the arms or legs of bipedal robots [33,35–37]; among these, the Kuka-LBR-iiwa models are the most adopted SRS-type redundant robots [8,9,11,38,39]. The experimentation on the proposed collision-avoidance strategy uses the Kuka-LBR-iiwa's kinematic model with seven degrees-of-freedom (DOFs), which is commonly used to test collision-avoidance strategies in collaborative cells [4,7,8,27]. Such a choice will allow comparisons and a more accurate assessment of the proposition's advantages.

Here, the proposed strategy is proven and discussed using a simulation experiment, where the collision-avoidance control acts to avoid the collisions between the links of the SRS-type redundant robot and two kinematic models of human arms that share the robot's safety volume [12] while performing tasks in an industrial simulation environment.

This paper is organized as follows. Section 2 presents the proposed collision-avoidance strategy. Section 3 presents the SRS-type redundant robot benchmark and a proposal for solving its inverse kinematics. Section 4 presents the new approach for defining the safe region around the links and calculating the distance between a point and the robot. Section 5 presents the graphical results, and Section 6 discusses the results. Finally, Section 7 draws the conclusions.

2. Critically Damped Collision Avoidance

Let P_c be a point with an imminent possibility of collision with a redundant robot. Such a point may belong to a human operator or to another apparatus occasionally inserted into the robot's safety volume [12,40,41]. The collision-avoidance strategy must avoid the collision by acting in the null space of a redundant robot so that the robot's configuration changes using movements, chosen according to the degree of redundancy [11,30], that do not affect the task performed by the robot. In short, without redundancy, avoiding a possible collision can be achieved only by changing the motion task of the robot.

The collision-avoidance method proposed in this paper holds for robots with only one degree of redundancy. So, we can only exploit a movement with one DOF to avoid a collision. In the literature, for the case of the SRS-type redundant robot of Figure 1, which is a benchmark for this problem, angle α is usually chosen for controlling the distance between the elbow point $\boldsymbol{P}_\mathrm{e}$ and the collision point $\boldsymbol{P}_\mathrm{c}$ using criteria such as geometric approaches [1,38], differential maps [17,29], and energy [5,25,26,32]. The problem reduces itself to ideate an inverse-kinematics algorithm that makes angle α perform the internal movements (null-space) of the kinematic chain without changing the position and orientation of the end-effector.

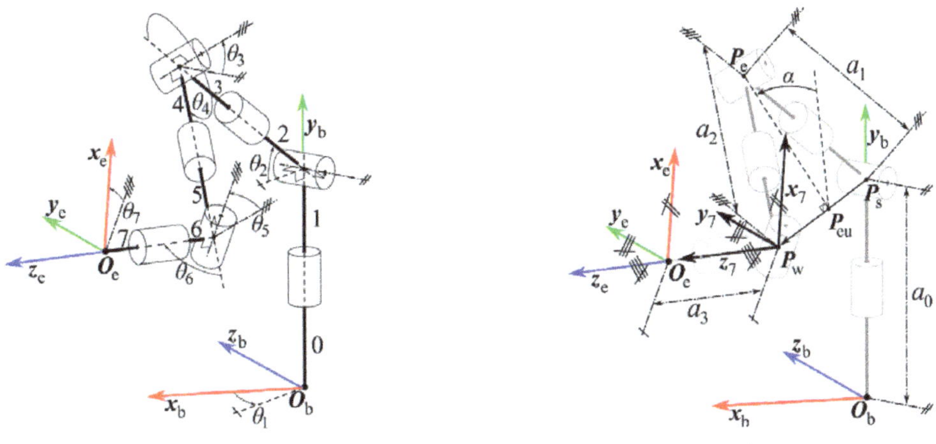

Figure 1. 7-DOF SRS-type redundant robot: (**a**) joint-variables' phase-references; (**b**) geometrical parameters.

From a geometric point of view, the problem consists of verifying the *minimum distance*, d_min, between the robot and the nearest point, $\boldsymbol{P}_\mathrm{c}$, of a mobile obstacle and then adjusting the value of angle α to guarantee a *minimum desired distance*, dr_min. Considering that point $\boldsymbol{P}_\mathrm{c}$ may change its coordinates randomly or abruptly in the space at each measurement of its coordinates (e.g., obtained through sensors or video cameras [42]), the respective change of angle α may produce abrupt oscillations of the joint positions, calculated with inverse-kinematics algorithms, which is an undesirable and often, unpractical result.

To avoid abrupt oscillations in the values of the joint positions, it is interesting to smooth angle-α transients before using them as input data for the inverse-kinematics algorithms. Smoothing improves continuity in signal transitions and variations based on polynomial numerical interpolations [43]. Such an approach has limitations when the order of signals' derivatives is greater than polynomials' order; indeed, in this case, continuity is no longer guaranteed.

This work proposes that angle α be numerically filtered by using a time-continuous second-order transfer function, *G(s)*, defined as follows, which is critically damped and has a static gain equal to 1 [28]:

$$\frac{\alpha_\mathrm{f}(s)}{\alpha_\mathrm{c}(s)} = G(s) = \frac{p^2}{(s+p)^2} \tag{1}$$

In Equation (1), $\alpha_c(s)$ corresponds to a time-continuous signal input for the α values on the complex s-domain calculated to avoid the collision; $\alpha_f(s)$ is the respective continuous output, signal-filtered; and p is an adjustable pole, which allows setting the speed of the transients [28]. The critically damped dynamics consist of second-order systems that have a time response with faster transients without overshoot, where the damping coefficient $\xi = 1$ [28].

Using the filtered signal, α_f, in a trajectory-generator algorithm needs the discretization of the transfer function, $G(s)$, defined by Equation (1). The discretization of $G(s)$ is given by its z-transform associated with a *zero-order hold* (ZOH) [28] as follows:

$$\frac{\alpha_f(s)}{\alpha_c(s)} = G(z) = \left(1 - z^{-1}\right) Z\left\{\frac{G(s)}{s}\right\} = \frac{(1 - e^{-m}(1+m))z^{-1} + e^{-m}(m - 1 + e^{-m})z^{-2}}{1 - 2e^{-m}z^{-1} + e^{-2m}z^{-2}} = \frac{h_1 z^{-1} + h_2 z^{-2}}{1 - b_1 z^{-1} - b_2 z^{-2}} \tag{2}$$

where T is the desired sample time (in general defined as a function of the computer clock, or interfaces sample times); m is the product Tp; $h_1 = (1 - e^{-m}(1 + m))$; $h_2 = e^{-m}(m - 1 + e^{-m})$; $b_1 = 2e^{-m}$; and $b_2 = -e^{-2m}$.

Using the right-shift or delay theorem [28], the variable z behaves like a delay operator for the input and output variables of the discrete transfer function presented in Equation (2). In this way, it is possible to obtain a command line for the algorithm that filters the α angle as follows:

$$\alpha_f(n) = h_1\alpha_c(n-1) + h_2\alpha_c(n-2) + b_1\alpha_f(n-1) + b_2\alpha_f(n-2) \tag{3}$$

where "n" is the discrete parameter, that is, $n = 0, 1, \ldots$, whereas α_c and α_f are the computed-input and the filtered-output angle α calculated at the instant "n", respectively. It is worth noting that since the signal starts at $n = 0$, it is necessary to define the initial conditions of α_f and α_c for $n = -2, -1$.

3. SRS-Type Redundant Robot Benchmark

3.1. Geometric Description

The SRS-type robot is a redundant robot with seven revolute joints (7-DOF) with the possibility of elbow movement, adapted mainly for collaborative tasks between robots or robots and humans [9,44,45]. The scope of applications has encouraged several authors to develop research and proposals for the kinematic modeling of this robot [8,9,26,35,37,46]. A typical SRS-type redundant robot is the Kuka-LBR-iiwa [4,44], whose kinematic model is shown in Figure 1.

With reference to Figure 1, the links are numbered from 0 (the frame) to 7 (the end effector), and the revolute(R)-joint variables are $\mathbf{q} = [\theta_1 \;\; \theta_2 \;\; \theta_3 \;\; \theta_4 \;\; \theta_5 \;\; \theta_6 \;\; \theta_7]^T$ (see Figure 1a), whereas the constant lengths are $\mathbf{a} = [a_0 \;\; a_1 \;\; a_3 \;\; a_4]^T$ (see Figure 1b). Link 0 is the frame (base), and the Cartesian reference $\boldsymbol{O}_b$-$x_b y_b z_b$ is fixed to the base, whereas link 7 is the end-effector and the Cartesian reference $\boldsymbol{O}_e$-$x_e y_e z_e$ is fixed to the end effector. Here, without losing generality, for the sake of simplicity, the assumption that the tool is displaced a_3 from the center of the spherical wrist is introduced. The seven R-joints are numbered from 1 to 7 following the order they are encountered by moving from the base to the end effector along the 7R kinematic chain of the robot.

The 7-DOF Kuka-LBR-iiwa robot is a serial chain of SRS-type. Indeed, the axes of the first three R-joints, with joint variables (θ_1, θ_2, θ_3), intersect at $\boldsymbol{P}_s$, which is the center of the first S-pair, and the axes of the last three R-joints, with joint variables (θ_5, θ_6, θ_7), intersect themselves at $\boldsymbol{P}_w$, which is the center of the second S-pair (i.e., the wrist of the robot). The axis of the first R-joint coincides with the $\mathbf{y}_b$-axis, whereas the axis of the second R-joint is perpendicular to the y_b-axis and passes through $\boldsymbol{P}_s$. The axis of the third R-joint is perpendicular to both the axes of the second and the fourth R-joints and intersects the axis of the fourth R-joint at $\boldsymbol{P}_e$. Points $\boldsymbol{O}_b$, $\boldsymbol{P}_s$, and $\boldsymbol{P}_w$ lie on a plane, hereafter named plane-π_1, which contains the $\mathbf{y}_b$-axis. Points $\boldsymbol{P}_s$, $\boldsymbol{P}_e$, and $\boldsymbol{P}_w$ lie on another plane, hereafter named plane-π_2, whose normal is parallel to the axis of the fourth R-joint. The axis of the fifth

R-joint passes through P_e, is perpendicular to both the axes of the fourth and the sixth R-joints, and intersects the axis of the sixth R-joint at P_w: θ_5 is the angle between the axes of the fourth and the sixth R-joints. Eventually, the axis of the seventh R-joint passes through P_w and is perpendicular to the axis of the sixth R-joint: θ_6 is the angle between the axes of the fifth and the seventh R-joints.

The constant lengths, shown in Figure 1b, are defined as follows: a_0 is the distance between points O_b and P_s; a_1 is the distance between points P_s and P_e; a_2 is the distance between points P_e and P_w; and a_3 is the distance between points P_w and O_e. In addition, three *characteristic internal points* are defined (see Figure 1b): P_s–*shoulder* point—in the center of the first S-pair; P_e–*elbow* point—in the crossing point of the joint axes 3, 4, and 5; and, lastly, P_w in the center of the second S-pair (the robot wrist). When $\theta_3 = 0$ rad, the points O_b, P_s, P_e, and P_w belong to the same plane (i.e., plane-π_1 and plane-π_2 coincide with one another), making the kinematic chain like a classical 6-DOF anthropomorphic robot. Differently, when $\theta_3 \neq 0$ rad, point P_e goes out of plane-π_1, and the line passing through points P_s and P_w becomes the common intersection between plane-π_1 and plane-π_2, which P_e lies on. The angle α between these two planes (see Figure 1b) can be chosen as the seventh generalized coordinate that, together with the coordinates ${}^b\mathbf{p}_7 = [x_7\ y_7\ z_7]^T$ of point P_w in O_b-$x_by_bz_b$ and the end-effector's orientation parameters ${}^b\boldsymbol{\varphi}_e = [\varphi_e\ \vartheta_e\ \psi_e\ ]^T$ with respect to O_b-$x_by_bz_b$, uniquely identifies the configuration of the 7-DOF robot in the operational space.

From a geometric point of view, angle α is a rotation angle around the line segment P_sP_w. In short, we have $\mathbf{x} = [{}^b\mathbf{p}_7{}^T\ {}^b\boldsymbol{\varphi}_e{}^T\ \alpha]\ ^T$ as operational-space coordinates and $\mathbf{q} = [\theta_1\ \ \theta_2\ \ \theta_3\ \ \theta_4\ \ \theta_5\ \ \theta_6\ \ \theta_7]^T$ as joint-space coordinates.

Mathematically the direct kinematics model: $\mathbf{q}\rightarrow\mathbf{x}$, uses a common methodology such as Denavit–Hartenberg [10,11,30] or successive screws [14,31], among others. Such methodologies are well known and allow the systematic deduction of the kinematics equation. For the inverse kinematics: $\mathbf{x}\rightarrow\mathbf{q}$, the angle α and the end-effector's pose parameters, ${}^b\mathbf{p}_7{}^T$ and ${}^b\boldsymbol{\varphi}_e{}^T$, are known and, in this case [9], the inverse kinematics has a finite number of solutions [14,30,31]. The next section presents a particular solution for the inverse kinematics, refining other propositions [8,11,46] for α Î $[0, 2\pi]$.

3.2. *An Analytical Approach to Solve the Inverse Kinematics of SRS-Type Redundant Robots*

Let ${}^b\mathbf{T}_e$ be a homogeneous transformation matrix built from the end-effector's pose parameters, ${}^b\mathbf{p}_e{}^T$ and ${}^b\boldsymbol{\varphi}_e{}^T$, which locates a Cartesian reference P_w-$x_7y_7z_7$ fixed to link 7 with respect to O_b-$x_by_bz_b$. Additionally, consider, without losing generality, that the tool action point of the end-effector is in the O_e-$x_ey_ez_e$ frame, as shown in Figure 1a, and has its pose with respect to the Cartesian reference P_w-$x_7y_7z_7$ identified by a known and constant homogeneous matrix ${}^7\mathbf{T}_e$. Thus, we have the two known homogeneous matrices (${}^j\mathbf{R}_i$ and ${}^j\mathbf{p}_i$ stand for the rotation matrix and the position vector, respectively, associated with the homogeneous transformation matrix ${}^j\mathbf{T}_i$):

$$
{}^b\mathbf{T}_e = \begin{bmatrix} & {}^b\mathbf{R}_e & & {}^b\mathbf{p}_e \\ 0 & 0 & 0 & 1 \end{bmatrix} \text{ and } {}^7\mathbf{T}_e = \begin{bmatrix} & {}^7\mathbf{R}_e & & {}^7\mathbf{p}_e \\ 0 & 0 & 0 & 1 \end{bmatrix} \tag{4}
$$

The first step consists of setting $\theta_3 = 0$ rad to make the SRS-type robot's kinematic chain like a classical 6-DOF anthropomorphic robot. Considering the resultant 6-DOF sub-structure, the solution of the inverse kinematics calculates an intermediary joint position vector $\bar{\mathbf{q}} = [\theta_1\ \ \theta_2\ \ \theta_4\ \ \theta_5\ \ \theta_6\ \ \theta_7]^T$, with $\theta_3 = 0$ rad through an analytical procedure [30,31,39].

With the values obtained for $\bar{\mathbf{q}}$, it is possible to calculate the coordinates of P_s, P_{eo}, and P_w, where P_{eo} is the coordinates of P_e when $\alpha = 0$ rad (or $\theta_3 = 0$ rad). The next step consists of rotating P_{eo} around the axis P_sP_w by the known angle α for calculating the actual coordinates of P_e, as shown in Figure 1b.

Using Rodrigues's rotation formula [14,31] to perform this rotation yields the following relationship:

$$\boldsymbol{P}_{\rm e} = \boldsymbol{P}_{\rm s} + \boldsymbol{R}_{\rm u}(\boldsymbol{P}_{\rm eo} - \boldsymbol{P}_{\rm s}) \tag{5}$$

with

$$\boldsymbol{R}_{\rm u} = \begin{bmatrix} u_x^2(1-c\alpha)+c\alpha & u_xu_y(1-c\alpha)-u_zs\alpha & u_xu_z(1-c\alpha)+u_ys\alpha \\ u_xu_y(1-c\alpha)+u_zs\alpha & u_y^2(1-c\alpha)+c\alpha & u_yu_z(1-c\alpha)-u_xs\alpha \\ u_xu_z(1-c\alpha)-u_ys\alpha & u_yu_z(1-c\alpha)+u_xs\alpha & u_z^2(1-c\alpha)+c\alpha \end{bmatrix}$$

where $\mathbf{u} = \begin{bmatrix} u_x & u_y & u_z \end{bmatrix}^T = (\boldsymbol{P}_{\rm w} - \boldsymbol{P}_{\rm s})/\|\boldsymbol{P}_{\rm w} - \boldsymbol{P}_{\rm s}\|$ is the unit vector that identifies the rotation axis with reference to which angle α is defined, whereas $c\alpha$ and $s\alpha$ stand for $\cos\alpha$ and $\sin\alpha$, respectively.

The above-described two steps calculated the coordinates of the characteristic internal points $\boldsymbol{P}_{\rm s}$, $\boldsymbol{P}_{\rm e}$ and $\boldsymbol{P}_{\rm w}$. Now, these computed data can be used in the third step for adjusting the joint-space vector according to the actual value of angle α of the final configuration of the robot. Note that after the rotation of angle α, only the angle θ_4 remains the same, and all the other joint variables require a correction that must be computed.

Joints θ_1, θ_2, and θ_3 compose the first spherical wrist with orientation defined by the consecutive elemental rotations of the YZX Tait–Bryan angles where the directions of the $\boldsymbol{y}$, $\boldsymbol{z}$, and $\boldsymbol{x}$- axes coincide with the R-pair axes of the wrist (see Figure 1). This definition yields the following rotation matrix of link 3 with respect to the base:

$${}^b\boldsymbol{R}_3 = \begin{bmatrix} c_1c_2 & -c_1s_2c_3 + s_1s_3 & c_1s_2s_3 + s_1c_3 \\ s_2 & c_2c_3 & -c_2s_3 \\ -s_1c_2 & s_1s_2c_3 + c_1s_3 & -s_1s_2s_3 + c_1c_3 \end{bmatrix} \tag{6}$$

where $c_{\rm i}$ and $s_{\rm i}$ denote $\cos\theta_{\rm i}$ and $\sin\theta_{\rm i}$, for i = 1, 2, 3, respectively.

Additionally, the same ${}^b\boldsymbol{R}_3$ rotation matrix can be obtained numerically using the coordinates of the characteristic internal points as follows:

$${}^b\boldsymbol{R}_3 = \begin{bmatrix} {}^b\boldsymbol{x}_3 & {}^b\boldsymbol{y}_3 & {}^b\boldsymbol{z}_3 \end{bmatrix} = \begin{bmatrix} r_{11} & r_{12} & r_{13} \\ r_{21} & r_{22} & r_{23} \\ r_{31} & r_{32} & r_{33} \end{bmatrix} \tag{7}$$

where, considering that (see Figure 1b) $\boldsymbol{P}_{\rm eu} = \boldsymbol{P}_{\rm s} + ((\boldsymbol{P}_{\rm e} - \boldsymbol{P}_{\rm s}) \cdot \mathbf{u})\mathbf{u}$ is the projection [31,43,47] of $\boldsymbol{P}_{\rm e}$ onto the line defined by the segment $\boldsymbol{P}_{\rm s}\boldsymbol{P}_{\rm w}$ and $\mathbf{v} = \boldsymbol{P}_{\rm e} - \boldsymbol{P}_{\rm eu}$:

$${}^b\mathbf{x}_3 = \begin{bmatrix} r_{11} \\ r_{21} \\ r_{31} \end{bmatrix} = \frac{(\boldsymbol{P}_{\rm e} - \boldsymbol{P}_{\rm s})}{\|\boldsymbol{P}_{\rm e} - \boldsymbol{P}_{\rm s}\|} \; {}^b\mathbf{z}_3 = \begin{bmatrix} r_{13} \\ r_{23} \\ r_{33} \end{bmatrix} = \frac{(\mathbf{u} \times \mathbf{v})}{\|\mathbf{u} \times \mathbf{v}\|} \; {}^b\mathbf{y}_3 = \begin{bmatrix} r_{12} \\ r_{22} \\ r_{32} \end{bmatrix} = {}^b\mathbf{z}_3 \times {}^b\mathbf{x}_3 \tag{8}$$

The comparison of Equations (6) and (7) allows the deduction of the following explicit expressions for θ_1, θ_2, and θ_3:

$$\begin{aligned} \theta_1 &= Atan2(-\text{sign}(\theta_2)r_{31}, \text{sign}(\theta_2)r_{11}) \\ \theta_2 &= Atan2\left(r_{21}, \text{sign}(\theta_2)\sqrt{r_{31}^2 + r_{11}^2}\right) \\ \theta_3 &= Atan2(-\text{sign}(\theta_2)r_{23}, \text{sign}(\theta_2)r_{22}) \end{aligned} \tag{9}$$

where sign(θ_2) is the chosen *sign* for θ_2 determining if $\theta_2 \in \left(\frac{-\pi}{2}, \frac{\pi}{2}\right)$ rad for sign(θ_2) > 0 or if $\theta_2 \in \left(\frac{\pi}{2}, \frac{3\pi}{2}\right)$ rad for sign(θ_2) < 0.

Note that for $\theta_2 = k\pi \pm \frac{\pi}{2}$ rad, for k = 0, 1, ... , it configures a singular posture in the first S-pair since, in this case, the axes of the R-joints 1 and 3 are aligned. Such a singularity is additional to the others, which are the alignment of links 3 and 4, when θ_4 = 0, and the axis alignment of joints 5 and 7 in the second S-pair (the robot wrist), both common to anthropomorphic robots with six DOFs.

Knowing the angles θ_1, θ_2, θ_3, and θ_4, the rotation matrix ${}^{b}\mathbf{R}_4$ describes the rotation of the frame of link 4 with respect to the base frame as follows:

$$ {}^{b}\boldsymbol{R}_4 = {}^{b}\boldsymbol{R}_3{}^{3}\boldsymbol{R}_4 \tag{10} $$

where ${}^{3}\boldsymbol{R}_4$ is the rotation matrix that relates the orientation of link 4 with respect to link 3 described by an elementary rotation around the z-axis by the angle θ_4.

The known rotation matrices ${}^{b}\boldsymbol{R}_{\text{e}}$ and ${}^{7}\boldsymbol{R}_{\text{e}}$ that appear in Equation (4) and the rotation matrix ${}^{b}\boldsymbol{R}_4$ computed through Equation (10) are related by the following relationship, which enables the computation of the remaining joint variables θ_5, θ_6, and θ_7:

$$ {}^{b}\boldsymbol{R}_{\text{e}} = {}^{b}\boldsymbol{R}_4{}^{4}\boldsymbol{R}_7{}^{7}\boldsymbol{R}_{\text{e}} \tag{11} $$

Indeed, isolating ${}^{4}\boldsymbol{R}_7$ in Equation (11) yields the following result:

$$ {}^{4}\boldsymbol{R}_7 = {}^{b}\boldsymbol{R}_4^{T}{}^{b}\boldsymbol{R}_{\text{e}}{}^{7}\boldsymbol{R}_{\text{e}}^{T} = \begin{bmatrix} n_{11} & n_{12} & n_{13} \\ n_{21} & n_{22} & n_{23} \\ n_{31} & n_{32} & n_{33} \end{bmatrix} \tag{12} $$

The second S-pair orientates link 7 with respect to link 4 through the angles θ_5, θ_6, and θ_7 defined as the consecutive elemental rotations of the XZX proper Euler angles, where the directions of the $\boldsymbol{x}$, $\boldsymbol{z}$, and $\boldsymbol{x}$-axes coincide with the R-pair axes of the S-pair (see Figure 1). Therefore, the following expression of rotation matrix ${}^{4}\boldsymbol{R}_7$ holds:

$$ {}^{4}\boldsymbol{R}_7 = \begin{bmatrix} c_6 & -s_6c_7 & s_6s_7 \\ c_5s_6 & c_5c_6c_7 - s_5s_7 & -c_5c_6s_7 - s_5c_7 \\ s_5s_6 & s_5c_6c_7 + c_5s_7 & -s_5c_6s_7 + c_5c_7 \end{bmatrix} \tag{13} $$

Comparing the rotation matrices presented in Equations (12) and (13) reaches the expressions for θ_5, θ_6, and θ_7 as follows:

$$ \begin{aligned} \theta_5 &= \text{Atan2}(\text{sign}(\theta_6)n_{31}, \text{sign}(\theta_6)n_{21}) \\ \theta_6 &= \text{Atan2}\left(\text{sign}(\theta_6)\sqrt{n_{12}^2 + n_{13}^2}, n_{11}\right) \\ \theta_7 &= \text{Atan2}(\text{sign}(\theta_6)n_{13}, -\text{sign}(\theta_6)n_{12}) \end{aligned} \tag{14} $$

where $\text{sign}(\theta_6)$ is the chosen *sign* for θ_6 determining if $\theta_6 \in (0, \pi)$ rad for $\text{sign}(\theta_6) > 0$ or if $\theta_6 \in (-\pi, 0)$ rad for $\text{sign}(\theta_6) < 0$.

The joint variables computed with Equations (9) and (14), together with the value of θ_4 obtained in the inverse kinematics of the intermediate 6-DOF sub-model, yield the sought-after joint-space vector $\mathbf{q}$ that solves the inverse kinematics.

4. Proposition of an Algorithm for Calculating the Distance between a Point and the Robot

4.1. Typical Strategy for Collision Avoidance

When an obstacle enters the robot's safety volume, the collision-avoidance strategy must impose additional movements for the robot that prevent the possibility of a collision [14,30].

The typical collision-avoidance approach firstly measures the distance, d, between the nearest point of the obstacle (i.e., point $\boldsymbol{P}_{\text{c}}$) and a relevant point of the robot, in general, the *"elbow"*, which, in Figure 1, is point $\boldsymbol{P}_{\text{e}}$. Then, by moving point $\boldsymbol{P}_{\text{e}}$, it tries to keep, in the worst case, point $\boldsymbol{P}_{\text{c}}$ on the spherical surface with the center at point $\boldsymbol{P}_{\text{e}}$ and radius equal to the minimum desired distance, dr_{min}; that is, it imposes $d \geq dr_{min}$.

In an SRS-type robot, if the value of dr_{min} is less than the lengths of the segments $\boldsymbol{P}_{\text{s}}\boldsymbol{P}_{\text{e}}$ and/or $\boldsymbol{P}_{\text{e}}\boldsymbol{P}_{\text{w}}$, point $\boldsymbol{P}_{\text{c}}$ can collide with one of the links belonging to the kinematic chain, thus making the strategy unfeasible.

4.2. Proposition of a New Approach for Obstacle Measurement

Let the kinematic structure of a robot be defined through a set of *relevant points*, such as intersections of R-pair axes and centers of spherical wrists. Hereafter, the segments between any two relevant points will be called *referential links*. It is worth noting that the referential links are not constrained to keep their lengths constant during the robot's motion.

This paper proposes the use of axis-symmetric solids, such as *right circular cylinders* and *truncated right-cones* (i.e., a right-cone whose top is cut by a plane parallel to the base), whose axes of symmetry are the referential links, to define the boundaries of safety regions around the links of the robot. In addition, for the relevant points, if necessary, the use of a sphere determines the boundaries of a safety region, as used in conventional approaches [4,9]. The objective of collision avoidance is to keep the points belonging to the obstacles out, or on the boundaries of the safety regions defined by all the above-defined solids. For the SRS-type model, therefore, we have $\boldsymbol{P}_{\mathrm{s}}$, $\boldsymbol{P}_{\mathrm{e}}$, and $\boldsymbol{P}_{\mathrm{w}}$ as relevant points, and the segments $\boldsymbol{P}_{\mathrm{s}}\boldsymbol{P}_{\mathrm{e}}$ and $\boldsymbol{P}_{\mathrm{e}}\boldsymbol{P}_{\mathrm{w}}$ [11] as the respective referential links.

Considering the referential links$\boldsymbol{P}_{\mathrm{s}}\boldsymbol{P}_{\mathrm{e}}$ and $\boldsymbol{P}_{\mathrm{e}}\boldsymbol{P}_{\mathrm{w}}$, we propose the use of truncated right cones to define the boundary of a safety volume around the robot links. For the line segment $\boldsymbol{P}_{\mathrm{s}}\boldsymbol{P}_{\mathrm{e}}$, the truncated right cone has its smaller base circle with radius d_{cmin} centered at point $\boldsymbol{P}_{\mathrm{s}}$ and the larger base circle with radius d_{cmax} centered at point $\boldsymbol{P}_{\mathrm{e}}$. Similarly, for the line segment $\boldsymbol{P}_{\mathrm{e}}\boldsymbol{P}_{\mathrm{w}}$, the truncated right cone has its larger base circle with radius d_{cmax} centered at point $\boldsymbol{P}_{\mathrm{e}}$ and the smaller base circle with radius d_{cmin} centered at point $\boldsymbol{P}_{\mathrm{w}}$. Eventually, as the elbow is a relevant point that is exposed, a sphere, centered at point $\boldsymbol{P}_{\mathrm{e}}$ with radius d_{cmax}, defines the boundary of its safety volume. Figure 2 shows the SRS-type robot together with its safety volume defined as explained above.

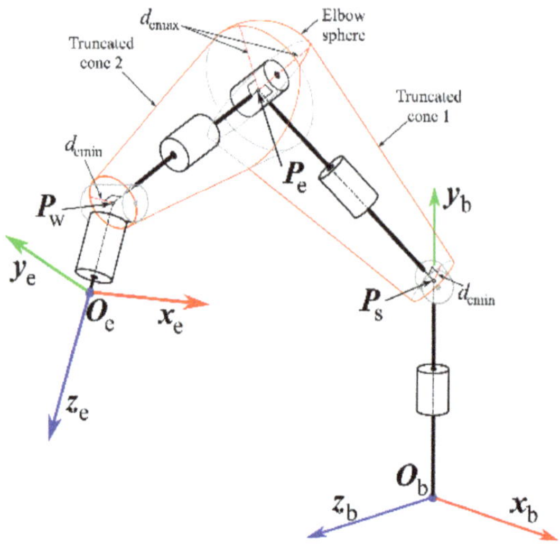

Figure 2. SRS-type robot with its safety volume.

It is worth stressing that points $\boldsymbol{P}_{\mathrm{s}}$ and $\boldsymbol{P}_{\mathrm{w}}$, when under the imminence of a collision, cannot be moved, since $\boldsymbol{P}_{\mathrm{s}}$ is fixed with respect to the base coordinate system and a displacement of point $\boldsymbol{P}_{\mathrm{w}}$ implies changes in the position or direction of the end-effector. In these cases, it is interesting to use $d_{\mathrm{cmin}} = 0$, for real applications featuring a right cone, as a safety volume.

4.3. Computing the Minimum Distance and Respective Collision-Avoidance Angle α

Concerning Figure 3, let us consider a point, $\boldsymbol{P}_{\mathrm{c}}$, belonging to an obstacle located inside the space region where the robot moves and placed on one side of plane-π_2, which is identified by the sign g. Additionally, let us consider the $\boldsymbol{P}_{\mathrm{c}}$ projection points $\boldsymbol{P}_{\mathrm{c1}}$ and

P_{c2} onto the lines defined by the referential links P_sP_e and P_eP_w, respectively [47]. The following relationships hold:

$$\begin{aligned} P_{c1} &= P_s + (P_c - P_s) \cdot (P_e - P_s)\left(\frac{(P_e - P_s)}{\|P_e - P_s\|^2}\right) \\ P_{c2} &= P_e + (P_c - P_e) \cdot (P_w - P_e)\left(\frac{(P_w - P_e)}{\|P_w - P_e\|^2}\right) \end{aligned} \quad (15)$$

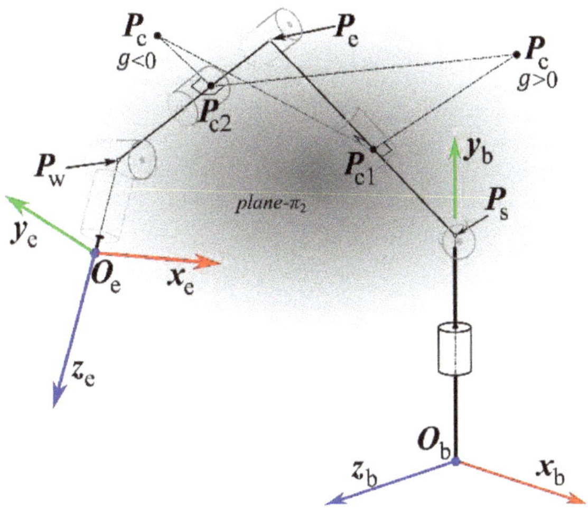

Figure 3. Obstacle point P_c on left or right of plane-π_2 and its projection points P_{c1} and P_{c2}.

In the proposed algorithm, evaluating whether point P_{c1} (P_{c2}) is inside the segment P_sP_e (segment P_eP_w) is the first step. Both, just one, or even none of the two points may belong to their respective segments. Such an evaluation will be true if it is true that for the two parameters u_1 and u_2, defined as follows:

$$u_1 = \frac{(P_{c1} - P_s) \cdot (P_e - P_s)\|P_{c1} - P_s\|}{|(P_{c1} - P_s) \cdot (P_e - P_s)|\|P_e - P_s\|} \quad u_2 = \frac{(P_{c2} - P_e) \cdot (P_w - P_e)\|P_{c2} - P_e\|}{|(P_{c2} - P_e) \cdot (P_w - P_e)|\|P_w - P_e\|}, \quad (16)$$

where the conditions $u_1 \in [0,1]$ and $u_2 \in [0,1]$ hold.

If P_{c1} (P_{c2}) belongs to P_sP_e (to P_eP_w), the next step consists of the computation of the three distances: $d_a = \|P_c - P_{c1}\|$ if $u_1 \in [0,1]$, $d_b = \|P_c - P_e\|$ and $d_c = \|P_c - P_{c2}\|$ if $u_2 \in [0,1]$, which are the distances of P_c, respectively, from the referential link P_sP_e, the relevant point P_e, and the referential link P_eP_w. Considering the minimum and maximum radii of the truncated cones (see Figure 2), the minimum allowed safety distance d_{m1} and d_{m2}, for each cone of each link are calculated as follows:

$$\begin{aligned} d_{m1} &= (1 - u_1)d_{cmin} + u_1 d_{cmax} \\ d_{m2} &= (1 - u_2)d_{cmax} + u_2 d_{cmin} \end{aligned} \quad (17)$$

The collision will be imminent if $d_a < d_{m1}$ or $d_b < d_{cmax}$ or $d_c < d_{m2}$. The three inequalities can be false, which implies that there is no possibility of collision; that is, $\alpha_c(n) = 0$ rad (see Equations (2) and (3)). Differently, if one or more inequalities are true, one or more points on the robot's kinematic chain are under collision imminence. In this case, with only one degree of redundancy and, therefore, only one possible additional movement, the smallest among the minimum distances exceeded (d_a or d_b or d_c) will be identified and will serve as a reference for calculating angle $\alpha_c(n)$ (see Figure 4).

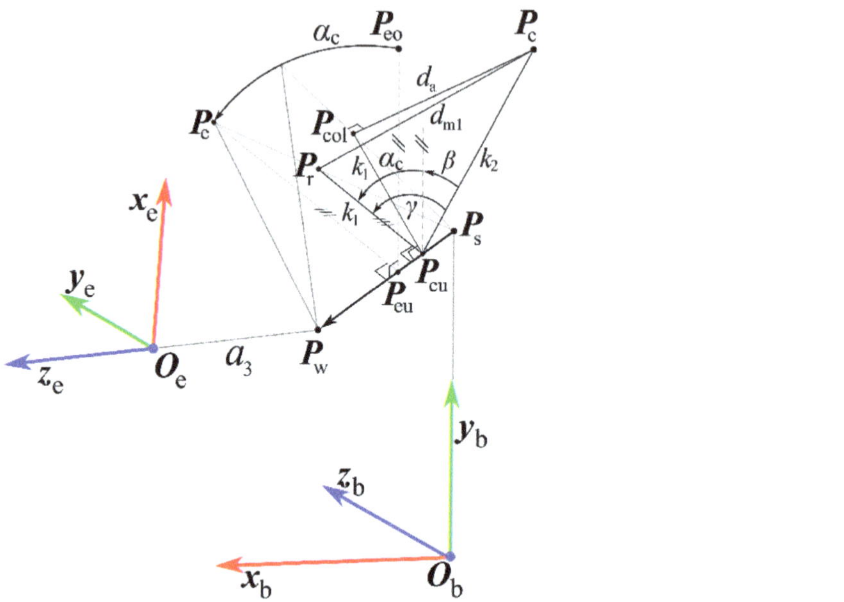

Figure 4. Case with the collision point, $\boldsymbol{P}_{\mathrm{col}}$, belonging to the segment $\boldsymbol{P}_{\mathrm{s}}\boldsymbol{P}_{\mathrm{e}}$: computation of the α_{c} value for avoiding collision.

Let us consider the case (see Figure 4) where the minimum distance between the point $\boldsymbol{P}_{\mathrm{c}}$ and the robot link $\boldsymbol{P}_{\mathrm{s}}\boldsymbol{P}_{\mathrm{e}}$ is less than the allowed distance d_{m1}. Angle $\alpha_{\mathrm{c}}(n)$ must make the distance between $\boldsymbol{P}_{\mathrm{c}}$ and its projection, $\boldsymbol{P}_{\mathrm{col}}$, onto $\boldsymbol{P}_{\mathrm{s}}\boldsymbol{P}_{\mathrm{e}}$ at least equal to d_{m1}, which implies moving $\boldsymbol{P}_{\mathrm{col}}$ from its initial position to at least point $\boldsymbol{P}_{\mathrm{r}}$ of Figure 4 by changing α (see Figure 1b).

Let $\boldsymbol{P}_{\mathrm{cu}}$ be the projection point of $\boldsymbol{P}_{\mathrm{c}}$ onto the line segment $\boldsymbol{P}_{\mathrm{s}}\boldsymbol{P}_{\mathrm{w}}$, calculated as follows:

$$\boldsymbol{P}_{\mathrm{cu}} = \boldsymbol{P}_{\mathrm{s}} + (\boldsymbol{P}_{\mathrm{c}} - \boldsymbol{P}_{\mathrm{s}}) \cdot (\boldsymbol{P}_{\mathrm{w}} - \boldsymbol{P}_{\mathrm{s}}) \left(\frac{(\boldsymbol{P}_{\mathrm{w}} - \boldsymbol{P}_{\mathrm{s}})}{\|\boldsymbol{P}_{\mathrm{w}} - \boldsymbol{P}_{\mathrm{s}}\|^2} \right) \tag{18}$$

Thus, inspecting Figure 4, one notes that the angle $\alpha_{\mathrm{c}}(n) = \gamma - \beta$, with

$$\begin{aligned} \gamma &= \mathrm{Atan2}\left(\mathrm{sign}(g)\sqrt{1 - \left(\frac{k_1^2 + k_2^2 - d_{\mathrm{m1}}^2}{2k_1k_2} \right)^2}, \frac{k_1^2 + k_2^2 - d_{\mathrm{m1}}^2}{2k_1k_2} \right) \\ \beta &= \mathrm{Atan2}\left(\mathrm{sign}(g)\sqrt{1 - (\mathbf{v}_1 \cdot \mathbf{v}_2)^2}, \mathbf{v}_1 \cdot \mathbf{v}_2 \right) \end{aligned} \tag{19}$$

where $k_1 = \|\boldsymbol{P}_{\mathrm{col}} - \boldsymbol{P}_{\mathrm{cu}}\|$, and $k_2 = \|\boldsymbol{P}_{\mathrm{c}} - \boldsymbol{P}_{\mathrm{cu}}\|$, $\mathbf{v}_1 = \frac{\boldsymbol{P}_{\mathrm{eo}} - \boldsymbol{P}_{\mathrm{eu}}}{\|\boldsymbol{P}_{\mathrm{eo}} - \boldsymbol{P}_{\mathrm{eu}}\|}$, $\mathbf{v}_2 = \frac{\boldsymbol{P}_{\mathrm{c}} - \boldsymbol{P}_{\mathrm{eu}}}{\|\boldsymbol{P}_{\mathrm{c}} - \boldsymbol{P}_{\mathrm{eu}}\|}$, and $g = ((\boldsymbol{P}_{\mathrm{c}} - \boldsymbol{P}_{\mathrm{cu}}) \times (\boldsymbol{P}_{\mathrm{col}} - \boldsymbol{P}_{\mathrm{cu}})) \cdot (\boldsymbol{P}_{\mathrm{w}} - \boldsymbol{P}_{\mathrm{s}})$. The sign of g indicates which side of plane-π_2 point $\boldsymbol{P}_{\mathrm{c}}$ belongs to (see Figure 3) and determines the rotation direction of the angles γ and β.

For the nth iteration of the trajectory generation, the previous steps can be summarized in Algorithm 1.

Algorithm 1. Computation of angle α_f at the nth iteration.

Input
Desired position and orientation of the end-effector at the iteration *n*;
Compute the inverse kinematics for $\alpha_c = 0$ rad;
Coordinates of the points belonging to the referential links and relevant points % *In this development* $\mathbf{P}_s$, $\mathbf{P}_e$ and $\mathbf{P}_w$
Retrieve the values of $\alpha_c(n-1)$, $\alpha_c(n-2)$, $\alpha_f(n-1)$, $\alpha_f(n-2)$
$\mathbf{P}_c$ % *Coordinates of a possible collision point*
find← false; % *Boolean variable to indicate if a collision point exists inside safety volume.*
d_{mc}← 0; % *Initializing the minimum of the minimal distances.*
$\mathbf{P}_{col}$← []; % *Point in the kinematic chain with the minor distance to* $\mathbf{P}_c$.
Compute u_1 and u_2;
if $u_1 \in [0,1]$ **then** % $\mathbf{P}_{c1}$ *is between* $\mathbf{P}_s$ *and* $\mathbf{P}_e$
  $d_a = \|\mathbf{P}_c - \mathbf{P}_{c1}\|$;
  $d_{m1} = (1-u_1)d_{cmin} + u_1 d_{cmax}$;
  if $d_a < d_{m1}$ **then**
    $d_{mc} \leftarrow d_{m1}$; % *saving the minimal allowed distance*
    $\mathbf{P}_{col} \leftarrow \mathbf{P}_{c1}$; % *saving the point on the kinematic chain*
    find← true;
  end
elseif $u_2 \in [0,1]$ **then** % $\mathbf{P}_{c2}$ is between $\mathbf{P}_e$ and $\mathbf{P}_w$
  $d_c = \|\mathbf{P}_c - \mathbf{P}_{c2}\|$;
  $d_{m2} = (1-u_2)d_{cmin} + u_2 d_{cmax}$;
  if $d_c < d_{m2}$ **then**
    if $d_c < d_a$ **then**
      $d_{mc} \leftarrow d_{m2}$; % *saving the minimal distance*
      $\mathbf{P}_{col} \leftarrow \mathbf{P}_{c2}$; % *saving the point on the kinematic chain*
      find← true;
    end if
  end if
end if
$d_b = \|\mathbf{P}_c - \mathbf{P}_e\|$; % *evaluating the collision distance from* $\mathbf{P}_c$ *to the point* $\mathbf{P}_e$
if $d_b < d_{cmax}$ **then** % *considering a sphere as safety region around* $\mathbf{P}_e$*–see* Figure 2
  if find=false **then**;
    $d_{mc} \leftarrow d_b$;
    $\mathbf{P}_{col} \leftarrow \mathbf{P}_e$;
    find← true;
  elseif $d_b < d_{mc}$ **then**
    $d_{mc} \leftarrow d_b$;
    $\mathbf{P}_{col} \leftarrow \mathbf{P}_e$;
  end if
end if
if find=false % *if there is no need to avoid collision*
  $\alpha_c(n) = 0$ rad;
  d_{mc}=0;
else % *computing the value of* α_c *to move the robot away from the collision point.*
  Compute γ and β;
  $\alpha_c(n) = \gamma - \beta$;
end if
$\alpha_f(n) = h_1\alpha_c(n-1) + h_2\alpha_c(n-2) + b_1\alpha_f(n-1) + b_2\alpha_f(n-2)$; % *Equation (3)*
Solve inverse kinematics using $\alpha_f(n)$;
Output Save joint positions;

5. Graphical Results

Numerical simulations were conducted to validate the proposed collision-avoidance strategy. In these simulations, the Kuka-LBR-iiwa real lengths [45], that is, a_0 = 340 mm, a_1 = 400 mm, a_2 = 400 mm, and a_3 = 126 mm, were selected for the geometric model of the SRS-type robot. Additionally, the chosen motion task for the robot is an end-effector path (see Figures 2 and 5) that keeps the z_e axis perpendicular to the x_bz_b coordinate-plane and makes point $\mathbf{O}_e$ perform five rides along the circular trajectory, with the center at point $[400\ 200\ 200]^T$ mm and radius 200 mm, that lies on the plane y_b = 200 mm. The robot motion is sampled in 1000 equidistant time-instants, and at every sampling instant, Algorithm 1 is executed.

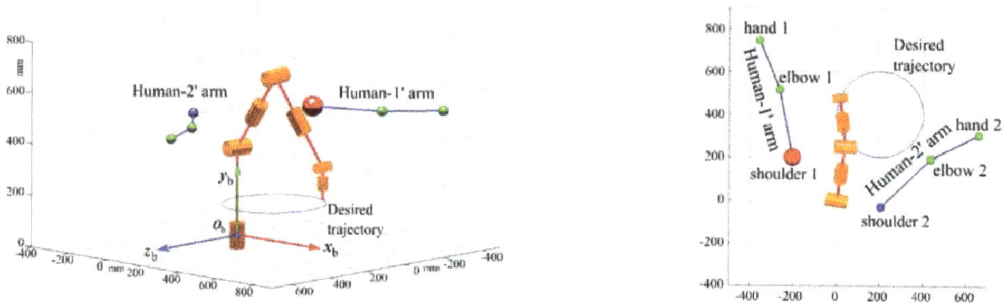

Figure 5. Virtual simulation environment at iteration 390. Blue ball—shoulder coordinates; green ball—coordinates of the elbows and hands; red ball—point under imminent collision.

The simulation environment, over the robot, includes two human arms identified as *Human-arm 1* and *Human-arm 2*, composed of two links and located in the space by three points named *shoulder*, *elbow*, and *hand*, thus improving the experiments presented in [48], where only one arm is used. The links shoulder–elbow and elbow–hand have lengths [49] of 319.6 mm and 246.5 mm with angle displacement defined with respect to the x_b-axis. The chosen motion for these two arms makes the three above-mentioned points perform oscillatory movements with the motion parameters reported in Table 1. Human-arm motion takes place while the robot's end-effector performs the above-defined five circular paths with the center at $[400\ 200\ 200]^T$ mm and a radius of 200 mm.

Table 1. Human-arms simulation parameters.

	Human-Arm 1	Human-Arm 2
Base (shoulder)	$[-50\ 500\ 250]^T$ mm	$[150\ 500\ -200]^T$ mm
Base rotation radius (around y-axis)	50 mm	50 mm
Base number of turns	5	11
Shoulder angle range	0.4 rad	0.4 rad
Shoulder angle offset	0.4 rad	0.2 rad
Shoulder number of turns	9	13
Elbow angle range	0.4 rad	0.4 rad
Elbow angle offset	0.4 rad	0 rad
Elbow number of turns	7	9

The tests were developed on a laptop with an Intel Core i7 CPU (2.7 GHz, 800 Gb RAM), spending 3.68 s in execution. All the simulations were developed by using MATLAB R2021a. Eventually, the second-order transfer function, the reference for programming the filter, was adjusted with $p = 0.2\ s^{-1}$, and the sampling time was adjusted to $T = 0.06$ s, which corresponds to $m = pT = 0.012$ and a simulation cycle of 60 s. It is worth stressing that, to avoid muscular injuries, a worker should not execute motion cycles with a period lower than 60 s; so, in these simulations, the human-arm motion is much faster than the admitted one (i.e., the chosen parameters make the simulation environment much faster than a realistic one).

The graphical simulation environment presented in Figure 5 enables us to observe the evolution of the proposed algorithm and the SRS-type's end-effector performance following the desired trajectory while dealing with collision avoidance.

A video that presents the complete simulation experiment of the proposed algorithm is available at [50] and as "Supplementary Materials" accompanying this paper. Figure 6 presents three frames from the above-mentioned video showing the SRS-type robot while avoiding a collision. The video shows that, in this simulation environment, a sampling rate (=1/T) of 16.7 Hz is sufficient to avoid collision with the proposed strategy. It is worth stressing that since this sampling rate corresponds to a "reaction time" of 60 ms for the robot and the human reaction time is about 200 ms, the proposed strategy can enable the robot to react three times faster than a human being.

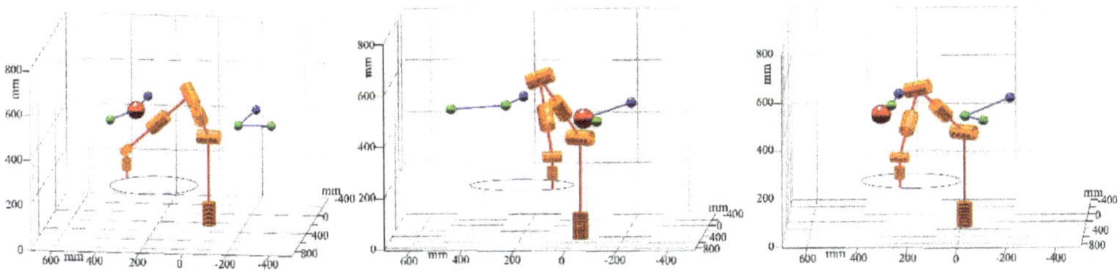

(a) (b) (c)

Figure 6. Frames from the simulation video at iteration (**a**) 76; (**b**) 179; (**c**) 407.

Figure 7 shows the computed angle α_c and its filtered counterpart α_f, used as a reference for the inverse kinematics, during the simulation cycle. The analysis of Figure 7 reveals that α_f is a regular curve.

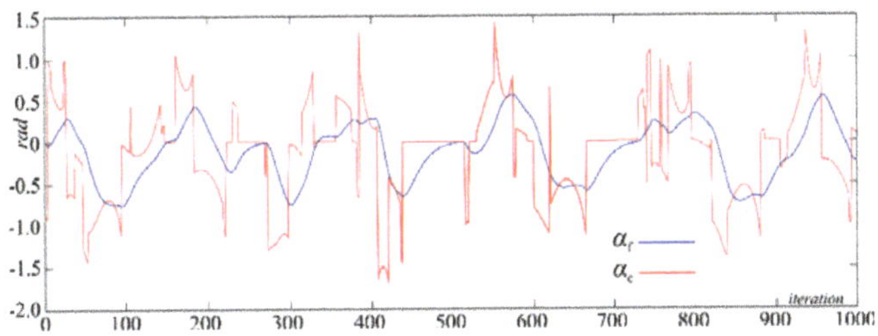

Figure 7. Comparison between the computed angle α_c and its filtered α_f used to solve the inverse kinematics.

Figures 8 and 9 show the rates of angle and angular speed as a function of the sampling period T to evaluate the continuity and smoothness of the trajectory for the filtered angle α_f.

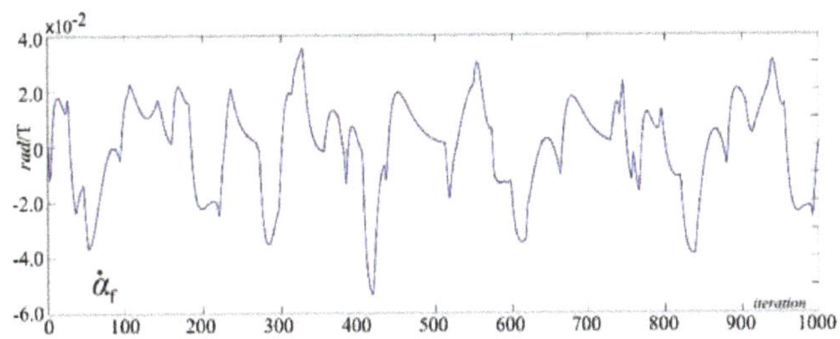

Figure 8. Computed rate of change of α_f.

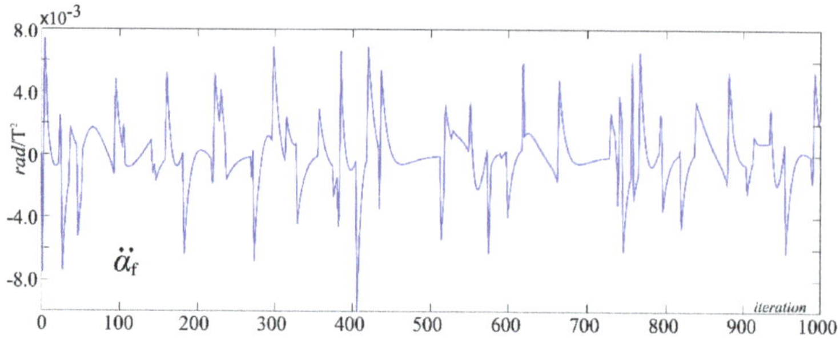

Figure 9. Computed rate of change of $\dot{\alpha}_f$.

The inverse-kinematics solution of the SRS-type robot yields the position profile of joints 1, 2, and 3, which are θ_1, θ_2, and θ_3, respectively, presented in Figure 10; as well as joints 4, 5, 6, and 7, which are θ_4, θ_5, θ_6, and θ_7, respectively, presented in Figure 11.

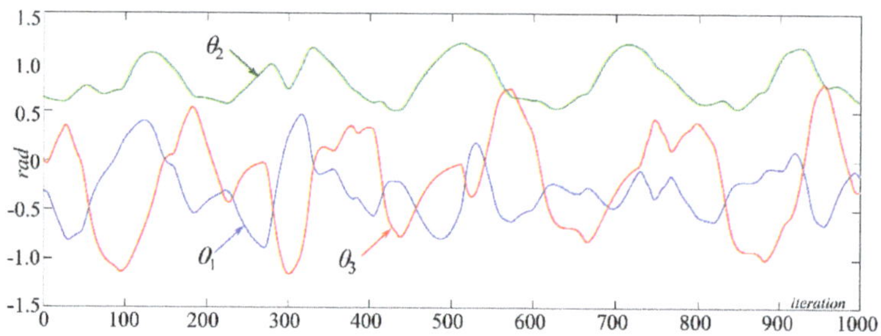

Figure 10. Result of the inverse kinematic for SRS-type robot joints 1, 2, and 3.

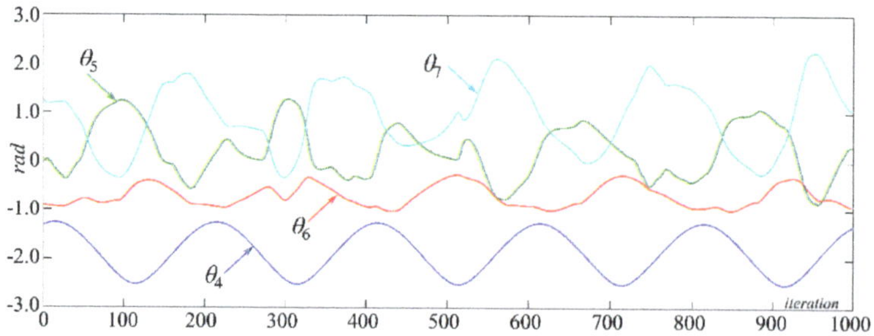

Figure 11. Result of the inverse kinematic for SRS-type robot joints 4, 5, 6, and 7.

For each iteration, and considering the known sampling time T, the rates of changes of joint variables and corresponding accelerations were calculated for each joint by first-order approximations. Thus, taking the maximum and minimum values obtained, Figures 12 and 13 show the range achieved for the rate of change of position and acceleration, respectively, in each iteration of all robot joint variables. The rates of joint positions and acceleration behavior did not show abrupt variations, remaining in a limited range with similar magnitudes.

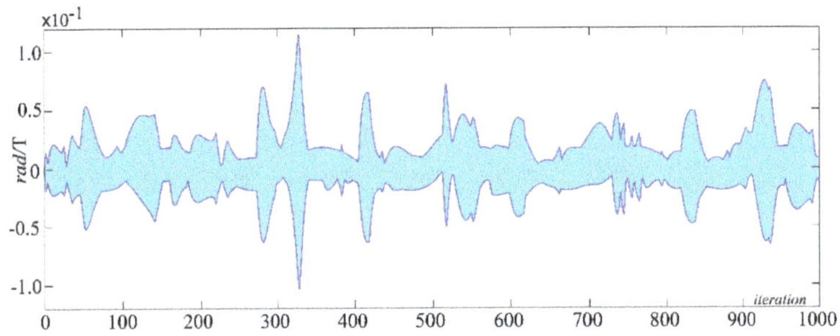

Figure 12. Range of computed maximum and minimum rates of change of joint variables.

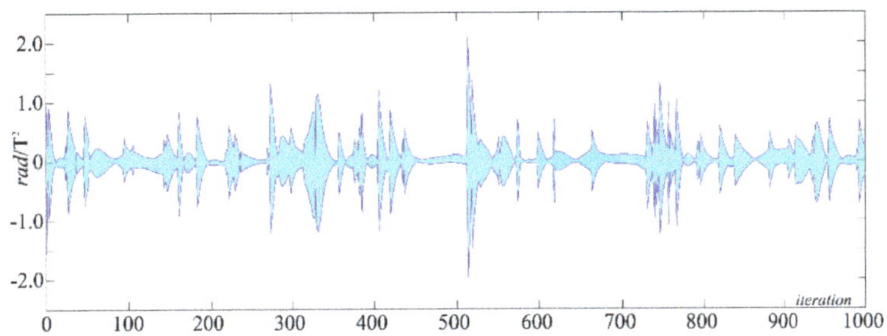

Figure 13. Range of computed maximum and minimum accelerations of joint.

The angle α_f smooths the resultant distances between the robot and the collision point. Figure 14 shows the behavior of the distance between the SRS-type robot and the collision point (red) along the simulation, compared with the distances when under imminent collision (blue). It is worth mentioning that when the robot is out of collision avoidance, the distance d_{mc} used to calculate the angle α_f is adjusted to 0, as seen in line 37 of Algorithm 1.

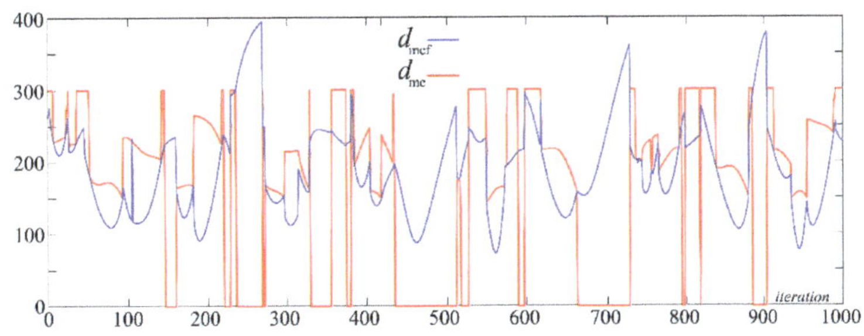

Figure 14. Computed d_{mc} distances from obstacle and the SRS-type robot along the simulation and the respective distance d_{mcf} used to compute the α_f when under collision imminence.

6. Discussions

The results presented in the test showed that the proposed algorithm smoothed all the joints' position and velocity profiles in a dynamic environment.

The test aims at replicating the relationship between a collision point crossing a circle and the consequent problem of variation in the resulting distances projected on the robot's links. Although the collision curve is deterministic in the test, the procedure applies to random variations in the collision point coordinates or a more significant number of collision points, including the number of links and relevant robot points. The algorithm can deal with multiple collision points, attempting to move the robot away from that nearest collision point. For the collision avoidance on k simultaneous collision points, the robot should have a redundancy index $r = k$; for example, if it is necessary to have three collision points or control the three distances, such as d_a, d_b, and d_c discussed in Section 4.3, the robot should have degrees of redundancy $r = 3$. In this way, the algorithm should be executed three times for each collision point featuring its expandability.

The presented inverse kinematics, which solves the α angle for the four quadrants, is based on review propositions [8,39,48] and refines them, thus improving the formulation.

Additionally, in the presented simulation, the available degree of redundancy is 1, and it is strictly sufficient to solve the collision-avoidance problem without including the joint limits control, as was achieved in other papers [15,20,48,51–53] where the introduced redundancy was exploited by optimization procedures to find the best α that avoids reaching joint-position limits. The possible involvement of joint-position limits' control, over collision avoidance, in programming trajectories and kinematic algorithms solely depends on the availability of redundancy degrees higher than one, which, if available, can be exploited by implementing one of the strategies proposed in the literature, even with the proposed algorithm.

Similar algorithms present safety volumes described by a limited number of points [44,54] or based on numerical approaches [4,20]. The here-discussed proposal uses a parametric surface around the links, instead of limited volumes as discussed in [4], to define the safety volume in the computation of the collision distance for any point of the link, which makes this new approach more realistic and reliable.

At the same time, the use of a second-order filter makes the strategy feasible, adjustable, and replicable, ensuring, together with the description of the safety volume, smooth transients in the trajectories as well as an analytical collision-avoidance algorithm without numerical procedures that increase the computational burden, as found in [23]. The smoothing of the collision-avoidance transients appears in the bounded rate of positions and acceleration profiles for the robot joints (see Figures 12 and 13), continuously ensuring that discontinuities, such as those shown in [4,54,55], do not arise despite the dynamic environment.

Using a digital filter in the transition between paths to be followed by the joints additionally guarantees the smoothness of the programmed trajectories.

It is important to note that the collision-avoidance algorithm's proposition has limitations when several collision points or several links lead to opposite displacements generating conflicts of movements characterizing singularities. Concerning the singularities, it is observed that collision avoidance does not occur when g = 0, or the collision point $\boldsymbol{P}_c$ belongs to the plane formed by the points $\boldsymbol{P}_s$, $\boldsymbol{P}_w$, and $\boldsymbol{P}_e$, as can be seen in Figures 3 and 4; that is,

$$D = \frac{|(\boldsymbol{P}_c - \boldsymbol{P}_s) \cdot \boldsymbol{n}_p|}{\|\boldsymbol{n}_p\|} = 0 \tag{20}$$

where D corresponds to the distance between the point $\boldsymbol{P}_c$ and the plane formed by the points P_s, P_w, and P_e, and n_p is the result of a cross-product as follows:

$$\boldsymbol{n}_p = \frac{(\boldsymbol{P}_e - \boldsymbol{P}_s)}{\|(\boldsymbol{P}_e - \boldsymbol{P}_s)\|} \times \frac{(\boldsymbol{P}_w - \boldsymbol{P}_e)}{\|(\boldsymbol{P}_w - \boldsymbol{P}_e)\|} \tag{21}$$

It is interesting to note that angle α_f does not produce errors of position and orientation of the end effector, despite the need for a time of accommodation or convergence to the final values following trajectories, since its primary task is to control the distance from the collision.

However, the need for transitory times for collision avoidance means suitable adjustments to the filter time constant according to the desired performance and control specifications.

The proposed algorithm can deal with collision avoidance between multiple robots simultaneously. In this case, the distance between the robot's links is controlled by calculating the geometric distance between the line segments to which the links belong. Thus, taking a robot as a reference, the $\mathbf{P}_c$ points belong to the other robots, and running the proposed algorithm in each robot's trajectory generator allows several robots to share the same space without generating collisions.

7. Conclusions

Workspaces shared between robots and humans in collaborative tasks require strategies that include collision avoidance and the programming of smooth transitional displacements.

This paper presented an algorithm for the treatment of collision and posture adjustment of robots based on analytical spatial geometry and the use of digital second-order critically damped filters. The strategy includes a posture adjustment of the robot based on an additional movement, defined as α, available for the redundant 7-DOF SRS-type robot. The simulations showed the smoothing of the joints' trajectory profiles and collision avoidance effectiveness without collisions in a dynamic environment.

The geometric approach allowed us to present a proposition for inverse kinematics considering the posture adjustment and the projection of the collision point on any point on the links of the robot's kinematic chain.

Spatial surfaces were used as an example for the determination of the safety regions of the robot. For relevant points and links (under imminent collision), we used the formulation of spheres and truncated cones, which allows us to identify the possibility of collision and activate the collision-avoidance strategy. According to the geometry of links and components that form the environment where the robot is inserted, other formulations can be adapted and used as safety surfaces.

The proposed collision-avoidance solution is feasible and straightforward, providing a quick implementation solution for collaborative robots. The fact that the proposal is based only on analytical formulations reduces the computational effort, thus increasing the reliability of the solutions for trajectories and the efficiency of collision control and other processing in each sampling time.

Supplementary Materials: The following supporting information can be downloaded at: https://www.mdpi.com/article/10.3390/robotics11050093/s1.

Author Contributions: Conceptualization, H.S. and R.D.G.; Funding acquisition, R.D.G.; Methodology, H.S. and R.D.G.; Software, H.S.; Supervision, R.D.G.; Validation, H.S.; Writing–original draft, H.S.; Writing–review & editing, R.D.G. All authors have read and agreed to the published version of the manuscript.

Funding: This research was developed at the Laboratory of Mechatronics and Virtual Prototyping (LaMaViP) of Ferrara Technopole, supported by the UNIFE FAR2020 fund, in partnership with the Raul Guenther Laboratory of Applied Robotics of the Federal University of Santa Catarina, supported by CNPq–Conselho Nacional de Desenvolvimento Científico e Tecnológico (National Council for Scientific and Technological Development) project 307249/2021-2, Brazil.

Institutional Review Board Statement: Not applicable.

Informed Consent Statement: Not applicable.

Data Availability Statement: This work does not use experimental data. The data necessary to replicate the computations illustrated in the paper are included in the text of the paper.

Conflicts of Interest: The authors declare no conflict of interest.

References

1. Mihelj, M.; Bajd, T.; Ude, A.; Lenarcic, J.; Stanovnik, A.; Munih, M.; Rejc, J.; Šlajpah, S. *Robotics*, 2nd ed.; Springer: Cham, Switzerland, 2018. [CrossRef]
2. Matheson, E.; Minto, R.; Zampieri, E.G.G.; Faccio, M.; Rosati, G. Human–Robot Collaboration in Manufacturing Applications: A Review. *Robotics* **2019**, *8*, 100. [CrossRef]
3. El Zaatari, S.; Marei, M.; Li, W.; Usman, Z. Cobot Programming for Collaborative Industrial Tasks: An Overview. *Robot. Auton. Syst.* **2019**, *116*, 162–180. [CrossRef]
4. Safeea, M.; Neto, P.; Bearee, R. On-Line Collision Avoidance for Collaborative Robot Manipulators by Adjusting off-Line Generated Paths: An Industrial Use Case. *Robot. Auton. Syst.* **2019**, *119*, 278–288. [CrossRef]

5. Alebooyeh, M.; Urbanic, R.J. Neural Network Model for Identifying Workspace, Forward and Inverse Kinematics of the 7-DOF YuMi 14000 ABB Collaborative Robot. *IFAC-PapersOnLine* **2019**, *52*, 176–181. [CrossRef]
6. IFR. *A Positioning Paper: Demystifying Collaborative Industrial Robots*; International Federation of Robotics: Frankfurt, Germany, 2018; pp. 1–5.
7. *ISO/TS 15066:2016*; Robots and Robotic Devices—Collaborative Robots. ISO: Geneva, Switzerland, 2016.
8. Busson, D.; Bearee, R.; Olabi, A. Task-Oriented Rigidity Optimization for 7 DOF Redundant Manipulators. *IFAC-PapersOnLine* **2017**, *50*, 14588–14593. [CrossRef]
9. Liu, W.; Chen, D.; Steil, J. Analytical Inverse Kinematics Solver for Anthropomorphic 7-DOF Redundant Manipulators with Human-Like Configuration Constraints. *J. Intell. Robot. Syst. Theory Appl.* **2017**, *86*, 63–79. [CrossRef]
10. Zhang, L.; Xiao, N. A Novel Artificial Bee Colony Algorithm for Inverse Kinematics Calculation of 7-DOF Serial Manipulators. *Soft Comput.* **2019**, *23*, 3269–3277. [CrossRef]
11. Liu, M.C.; Tsai, H.H.; Hsiao, T. Kinematics-Based Studies on a 7-DOF Redundant Manipulator. In Proceedings of the 2014 CACS International Automatic Control Conference (CACS 2014), Kaohsiung, Taiwan, 26–28 November 2014; pp. 228–231. [CrossRef]
12. Di Gregorio, R.; Simas, H. Dimensional Synthesis of the Single-Loop Translational Parallel Manipulator PRRR-PRPU. *Meccanica* **2018**, *53*, 481–495. [CrossRef]
13. Simas, H.; Martins, D.; Di Gregorio, R. Smooth Path Planning for Redundant Robots on Collision Avoidance. In *Mechanisms and Machine Science*; Springer: Cham, Switzerland, 2019; Volume 73, pp. 1869–1878. [CrossRef]
14. Simas, H.; Guenther, R.; Da Cruz, D.F.M.; Martins, D. A New Method to Solve Robot Inverse Kinematics Using Assur Virtual Chains. *Robotica* **2009**, *27*, 1017–1026. [CrossRef]
15. Rouhani, M.; Ebrahimabadi, S. Inverse Kinematics of a 7-DOF Redundant Robot Manipulator Using the Active Set Approach under Joint Physical Limits. *Turk. J. Electr. Eng. Comput. Sci.* **2017**, *25*, 3920–3931. [CrossRef]
16. Osa, T. Multimodal Trajectory Optimization for Motion Planning. *Int. J. Robot. Res.* **2020**, *39*, 983–1001. [CrossRef]
17. Dereli, S.; Köker, R. Calculation of the Inverse Kinematics Solution of the 7-DOF Redundant Robot Manipulator by the Firefly Algorithm and Statistical Analysis of the Results in Terms of Speed and Accuracy. *Inverse Probl. Sci. Eng.* **2020**, *28*, 601–613. [CrossRef]
18. Ayten, K.K.; Sahinkaya, M.N.; Dumlu, A. Real Time Optimum Trajectory Generation for Redundant/Hyper-Redundant Serial Industrial Manipulators. *Int. J. Adv. Robot. Syst.* **2017**, *14*, 172988141773724. [CrossRef]
19. Siciliano, B. Kinematic Control of Redundant Robot Manipulators: A Tutorial. *J. Intell. Robot. Syst.* **1990**, *3*, 201–212. [CrossRef]
20. Quispe, A.H.; Stilman, M. Deterministic Motion Planning for Redundant Robots along End-Effector Paths. In Proceedings of the 2012 12th IEEE-RAS International Conference on Humanoid Robots (Humanoids 2012), Osaka, Japan, 29 November–1 December 2012; pp. 785–790. [CrossRef]
21. Zanchettin, A.M.; Rocco, P. A General User-Oriented Framework for Holonomic Redundancy Resolution in Robotic Manipulators Using Task Augmentation. *IEEE Trans. Robot.* **2012**, *28*, 514–521. [CrossRef]
22. Klein, C.A.; Huang, C.H. Review of Pseudoinverse Control for Use with Kinematically Redundant Manipulators. *IEEE Trans. Syst. Man Cybern.* **1983**, *SMC-13*, 245–250. [CrossRef]
23. Kalakrishnan, M.; Chitta, S.; Theodorou, E.; Pastor, P.; Schaal, S. STOMP: Stochastic Trajectory Optimization for Motion Planning. In Proceedings of the 2011 IEEE International Conference on Robotics and Automation, Shanghai, China, 9–13 May 2011; pp. 4569–4574. [CrossRef]
24. Besset, P.; Béarée, R. FIR Filter-Based Online Jerk-Constrained Trajectory Generation. *Control Eng. Pract.* **2017**, *66*, 168–180. [CrossRef]
25. Gerelli, O.; Guarino Lo Bianco, C. A Discrete-Time Filter for the on-Line Generation of Trajectories with Bounded Velocity, Acceleration, and Jerk. In Proceedings of the 2010 IEEE International Conference on Robotics and Automation, Anchorage, AK, USA, 3–7 May 2010; pp. 3989–3994. [CrossRef]
26. Biagiotti, L.; Melchiorri, C. Online Trajectory Planning and Filtering for Robotic Applications via B-Spline Smoothing Filters. In Proceedings of the 2013 IEEE/RSJ International Conference on Intelligent Robots and Systems, Tokyo, Japan, 3–7 November 2013; pp. 5668–5673. [CrossRef]
27. Preiss, J.A.; Hausman, K.; Sukhatme, G.S.; Weiss, S. Simultaneous Self-Calibration and Navigation Using Trajectory Optimization. *Int. J. Robot. Res.* **2018**, *37*, 1573–1594. [CrossRef]
28. Franklin, G.F.; Powell, J.D.; Emami-Naeini, A. *Feedback Control of Dynamic Systems*, 8th ed.; Pearson: London, UK, 2019.
29. Colome, A.; Torras, C. Redundant Inverse Kinematics: Experimental Comparative Review and Two Enhancements. In Proceedings of the 2012 IEEE/RSJ International Conference on Intelligent Robots and Systems, Vilamoura-Algarve, Portugal, 7–12 October 2012; pp. 5333–5340. [CrossRef]
30. Siciliano, B.; Lorenzo, S.; Villani, L.; Orilo, G. *Robotics: Modelling, Planning and Control*, 2nd ed.; Springer: London, UK, 2010.
31. Tsai, L.-W. *Robot Analysis and Design: The Mechanics of Serial and Parallel Manipulators*, 1st ed.; John Wiley & Sons, Inc.: Hoboken, NJ, USA, 1999.
32. Dereli, S.; Köker, R. A Meta-Heuristic Proposal for Inverse Kinematics Solution of 7-DOF Serial Robotic Manipulator: Quantum Behaved Particle Swarm Algorithm. *Artif. Intell. Rev.* **2020**, *53*, 949–964. [CrossRef]
33. Mooney, J.G.; Johnson, E.N. A Comparison of Automatic Nap-of-the-Earth Guidance Strategies for Helicopters. *J. Field Robot.* **2014**, *33*, 1–17. [CrossRef]

34. Zhao, S.; Zhu, Z.; Luo, J. Multitask-Based Trajectory Planning for Redundant Space Robotics Using Improved Genetic Algorithm. *Appl. Sci.* **2019**, *9*, 2226. [CrossRef]
35. Ogura, Y.; Aikawa, H.; Shimomura, K.; Kondo, H.; Morishima, A.; Lim, H.O.; Takanishi, A. Development of a New Humanoid Robot WABIAN-2. In Proceedings of the 2006 IEEE International Conference on Robotics and Automation, 2006. ICRA 2006, Orlando, FL, USA, 15–19 May 2006; pp. 76–81. [CrossRef]
36. Starke, S.; Hendrich, N.; Krupke, D.; Zhang, J. Evolutionary Multi-Objective Inverse Kinematics on Highly Articulated and Humanoid Robots. In Proceedings of the 2017 IEEE/RSJ International Conference on Intelligent Robots and Systems (IROS), Vancouver, BC, Canada, 24–28 September 2017; pp. 6959–6966. [CrossRef]
37. Lim, J.; Oh, J.-H. Backward Ladder Climbing Locomotion of Humanoid Robot with Gain Overriding Method on Position Control. *J. Field Robot.* **2016**, *33*, 687–705. [CrossRef]
38. Faria, C.; Ferreira, F.; Erlhagen, W.; Monteiro, S.; Bicho, E. Position-Based Kinematics for 7-DoF Serial Manipulators with Global Configuration Control, Joint Limit and Singularity Avoidance. *Mech. Mach. Theory* **2018**, *121*, 317–334. [CrossRef]
39. Wang, Y.; Artemiadis, P. Closed-Form Inverse Kinematic Solution for Anthropomorphic Motion in Redundant Robot Arms. *Adv. Robot. Autom.* **2013**, *2*, 1000110. [CrossRef]
40. Simas, H.; Di Gregorio, R. Position Analysis, Singularity Loci and Workspace of a Novel 2PRPU Schoenflies-Motion Generator. *Robotica* **2019**, *37*, 141–160. [CrossRef]
41. Simas, H.; Di Gregorio, R. Kinetostatics and Optimal Design of a 2PRPU Shoenflies-Motion Generator. In *Mechanisms and Machine Science*; Springer Cham: Cham, Switzerland, 2017; Volume 54. [CrossRef]
42. Mohammed, A.; Schmidt, B.; Wang, L. Active Collision Avoidance for Human–Robot Collaboration Driven by Vision Sensors. *Int. J. Comput. Integr. Manuf.* **2017**, *30*, 970–980. [CrossRef]
43. Süli, E.; Mayers, D.F. *An Introduction to Numerical Analysis*; Cambridge University Press: New York, NY, USA, 2003. [CrossRef]
44. Luo, R.C.; Ko, M.C.; Chung, Y.T.; Chatila, R. Repulsive Reaction Vector Generator for Whole-Arm Collision Avoidance of 7-DoF Redundant Robot Manipulator. In Proceedings of the 2014 IEEE/ASME International Conference on Advanced Intelligent Mechatronics, Besacon, France, 8–11 July 2014; pp. 1036–1041. [CrossRef]
45. KUKA Sensitive Robotics: LBR iiwa. Catalog published by KUKA Roboter GmbH: Augsburg, Germany. 2016. Available online: https://pdf.directindustry.com/pdf/kuka-ag/kuka-sensitive-robotics-lbr-iiwa/17587-724449.html (accessed on 9 July 2022).
46. Huber, G.; Wollherr, D. Efficient Closed-Form Task Space Manipulability for a 7-DOF Serial Robot. *Robotics* **2019**, *8*, 98. [CrossRef]
47. Larson, R.; Edwards, B.H. *Multivariable Calculus*, 9th ed.; Cengage Learning: Boston, MA, USA, 2009.
48. Herrera Pineda, J.C.; Mejia Rincon, L.; Simoni, R.; Simas, H. Maximum Isotropic Force Capability Maps in Planar Cooperative Systems: A Practical Case Study. In *Mechanisms and Machine Science*; Springer: Cham, Switzerland, 2018; Volume 54, pp. 160–170. [CrossRef]
49. Oh, J.; Bae, H.; Oh, J.H. Analytic Inverse Kinematics Considering the Joint Constraints and Self-Collision for Redundant 7DOF Manipulator. In Proceedings of the 2017 First IEEE International Conference on Robotic Computing (IRC), Taichung, Taiwan, 10–12 April 2017; pp. 123–128. [CrossRef]
50. Video that Presents the Complete Simulation Experiment. Available online: https://youtu.be/BfRjA8dgnq8 (accessed on 9 July 2022).
51. Frantz, J.C.; Rincon, L.M.; Simas, H.; Martins, D. Wrench Distribution of a Cooperative Robotic System Using a Modified Scaling Factor Method. *J. Braz. Soc. Mech. Sci. Eng.* **2018**, *40*, 177. [CrossRef]
52. Zhou, D.; Ji, L.; Zhang, Q.; Wei, X. Practical Analytical Inverse Kinematic Approach for 7-DOF Space Manipulators with Joint and Attitude Limits. *Intell. Serv. Robot.* **2015**, *8*, 215–224. [CrossRef]
53. Scimmi, L.S.; Melchiorre, M.; Mauro, S.; Pastorelli, S. Multiple Collision Avoidance between Human Limbs and Robot Links Algorithm in Collaborative Tasks. In Proceedings of the 15th International Conference on Informatics in Control, Automation and Robotics, Porto, Portugal, 29–31 July 2018; pp. 291–298. [CrossRef]
54. Karimi, G.; Jahanian, O. Genetic Algorithm Application in Swing Phase Optimization of AK Prosthesis with Passive Dynamics and Biomechanics Considerations. In *Genetic. Algorithms Application*; Intechopen: London, UK, 2012; pp. 71–88. [CrossRef]
55. Xu, Z.; Gan, Y.; Dai, X. Obstacle Avoidance of 7-DOF Redundant Manipulators. In Proceedings of the 2019 Chinese Control And Decision Conference (CCDC), Nanchang, China, 3–5 June 2019; pp. 4184–4189. [CrossRef]

robotics

MDPI

Article

Inverse Kinematics of a Class of 6R Collaborative Robots with Non-Spherical Wrist

Luca Carbonari †, Matteo-Claudio Palpacelli † and Massimo Callegari *,†

Department of Industrial Engineering and Mathematical Sciences, Polytechnic University of Marche, 60131 Ancona, Italy
* Correspondence: m.callegari@univpm.it
† These authors contributed equally to this work.

Abstract: The spread of cobotsin common industrial practice has led constructors to prefer the development of collaborative features that are necessary to prevent injuries to operators over the realization of simple kinematic structures for which the joints-to-workspace mapping is well known. An example is given by the replacement in serial robots of spherical wrists with safer solutions, where the danger of crushing and shearing is intrinsically avoided. Despite this tendency, the kinematic map between actuated joints and the Cartesian workspace remains of paramount importance for robot analysis and programming, deserving the attention of the research community. This paper proposes a closed-form solution for the inverse kinematics of a class of 6R robotic arms with six degrees of freedom and non-spherical wrists. The solutions are worked out by a single polynomial, of minimum degree, in terms of one of the positioning parameters chosen for the description of the robot posture. The roots of such a polynomial are then back-substituted to determine all the remaining unknowns. A numerical example is finally shown to verify the validity of the proposed implementation for a commercial collaborative robot.

Keywords: inverse kinematics; collaborative robot; non-spherical wrist

Citation: Carbonari, L.; Palpacelli, M.-C.; Callegari, M. Inverse Kinematics of a Class of 6R Collaborative Robots with Non-Spherical Wrist. *Robotics* **2023**, *12*, 36. https://doi.org/10.3390/robotics12020036

Academic Editor: Zhangguo Yu

Received: 2 February 2023
Revised: 16 February 2023
Accepted: 1 March 2023
Published: 3 March 2023

1. Introduction

The last decade has witnessed the spread of collaborative robotics in many aspects of everyday life, from common industrial practice [1] to home assistance and healthcare [2,3], and service robotics in general [4]. The main reason for such an impact lies in the flexible use of cobots (collaborative robots). Their success in industry, in fact, is mostly attributed to the possibility to implement workflows where humans and robots safely cooperate in a shared environment [5–7]. This feature has recently being exploited in fields of applications different from bare production, where the use of classical industrial robots would be impossible for obvious safety reasons.

From the point of view of robot producers, such an unprecedented diffusion has required the development of innovative safety features, aimed at making the use of cobots increasingly secure and accessible to the wider public. Therefore, on the one hand, producers have been required to implement expensive sensors and use appropriate materials to assess the risk of injury in using cobots, and, on the other, they have been forced to maintain competitive production costs to make their machines accessible. The first consequence, as visible in many well-known commercial cobots, has been the adoption of serial kinematic structures that are simple and inexpensive to realize, such as those characterized by non-serial wrists.

The class of collaborative robots produced by FANUC, among others, belongs to such a category. Their kinematic structure, described in the following in detail, is characterized by a non-spherical roll–pitch–roll wrist that makes inverse kinematics mapping quite challenging. Recent literature shows different approaches to the issue. Trinh et al. [8]

proposed a geometrical approach consisting of the solution of four separate univariate polynomials. In [9], the authors proposed a solution via numerical optimization of a problem formulated by means of polynomials for a similar robot (namely the Kinova Gen-3 Lite). In [10], a solution for a different non-spherical collaborative robot is tackled as well, although the wrist topology in this case allows a far simpler approach based on the computation of the first joint rotation at the very first step of the algorithm.

In less recent years, the problem has been approached in more general ways. Among others, Raghavan and Roth approached the problem of general linkages [11,12], demonstrating the existence of a maximum number of 16 solutions for the inverse kinematics of 6R (i.e., mechanisms owning a kinematic chain of six revolute joints) linkages. Their work, based on the solution via dialytic elimination of joints variables, inspired many other polynomial approaches, such as the Groebner basis work proposed by Wang et al. [13], or the eigenvalue approaches of Fu et al. [14] and Ghazvini [15].

Aside from such general solutions, the research community has provided many others for manipulators with spherical wrists. Among them, the method proposed by Xiao et al. [16] proposed to reach a solution by applying two cutting points within the kinematic chain and comparing the equations coming from their re-connection. Li et al. [17] considered the effect of compensation of the link lengths, so that their approach became effective after kinematic calibration.

Of course, the literature also offers plenty of numerical approaches to the problem, commonly used to find a solution by letting a starting configuration converge towards an end pose. Such methods are often based on the robot Jacobian matrix [18], differing one from the other for the management of singularities within the solution path. Among them, the Damped Least-Squares is the iterative approach most widely used [19–22], although convergence is strictly dependent on the used damping factor. Other more recent approaches have also been experimented with, such as the training of specific artificial intelligence as done in [23]. In any case, all of these approaches only allow the discovery of one solution to the problem (the closest to a first guess, usually), disregarding the large number of postures a 6R kinematic chain can exhibit to reach a given pose.

To overcome the limits of numerical approaches, a specific closed-form solution to the inverse kinematics of the FANUC CRX family of collaborative robots is proposed in this manuscript. As discussed in the following, the solution is worked out in the form of a univariate polynomial in one of the orientation parameters of the first body of the wrist. Such an approach allows the obtaining of the solving polynomial (of degree 16) with relative ease, starting from a system of six constraint equations. The backwards substitution needed for the computation of the remainder of the unknowns is then described in detail. Finally, a numerical example to verify the correctness of the solution is also proposed.

2. Robot Description

A robot of great interest in the collaborative robotics scene is the FANUC CRX-10iA/L (see Figure 1a), which meets industrial reliability requirements and provides all the necessary functionality for human–robot collaboration. In this paper, it is taken as a reference from the family of serial robots with non-spherical wrists, without loss of generality. The kinematic chain of the FANUC CRX-10iA/L is made of six revolute joints (6R) arranged in serial configuration. The first three joints (J_1, J_2 and J_3 in Figure 1b) are arranged in a classic kinematic sequence typical of robotic arms while the roll–pitch–roll wrist has a non-spherical configuration. The last three joints (J_4, J_5 and J_6), in fact, have axes that are two-by-two perpendicular and incident. Moreover, the lack of a common intersection point obviously makes the wrist non-spherical, making it impossible to approach the inverse kinematic problem with the usual methods for serial anthropomorphic manipulators, such as the solution proposed by Pieper [24].

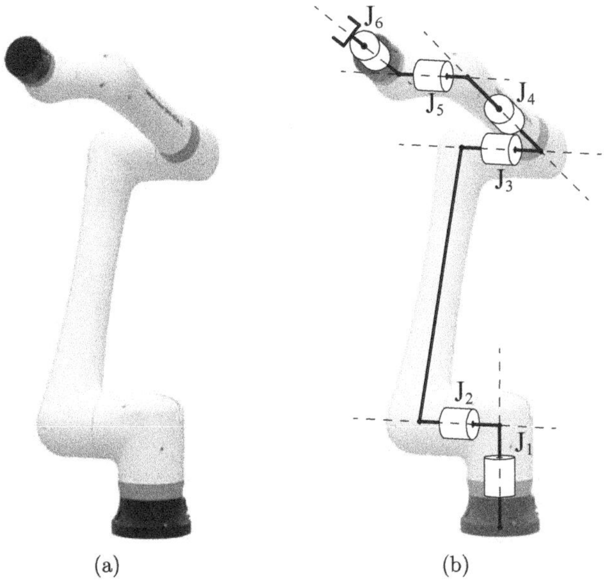

Figure 1. (**a**) Collaborative Robot FANUC CRX-10iA/L and (**b**) its kinematic architecture.

The position and orientation of robot bodies in space are described here with the usual Denavit–Hartenberg notation. It is worth remembering that the mutual position between two frames can be represented by a homogeneous transformation, whose expression is given by:

$$^{i}\mathbf{T}_{j} = \begin{bmatrix} \cos\theta_i & -\sin\theta_i & 0 & 0 \\ \sin\theta_i & \cos\theta_i & 0 & 0 \\ 0 & 0 & 1 & d_i \\ 0 & 0 & 0 & 1 \end{bmatrix} \begin{bmatrix} 1 & 0 & 0 & a_i \\ 0 & \cos\alpha_i & -\sin\alpha_i & 0 \\ 0 & \sin\alpha_i & \cos\alpha_i & 0 \\ 0 & 0 & 0 & 1 \end{bmatrix} \quad (1)$$

where $\theta_i, d_i, \alpha_i, a_i$ are the well-known Denavit–Hartenberg parameters. In the present case, the actuated joint variables q_i are the rotations θ_i around the local z-axes.

The FANUC CRX-10iA/L frames are arranged as shown in Figure 2a, while the parameters collected in Table 1 complete their description. The configuration in space of the tool frame $\{6\}$ with respect to the global reference frame $\{0\}$ is obtained by composing the local transformations $^{i}\mathbf{T}_{j}$ according to the convention of successive transformation on mobile axes:

$$^{0}\mathbf{T}_{6} = \prod_{i=1}^{6} {}^{i-1}\mathbf{T}_{i} = {}^{0}\mathbf{T}_{1}{}^{1}\mathbf{T}_{2}{}^{2}\mathbf{T}_{3}{}^{3}\mathbf{T}_{4}{}^{4}\mathbf{T}_{5}{}^{5}\mathbf{T}_{6} \quad (2)$$

Table 1. Denavit–Hartenberg parameters of the FANUC CRX-10iA/L.

i	α_i (rad)	a_i (mm)	d_i (mm)	θ_i (rad)	Motion Range (rad)
1	$\pi/2$	0	250.3	q_1	$\pm\pi$
2	$-\pi$	710.0	260.4	q_2	$\pm\pi$
3	$-\pi/2$	0	260.4	q_3	$\pm1.5\pi$
4	$-\pi/2$	0	540.0	q_4	$\pm1.06\pi$
5	$\pi/2$	0	150.0	q_5	$\pm\pi$
6	0	0	160.0	q_6	$\pm1.25\pi$

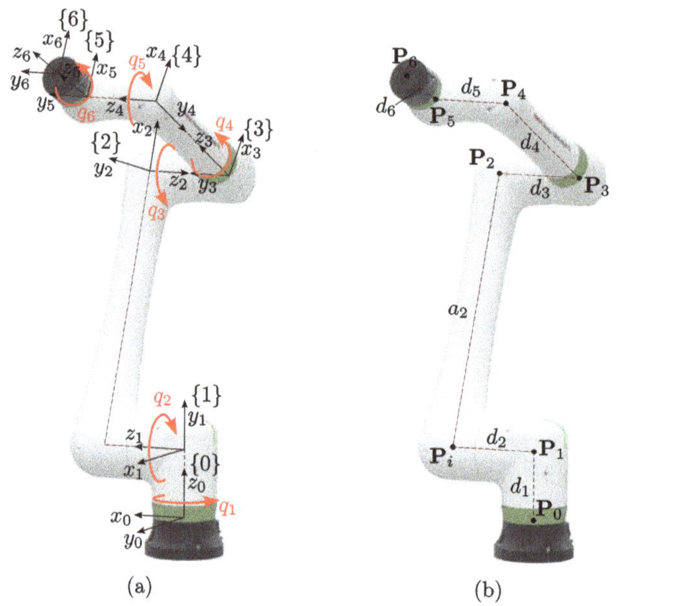

Figure 2. (**a**) Frames attached to the robot bodies and (**b**) details of their lengths.

3. Inverse Kinematics Problem

As is well known, the inverse kinematics mapping of a serial chain consists of determining the joint parameters (q_i in the previous notation) able to provide a given pose for the terminal body of the chain (i.e., body 6 in this manuscript, described by frame $\{6\}$). It is also well known that such a problem has many real solutions corresponding to the many postures that the robot can exhibit while reaching a given position and orientation of the terminal body.

A usual approach to this problem consists of simply comparing the elements of the final transformation ${}^0\mathbf{T}_6$ with a symbolic expression of the same matrix obtained considering unknown the six joint parameters q_i:

$$ {}^0\mathbf{T}_6 \quad \xrightarrow[kinematics]{inverse} \quad \mathbf{q} = \begin{bmatrix} q_1 & q_2 & q_3 & q_4 & q_5 & q_6 \end{bmatrix} \tag{3} $$

This allows the obtaining of a set of 12 equations that can be reduced to six considering that only three elements of the rotational part of the homogeneous transformation matrix are actually independent. Additionally, a common issue in approaching this kind of equation is given by the presence of sines and cosines, which add complexity to the problem. The usual solution is to apply the Weierstrass substitution to have a common parameterization without trigonometric functions, or to treat them (in this case, 12 expressions of $\cos q_i$ and $\sin q_i$ for $i = 1, \ldots, 6$) as distinct unknowns related by six further homogeneous equations ($\cos^2 q_i + \sin^2 q_i - 1 = 0$). Both strategies lead to a set of polynomial equations of degree 2, whose solution yields to a univariate polynomial: nonetheless, the reduction of such a system is often computationally burdening and the possibility of obtaining a final degree higher than the number of expected results is also significant.

The approach proposed in this manuscript takes into account a different set of unknown variables, hereby collected in a vector called $\mathbf{p}$, which locates the reference frame of one of the robot bodies (frame $\{4\}$ in particular) and, as a result, allows the computation of the coordinates in space of the robot nodal points $\mathbf{P}_i$ (shown in Figure 2). From such points, then, the computation of the joint variables becomes a trivial issue.

In more detail, the parameterization chosen to express the rotation matrix ${}^0\mathbf{R}_4$ between frame $\{4\}$ and the fixed frame $\{0\}$ avoiding trigonometric functions is the Cayley transform, which maps the skew–symmetric matrix $\mathbf{C}$ given by:

$$\mathbf{C} = \begin{bmatrix} 0 & -c_3 & c_2 \\ c_3 & 0 & -c_1 \\ -c_2 & c_1 & 0 \end{bmatrix} \tag{4}$$

into a rotation matrix:

$${}^0\mathbf{R}_4 = (\mathbf{I} + \mathbf{C})(\mathbf{I} - \mathbf{C})^{-1} \tag{5}$$

Therefore, the shape assumed by the resulting rotation matrix ${}^0\mathbf{R}_4$ in terms of the three parameters c_i (with $i = 1, 2, 3$) is:

$${}^0\mathbf{R}_4 = \frac{1}{c_1^2 + c_2^2 + c_3^2 + 1} \begin{bmatrix} c_1^2 - c_2^2 - c_3^2 + 1 & 2c_1c_2 - 2c_3 & 2c_2 + 2c_1c_3 \\ 2c_3 + 2c_1c_2 & -c_1^2 + c_2^2 - c_3^2 + 1 & 2c_2c_3 - 2c_1 \\ 2c_1c_3 - 2c_2 & 2c_1 + 2c_2c_3 & -c_1^2 - c_2^2 + c_3^2 + 1 \end{bmatrix} \tag{6}$$

Based on such a choice, the Inverse Kinematics Problem actually consists of finding the parameters that allow the determination of the pose of $\{4\}$ (expressed by means of c_1, c_2, c_3 and coordinates of $\mathbf{P}_4$) starting from $\{6\}$, and it can be stated as:

$${}^0\mathbf{T}_6 \xrightarrow[kinematics]{inverse} \mathbf{p} = \begin{bmatrix} c_1 & c_2 & c_3 & P_{4,x} & P_{4,y} & P_{4,z} \end{bmatrix} \tag{7}$$

where the variables c_i can be used to describe the orientation of frame $\{4\}$, as mentioned above, and $P_{4,x}$, $P_{4,y}$, $P_{4,z}$ are the coordinates of its origin, gathered in the column vector $\mathbf{P}_4$. The 6 parameters are then used to build up the needed transformation matrix:

$${}^0\mathbf{T}_4 = \begin{bmatrix} {}^0\mathbf{R}_4 & \mathbf{P}_4 \\ \mathbf{0}_{1\times 3} & 1 \end{bmatrix} \tag{8}$$

with obvious meaning of the terms involved. In the following, the system of equations used to work out the solution of the kinematic problem is shown in detail together with its solution path.

3.1. System of Equations

To find the configuration of frame $\{4\}$ once given a value for ${}^0\mathbf{R}_6$, a set of six equations must be provided. To achieve such an aim, it is possible to exploit both the loop-closure of the kinematic chain and the mobility provided by the joint topology. The former can be obtained by considering as known the position in space of points $\mathbf{P}_5$ and $\mathbf{P}_6$, while the latter is achieved through the mobility of the non-spherical wrist and the anthropomorphic arm.

Starting from the kinematic closure, the coordinates of $\mathbf{P}_5$ are easily obtained via:

$$\begin{bmatrix} \mathbf{P}_5 \\ 1 \end{bmatrix} = {}^0\mathbf{T}_6 \begin{bmatrix} 0 \\ 0 \\ -d_6 \\ 1 \end{bmatrix} = \begin{bmatrix} P_{5,x} \\ P_{5,y} \\ P_{5,z} \\ 1 \end{bmatrix} \tag{9}$$

The same point can be written starting from $\{4\}$ in terms of the problem unknowns, namely:

$$\begin{bmatrix} \mathbf{P}_5 \\ 1 \end{bmatrix} = {}^0\mathbf{T}_4 \begin{bmatrix} 0 \\ 0 \\ d_5 \\ 1 \end{bmatrix} \tag{10}$$

By comparing (9) and (10), the first three equations are obtained. After simplification and elimination of the non-vanishing terms, the three equations obtained are three polynomials in terms of the six unknowns collected in $\mathbf{p}$:

$$\begin{aligned}
\Phi_1: &\quad (P_{4,x} - P_{5,x})(c_1^2 + c_2^2 + c_3^2 + 1) + 2d_5(c_1c_3 + c_2) = 0 \\
\Phi_2: &\quad (P_{4,y} - P_{5,y})(c_1^2 + c_2^2 + c_3^2 + 1) + 2d_5(c_2c_3 - c_1) = 0 \\
\Phi_3: &\quad (P_{4,z} - P_{5,z})(c_1^2 + c_2^2 + c_3^2 + 1) - d_5(c_1^2 + c_2^2 - c_3^2 - 1) = 0
\end{aligned} \tag{11}$$

Three additional equations can be worked out from the mobility of frame $\{4\}$. First, it can be noticed that the y_4-axis of the frame is constrained to intersect the z_0-axis of the global coordinate system (since $d_2 = d_3$), except for the case where the two axes are parallel and not coincident, i.e., when the y_4-axis is directed along vertical and does not pass through the origin of the fixed frame. The latter condition, by the way, is far from the operating conditions of the robot. A more general condition can be stated as the coplanarity of the four points $\mathbf{P}_0$, $\mathbf{P}_1$, $\mathbf{P}_3$ and $\mathbf{P}_4$ or better as the linear dependency among the three vectors $\mathbf{z}_0$, $\mathbf{y}_4$ and $(\mathbf{P}_4 - \mathbf{P}_0)$. Such a condition is fulfilled when:

$$\det\begin{bmatrix}\mathbf{z}_0 & \mathbf{y}_4 & (\mathbf{P}_4 - \mathbf{P}_0)\end{bmatrix} = 0 \tag{12}$$

where $\mathbf{z}_0 = \begin{bmatrix}0 & 0 & 1\end{bmatrix}^T$, $\mathbf{y}_4 = {}^0\mathbf{R}_4\begin{bmatrix}0 & 1 & 0\end{bmatrix}^T$ and $(\mathbf{P}_4 - \mathbf{P}_0) = \begin{bmatrix}P_{4,x} & P_{4,y} & P_{4,z}\end{bmatrix}^T$. After some manipulation and elimination of non-vanishing components, the determinant in (12) can be expanded to obtain the fourth equation of the system:

$$\Phi_4: \quad (c_1^2 - c_2^2 + c_3^2 - 1)P_{4,x} + (2c_1c_2 - 2c_3)P_{4,y} = 0 \tag{13}$$

The mobility of the kinematic chain, of the wrist in particular, also constrains frame $\{4\}$ to maintain an axis perpendicular to the assigned coordinate system $\{6\}$, i.e., the axes identified by vectors $\mathbf{z}_4$ and $\mathbf{z}_6$. The perpendicularity is fulfilled when:

$$\mathbf{z}_4^T\mathbf{z}_6 = 0 \tag{14}$$

where $\mathbf{z}_4 = {}^0\mathbf{R}_4\begin{bmatrix}0 & 0 & 1\end{bmatrix}^T$ and $\mathbf{z}_6$ is the unit vector of $(\mathbf{P}_5 - \mathbf{P}_6)$ (for the constraint purpose, they can be used interchangeably). After substitution and simplification, the following polynomial is obtained:

$$\begin{aligned}
\Phi_5: &\quad 2(c_1c_3 + c_2)(P_{5,x} - P_{6,x}) + 2(c_2c_3 - c_1)(P_{5,y} - P_{6,y}) + \\
&\quad -(c_1^2 + c_2^2 - c_3^2 - 1)(P_{5,z} - P_{6,z}) = 0
\end{aligned} \tag{15}$$

Finally, it must be noted that point $\mathbf{P}_3$ maintains a constant distance with respect to the center of the robot shoulder, i.e., to the point $\mathbf{P}_1$. Thus, it is:

$$(\mathbf{P}_3 - \mathbf{P}_1)^T(\mathbf{P}_3 - \mathbf{P}_1) - a_2^2 = 0 \tag{16}$$

where $\mathbf{P}_3 = \mathbf{P}_4 + {}^0\mathbf{R}_4\begin{bmatrix}0 & d_4 & 0\end{bmatrix}^T$ and $\mathbf{P}_1 = \begin{bmatrix}0 & 0 & d_1\end{bmatrix}^T$, yielding the simplified form:

$$\begin{aligned}
\Phi_6: &\quad \left(P_{4,x}(c_1^2 + c_2^2 + c_3^2 + 1) + 2d_4(c_1c_2 - c_3)\right)^2 + \\
&\quad + \left(P_{4,y}(c_1^2 + c_2^2 + c_3^2 + 1) + d_4(c_1^2 - c_2^2 + c_3^2 - 1)\right)^2 + \\
&\quad + \left((P_{4,z} - d_1)(c_1^2 + c_2^2 + c_3^2 + 1) + 2d_4(c_2c_3 + c_1)\right)^2 - a_2^2 = 0
\end{aligned} \tag{17}$$

The six polynomial equations Φ_i must now be solved in the unknown variables of $\mathbf{p}$ to obtain a single polynomial in terms of c_1 only. The solution path is detailed in the following section.

3.2. System Solution

At first, it is possible to remove the three Cartesian variables $P_{4,x}$, $P_{4,y}$ and $P_{4,z}$ exploiting the three equations Φ_1, Φ_2 and Φ_3 in which they only appear in linear form. Thus, the following expressions can be found in terms of c_1, c_2 and c_3:

$$\begin{aligned} P_{4,x}(c_1,c_2,c_3) &= P_{5,x} - d_5\frac{2(c_1c_3+c_2)}{c_1^2+c_2^2+c_3^2+1} \\ P_{4,y}(c_1,c_2,c_3) &= P_{5,y} - d_5\frac{2(c_2c_3-c_1)}{c_1^2+c_2^2+c_3^2+1} \\ P_{4,z}(c_1,c_2,c_3) &= P_{5,z} + d_5\frac{c_1^2+c_2^2-c_3^2-1}{c_1^2+c_2^2+c_3^2+1} \end{aligned} \tag{18}$$

Substitution of (18) into Φ_4, Φ_5 and Φ_6 yields to a formulation of the robot kinematics in terms of only c_1, c_2 and c_3. The resulting polynomials own a maximum degree of 2 and can be expressed as:

$$\Phi_h : \sum_{i,j,k=0}^{i+j+k\leq 2} \varphi_{h,ijk}c_1^i c_2^j c_3^k = 0 \tag{19}$$

where the coefficients $\varphi_{h,ijk}$ are function of the robot geometric parameters and $h = 4,5,6$. Appendix A shows in detail the value of such coefficients.

To reduce the system of equations and obtain a polynomial in terms of just one of the three unknowns (namely c_1), the Equation (19) can be rewritten consequently. To remove c_3 at first, the following shape can be worked out:

$$X_h : \sum_{k=0}^{2} \chi_{h,k}c_3^k = 0 \tag{20}$$

where the coefficients $\chi_{h,k}$ (detailed in Appendix B) are functions of c_1 and c_2. It is worth remarking that the formulations (19) and (20) coincide, thus $\Phi_h = X_h$. However, polynomials (20) can be used to remove c_3 by means of two Sylvester matrices, whose determinants provide two further reduced equations:

$$\begin{aligned} \Psi_1 : \det\begin{bmatrix} \chi_{4,2} & \chi_{4,1} & \chi_{4,0} & 0 \\ 0 & \chi_{4,2} & \chi_{4,1} & \chi_{4,0} \\ \chi_{6,2} & \chi_{6,1} & \chi_{6,0} & 0 \\ 0 & \chi_{6,2} & \chi_{6,1} & \chi_{6,0} \end{bmatrix} = 0 \\ \Psi_2 : \det\begin{bmatrix} \chi_{5,2} & \chi_{5,1} & \chi_{5,0} & 0 \\ 0 & \chi_{5,2} & \chi_{5,1} & \chi_{5,0} \\ \chi_{6,2} & \chi_{6,1} & \chi_{6,0} & 0 \\ 0 & \chi_{6,2} & \chi_{6,1} & \chi_{6,0} \end{bmatrix} = 0 \end{aligned} \tag{21}$$

The vanishing set of the determinants Ψ_1 and Ψ_2 (which are polynomials of c_1 and c_2) represent the solution to the inverse kinematics problem. The maximum degree of the polynomials is 4 and in compact form they can be formulated as:

$$\Psi_h : \sum_{i,j=0}^{i+j\leq 4} \psi_{h,ij}c_1^i c_2^j = 0 \tag{22}$$

The elimination via Sylvester matrix can be adopted also to remove the unknown c_2 by rewriting (22) as:

$$\Omega_h : \sum_{j=0}^{4} \omega_{h,j}c_2^j = 0 \tag{23}$$

Again, it is remarked that (23) is a different formulation of (22), nevertheless $\Psi_h = \Omega_h$. A complete formulation of coefficients $\omega_{h,j}$ is provided in Appendix C, where they are presented already as polynomials of the only unknown c_1.

The last step of the solution is made using again the Sylvester method for reduction of variable c_2 by means of (23). A further matrix is built, whose determinant is a polynomial of degree 16 in terms of the only unknown c_1:

$$\Lambda : \det \begin{bmatrix} \omega_{1,4} & \omega_{1,3} & \omega_{1,2} & \omega_{1,1} & \omega_{1,0} & 0 & 0 & 0 \\ 0 & \omega_{1,4} & \omega_{1,3} & \omega_{1,2} & \omega_{1,1} & \omega_{1,0} & 0 & 0 \\ 0 & 0 & \omega_{1,4} & \omega_{1,3} & \omega_{1,2} & \omega_{1,1} & \omega_{1,0} & 0 \\ 0 & 0 & 0 & \omega_{1,4} & \omega_{1,3} & \omega_{1,2} & \omega_{1,1} & \omega_{1,0} \\ \omega_{2,4} & \omega_{2,3} & \omega_{2,2} & \omega_{2,1} & \omega_{2,0} & 0 & 0 & 0 \\ 0 & \omega_{2,4} & \omega_{2,3} & \omega_{2,2} & \omega_{2,1} & \omega_{2,0} & 0 & 0 \\ 0 & 0 & \omega_{2,4} & \omega_{2,3} & \omega_{2,2} & \omega_{2,1} & \omega_{2,0} & 0 \\ 0 & 0 & 0 & \omega_{2,4} & \omega_{2,3} & \omega_{2,2} & \omega_{2,1} & \omega_{2,0} \end{bmatrix} = 0 \tag{24}$$

In more compact form, the solving equation Λ is:

$$\Lambda : \sum_{i=0}^{16} \lambda_i c_1^i = 0 \tag{25}$$

where the coefficients λ_i for $i = 1, \ldots, 16$ are finally functions of only geometric parameters. Unfortunately, explicit expressions for the coefficients are too large to be shown in this document. Nonetheless, Appendix D shows (25) in terms of the polynomials $\omega_{h,i}$ appearing in matrix (24). The polynomial provides up to 16 real solutions to the problem, corresponding to its roots. The number of solutions does not change depending on proximity to singular poses or joint limits. However, it obviously becomes equal to 0 for singular configurations (such as any point out of the robot workspace) for which only complex roots can be found (25). In the remainder of the manuscript, details are provided about the back-substitution of such values and the numeric implementation for the computation of the other variables of the problem.

4. Implementation and Verification

As aforementioned, the vanishing set of the polynomial (25) allows the discovery of up to 16 real solutions for the inverse kinematics problem, corresponding to the roots of Λ. The expanded equations shown in the appendices provide an impression of the coefficient dimensions that cannot be further reduced. For this reason, the products of polynomials $\omega_{h,i}$ have not been made explicit during the implementation of the inverse kinematics algorithm. On the contrary, the coefficients of Λ have been obtained as a sum of the coefficients of each addendum appearing in Appendix D, and each of them was worked out via the convolution of the corresponding discrete sequence of coefficients.

As is known, in fact, given two polynomials $f(x)$ and $g(x)$of degree m and n, respectively:

$$f(x) = \Sigma_{i=0}^{m} f_i x^i \qquad g(x) = \Sigma_{i=0}^{n} g_i x^i \tag{26}$$

the product $f(x)g(x)$ is a $m+n$ degree polynomial given by:

$$h(x) = f(x)g(x) = \Sigma_{i=0}^{m+n} h_i x^i \qquad \text{with} \qquad h_i = \Sigma_{k=0}^{i} f_k g_{i-k} \tag{27}$$

The definition of h_i can be easily extended to a $-\infty, +\infty$ sum, and then to a definition of the convolution of a discrete series. Therefore, the coefficients h_i of polynomial $h(x) = f(x)g(x)$ are computable as the convolution $f_i \circledast g_i$. Let us take, for example, the multiplication $\omega_{1,3}\omega_{2,2}$. The coefficients of the resulting polynomial will be provided by the convolution of the coefficients of the factors, thus:

$$\omega_{1,3}\omega_{2,2} = \left(\begin{bmatrix} B_1, C_1 \end{bmatrix} \begin{bmatrix} c_1 \\ 1 \end{bmatrix} \right) \left(\begin{bmatrix} D_2, E_2, F_2 \end{bmatrix} \begin{bmatrix} c_1^2 \\ c_1 \\ 1 \end{bmatrix} \right) = \left(\begin{bmatrix} B_1, C_1 \end{bmatrix} \circledast \begin{bmatrix} D_2, E_2, F_2 \end{bmatrix} \right) \begin{bmatrix} c_1^3 \\ c_1^2 \\ c_1 \\ 1 \end{bmatrix} \tag{28}$$

where expressions in terms of the robot geometric parameters are provided in the appendices for A_i, B_i, C_i, etc.

Using convolution rather than multiplication allows computation with relative ease of the coefficients of Λ and therefore its 16 roots. Then, the remaining unknowns can be computed following the solution path backwards and substituting the values obtained time by time. Starting from (24), a linear system of equation can be built as:

$$\begin{bmatrix} \omega_{1,4} & \omega_{1,3} & \omega_{1,2} & \omega_{1,1} & \omega_{1,0} & 0 \\ 0 & \omega_{1,4} & \omega_{1,3} & \omega_{1,2} & \omega_{1,1} & \omega_{1,0} \\ 0 & 0 & \omega_{1,4} & \omega_{1,3} & \omega_{1,2} & \omega_{1,1} \\ \omega_{2,4} & \omega_{2,3} & \omega_{2,2} & \omega_{2,1} & \omega_{2,0} & 0 \\ 0 & \omega_{2,4} & \omega_{2,3} & \omega_{2,2} & \omega_{2,1} & \omega_{2,0} \\ 0 & 0 & \omega_{2,4} & \omega_{2,3} & \omega_{2,2} & \omega_{2,1} \end{bmatrix} \begin{bmatrix} c_2^6 \\ c_2^5 \\ c_2^4 \\ c_2^3 \\ c_2^2 \\ c_2 \end{bmatrix} = \begin{bmatrix} 0 \\ 0 \\ -\omega_{1,0} \\ 0 \\ 0 \\ -\omega_{2,0} \end{bmatrix} \tag{29}$$

Substituting one at a time the 16 values of c_1 (corresponding to the roots of Λ) to evaluate the polynomials $\omega_{h,i}$, the respective values of c_2 can be found from a solution of (29). As c_1 and c_2 are now known, the coefficients $\chi_{h,k}$ of (20) can be computed and used to obtain c_3 picking two equations from the three X_h. For example, it is possible to formulate the linear system:

$$\begin{bmatrix} \chi_{4,2} & \chi_{4,1} \\ \chi_{5,2} & \chi_{5,1} \end{bmatrix} \begin{bmatrix} c_3^2 \\ c_3 \end{bmatrix} = \begin{bmatrix} -\chi_{4,0} \\ -\chi_{5,0} \end{bmatrix} \tag{30}$$

Lastly, the three orientation parameters c_1, c_2 and c_3 can be substituted into (18) to obtain the coordinates $P_{4,x}$, $P_{4,y}$, $P_{4,z}$ of the origin of the reference frame $\{4\}$. This passage almost closes the inverse kinematics problem, although the last few steps of joint variables computation are still at stake. To provide a complete mapping ${}^0\mathbf{T}_6 \longrightarrow \mathbf{q}$, the coordinates of the points $\mathbf{P}_4$ and $\mathbf{P}_3 = \mathbf{P}_4 + {}^0\mathbf{R}_4 \begin{bmatrix} 0 & d_4 & 0 \end{bmatrix}^T$ can be exploited. The passages to obtain the values of the joint variables, which are quite trivial, are not shown here for the sake of conciseness, although their full formulation is shown in Appendix E. However, it is worth remarking that their values are calculated in cascade, following the order q_2, q_1, q_3, q_4, q_5, q_6.

It is also worth noting that the very first computation, i.e., $q_2 = \arcsin{(P_{3,x} - P_{1,x})}/a_2$, doubles the number of available solutions (being acceptable both q_2 and $\pi - q_2$). Actually, the solutions obtained considering the full domain of the function arcsin are already present within the set of points coordinates previously found. In practice, applying in sequence the mapping ${}^0\mathbf{T}_6 \longrightarrow \mathbf{p}$ and $\mathbf{p} \longrightarrow \mathbf{q}$ duplicated solutions are added to the available ones (that can be easily eliminated).

Numerical Example

Finally, to verify the effectiveness of the proposed solution, a numerical example is shown. A random set of joint angles is picked to compute the robot forward kinematics by (2). In particular, in degrees:

$$\mathbf{q} = \begin{bmatrix} 78^\circ & 131^\circ & 24^\circ & 42^\circ & -60^\circ & -10^\circ \end{bmatrix}^T \tag{31}$$

which provide the transformation matrix:

$$
{}^{0}\mathbf{T}_6 = \begin{bmatrix} 0.3363 & 0.8387 & -0.4283 & 0.0571 \\ 0.6182 & 0.1464 & 0.7722 & 0.1786 \\ 0.7104 & -0.5245 & -0.4693 & 0.7677 \\ 0 & 0 & 0 & 1 \end{bmatrix} \tag{32}
$$

The resulting known points useful for inverse kinematics computations and for results check are (in mm):

$$
\begin{array}{lll} \mathbf{P}_0 = \begin{bmatrix} 0.0 \\ 0.0 \\ 0.0 \end{bmatrix} & \mathbf{P}_1 = \begin{bmatrix} 0.0 \\ 0.0 \\ 250.3 \end{bmatrix} & \mathbf{P}_3 = \begin{bmatrix} 125.7 \\ 55.0 \\ 842.7 \end{bmatrix} \\ \mathbf{P}_4 = \begin{bmatrix} 0.0 \\ 0.0 \\ 250.3 \end{bmatrix} & \mathbf{P}_5 = \begin{bmatrix} 125.7 \\ 55.0 \\ 842.7 \end{bmatrix} & \mathbf{P}_6 = \begin{bmatrix} 57.1 \\ 178.6 \\ 767.7 \end{bmatrix} \end{array} \tag{33}
$$

Such coordinates, together with the geometrical parameters already introduced in Table 1, can be substituted into the explicit expressions of the coefficients of polynomials $\omega_{h,i}$. Such polynomials allow computing, via convolution, the coefficients of Λ, whose roots represent the values of c_1 which are the solution of the inverse kinematics problem. Substituting in cascade into (29), (30) and (18), it is possible to obtain the values of the remaining unknowns, which are shown in Table 2. As is shown, the polynomial Λ provides in this case only eight real solutions, for which it is possible to evaluate eight poses for the reference frame $\{4\}$.

Table 2. Solutions of the inverse kinematics for the given joint variables (31): eight real solutions for c_1 out of 16 roots of Λ.

c_1	c_2	c_3	$P_{4,x}$	$P_{4,y}$	$P_{4,z}$
−13.105	−0.262	−22.831	−3.5	46.7	766.9
−9.673	1.672	−4.490	10.5	49.4	938.7
−7.730	−18.981	8.395	176.8	147.5	949.1
−7.603	0.844	−3.933	2.6	37.9	926.8
−6.135	0.372	−3.417	−1.2	26.1	917.3
−3.738	−0.223	−2.270	2.7	−8.0	901.1
−3.230	−0.296	−4.901	−5.4	15.5	781.5
−0.667 + 0.50i	–	–	–	–	–
−0.667 − 0.50i	–	–	–	–	–
−0.480	1.417	−0.041	−7.3	16.0	900.0
−0.233 + 0.01i	–	–	–	–	–
−0.233 − 0.01i	–	–	–	–	–
−0.088 + 0.04i	–	–	–	–	–
−0.088 − 0.04i	–	–	–	–	–
−0.041 + 0.02i	–	–	–	–	–
−0.041 − 0.02i	–	–	–	–	–

The results offered by the algorithm in terms of pose of $\{4\}$ can be then mapped into sets of joint variables by means of the equations presented in Appendix D. As aforementioned, the eight real solutions of c_1 are doubled by the angle computations (see Table 3), although four of the resulting configurations (shown in Figure 3) are repeated.

Table 3. Computedjoints variables corresponding to the eight real roots of Λ: total of 16 configurations, four of which repeated.

#	c_1	q_1	q_2	q_3	q_4	q_5	q_6
1	−13.105	−30.30	33.99	136.01	143.27	−48.79	77.61
2		149.69	146.00	43.98	−36.72	−48.79	77.61
3	−9.673	−101.99	48.99	155.99	−138.00	−59.99	−9.99
4		78.00	131.00	24.00	42.00	−60.00	−10.00
5	−7.730	39.90	28.21	161.19	−75.16	116.22	−90.34
6		−140.09	151.78	18.80	104.83	116.22	−90.34
7	−7.603	−93.98	47.62	154.56	−144.11	−55.12	−2.97
8		86.01	132.37	25.43	35.88	−55.12	−2.97
9	−6.135	−93.98	47.62	154.56	−144.11	−55.12	−2.97
10		86.01	132.37	25.43	35.88	−55.12	−2.97
11	−3.738	−93.98	47.62	154.56	−144.11	−55.12	−2.97
12		86.01	132.37	25.43	35.88	−55.12	−2.97
13	−3.230	−29.41	33.90	135.67	142.25	−49.24	78.79
14		150.58	146.09	44.32	−37.74	−49.24	78.79
15	−0.480	114.69	42.07	151.46	−23.88	170.53	10.81
16		−65.30	137.92	28.53	156.11	170.53	10.81

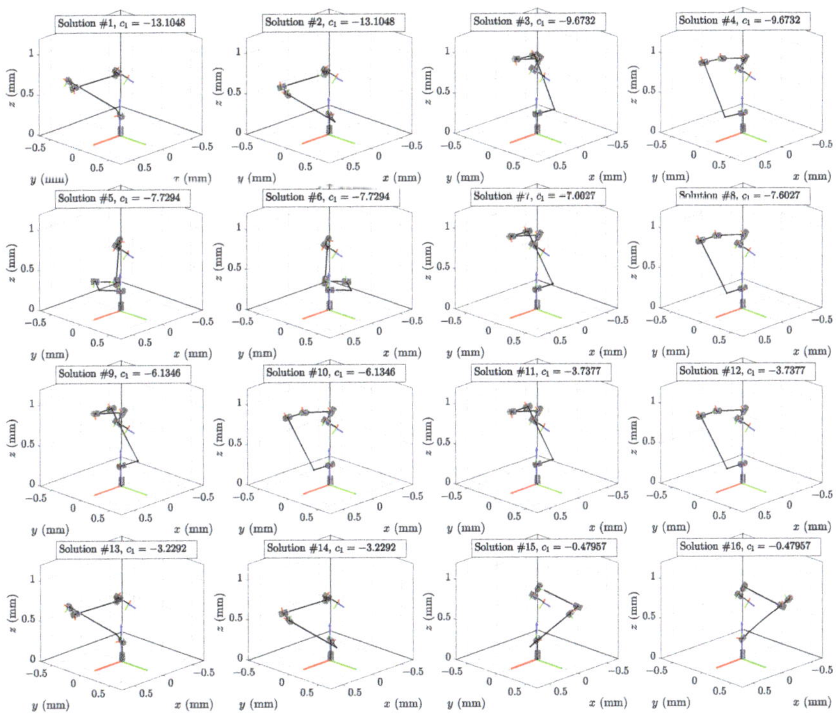

Figure 3. Representation of the 16 configurations resulting from the inverse kinematics problem solution.

5. Conclusions

This manuscript proposes a closed-form solution for the inverse kinematics mapping of a class of 6R robotic manipulators, characterized by a non-spherical wrist. On the one hand, such a feature improves the robot mechanical simplicity and provides some advantage in terms of dexterity; on the other, it prevents the use of common methods developed for anthropomorphic arms. For this reason, a specific approach has been worked out based on the constraint equations that characterize the mobility of an inner body of the kinematic chain (i.e., the first body of the non-spherical wrist). The solution path eliminates one at a time five out of the six unknowns of the problem and provides a univariate polynomial of degree 16, whose roots represent the problem solution. Backward substitution and computation of the actuated joint angles are shown to close the solution path and provide a full map of the end-effector configurations space and the motors space. In conclusion, the manuscript provides a complete walk-through to compute all the possible postures that the FANUC CRX family of cobots can exhibit to reach a given pose. Obviously, the proposed method is effective also for all the manipulators sharing their joint topology with the FANUC CRX family, i.e., for those manipulators for which the written equations are valid. Future developments will involve the analysis of the robot workspace to also provide a complete map of robot workspace boundaries and inner singularities, the analysis of representation singularities, and the impact of the problem of numerical conditioning on the accuracy of the kinematics solution.

Author Contributions: Conceptualization, L.C. and M.-C.P.; methodology, L.C.; software, L.C.; validation, L.C.; formal analysis, L.C., M.-C.P. and M.C.; investigation, L.C.; resources, M.C.; data curation, L.C.; writing—original draft preparation, L.C.; writing—review and editing, M.-C.P. and M.C.; visualization, M.-C.P.; supervision, M.C.; project administration, L.C.; funding acquisition, M.-C.P. and M.C. All authors have read and agreed to the published version of the manuscript.

Funding: This research received no external funding.

Institutional Review Board Statement: Not applicable.

Informed Consent Statement: Not applicable.

Data Availability Statement: Not applicable.

Conflicts of Interest: The authors declare no conflict of interest.

Appendix A. Coefficients of Polynomials Φ_h

$$
\begin{aligned}
\varphi_{4,002} &= P_{5,x} & \varphi_{5,002} &= P_{5,z} - P_{6,z} \\
\varphi_{4,011} &= 0 & \varphi_{5,011} &= 2P_{5,y} - 2P_{6,y} \\
\varphi_{4,101} &= -2d_5 & \varphi_{5,101} &= 2P_{5,x} - 2P_{6,x} \\
\varphi_{4,001} &= -2P_{5,y} & \varphi_{5,001} &= 0 \\
\varphi_{4,020} &= -P_{5,x} & \varphi_{5,020} &= P_{6,z} - P_{5,z} \\
\varphi_{4,110} &= 2P_{5,y} & \varphi_{5,110} &= 0 \\
\varphi_{4,010} &= 2d_5 & \varphi_{5,010} &= 2P_{5,x} - 2P_{6,x} \\
\varphi_{4,200} &= P_{5,x} & \varphi_{5,200} &= P_{6,z} - P_{5,z} \\
\varphi_{4,100} &= 0 & \varphi_{5,100} &= 2P_{6,y} - 2P_{5,y} \\
\varphi_{4,000} &= -P_{5,x} & \varphi_{5,000} &= P_{5,z} - P_{6,z}
\end{aligned}
$$

$$
\begin{aligned}
\varphi_{6,002} &= P_{5,x}^2 + P_{5,y}^2 + P_{5,z}^2 - 2P_{5,y}d_4 - 2P_{5,z}(d_1 + d_5) - d_2^2 + d_1^2 + 2d_1d_5 + d_4^2 + d_5^2 \\
\varphi_{6,011} &= 4P_{5,z}d_4 - 4P_{5,y}d_5 - 4d_1d_4 \\
\varphi_{6,101} &= -4P_{5,x}d_5 \\
\varphi_{6,001} &= -4P_{5,x}d_4 \\
\varphi_{6,020} &= P_{5,x}^2 + P_{5,y}^2 + P_{5,z}^2 + 2P_{5,y}d_4 - 2P_{5,z}(d_1 - d_5) - d_2^2 + d_1^2 - 2d_1d_5 + d_4^2 + d_5^2 \\
\varphi_{6,110} &= 4P_{5,x}d_4 \\
\varphi_{6,010} &= -4P_{5,x}d_5 \\
\varphi_{6,200} &= P_{5,x}^2 + P_{5,y}^2 + P_{5,z}^2 - 2P_{5,y}d_4 - 2P_{5,z}(d_1 + d_5) - d_2^2 + d_1^2 - 2d_1d_5 + d_4^2 + d_5^2 \\
\varphi_{6,100} &= 4P_{5,y}d_5 + 4P_{5,z}d_4 - 4d_1d_4 \\
\varphi_{6,000} &= P_{5,x}^2 + P_{5,y}^2 + P_{5,z}^2 + 2P_{5,y}d_4 - 2P_{5,z}(d_1 + d_5) - d_2^2 + d_1^2 + 2d_1d_5 + d_4^2 + d_5^2
\end{aligned}
$$

Appendix B. Coefficients of Polynomials X_h

$$
\begin{aligned}
\chi_{h,2} &= \varphi_{h,002} \\
\chi_{h,1} &= \varphi_{h,001} + \varphi_{h,101}c_1 + \varphi_{h,011}c_2 \\
\chi_{h,0} &= \varphi_{h,200}c_1^2 + \varphi_{h,110}c_1c_2 + \varphi_{h,100}c_1 + \varphi_{h,020}c_2^2 + \varphi_{h,010}c_2 + \varphi_{h,000}
\end{aligned}
$$

Appendix C. Coefficients of Polynomials Ω_h

$$
\begin{aligned}
\omega_{h,4} &= A_h \\
\omega_{h,3} &= B_hc_1 + C_h \\
\omega_{h,2} &= D_hc_1^2 + E_hc_1 + F_h \\
\omega_{h,1} &= G_hc_1^3 + H_hc_1^2 + I_hc_1 + J_h \\
\omega_{h,0} &= K_hc_1^4 + L_hc_1^3 + M_hc_1^2 + N_hc_1 + O_h
\end{aligned}
$$

In the following, $h^* = 4$ for $h = 1$ and $h^* = 5$ for $h = 2$.

$$
\begin{aligned}
A_h = \; & \varphi_{h^*,002}^2\varphi_{6,020}^2 - \varphi_{h^*,002}\varphi_{h^*,011}\varphi_{6,011}\varphi_{6,020} - 2\varphi_{h^*,002}\varphi_{h^*,020}\varphi_{6,002}\varphi_{6,020} \\
& + \varphi_{h^*,002}\varphi_{h^*,020}\varphi_{6,011}^2 + \varphi_{h^*,011}^2\varphi_{6,002}\varphi_{6,020} - \varphi_{h^*,011}\varphi_{h^*,020}\varphi_{6,002}\varphi_{6,011} + \varphi_{h^*,020}^2\varphi_{6,002}^2 \\
B_h = \; & \varphi_{h^*,002}\varphi_{h^*,110}\varphi_{6,011}^2 + 2\varphi_{h^*,020}\varphi_{h^*,110}\varphi_{6,002}^2 + \varphi_{h^*,011}^2\varphi_{6,002}\varphi_{6,110} + 2\varphi_{h^*,002}^2\varphi_{6,020}\varphi_{6,110} \\
& - \varphi_{h^*,002}\varphi_{h^*,011}\varphi_{6,011}\varphi_{6,110} - \varphi_{h^*,002}\varphi_{h^*,011}\varphi_{6,020}\varphi_{6,101} - 2\varphi_{h^*,002}\varphi_{h^*,020}\varphi_{6,002}\varphi_{6,110} \\
& + 2\varphi_{h^*,002}\varphi_{h^*,020}\varphi_{6,011}\varphi_{6,101} - \varphi_{h^*,002}\varphi_{h^*,101}\varphi_{6,011}\varphi_{6,020} - 2\varphi_{h^*,002}\varphi_{h^*,110}\varphi_{6,002}\varphi_{6,020} \\
& - \varphi_{h^*,011}\varphi_{h^*,020}\varphi_{6,002}\varphi_{6,101} + 2\varphi_{h^*,011}\varphi_{h^*,101}\varphi_{6,002}\varphi_{6,020} - \varphi_{h^*,011}\varphi_{h^*,110}\varphi_{6,002}\varphi_{6,011} \\
& - \varphi_{h^*,020}\varphi_{h^*,101}\varphi_{6,002}\varphi_{6,011} \\
C_h = \; & \varphi_{h^*,002}\varphi_{h^*,010}\varphi_{6,011}^2 + 2\varphi_{h^*,010}\varphi_{h^*,020}\varphi_{6,002}^2 + \varphi_{h^*,011}^2\varphi_{6,002}\varphi_{6,010} + 2\varphi_{h^*,002}^2\varphi_{6,010}\varphi_{6,020} \\
& - \varphi_{h^*,001}\varphi_{h^*,002}\varphi_{6,011}\varphi_{6,020} + 2\varphi_{h^*,001}\varphi_{h^*,011}\varphi_{6,002}\varphi_{6,020} - \varphi_{h^*,001}\varphi_{h^*,020}\varphi_{6,002}\varphi_{6,011} \\
& - 2\varphi_{h^*,002}\varphi_{h^*,010}\varphi_{6,002}\varphi_{6,020} - \varphi_{h^*,002}\varphi_{h^*,011}\varphi_{6,001}\varphi_{6,020} - \varphi_{h^*,002}\varphi_{h^*,011}\varphi_{6,010}\varphi_{6,011} \\
& + 2\varphi_{h^*,002}\varphi_{h^*,020}\varphi_{6,001}\varphi_{6,011} - 2\varphi_{h^*,002}\varphi_{h^*,020}\varphi_{6,002}\varphi_{6,010} - \varphi_{h^*,010}\varphi_{h^*,011}\varphi_{6,002}\varphi_{6,011} \\
& - \varphi_{h^*,011}\varphi_{h^*,020}\varphi_{6,001}\varphi_{6,002} \\
D_h = \; & \varphi_{h^*,002}^2\varphi_{6,110}^2 + \varphi_{h^*,110}^2\varphi_{6,002}^2 + \varphi_{h^*,002}\varphi_{h^*,020}\varphi_{6,101}^2 + \varphi_{h^*,002}\varphi_{h^*,200}\varphi_{6,011}^2 \\
& + 2\varphi_{h^*,020}\varphi_{h^*,200}\varphi_{6,002}^2 + \varphi_{h^*,101}^2\varphi_{6,002}\varphi_{6,020} + \varphi_{h^*,011}^2\varphi_{6,002}\varphi_{6,200} + 2\varphi_{h^*,002}^2\varphi_{6,020}\varphi_{6,200} \\
& - \varphi_{h^*,002}\varphi_{h^*,011}\varphi_{6,011}\varphi_{6,200} - \varphi_{h^*,002}\varphi_{h^*,011}\varphi_{6,101}\varphi_{6,110} - 2\varphi_{h^*,002}\varphi_{h^*,020}\varphi_{6,002}\varphi_{6,200} \\
& - \varphi_{h^*,002}\varphi_{h^*,101}\varphi_{6,011}\varphi_{6,110} - \varphi_{h^*,002}\varphi_{h^*,101}\varphi_{6,020}\varphi_{6,101} - 2\varphi_{h^*,002}\varphi_{h^*,110}\varphi_{6,002}\varphi_{6,110} \\
& + 2\varphi_{h^*,002}\varphi_{h^*,110}\varphi_{6,011}\varphi_{6,101} - 2\varphi_{h^*,002}\varphi_{h^*,200}\varphi_{6,002}\varphi_{6,020} + 2\varphi_{h^*,011}\varphi_{h^*,101}\varphi_{6,002}\varphi_{6,110} \\
& - \varphi_{h^*,011}\varphi_{h^*,110}\varphi_{6,002}\varphi_{6,101} - \varphi_{h^*,011}\varphi_{h^*,200}\varphi_{6,002}\varphi_{6,011} - \varphi_{h^*,020}\varphi_{h^*,101}\varphi_{6,002}\varphi_{6,101} \\
& - \varphi_{h^*,101}\varphi_{h^*,110}\varphi_{6,002}\varphi_{6,011} \\
E_h = \; & \varphi_{h^*,002}\varphi_{h^*,100}\varphi_{6,011}^2 + 2\varphi_{h^*,010}\varphi_{h^*,110}\varphi_{6,002}^2 + 2\varphi_{h^*,020}\varphi_{h^*,100}\varphi_{6,002}^2 \\
& + \varphi_{h^*,011}^2\varphi_{6,002}\varphi_{6,100} + 2\varphi_{h^*,002}^2\varphi_{6,010}\varphi_{6,110} + 2\varphi_{h^*,002}^2\varphi_{6,020}\varphi_{6,100} \\
& - \varphi_{h^*,001}\varphi_{h^*,002}\varphi_{6,011}\varphi_{6,110} - \varphi_{h^*,001}\varphi_{h^*,002}\varphi_{6,020}\varphi_{6,101} + 2\varphi_{h^*,001}\varphi_{h^*,011}\varphi_{6,002}\varphi_{6,110} \\
& - \varphi_{h^*,001}\varphi_{h^*,020}\varphi_{6,002}\varphi_{6,101} + 2\varphi_{h^*,001}\varphi_{h^*,101}\varphi_{6,002}\varphi_{6,020} - \varphi_{h^*,001}\varphi_{h^*,110}\varphi_{6,002}\varphi_{6,011} \\
& - 2\varphi_{h^*,002}\varphi_{h^*,010}\varphi_{6,002}\varphi_{6,110} + 2\varphi_{h^*,002}\varphi_{h^*,010}\varphi_{6,011}\varphi_{6,101} - \varphi_{h^*,002}\varphi_{h^*,011}\varphi_{6,001}\varphi_{6,110} \\
& - \varphi_{h^*,002}\varphi_{h^*,011}\varphi_{6,010}\varphi_{6,101} - \varphi_{h^*,002}\varphi_{h^*,011}\varphi_{6,011}\varphi_{6,100} + 2\varphi_{h^*,002}\varphi_{h^*,020}\varphi_{6,001}\varphi_{6,101} \\
& - 2\varphi_{h^*,002}\varphi_{h^*,020}\varphi_{6,002}\varphi_{6,100} - 2\varphi_{h^*,002}\varphi_{h^*,100}\varphi_{6,002}\varphi_{6,020} - \varphi_{h^*,002}\varphi_{h^*,101}\varphi_{6,001}\varphi_{6,020} \\
& - \varphi_{h^*,002}\varphi_{h^*,101}\varphi_{6,010}\varphi_{6,011} + 2\varphi_{h^*,002}\varphi_{h^*,110}\varphi_{6,001}\varphi_{6,011} - 2\varphi_{h^*,002}\varphi_{h^*,110}\varphi_{6,002}\varphi_{6,010} \\
& - \varphi_{h^*,010}\varphi_{h^*,011}\varphi_{6,002}\varphi_{6,101} - \varphi_{h^*,010}\varphi_{h^*,101}\varphi_{6,002}\varphi_{6,011} - \varphi_{h^*,011}\varphi_{h^*,100}\varphi_{6,002}\varphi_{6,011} \\
& + 2\varphi_{h^*,011}\varphi_{h^*,101}\varphi_{6,002}\varphi_{6,010} - \varphi_{h^*,011}\varphi_{h^*,110}\varphi_{6,001}\varphi_{6,002} - \varphi_{h^*,020}\varphi_{h^*,101}\varphi_{6,001}\varphi_{6,002} \\
F_h = \; & \varphi_{h^*,002}^2\varphi_{6,010}^2 + \varphi_{h^*,010}^2\varphi_{6,002}^2 + \varphi_{h^*,000}\varphi_{h^*,002}\varphi_{6,011}^2 + 2\varphi_{h^*,000}\varphi_{h^*,020}\varphi_{6,002}^2 \\
& + \varphi_{h^*,002}\varphi_{h^*,020}\varphi_{6,001}^2 + \varphi_{h^*,011}^2\varphi_{6,000}\varphi_{6,002} + 2\varphi_{h^*,002}^2\varphi_{6,000}\varphi_{6,020} + \varphi_{h^*,001}^2\varphi_{6,002}\varphi_{6,020} \\
& - 2\varphi_{h^*,000}\varphi_{h^*,002}\varphi_{6,002}\varphi_{6,020} - \varphi_{h^*,000}\varphi_{h^*,011}\varphi_{6,002}\varphi_{6,011} - \varphi_{h^*,001}\varphi_{h^*,002}\varphi_{6,001}\varphi_{6,020} \\
& - \varphi_{h^*,001}\varphi_{h^*,002}\varphi_{6,010}\varphi_{6,011} - \varphi_{h^*,001}\varphi_{h^*,010}\varphi_{6,002}\varphi_{6,011} + 2\varphi_{h^*,001}\varphi_{h^*,011}\varphi_{6,002}\varphi_{6,010} \\
& - \varphi_{h^*,001}\varphi_{h^*,020}\varphi_{6,001}\varphi_{6,002} + 2\varphi_{h^*,002}\varphi_{h^*,010}\varphi_{6,001}\varphi_{6,011} - 2\varphi_{h^*,002}\varphi_{h^*,010}\varphi_{6,002}\varphi_{6,010} \\
& - \varphi_{h^*,002}\varphi_{h^*,011}\varphi_{6,000}\varphi_{6,011} - \varphi_{h^*,002}\varphi_{h^*,011}\varphi_{6,001}\varphi_{6,010} - 2\varphi_{h^*,002}\varphi_{h^*,020}\varphi_{6,000}\varphi_{6,002} \\
& - \varphi_{h^*,010}\varphi_{h^*,011}\varphi_{6,001}\varphi_{6,002}
\end{aligned}
$$

$$
\begin{aligned}
G_h = \; & \varphi_{h^*,002}\varphi_{h^*,110}\varphi_{6,101}^2 + 2\varphi_{h^*,110}\varphi_{h^*,200}\varphi_{6,002}^2 + \varphi_{h^*,101}^2\varphi_{6,002}\varphi_{6,110} + 2\varphi_{h^*,002}^2\varphi_{6,110}\varphi_{6,200} \\
& -\varphi_{h^*,002}\varphi_{h^*,011}\varphi_{6,101}\varphi_{6,200} - \varphi_{h^*,002}\varphi_{h^*,101}\varphi_{6,011}\varphi_{6,200} - \varphi_{h^*,002}\varphi_{h^*,101}\varphi_{6,101}\varphi_{6,110} \\
& -2\varphi_{h^*,002}\varphi_{h^*,110}\varphi_{6,002}\varphi_{6,200} - 2\varphi_{h^*,002}\varphi_{h^*,200}\varphi_{6,002}\varphi_{6,110} + 2\varphi_{h^*,002}\varphi_{h^*,200}\varphi_{6,011}\varphi_{6,101} \\
& +2\varphi_{h^*,011}\varphi_{h^*,101}\varphi_{6,002}\varphi_{6,200} - \varphi_{h^*,011}\varphi_{h^*,200}\varphi_{6,002}\varphi_{6,101} - \varphi_{h^*,101}\varphi_{h^*,110}\varphi_{6,002}\varphi_{6,101} \\
& -\varphi_{h^*,101}\varphi_{h^*,200}\varphi_{6,002}\varphi_{6,011} \\
H_h = \; & \varphi_{h^*,002}\varphi_{h^*,010}\varphi_{6,101}^2 + 2\varphi_{h^*,010}\varphi_{h^*,200}\varphi_{6,002}^2 + 2\varphi_{h^*,100}\varphi_{h^*,110}\varphi_{6,002}^2 + \varphi_{h^*,101}^2\varphi_{6,002}\varphi_{6,010} \\
& +2\varphi_{h^*,002}^2\varphi_{6,010}\varphi_{6,200} + 2\varphi_{h^*,002}^2\varphi_{6,100}\varphi_{6,110} - \varphi_{h^*,001}\varphi_{h^*,002}\varphi_{6,011}\varphi_{6,200} \\
& -\varphi_{h^*,001}\varphi_{h^*,002}\varphi_{6,101}\varphi_{6,110} + 2\varphi_{h^*,001}\varphi_{h^*,011}\varphi_{6,002}\varphi_{6,200} + 2\varphi_{h^*,001}\varphi_{h^*,101}\varphi_{6,002}\varphi_{6,110} \\
& -\varphi_{h^*,001}\varphi_{h^*,110}\varphi_{6,002}\varphi_{6,101} - \varphi_{h^*,001}\varphi_{h^*,200}\varphi_{6,002}\varphi_{6,011} - 2\varphi_{h^*,002}\varphi_{h^*,010}\varphi_{6,002}\varphi_{6,200} \\
& -\varphi_{h^*,002}\varphi_{h^*,011}\varphi_{6,001}\varphi_{6,200} - \varphi_{h^*,002}\varphi_{h^*,011}\varphi_{6,100}\varphi_{6,101} - 2\varphi_{h^*,002}\varphi_{h^*,100}\varphi_{6,002}\varphi_{6,110} \\
& +2\varphi_{h^*,002}\varphi_{h^*,100}\varphi_{6,011}\varphi_{6,101} - \varphi_{h^*,002}\varphi_{h^*,101}\varphi_{6,001}\varphi_{6,110} - \varphi_{h^*,002}\varphi_{h^*,101}\varphi_{6,010}\varphi_{6,101} \\
& -\varphi_{h^*,002}\varphi_{h^*,101}\varphi_{6,011}\varphi_{6,100} + 2\varphi_{h^*,002}\varphi_{h^*,110}\varphi_{6,001}\varphi_{6,101} - 2\varphi_{h^*,002}\varphi_{h^*,110}\varphi_{6,002}\varphi_{6,100} \\
& +2\varphi_{h^*,002}\varphi_{h^*,200}\varphi_{6,001}\varphi_{6,011} - 2\varphi_{h^*,002}\varphi_{h^*,200}\varphi_{6,002}\varphi_{6,010} - \varphi_{h^*,010}\varphi_{h^*,101}\varphi_{6,002}\varphi_{6,101} \\
& -\varphi_{h^*,011}\varphi_{h^*,100}\varphi_{6,002}\varphi_{6,101} + 2\varphi_{h^*,011}\varphi_{h^*,101}\varphi_{6,002}\varphi_{6,100} - \varphi_{h^*,011}\varphi_{h^*,200}\varphi_{6,001}\varphi_{6,002} \\
& -\varphi_{h^*,100}\varphi_{h^*,101}\varphi_{6,002}\varphi_{6,011} - \varphi_{h^*,101}\varphi_{h^*,110}\varphi_{6,001}\varphi_{6,002} \\
I_h = \; & 2\varphi_{h^*,000}\varphi_{h^*,110}\varphi_{6,002}^2 + 2\varphi_{h^*,010}\varphi_{h^*,100}\varphi_{6,002}^2 + \varphi_{h^*,002}\varphi_{h^*,110}\varphi_{6,001}^2 + 2\varphi_{h^*,002}^2\varphi_{6,000}\varphi_{6,110} \\
& +2\varphi_{h^*,002}^2\varphi_{6,010}\varphi_{6,100} + \varphi_{h^*,001}^2\varphi_{6,002}\varphi_{6,110} - 2\varphi_{h^*,000}\varphi_{h^*,002}\varphi_{6,002}\varphi_{6,110} \\
& +2\varphi_{h^*,000}\varphi_{h^*,002}\varphi_{6,011}\varphi_{6,101} - \varphi_{h^*,000}\varphi_{h^*,011}\varphi_{6,002}\varphi_{6,101} - \varphi_{h^*,000}\varphi_{h^*,101}\varphi_{6,002}\varphi_{6,011} \\
& -\varphi_{h^*,001}\varphi_{h^*,002}\varphi_{6,001}\varphi_{6,110} - \varphi_{h^*,001}\varphi_{h^*,002}\varphi_{6,010}\varphi_{6,101} - \varphi_{h^*,001}\varphi_{h^*,002}\varphi_{6,011}\varphi_{6,100} \\
& -\varphi_{h^*,001}\varphi_{h^*,010}\varphi_{6,002}\varphi_{6,101} + 2\varphi_{h^*,001}\varphi_{h^*,011}\varphi_{6,002}\varphi_{6,100} - \varphi_{h^*,001}\varphi_{h^*,100}\varphi_{6,002}\varphi_{6,011} \\
& +2\varphi_{h^*,001}\varphi_{h^*,101}\varphi_{6,002}\varphi_{6,010} - \varphi_{h^*,001}\varphi_{h^*,110}\varphi_{6,001}\varphi_{6,002} + 2\varphi_{h^*,002}\varphi_{h^*,010}\varphi_{6,001}\varphi_{6,101} \\
& -2\varphi_{h^*,002}\varphi_{h^*,010}\varphi_{6,002}\varphi_{6,100} - \varphi_{h^*,002}\varphi_{h^*,011}\varphi_{6,000}\varphi_{6,101} - \varphi_{h^*,002}\varphi_{h^*,011}\varphi_{6,001}\varphi_{6,100} \\
& +2\varphi_{h^*,002}\varphi_{h^*,100}\varphi_{6,001}\varphi_{6,011} - 2\varphi_{h^*,002}\varphi_{h^*,100}\varphi_{6,002}\varphi_{6,010} - \varphi_{h^*,002}\varphi_{h^*,101}\varphi_{6,000}\varphi_{6,011} \\
& -\varphi_{h^*,002}\varphi_{h^*,101}\varphi_{6,001}\varphi_{6,010} - 2\varphi_{h^*,002}\varphi_{h^*,110}\varphi_{6,000}\varphi_{6,002} - \varphi_{h^*,010}\varphi_{h^*,101}\varphi_{6,001}\varphi_{6,002} \\
& -\varphi_{h^*,011}\varphi_{h^*,100}\varphi_{6,001}\varphi_{6,002} + 2\varphi_{h^*,011}\varphi_{h^*,101}\varphi_{6,000}\varphi_{6,002} \\
J_h = \; & 2\varphi_{h^*,000}\varphi_{h^*,010}\varphi_{6,002}^2 + \varphi_{h^*,002}\varphi_{h^*,010}\varphi_{6,001}^2 + 2\varphi_{h^*,002}^2\varphi_{6,000}\varphi_{6,010} + \varphi_{h^*,001}^2\varphi_{6,002}\varphi_{6,010} \\
& -\varphi_{h^*,000}\varphi_{h^*,001}\varphi_{6,002}\varphi_{6,011} + 2\varphi_{h^*,000}\varphi_{h^*,002}\varphi_{6,001}\varphi_{6,011} - 2\varphi_{h^*,000}\varphi_{h^*,002}\varphi_{6,002}\varphi_{6,010} \\
& -\varphi_{h^*,000}\varphi_{h^*,011}\varphi_{6,001}\varphi_{6,002} - \varphi_{h^*,001}\varphi_{h^*,002}\varphi_{6,000}\varphi_{6,011} - \varphi_{h^*,001}\varphi_{h^*,002}\varphi_{6,001}\varphi_{6,010} \\
& -\varphi_{h^*,001}\varphi_{h^*,010}\varphi_{6,001}\varphi_{6,002} + 2\varphi_{h^*,001}\varphi_{h^*,011}\varphi_{6,000}\varphi_{6,002} - 2\varphi_{h^*,002}\varphi_{h^*,010}\varphi_{6,000}\varphi_{6,002} \\
& -\varphi_{h^*,002}\varphi_{h^*,011}\varphi_{6,000}\varphi_{6,001} \\
K_h = \; & \varphi_{h^*,002}^2\varphi_{6,200}^2 - \varphi_{h^*,002}\varphi_{h^*,101}\varphi_{6,101}\varphi_{6,200} - 2\varphi_{h^*,002}\varphi_{h^*,200}\varphi_{6,002}\varphi_{6,200} \\
& +\varphi_{h^*,002}\varphi_{h^*,200}\varphi_{6,101}^2 + \varphi_{h^*,101}^2\varphi_{6,002}\varphi_{6,200} - \varphi_{h^*,101}\varphi_{h^*,200}\varphi_{6,002}\varphi_{6,101} \\
& +\varphi_{h^*,200}^2\varphi_{6,002}^2 \\
L_h = \; & \varphi_{h^*,002}\varphi_{h^*,100}\varphi_{6,101}^2 + 2\varphi_{h^*,100}\varphi_{h^*,200}\varphi_{6,002}^2 + \varphi_{h^*,101}^2\varphi_{6,002}\varphi_{6,100} + 2\varphi_{h^*,002}^2\varphi_{6,100}\varphi_{6,200} \\
& -\varphi_{h^*,001}\varphi_{h^*,002}\varphi_{6,101}\varphi_{6,200} + 2\varphi_{h^*,001}\varphi_{h^*,101}\varphi_{6,002}\varphi_{6,200} - \varphi_{h^*,001}\varphi_{h^*,200}\varphi_{6,002}\varphi_{6,101} \\
& -2\varphi_{h^*,002}\varphi_{h^*,100}\varphi_{6,002}\varphi_{6,200} - \varphi_{h^*,002}\varphi_{h^*,101}\varphi_{6,001}\varphi_{6,200} - \varphi_{h^*,002}\varphi_{h^*,101}\varphi_{6,100}\varphi_{6,101} \\
& +2\varphi_{h^*,002}\varphi_{h^*,200}\varphi_{6,001}\varphi_{6,101} - 2\varphi_{h^*,002}\varphi_{h^*,200}\varphi_{6,002}\varphi_{6,100} - \varphi_{h^*,100}\varphi_{h^*,101}\varphi_{6,002}\varphi_{6,101} \\
& -\varphi_{h^*,101}\varphi_{h^*,200}\varphi_{6,001}\varphi_{6,002} \\
M_h = \; & \varphi_{h^*,002}^2\varphi_{6,100}^2 + \varphi_{h^*,100}^2\varphi_{6,002}^2 + \varphi_{h^*,000}\varphi_{h^*,002}\varphi_{6,101}^2 + 2\varphi_{h^*,000}\varphi_{h^*,200}\varphi_{6,002}^2 \\
& +\varphi_{h^*,002}\varphi_{h^*,200}\varphi_{6,001}^2 + \varphi_{h^*,101}^2\varphi_{6,000}\varphi_{6,002} + 2\varphi_{h^*,002}^2\varphi_{6,000}\varphi_{6,200} + \varphi_{h^*,001}^2\varphi_{6,002}\varphi_{6,200} \\
& -2\varphi_{h^*,000}\varphi_{h^*,002}\varphi_{6,002}\varphi_{6,200} - \varphi_{h^*,000}\varphi_{h^*,101}\varphi_{6,002}\varphi_{6,101} - \varphi_{h^*,001}\varphi_{h^*,002}\varphi_{6,001}\varphi_{6,200} \\
& -\varphi_{h^*,001}\varphi_{h^*,002}\varphi_{6,100}\varphi_{6,101} - \varphi_{h^*,001}\varphi_{h^*,100}\varphi_{6,002}\varphi_{6,101} + 2\varphi_{h^*,001}\varphi_{h^*,101}\varphi_{6,002}\varphi_{6,100} \\
& -\varphi_{h^*,001}\varphi_{h^*,200}\varphi_{6,001}\varphi_{6,002} + 2\varphi_{h^*,002}\varphi_{h^*,100}\varphi_{6,001}\varphi_{6,101} - 2\varphi_{h^*,002}\varphi_{h^*,100}\varphi_{6,002}\varphi_{6,100} \\
& -\varphi_{h^*,002}\varphi_{h^*,101}\varphi_{6,000}\varphi_{6,101} - \varphi_{h^*,002}\varphi_{h^*,101}\varphi_{6,001}\varphi_{6,100} - 2\varphi_{h^*,002}\varphi_{h^*,200}\varphi_{6,000}\varphi_{6,002} \\
& -\varphi_{h^*,100}\varphi_{h^*,101}\varphi_{6,001}\varphi_{6,002} \\
N_h = \; & 2\varphi_{h^*,000}\varphi_{h^*,100}\varphi_{6,002}^2 + \varphi_{h^*,002}\varphi_{h^*,100}\varphi_{6,001}^2 + 2\varphi_{h^*,002}^2\varphi_{6,000}\varphi_{6,100} + \varphi_{h^*,001}^2\varphi_{6,002}\varphi_{6,100} \\
& -\varphi_{h^*,000}\varphi_{h^*,001}\varphi_{6,002}\varphi_{6,101} + 2\varphi_{h^*,000}\varphi_{h^*,002}\varphi_{6,001}\varphi_{6,101} - 2\varphi_{h^*,000}\varphi_{h^*,002}\varphi_{6,002}\varphi_{6,100} \\
& -\varphi_{h^*,000}\varphi_{h^*,101}\varphi_{6,001}\varphi_{6,002} - \varphi_{h^*,001}\varphi_{h^*,002}\varphi_{6,000}\varphi_{6,101} - \varphi_{h^*,001}\varphi_{h^*,002}\varphi_{6,001}\varphi_{6,100} \\
& -\varphi_{h^*,001}\varphi_{h^*,100}\varphi_{6,001}\varphi_{6,002} + 2\varphi_{h^*,001}\varphi_{h^*,101}\varphi_{6,000}\varphi_{6,002} - 2\varphi_{h^*,002}\varphi_{h^*,100}\varphi_{6,000}\varphi_{6,002} \\
& -\varphi_{h^*,002}\varphi_{h^*,101}\varphi_{6,000}\varphi_{6,001} \\
O_h = \; & \varphi_{h^*,000}^2\varphi_{6,002}^2 - \varphi_{h^*,000}\varphi_{h^*,001}\varphi_{6,001}\varphi_{6,002} - 2\varphi_{h^*,000}\varphi_{h^*,002}\varphi_{6,000}\varphi_{6,002} \\
& +\varphi_{h^*,000}\varphi_{h^*,002}\varphi_{6,001}^2 + \varphi_{h^*,001}^2\varphi_{6,000}\varphi_{6,002} - \varphi_{h^*,001}\varphi_{h^*,002}\varphi_{6,000}\varphi_{6,001} + \varphi_{h^*,002}^2\varphi_{6,000}^2
\end{aligned}
$$

Appendix D. Univariate Polynomial

$$
\begin{array}{l}
\Lambda : \\
\omega_{1,4}^4\omega_{2,0}^4 + \omega_{1,0}^4\omega_{2,4}^4 + \omega_{1,4}^2\omega_{1,0}^2\omega_{2,2}^4 + \omega_{1,2}^4\omega_{2,4}^2\omega_{2,0}^2 + \omega_{1,4}\omega_{1,0}^3\omega_{2,3}^4 + \omega_{1,4}^3\omega_{1,0}\omega_{2,1}^4 \\
+\omega_{1,3}^4\omega_{2,4}\omega_{2,0}^3 + \omega_{1,1}^4\omega_{2,4}^3\omega_{2,0} - \omega_{1,4}\omega_{1,3}^3\omega_{2,3}\omega_{2,0}^3 - \omega_{1,3}\omega_{1,0}^3\omega_{2,4}\omega_{2,3}^3 - \omega_{1,4}\omega_{1,1}^3\omega_{2,3}^3\omega_{2,0} \\
-4\omega_{1,4}\omega_{1,0}^3\omega_{2,4}^3\omega_{2,0} - 3\omega_{1,3}\omega_{1,0}^3\omega_{2,4}^3\omega_{2,1} - 2\omega_{1,2}\omega_{1,0}^3\omega_{2,4}^3\omega_{2,2} - \omega_{1,1}\omega_{1,0}^3\omega_{2,4}^3\omega_{2,3} \\
-\omega_{1,4}^3\omega_{1,3}\omega_{2,1}\omega_{2,0}^3 - 2\omega_{1,4}^3\omega_{1,2}\omega_{2,2}\omega_{2,0}^3 - 3\omega_{1,4}^3\omega_{1,1}\omega_{2,3}\omega_{2,0}^3 - 4\omega_{1,4}^3\omega_{1,0}\omega_{2,4}\omega_{2,0}^3 \\
-\omega_{1,3}^3\omega_{1,0}\omega_{2,4}\omega_{2,1}^3 - \omega_{1,4}^3\omega_{1,1}\omega_{2,1}^3\omega_{2,0} - \omega_{1,1}^3\omega_{1,0}\omega_{2,4}^3\omega_{2,1} + 2\omega_{1,4}\omega_{1,0}^3\omega_{2,4}^2\omega_{2,2}^2 \\
+2\omega_{1,4}^2\omega_{1,2}^2\omega_{2,4}\omega_{2,0}^3 + \omega_{1,4}\omega_{1,2}^3\omega_{2,3}^2\omega_{2,0}^2 + \omega_{1,2}\omega_{1,0}^3\omega_{2,4}^2\omega_{2,3}^2 + \omega_{1,4}^2\omega_{1,3}^2\omega_{2,2}\omega_{2,0}^3 \\
+\omega_{1,3}^2\omega_{1,0}^2\omega_{2,4}\omega_{2,2}^3 + \omega_{1,4}^2\omega_{1,1}^2\omega_{2,2}^3\omega_{2,0} + \omega_{1,4}^3\omega_{1,2}\omega_{2,1}^2\omega_{2,0}^2 + \omega_{1,2}^3\omega_{1,0}\omega_{2,4}^2\omega_{2,1}^2 \\
+\omega_{1,1}^2\omega_{1,0}^2\omega_{2,4}^3\omega_{2,2} + 2\omega_{1,4}^3\omega_{1,0}\omega_{2,2}^2\omega_{2,0}^2 + 2\omega_{1,2}^2\omega_{1,0}^2\omega_{2,4}^3\omega_{2,0} + \omega_{1,4}^2\omega_{1,2}^2\omega_{2,2}^2\omega_{2,0}^2 \\
+3\omega_{1,4}^2\omega_{1,1}^2\omega_{2,3}^2\omega_{2,0}^2 + 6\omega_{1,4}^2\omega_{1,0}^2\omega_{2,4}^2\omega_{2,0}^2 + 2\omega_{1,4}^2\omega_{1,0}^2\omega_{2,3}^2\omega_{2,1}^2 + 2\omega_{1,3}^2\omega_{1,1}^2\omega_{2,4}^2\omega_{2,0}^2 \\
+3\omega_{1,3}^2\omega_{1,0}^2\omega_{2,4}^2\omega_{2,1}^2 + \omega_{1,2}^2\omega_{1,0}^2\omega_{2,4}^2\omega_{2,2}^2 - 4\omega_{1,4}\omega_{1,3}^2\omega_{1,2}\omega_{2,4}\omega_{2,0}^3 \\
-\omega_{1,4}\omega_{1,3}\omega_{1,0}^2\omega_{2,3}\omega_{2,2}^3 - 2\omega_{1,4}\omega_{1,2}\omega_{1,0}^2\omega_{2,4}\omega_{2,2}^3 + 3\omega_{1,4}^2\omega_{1,3}\omega_{1,2}\omega_{2,3}\omega_{2,0}^3 \\
+4\omega_{1,4}^2\omega_{1,3}\omega_{1,1}\omega_{2,4}\omega_{2,0}^3 + \omega_{1,4}\omega_{1,3}^2\omega_{1,0}\omega_{2,3}\omega_{2,1}^3 - 3\omega_{1,4}\omega_{1,3}\omega_{1,0}^2\omega_{2,3}^3\omega_{2,0} \\
-2\omega_{1,4}\omega_{1,2}\omega_{1,0}^2\omega_{2,3}^3\omega_{2,1} - \omega_{1,4}\omega_{1,1}\omega_{1,0}^2\omega_{2,3}^3\omega_{2,2} - \omega_{1,4}^2\omega_{1,3}\omega_{1,0}\omega_{2,2}\omega_{2,1}^3 \\
-2\omega_{1,4}^2\omega_{1,2}\omega_{1,0}\omega_{2,3}\omega_{2,1}^3 - 3\omega_{1,4}^2\omega_{1,1}\omega_{1,0}\omega_{2,4}\omega_{2,1}^3 + \omega_{1,4}\omega_{1,1}^2\omega_{1,0}\omega_{2,3}^3\omega_{2,1} \\
+4\omega_{1,3}\omega_{1,1}\omega_{1,0}^2\omega_{2,4}^3\omega_{2,0} + 3\omega_{1,2}\omega_{1,1}\omega_{1,0}^2\omega_{2,4}^3\omega_{2,1} - 2\omega_{1,4}^2\omega_{1,2}\omega_{1,0}\omega_{2,2}^3\omega_{2,0} \\
-\omega_{1,4}^2\omega_{1,1}\omega_{1,0}\omega_{2,2}^3\omega_{2,1} - 4\omega_{1,2}\omega_{1,1}^2\omega_{1,0}\omega_{2,4}^3\omega_{2,0} - 4\omega_{1,4}\omega_{1,0}^3\omega_{2,4}\omega_{2,3}^2\omega_{2,2} \\
-2\omega_{1,4}\omega_{1,2}^3\omega_{2,4}\omega_{2,2}\omega_{2,0}^2 + 4\omega_{1,4}\omega_{1,0}^3\omega_{2,4}^2\omega_{2,3}\omega_{2,1} - \omega_{1,3}\omega_{1,2}^3\omega_{2,4}\omega_{2,3}\omega_{2,0}^2 \\
+3\omega_{1,3}\omega_{1,0}^3\omega_{2,4}^2\omega_{2,3}\omega_{2,2} + \omega_{1,3}\omega_{1,1}^3\omega_{2,4}\omega_{2,3}^2\omega_{2,0} - 3\omega_{1,4}\omega_{1,1}^3\omega_{2,4}^2\omega_{2,1}\omega_{2,0} \\
-2\omega_{1,3}\omega_{1,1}^3\omega_{2,4}^2\omega_{2,2}\omega_{2,0} - \omega_{1,2}\omega_{1,1}^3\omega_{2,4}^2\omega_{2,3}\omega_{2,0} - \omega_{1,3}^3\omega_{1,2}\omega_{2,4}\omega_{2,1}\omega_{2,0}^2 \\
-2\omega_{1,3}^3\omega_{1,1}\omega_{2,4}\omega_{2,2}\omega_{2,0}^2 - 3\omega_{1,3}^3\omega_{1,0}\omega_{2,4}\omega_{2,3}\omega_{2,0}^2 + \omega_{1,3}^3\omega_{1,1}\omega_{2,4}\omega_{2,1}^2\omega_{2,0} \\
+3\omega_{1,4}^3\omega_{1,1}\omega_{2,2}\omega_{2,1}\omega_{2,0}^2 + 4\omega_{1,4}^3\omega_{1,0}\omega_{2,3}\omega_{2,1}\omega_{2,0}^2 - \omega_{1,2}^3\omega_{1,1}\omega_{2,4}^2\omega_{2,1}\omega_{2,0} \\
-2\omega_{1,2}^3\omega_{1,0}\omega_{2,4}^2\omega_{2,2}\omega_{2,0} - 4\omega_{1,4}^3\omega_{1,0}\omega_{2,2}\omega_{2,1}^2\omega_{2,0} + 4\omega_{1,4}\omega_{1,2}\omega_{1,1}^2\omega_{2,4}^2\omega_{2,0}^2 \\
-3\omega_{1,4}\omega_{1,2}\omega_{1,0}^2\omega_{2,4}^2\omega_{2,1}^2 + \omega_{1,4}\omega_{1,2}\omega_{1,0}^2\omega_{2,3}^2\omega_{2,2}^2 + 3\omega_{1,4}\omega_{1,3}^2\omega_{1,0}\omega_{2,3}^2\omega_{2,0}^2 \\
-4\omega_{1,4}\omega_{1,2}^2\omega_{1,0}\omega_{2,4}^2\omega_{2,0}^2 + \omega_{1,4}\omega_{1,2}^2\omega_{1,0}\omega_{2,3}^2\omega_{2,1}^2 + 3\omega_{1,4}\omega_{1,1}^2\omega_{1,0}\omega_{2,4}^2\omega_{2,1}^2 \\
-2\omega_{1,3}\omega_{1,1}\omega_{1,0}^2\omega_{2,4}^2\omega_{2,2}^2 - 4\omega_{1,3}\omega_{1,2}^2\omega_{1,1}\omega_{2,4}^2\omega_{2,0}^2 - 2\omega_{1,4}^2\omega_{1,3}\omega_{1,1}\omega_{2,2}^2\omega_{2,0}^2 \\
-3\omega_{1,4}^2\omega_{1,2}\omega_{1,0}\omega_{2,3}^2\omega_{2,0}^2 + \omega_{1,4}^2\omega_{1,2}\omega_{1,0}\omega_{2,2}^2\omega_{2,1}^2 + 4\omega_{1,3}^2\omega_{1,2}\omega_{1,0}\omega_{2,4}^2\omega_{2,0}^2 \\
-3\omega_{1,4}^2\omega_{1,1}^2\omega_{2,4}\omega_{2,2}\omega_{2,0}^2 + 4\omega_{1,4}^2\omega_{1,0}^2\omega_{2,4}\omega_{2,2}\omega_{2,1}^2 + \omega_{1,3}^2\omega_{1,2}^2\omega_{2,4}\omega_{2,2}\omega_{2,0}^2 \\
-2\omega_{1,4}^2\omega_{1,2}^2\omega_{2,3}\omega_{2,1}\omega_{2,0}^2 + 3\omega_{1,4}^2\omega_{1,1}^2\omega_{2,4}\omega_{2,1}^2\omega_{2,0} - 4\omega_{1,4}^2\omega_{1,0}^2\omega_{2,4}\omega_{2,2}^2\omega_{2,0} \\
-4\omega_{1,4}^2\omega_{1,0}^2\omega_{2,3}\omega_{2,2}^2\omega_{2,1} + \omega_{1,3}^2\omega_{1,1}^2\omega_{2,4}\omega_{2,2}^2\omega_{2,0} + 3\omega_{1,3}^2\omega_{1,0}^2\omega_{2,4}\omega_{2,3}^2\omega_{2,0} \\
-2\omega_{1,2}^2\omega_{1,0}^2\omega_{2,4}^2\omega_{2,3}\omega_{2,1} + 4\omega_{1,4}^2\omega_{1,0}^2\omega_{2,3}^2\omega_{2,2}\omega_{2,0} - 3\omega_{1,3}^2\omega_{1,0}^2\omega_{2,4}^2\omega_{2,2}\omega_{2,0} \\
+\omega_{1,2}^2\omega_{1,1}^2\omega_{2,4}^2\omega_{2,2}\omega_{2,0} + 3\omega_{1,4}\omega_{1,3}\omega_{1,2}\omega_{1,0}\omega_{2,4}\omega_{2,1}^3 + 3\omega_{1,4}\omega_{1,2}\omega_{1,1}\omega_{1,0}\omega_{2,3}^3\omega_{2,0} \\
+3\omega_{1,4}\omega_{1,1}^3\omega_{2,4}\omega_{2,3}\omega_{2,2}\omega_{2,0} + 3\omega_{1,3}^3\omega_{1,0}\omega_{2,4}\omega_{2,2}\omega_{2,1}\omega_{2,0} - 3\omega_{1,4}\omega_{1,3}\omega_{1,2}\omega_{1,1}\omega_{2,3}^2\omega_{2,0}^2 \\
-8\omega_{1,4}\omega_{1,3}\omega_{1,1}\omega_{1,0}\omega_{2,4}^2\omega_{2,0}^2 - 2\omega_{1,4}\omega_{1,3}\omega_{1,1}\omega_{1,0}\omega_{2,3}^2\omega_{2,1}^2 - 3\omega_{1,3}\omega_{1,2}\omega_{1,1}\omega_{1,0}\omega_{2,4}^2\omega_{2,1}^2 \\
-5\omega_{1,4}\omega_{1,3}\omega_{1,1}^2\omega_{2,4}\omega_{2,3}\omega_{2,0}^2 - 5\omega_{1,4}\omega_{1,3}\omega_{1,0}^2\omega_{2,4}\omega_{2,3}\omega_{2,1}^2 + 3\omega_{1,4}\omega_{1,3}\omega_{1,2}^2\omega_{2,4}\omega_{2,1}\omega_{2,0}^2 \\
-\omega_{1,4}\omega_{1,3}\omega_{1,2}^2\omega_{2,3}\omega_{2,2}\omega_{2,0}^2 + \omega_{1,4}\omega_{1,3}\omega_{1,0}^2\omega_{2,4}\omega_{2,2}^2\omega_{2,1} + 3\omega_{1,4}\omega_{1,1}\omega_{1,0}^2\omega_{2,4}\omega_{2,3}\omega_{2,2}^2 \\
+\omega_{1,4}\omega_{1,2}^2\omega_{1,1}\omega_{2,4}\omega_{2,3}\omega_{2,0}^2 - \omega_{1,3}\omega_{1,2}\omega_{1,0}^2\omega_{2,4}\omega_{2,3}\omega_{2,2}^2 - \omega_{1,4}\omega_{1,3}\omega_{1,1}^2\omega_{2,3}\omega_{2,2}^2\omega_{2,0} \\
+3\omega_{1,4}\omega_{1,3}\omega_{1,0}^2\omega_{2,3}^2\omega_{2,2}\omega_{2,1} - 2\omega_{1,4}\omega_{1,2}\omega_{1,1}^2\omega_{2,4}\omega_{2,2}^2\omega_{2,0} + 2\omega_{1,4}\omega_{1,2}\omega_{1,0}^2\omega_{2,4}\omega_{2,3}^2\omega_{2,0} \\
+\omega_{1,4}\omega_{1,1}\omega_{1,0}^2\omega_{2,4}\omega_{2,3}^2\omega_{2,1} + \omega_{1,4}\omega_{1,3}^2\omega_{1,2}\omega_{2,3}\omega_{2,1}\omega_{2,0}^2 + \omega_{1,4}\omega_{1,3}^2\omega_{1,1}\omega_{2,4}\omega_{2,1}\omega_{2,0}^2 \\
+2\omega_{1,4}\omega_{1,3}^2\omega_{1,1}\omega_{2,3}\omega_{2,2}\omega_{2,0}^2 + 2\omega_{1,4}\omega_{1,3}^2\omega_{1,0}\omega_{2,4}\omega_{2,2}\omega_{2,0}^2 - 2\omega_{1,4}\omega_{1,2}^2\omega_{1,0}\omega_{2,4}\omega_{2,2}\omega_{2,1}^2 \\
+2\omega_{1,3}\omega_{1,2}\omega_{1,0}^2\omega_{2,4}\omega_{2,3}^2\omega_{2,1} + \omega_{1,3}\omega_{1,1}\omega_{1,0}^2\omega_{2,4}\omega_{2,3}^2\omega_{2,2} - \omega_{1,3}\omega_{1,2}^2\omega_{1,0}\omega_{2,4}\omega_{2,3}\omega_{2,1}^2 \\
+3\omega_{1,3}^2\omega_{1,2}\omega_{1,1}\omega_{2,4}\omega_{2,3}\omega_{2,0}^2 + 2\omega_{1,4}\omega_{1,3}\omega_{1,1}^2\omega_{2,3}^2\omega_{2,1}\omega_{2,0} + 5\omega_{1,4}\omega_{1,3}\omega_{1,0}^2\omega_{2,4}^2\omega_{2,1}\omega_{2,0} \\
+\omega_{1,4}\omega_{1,2}\omega_{1,1}^2\omega_{2,3}^2\omega_{2,2}\omega_{2,0} + 2\omega_{1,4}\omega_{1,2}\omega_{1,0}^2\omega_{2,4}^2\omega_{2,2}\omega_{2,0} - \omega_{1,4}\omega_{1,1}\omega_{1,0}^2\omega_{2,4}^2\omega_{2,3}\omega_{2,0} \\
-5\omega_{1,4}\omega_{1,1}\omega_{1,0}^2\omega_{2,4}^2\omega_{2,2}\omega_{2,1} - \omega_{1,4}\omega_{1,3}^2\omega_{1,1}\omega_{2,3}\omega_{2,1}^2\omega_{2,0} - \omega_{1,4}\omega_{1,3}^2\omega_{1,0}\omega_{2,4}\omega_{2,1}^2\omega_{2,0} \\
+4\omega_{1,4}\omega_{1,2}^2\omega_{1,0}\omega_{2,4}\omega_{2,2}^2\omega_{2,0} - \omega_{1,4}\omega_{1,1}^2\omega_{1,0}\omega_{2,4}\omega_{2,3}^2\omega_{2,0} - 5\omega_{1,3}\omega_{1,2}\omega_{1,0}^2\omega_{2,4}^2\omega_{2,3}\omega_{2,0} \\
+\omega_{1,3}\omega_{1,2}\omega_{1,0}^2\omega_{2,4}^2\omega_{2,2}\omega_{2,1} - \omega_{1,3}\omega_{1,1}\omega_{1,0}^2\omega_{2,4}^2\omega_{2,3}\omega_{2,1} - \omega_{1,3}\omega_{1,1}^2\omega_{1,0}\omega_{2,4}\omega_{2,3}^2\omega_{2,1} \\
-\omega_{1,2}\omega_{1,1}\omega_{1,0}^2\omega_{2,4}^2\omega_{2,3}\omega_{2,2} - \omega_{1,4}^2\omega_{1,3}\omega_{1,2}\omega_{2,2}\omega_{2,1}\omega_{2,0}^2 - \omega_{1,4}^2\omega_{1,3}\omega_{1,1}\omega_{2,3}\omega_{2,1}\omega_{2,0}^2
\end{array}
$$

$$
\begin{aligned}
&-\omega_{1,4}^2\omega_{1,3}\omega_{1,0}\omega_{2,4}\omega_{2,1}\omega_{2,0}^2 - 5\omega_{1,4}^2\omega_{1,3}\omega_{1,0}\omega_{2,3}\omega_{2,2}\omega_{2,0}^2 - 5\omega_{1,4}^2\omega_{1,2}\omega_{1,1}\omega_{2,4}\omega_{2,1}\omega_{2,0}^2 \\
&+\omega_{1,4}^2\omega_{1,2}\omega_{1,1}\omega_{2,3}\omega_{2,2}\omega_{2,0}^2 + 2\omega_{1,4}^2\omega_{1,2}\omega_{1,0}\omega_{2,4}\omega_{2,2}\omega_{2,0}^2 + 5\omega_{1,4}^2\omega_{1,1}\omega_{1,0}\omega_{2,4}\omega_{2,3}\omega_{2,0}^2 \\
&+\omega_{1,3}^2\omega_{1,2}\omega_{1,0}\omega_{2,4}\omega_{2,2}\omega_{2,1}^2 + 2\omega_{1,3}^2\omega_{1,1}\omega_{1,0}\omega_{2,4}\omega_{2,3}\omega_{2,1}^2 - \omega_{1,4}\omega_{1,2}^2\omega_{1,1}\omega_{2,3}^2\omega_{2,1}\omega_{2,0} \\
&-2\omega_{1,4}\omega_{1,2}^2\omega_{1,0}\omega_{2,3}^2\omega_{2,2}\omega_{2,0} + 2\omega_{1,4}\omega_{1,1}^2\omega_{1,0}\omega_{2,4}^2\omega_{2,2}\omega_{2,0} + 3\omega_{1,3}\omega_{1,2}\omega_{1,1}^2\omega_{2,4}^2\omega_{2,1}\omega_{2,0} \\
&+\omega_{1,3}\omega_{1,1}^2\omega_{1,0}\omega_{2,4}^2\omega_{2,3}\omega_{2,0} + 2\omega_{1,3}\omega_{1,1}^2\omega_{1,0}\omega_{2,4}^2\omega_{2,2}\omega_{2,1} + \omega_{1,2}\omega_{1,1}^2\omega_{1,0}\omega_{2,4}^2\omega_{2,3}\omega_{2,1} \\
&+\omega_{1,4}^2\omega_{1,3}\omega_{1,1}\omega_{2,2}\omega_{2,1}^2\omega_{2,0} + \omega_{1,4}^2\omega_{1,3}\omega_{1,0}\omega_{2,3}\omega_{2,1}^2\omega_{2,0} + 2\omega_{1,4}^2\omega_{1,2}\omega_{1,1}\omega_{2,3}\omega_{2,1}^2\omega_{2,0} \\
&+2\omega_{1,4}^2\omega_{1,2}\omega_{1,0}\omega_{2,4}\omega_{2,1}^2\omega_{2,0} + 3\omega_{1,4}^2\omega_{1,1}\omega_{1,0}\omega_{2,3}\omega_{2,2}\omega_{2,1}^2 - 2\omega_{1,3}^2\omega_{1,2}\omega_{1,0}\omega_{2,4}\omega_{2,2}^2\omega_{2,0} \\
&-\omega_{1,3}^2\omega_{1,1}\omega_{1,0}\omega_{2,4}\omega_{2,2}^2\omega_{2,1} + \omega_{1,3}\omega_{1,2}^2\omega_{1,0}\omega_{2,4}^2\omega_{2,1}\omega_{2,0} + 3\omega_{1,4}^2\omega_{1,3}\omega_{1,0}\omega_{2,2}^2\omega_{2,1}\omega_{2,0} \\
&-\omega_{1,4}^2\omega_{1,2}\omega_{1,1}\omega_{2,2}^2\omega_{2,1}\omega_{2,0} + \omega_{1,4}^2\omega_{1,1}\omega_{1,0}\omega_{2,3}\omega_{2,2}^2\omega_{2,0} + 3\omega_{1,2}^2\omega_{1,1}\omega_{1,0}\omega_{2,4}^2\omega_{2,3}\omega_{2,0} \\
&-\omega_{1,2}^2\omega_{1,1}\omega_{1,0}\omega_{2,4}^2\omega_{2,2}\omega_{2,1} - 5\omega_{1,4}^2\omega_{1,1}\omega_{1,0}\omega_{2,3}^2\omega_{2,1}\omega_{2,0} - 5\omega_{1,3}^2\omega_{1,1}\omega_{1,0}\omega_{2,4}^2\omega_{2,1}\omega_{2,0} \\
&-3\omega_{1,3}^2\omega_{1,0}^2\omega_{2,4}\omega_{2,3}\omega_{2,2}\omega_{2,1} - 8\omega_{1,4}^2\omega_{1,0}^2\omega_{2,4}\omega_{2,3}\omega_{2,1}\omega_{2,0} - 2\omega_{1,3}^2\omega_{1,1}^2\omega_{2,4}\omega_{2,3}\omega_{2,1}\omega_{2,0} \\
&-3\omega_{1,4}^2\omega_{1,1}^2\omega_{2,3}\omega_{2,2}\omega_{2,1}\omega_{2,0} + 4\omega_{1,4}\omega_{1,3}\omega_{1,2}\omega_{1,1}\omega_{2,4}\omega_{2,2}\omega_{2,0}^2 \\
&+2\omega_{1,4}\omega_{1,3}\omega_{1,2}\omega_{1,0}\omega_{2,4}\omega_{2,3}\omega_{2,0}^2 - 3\omega_{1,4}\omega_{1,3}\omega_{1,2}\omega_{1,1}\omega_{2,4}\omega_{2,1}^2\omega_{2,0} \\
&-\omega_{1,4}\omega_{1,3}\omega_{1,2}\omega_{1,0}\omega_{2,3}\omega_{2,2}\omega_{2,1}^2 - \omega_{1,4}\omega_{1,3}\omega_{1,1}\omega_{1,0}\omega_{2,4}\omega_{2,2}\omega_{2,1}^2 \\
&+\omega_{1,4}\omega_{1,2}\omega_{1,1}\omega_{1,0}\omega_{2,4}\omega_{2,3}\omega_{2,1}^2 + 2\omega_{1,4}\omega_{1,3}\omega_{1,2}\omega_{1,0}\omega_{2,3}\omega_{2,2}^2\omega_{2,0} \\
&+\omega_{1,4}\omega_{1,3}\omega_{1,1}\omega_{1,0}\omega_{2,3}\omega_{2,2}^2\omega_{2,1} + 2\omega_{1,4}\omega_{1,2}\omega_{1,1}\omega_{1,0}\omega_{2,4}\omega_{2,2}^2\omega_{2,1} \\
&+\omega_{1,4}\omega_{1,3}\omega_{1,2}\omega_{1,0}\omega_{2,3}^2\omega_{2,1}\omega_{2,0} - \omega_{1,4}\omega_{1,3}\omega_{1,1}\omega_{1,0}\omega_{2,3}^2\omega_{2,2}\omega_{2,0} \\
&-\omega_{1,4}\omega_{1,2}\omega_{1,1}\omega_{1,0}\omega_{2,3}^2\omega_{2,2}\omega_{2,1} - 3\omega_{1,3}\omega_{1,2}\omega_{1,1}\omega_{1,0}\omega_{2,4}\omega_{2,3}^2\omega_{2,0} \\
&+2\omega_{1,4}\omega_{1,2}\omega_{1,1}\omega_{1,0}\omega_{2,4}^2\omega_{2,1}\omega_{2,0} + 4\omega_{1,3}\omega_{1,2}\omega_{1,1}\omega_{1,0}\omega_{2,4}^2\omega_{2,2}\omega_{2,0} \\
&+2\omega_{1,4}\omega_{1,3}\omega_{1,0}^2\omega_{2,4}\omega_{2,3}\omega_{2,2}\omega_{2,0} + 4\omega_{1,4}\omega_{1,2}\omega_{1,0}^2\omega_{2,4}\omega_{2,3}\omega_{2,2}\omega_{2,1} \\
&+\omega_{1,4}\omega_{1,3}\omega_{1,1}^2\omega_{2,4}\omega_{2,2}\omega_{2,1}\omega_{2,0} - \omega_{1,4}\omega_{1,2}\omega_{1,1}^2\omega_{2,4}\omega_{2,3}\omega_{2,1}\omega_{2,0} \\
&-3\omega_{1,4}\omega_{1,1}^2\omega_{1,0}\omega_{2,4}\omega_{2,3}\omega_{2,2}\omega_{2,1} - \omega_{1,3}\omega_{1,2}\omega_{1,1}^2\omega_{2,4}\omega_{2,3}\omega_{2,2}\omega_{2,0} \\
&+2\omega_{1,4}\omega_{1,2}^2\omega_{1,1}\omega_{2,4}\omega_{2,2}\omega_{2,1}\omega_{2,0} + \omega_{1,3}\omega_{1,2}^2\omega_{1,1}\omega_{2,4}\omega_{2,3}\omega_{2,1}\omega_{2,0} \\
&+2\omega_{1,3}\omega_{1,2}^2\omega_{1,0}\omega_{2,4}\omega_{2,3}\omega_{2,2}\omega_{2,0} - 3\omega_{1,4}\omega_{1,3}^2\omega_{1,0}\omega_{2,3}\omega_{2,2}\omega_{2,1}\omega_{2,0} \\
&-\omega_{1,3}^2\omega_{1,2}\omega_{1,1}\omega_{2,4}\omega_{2,2}\omega_{2,1}\omega_{2,0} - \omega_{1,3}^2\omega_{1,2}\omega_{1,0}\omega_{2,4}\omega_{2,3}\omega_{2,1}\omega_{2,0} \\
&+\omega_{1,3}^2\omega_{1,1}\omega_{1,0}\omega_{2,4}\omega_{2,3}\omega_{2,2}\omega_{2,0} + 4\omega_{1,4}^2\omega_{1,2}\omega_{1,0}\omega_{2,3}\omega_{2,2}\omega_{2,1}\omega_{2,0} \\
&+2\omega_{1,4}^2\omega_{1,1}\omega_{1,0}\omega_{2,4}\omega_{2,2}\omega_{2,1}\omega_{2,0} + \omega_{1,4}\omega_{1,3}\omega_{1,2}\omega_{1,1}\omega_{2,3}\omega_{2,2}\omega_{2,1}\omega_{2,0} \\
&-8\omega_{1,4}\omega_{1,3}\omega_{1,2}\omega_{1,0}\omega_{2,4}\omega_{2,2}\omega_{2,1}\omega_{2,0} + 10\omega_{1,4}\omega_{1,3}\omega_{1,1}\omega_{1,0}\omega_{2,4}\omega_{2,3}\omega_{2,1}\omega_{2,0} \\
&-8\omega_{1,4}\omega_{1,2}\omega_{1,1}\omega_{1,0}\omega_{2,4}\omega_{2,3}\omega_{2,2}\omega_{2,0} + \omega_{1,3}\omega_{1,2}\omega_{1,1}\omega_{1,0}\omega_{2,4}\omega_{2,3}\omega_{2,2}\omega_{2,1} = 0
\end{aligned}
$$

Appendix E. Joint Variables

For the sake of conciseness, in the following the notation $\cos q_i = C_i$ and $\sin q_i = S_i$ has been used.

$$
\begin{aligned}
q_2 = &\quad \arcsin \tfrac{P_{3,x} - P_{1,x}}{a_2} \\
q_1 = &\quad \operatorname{atan2}\left(\tfrac{P_{3,y}}{a_2 C_2}, \tfrac{P_{3,x}}{a_2 C_2}\right) \\
q_3 = &\quad \operatorname{atan2}(S_3, C_3) \\
&\quad \text{where} \\
&\quad S_3 = \tfrac{P_{3,y}C_2 - P_{4,y}C_2 + P_{3,z}S_1S_2 - P_{4,z}S_1S_2}{d_4 S_1} \\
&\quad C_3 = -\tfrac{P_{3,y}S_2 - P_{4,y}S_2 - P_{3,z}C_2S_1 + P_{4,z}C_2S_1}{d_4 S_1} \\
q_4 = &\quad \operatorname{atan2}(S_4, C_4) \\
&\quad \text{where} \\
&\quad S_4 = \tfrac{P_{4,z} - P_{5,z}}{d_5 S_{2-3}} \\
&\quad C_4 = -\tfrac{P_{4,z}C_{1+2-3} + P_{4,z}C_{1-2+3} - P_{5,z}C_{1+2-3} - P_{5,z}C_{1-2+3} - 2P_{4,x}S_{2-3} + 2P_{5,x}S_{2-3}}{2d_5 S_{2-3} S_1}
\end{aligned}
$$

$$
\begin{aligned}
q_5 = \; & \operatorname{atan2}(S_5, C_5) \\
& \text{where} \\
& S_5 = \tfrac{1}{d_6}\left(P_{65,x} + d_2 S_4 - d_3 S_4 + a_2 C_3 C_4 - d_1 C_2 C_4 S_3 + d_1 C_3 C_4 S_2\right) \\
& C_5 = -\tfrac{1}{d_6}\left(P_{65,y} - d_4 + a_2 S_3 + d_1 C_{2-3}\right) \\
& \text{and} \\
& \begin{bmatrix} {}^4\mathbf{P}_{65} \\ 1 \end{bmatrix} = \begin{bmatrix} P_{65,x} \\ P_{65,y} \\ P_{65,z} \\ 1 \end{bmatrix} = {}^0\mathbf{T}_4^{-1} \begin{bmatrix} \mathbf{P}_6 - \mathbf{P}_5 \\ 1 \end{bmatrix} \\
q_6 = \; & \operatorname{atan2}(S_6, C_6) \\
& \text{where} \\
& S_6 = \begin{bmatrix} 0 & 1 & 0 & 0 \end{bmatrix} {}^5\mathbf{T_6} \begin{bmatrix} 1 & 0 & 0 & 0 \end{bmatrix}^T \\
& C_6 = \begin{bmatrix} 1 & 0 & 0 & 0 \end{bmatrix} {}^5\mathbf{T_6} \begin{bmatrix} 1 & 0 & 0 & 0 \end{bmatrix}^T \\
& \text{and} \\
& {}^5\mathbf{T_6} = {}^0\mathbf{T_5}^{-1}(q_1, q_2, q_3, q_4, q_5)\, {}^0\mathbf{T_6}
\end{aligned}
$$

References

1. Matheson, E.; Minto, R.; Zampieri, E.G.; Faccio, M.; Rosati, G. Human–robot collaboration in manufacturing applications: A review. *Robotics* **2019**, *8*, 100. [CrossRef]
2. Chiriatti, G.; Carbonari, L.; Costa, D.; Palmieri, G. Implementation of a Robot Assisted Framework for Rehabilitation Practices. In Proceedings of the Advances in Italian Mechanism Science: Proceedings of the 4th International Conference of IFToMM Italy, Naples, Italy, 7–9 September 2022; Springer: Berlin/Heidelberg, Germany, 2022; pp. 541–548.
3. Kyrarini, M.; Lygerakis, F.; Rajavenkatanarayanan, A.; Sevastopoulos, C.; Nambiappan, H.R.; Chaitanya, K.K.; Babu, A.R.; Mathew, J.; Makedon, F. A survey of robots in healthcare. *Technologies* **2021**, *9*, 8. [CrossRef]
4. Holland, J.; Kingston, L.; McCarthy, C.; Armstrong, E.; O'Dwyer, P.; Merz, F.; McConnell, M. Service robots in the healthcare sector. *Robotics* **2021**, *10*, 47. [CrossRef]
5. Galin, R.; Meshcheryakov, R. Automation and robotics in the context of Industry 4.0: The shift to collaborative robots. In Proceedings of the IOP Conference Series: Materials Science and Engineering, Prague, Czech Republic, 18–22 June 2018; IOP Publishing: Bristol, UK, 2019; Volume 537, p. 032073.
6. Maddikunta, P.K.R.; Pham, Q.V.; Prabadevi, B.; Deepa, N.; Dev, K.; Gadekallu, T.R.; Ruby, R.; Liyanage, M. Industry 5.0: A survey on enabling technologies and potential applications. *J. Ind. Inf. Integr.* **2021**, *26*, 100257. [CrossRef]
7. Chiriatti, G.; Palmieri, G.; Scoccia, C.; Palpacelli, M.C.; Callegari, M. Adaptive Obstacle Avoidance for a Class of Collaborative Robots. *Machines* **2021**, *9*, 113. [CrossRef]
8. Trinh, C.; Zlatanov, D.; Zoppi, M.; Molfino, R. A geometrical approach to the inverse kinematics of 6r serial robots with offset wrists. In Proceedings of the International Design Engineering Technical Conferences and Computers and Information in Engineering Conference, Anaheim, CA, USA, 18–21 August 2015; American Society of Mechanical Engineers: New York, NY, USA, 2015; Volume 57144, p. V05CT08A016.
9. Zohour, H.M.; Belzile, B.; St-Onge, D. Kinova Gen3-Lite manipulator inverse kinematics: Optimal polynomial solution. *arXiv* **2021**, arXiv:2102.01217.
10. Villalobos, J.; Sanchez, I.Y.; Martell, F. Singularity Analysis and Complete Methods to Compute the Inverse Kinematics for a 6-DOF UR/TM-Type Robot. *Robotics* **2022**, *11*, 137. [CrossRef]
11. Raghavan, M.; Roth, B. A general solution for the inverse kinematics of all series chains. In Proceedings of the 8th CISM-IFTOMM Symposium on Robots and Manipulators, Krakow, Poland, July 1990.
12. Raghavan, M.; Roth, B. Inverse Kinematics of the General 6R Manipulator and Related Linkages. *J. Mech. Des.* **1993**, *115*, 502–508. [CrossRef]
13. Wang, Y.; Hang, L.b.; Yang, T.l. Inverse kinematics analysis of general 6r serial robot mechanism based on Gröbner base. *Front. Mech. Eng.* **2006**, *1*, 115–124. [CrossRef]
14. Fu, Z.; Yang, W.; Yang, Z. Solution of inverse kinematics for 6R robot manipulators with offset wrist based on geometric algebra. *J. Mech. Robot.* **2013**, *5*, 031010. [CrossRef] [PubMed]
15. Ghazvini, M. Reducing the inverse kinematics of manipulators to the solution of a generalized eigenproblem. *Comput. Kinemat.* **1993**, *28*, 15–26.

16. Xiao, F.; Li, G.; Jiang, D.; Xie, Y.; Yun, J.; Liu, Y.; Huang, L.; Fang, Z. An effective and unified method to derive the inverse kinematics formulas of general six-DOF manipulator with simple geometry. *Mech. Mach. Theory* **2021**, *159*, 104265. [CrossRef]
17. Li, G.; Xiao, F.; Zhang, X.; Tao, B.; Jiang, G. An inverse kinematics method for robots after geometric parameters compensation. *Mech. Mach. Theory* **2022**, *174*, 104903. [CrossRef]
18. Sugihara, T. Solvability-unconcerned inverse kinematics by the Levenberg–Marquardt method. *IEEE Trans. Robot.* **2011**, *27*, 984–991. [CrossRef]
19. Cho, G.R.; Lee, M.J.; Kim, M.G.; Li, J.H. Inverse kinematics for autonomous underwater manipulations using weighted damped least squares. In Proceedings of the 2017 14th International Conference on Ubiquitous Robots and Ambient Intelligence (URAI), Jeju, Republic of Korea, 28 June–1 July 2017; IEEE: New York, NY, USA, 2017; pp. 765–770.
20. Zhao, J.; Xu, T.; Fang, Q.; Xie, Y.; Zhu, Y. A synthetic inverse kinematic algorithm for 7-DOF redundant manipulator. In Proceedings of the 2018 IEEE International Conference on Real-time Computing and Robotics (RCAR), Kandima, Maldives, 1–5 August 2018; IEEE: New York, NY, USA, 2018; pp. 112–117.
21. Callegari, M.; Carbonari, L.; Costa, D.; Palmieri, G.; Palpacelli, M.C.; Papetti, A.; Scoccia, C. Tools and Methods for Human Robot Collaboration: Case Studies at i-LABS. *Machines* **2022**, *10*, 997. [CrossRef]
22. Simas, H.; Di Gregorio, R. Collision Avoidance for Redundant 7-DOF Robots Using a Critically Damped Dynamic Approach. *Robotics* **2022**, *11*, 93. [CrossRef]
23. Malik, A.; Lischuk, Y.; Henderson, T.; Prazenica, R. A Deep Reinforcement-Learning Approach for Inverse Kinematics Solution of a High Degree of Freedom Robotic Manipulator. *Robotics* **2022**, *11*, 44. [CrossRef]
24. Pieper, D.; Roth, B. The Kinematics of Manipulators Under Computer Control. In Proceedings of the 2nd International Congress on Theory of Machines and Mechanisms, Zakopane, Poland, 23–27 September 1969; Volume 2, pp. 159–169.

robotics

MDPI

Article

Singularity Analysis and Complete Methods to Compute the Inverse Kinematics for a 6-DOF UR/TM-Type Robot

Jessica Villalobos [1,*], Irma Y. Sanchez [2] and Fernando Martell [1]

1 Technology Development Department, Centro de Investigaciones en Optica, Aguascalientes 20200, Mexico
2 Postgraduate Academic Department, Universidad Politecnica de Aguascalientes, Aguascalientes 20342, Mexico
* Correspondence: jessicavillalobos@cio.mx

Abstract: Improving the strategies employed to control robotic arms is of great importance because of the increase in their use in advanced supervisory control strategies, such as digital twins. The inverse kinematic (IK) control of manipulators requires an IK solution and an awareness of the singular configurations. This work presents a complete IK calculation system with singularity analysis for the UR5 robotic arm created by Universal Robots. For a specific robot pose, different angle solution sets are obtained, and one of these solution sets has to be selected to achieve movement continuity and avoid singularities. Two methods for this double purpose are proposed: one calculates all the solution possibilities, and the other obtains only one solution set by following a sequence of decisions and calculations clearly stated by a finite state machine (FSM). Both methods are effective in managing singularities. The FSM-based method complements the IK solution procedure with advantages in the number of computations and performance by producing results that would not lead the joints to move abruptly. The results prove that the presented methods select an IK solution that does not result in a singular configuration, and that most of the time, they lead to the same valid IK solution.

Keywords: inverse kinematics; singularity analysis; finite state machine; complete kinematic solution analysis

Citation: Villalobos, J.; Sanchez, I.Y.; Martell, F. Singularity Analysis and Complete Methods to Compute the Inverse Kinematics for a 6-DOF UR/TM-Type Robot. *Robotics* **2022**, *11*, 137. https://doi.org/10.3390/robotics11060137

Academic Editor: Raffaele Di Gregorio

Received: 4 November 2022
Accepted: 25 November 2022
Published: 29 November 2022

1. Introduction

Robots are used to increase production and improve product quality; because of this, Industry 4.0 uses technologies such as robotic manipulators and their digital twins to create cyber–physical systems that are advanced supervisory control systems that improve the overall performance of the physical systems. An inverse kinematic (IK) solution of a robot is required when controlling robots and their digital twins. The twin-in-the-loop architecture [1] is one example of how digital twins can be used. The extensive use of robots and the fact that they interact with their users make it necessary to improve their behavior; this can be achieved by using control strategies such as the IK control and the model reference adaptive control [2]; however, singularities must be avoided for the robot to be free to move in any direction within its workspace and with reasonable joint speed; the latter is necessary because joint velocities tend to infinity as a singular configuration is approached.

Improving the behavior of collaborative robots (cobots) is important because they should safely interact with their users. This is the case of the UR5 robotic arm, a 6-degree-of-freedom (DOF) manipulator created by Universal Robots, and other cobots that have been produced with the same geometry by companies such as Smokie Robotics, Techman Robot, AUBO Robotics, and Doosan Robotics. The kinematics of the UR5 robot has been studied in [3–9], some solutions have been statistically compared in [10], and a method to classify IK branches of the UR-type robot has been recently proposed [11]. The forward or direct (DK) and inverse kinematics (IK) can be studied with different methods;

a well-known technique uses the original or the modified Denavit–Hartenberg (DH) parameters [12]. An IK solution allows for computing the configurations that lead to a specific pose, which is necessary when the position and orientation of the end effector are controlled. Even though works such as [13–18] studied the IK of the general 6R serial manipulator that results in 16 solutions, robots with the UR5 geometry only have 8. However, in the analytical and numerical IK solutions of serial robot architectures, it is required to avoid singular configurations; therefore, singularity analysis needs to be considered as a part of the complete IK solution, as in [19]. Although singularity analyses for the UR robot geometry were performed [20,21], a simplified expression is more appropriate for implementations. A code that allows for computing the Jacobian and its determinant can be found in [22], and only a few changes are necessary to adapt it to other robots; in [23], the same authors presented the Denavit–Hartenberg parameters for the Open Unit Robot (OUR) manipulator created by Smokie Robotics.

To the best of the authors' knowledge, a complete IK solution that integrates the calculation of singularities and a strategy to circumvent them has not been published. Therefore, this work provides a practical IK solution that manages singularities besides the multiplicity of angle solutions inherent to the IK problem, and constitutes a ready-to-use tool for developing, simulating, and controlling robotic systems.

Additionally, two modalities or methods for dealing with singularities and multiple solution sets are proposed. A finite state machine (FSM) is used to tailor one of the methods for complementing the IK analysis with the direct production of an appropriate nonsingular solution without calculating all the possible solutions. An IK solution that avoids singular configurations is helpful in designing and controlling a robot and its digital twin [24], and in image-based visual servoing [25].

This paper presents a complete IK solution, a compact expression for the determinant of the Jacobian that can be easily implemented, and two effective methods with their corresponding algorithms for handling singularities and choosing a set of angles or calculating a single one. The rest of this paper is organized as follows: Section 2 shows a set of expressions used to compute the IK solution, and presents the singularity analysis for robots with the UR geometry and two algorithms that can be used to choose a set of angles. The results obtained with the two algorithms are compared in Section 3. Lastly, conclusions are presented in Section 4.

2. Materials and Methods

First, this section shows the IK solution on which the selection algorithms are based; then, it describes the singularity problem and presents the singularity analysis for robots with the same geometry as the UR5; lastly, the proposed selection algorithms are described.

2.1. Inverse Kinematic Solution

The UR5, which is the specific robot that was used for the experiments, is a 6-degree-of-freedom (DOF) cobot that only has rotational joints. Table 1 presents its DH parameters, which are illustrated in Figure 1. However, the IK solution and the singularity analysis only use variables, because this allows for using them for any other robot with the same geometry.

Table 1. Denavit–Hartenberg (DH) parameters of the UR5.

i	α_i(rad)	a_i(mm) [1]	d_i(mm) [1]	θ_i
1	$\pi/2$	0	$d_1 = 89.2$	θ_1
2	0	$a_2 = 425.0$	0	θ_2
3	0	$a_3 = 392.0$	0	θ_3
4	$\pi/2$	0	$d_4 = 109.3$	θ_4
5	$-\pi/2$	0	$d_5 = 94.75$	θ_5
6	–	–	$d_6 = 82.5$	θ_6

[1] The values for these parameters were obtained from [26].

From the different IK solutions found in the literature, only one was considered to define the algorithms presented here; however, these algorithms can be easily modified to use a solution that calculates the angles in a different order. The following notation is used in this work: $s_i = \sin\theta_i$, $c_i = \cos\theta_i$, $s_{ij...} = \sin(\theta_i + \theta_j + ...)$ and $c_{ij...} = \cos(\theta_i + \theta_j + ...)$. The homogeneous transformation matrix from Equation (1) is composed of a rotational submatrix (elements represented by r) and a position vector (p elements); these define the orientation and position of frame 6 with respect to the base (frame 0). Since the DK solution is required to compute the IK solution, the resulting transformation matrix and its elements are shown next [10]:

$$ {}^{0}_{6}T = \begin{bmatrix} r_{11} & r_{12} & r_{13} & p_x \\ r_{21} & r_{22} & r_{23} & p_y \\ r_{31} & r_{32} & r_{33} & p_z \\ 0 & 0 & 0 & 1 \end{bmatrix} \tag{1} $$

$$ r_{11} = c_1 c_{234} c_5 c_6 + c_6 s_1 s_5 - c_1 s_{234} s_6 \tag{2} $$

$$ r_{21} = c_{234} c_5 c_6 s_1 - c_1 c_6 s_5 - s_1 s_{234} s_6 \tag{3} $$

$$ r_{31} = c_5 c_6 s_{234} + c_{234} s_6 \tag{4} $$

$$ r_{12} = -c_1 c_{234} c_5 s_6 - s_1 s_5 s_6 - c_1 c_6 s_{234} \tag{5} $$

$$ r_{22} = -c_{234} c_5 s_1 s_6 + c_1 s_5 s_6 - c_6 s_1 s_{234} \tag{6} $$

$$ r_{32} = -c_5 s_{234} s_6 + c_{234} c_6 \tag{7} $$

$$ r_{13} = -c_1 c_{234} s_5 + c_5 s_1 \tag{8} $$

$$ r_{23} = -c_{234} s_1 s_5 - c_1 c_5 \tag{9} $$

$$ r_{33} = -s_{234} s_5 \tag{10} $$

$$ p_x = r_{13} d_6 + c_1 (s_{234} d_5 + c_{23} a_3 + c_2 a_2) + s_1 d_4 \tag{11} $$

$$ p_y = r_{23} d_6 + s_1 (s_{234} d_5 + c_{23} a_3 + c_2 a_2) - c_1 d_4 \tag{12} $$

$$ p_z = r_{33} d_6 - c_{234} d_5 + s_{23} a_3 + s_2 a_2 + d_1 \tag{13} $$

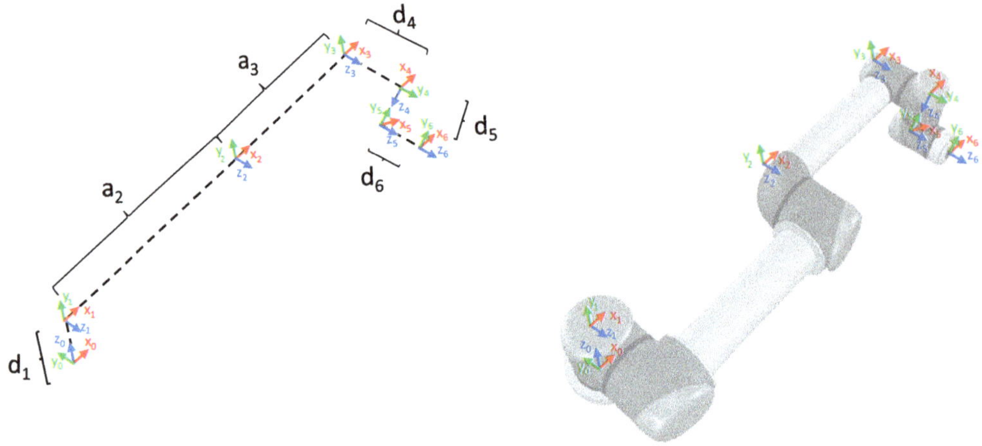

Figure 1. Coordinate frame assignment using the original DH convention ($\theta_i = 0$ for $i = 0, 1, 2, 3, 4, 5, 6$).

Although some equations are expressed differently, the IK solution used in this work consists of the following expressions that were proposed in [8]:

$$A = p_y - d_6 r_{23} \tag{14}$$

$$B = p_x - d_6 r_{13} \tag{15}$$

$$C = c_1 r_{11} + s_1 r_{21} \tag{16}$$

$$D = c_1 r_{22} - s_1 r_{12} \tag{17}$$

$$E = s_1 r_{11} - c_1 r_{21} \tag{18}$$

$$F = c_5 c_6 \tag{19}$$

$$\theta_1 = \pm\,\text{atan2}\left(\sqrt{B^2 + (-A)^2 - d_4^2}, d_4\right) + \text{atan2}(B, -A) \tag{20}$$

$$\theta_5 = \pm\,\text{atan2}\left(\sqrt{E^2 + D^2}, s_1 r_{13} - c_1 r_{23}\right) \tag{21}$$

$$\theta_6 = \text{atan2}\left(\frac{D}{s_5}, \frac{E}{s_5}\right) \tag{22}$$

$$\theta_{234} = \text{atan2}\left(r_{31} F - s_6 C, FC + s_6 r_{31}\right) \tag{23}$$

$$K_C = c_1 p_x + s_1 p_y - s_{234} d_5 + c_{234} s_5 d_6 \tag{24}$$

$$K_S = p_z - d_1 + c_{234} d_5 + s_{234} s_5 d_6 \tag{25}$$

$$c_3 = \frac{K_S^2 + K_C^2 - a_2^2 - a_3^2}{2 a_2 a_3} \tag{26}$$

$$s_3 = \sqrt{1 - c_3^2} \quad (27)$$

$$\theta_3 = \pm\,\text{atan2}(s_3, c_3) \quad (28)$$

$$\theta_2 = \text{atan2}(K_S, K_C) - \text{atan2}(s_3 a_3, c_3 a_3 + a_2) \quad (29)$$

$$\theta_4 = \theta_{234} - \theta_2 - \theta_3 \quad (30)$$

2.2. Singularities

The Jacobian is a matrix that relates the joint velocities to the linear and angular velocities of the end effector. This matrix is used to find singularities because, for nonredundant robots such as the UR5, they exist when $\det(\boldsymbol{J}) = 0$; the Jacobian can be computed using different methods, one of which is shown in [27].

In [28], the authors mentioned that a robot is at a singularity or singular configuration when it is impossible to move the end effector in at least one direction; there, the Jacobian was computed as in Equations (31) and (32) for prismatic and rotational joints, respectively.

$$\boldsymbol{J} = \begin{bmatrix} \boldsymbol{J}_{L_i} \\ \boldsymbol{J}_{A_i} \end{bmatrix} = \begin{bmatrix} \boldsymbol{b}_{i-1} \\ \boldsymbol{0} \end{bmatrix} \quad (31)$$

$$\boldsymbol{J} = \begin{bmatrix} \boldsymbol{J}_{L_i} \\ \boldsymbol{J}_{A_i} \end{bmatrix} = \begin{bmatrix} \boldsymbol{b}_{i-1} \times \boldsymbol{r}_{i-1,e} \\ \boldsymbol{b}_{i-1} \end{bmatrix} \quad (32)$$

In the previous equations, $\boldsymbol{b}_{i-1}$ is the unit vector representing the z-axis of joint i 1 with respect to the base (frame 0), $\boldsymbol{r}_{i-1,e}$ is the end-effector position with respect to frame $i-1$, and $\boldsymbol{J}_{L_i}$ and $\boldsymbol{J}_{A_i}$ represent the parts of the Jacobian that relate the joint velocities to the linear and angular ones, respectively. An example of how the Jacobian matrix is computed using this method can be found in [28].

Vectors $\boldsymbol{b}_{i-1}$ and $\boldsymbol{r}_{i-1,e}$ can be obtained from the DK study because $\boldsymbol{b}_{i-1}$ is equal to ${}^{0}\hat{\boldsymbol{z}}_{i-1}$ (third column of rotational matrix ${}_{i-1}^{0}\boldsymbol{R}$), and $\boldsymbol{r}_{i-1,e}$ can be obtained by subtracting the translation vector ${}_{i-1}^{0}\boldsymbol{P}$ from the end-effector position.

Since joint velocities tend to infinity as the robot approaches a singular configuration, studying and avoiding singularities is necessary to help in rendering the interaction between robot and user safer.

2.3. Singularity Analysis

Only Equation (32) is necessary to compute the Jacobian because the UR5 consists exclusively of rotational joints. Adapting the Jacobian matrix with the corresponding UR5 parameters, the following can be expressed:

$$\boldsymbol{J}_A = \begin{bmatrix} 0 & s_1 & s_1 & s_1 & c_1 s_{234} & r_{13} \\ 0 & -c_1 & -c_1 & -c_1 & s_1 s_{234} & r_{23} \\ 1 & 0 & 0 & 0 & -c_{234} & r_{33} \end{bmatrix} \quad (33)$$

$$\boldsymbol{r}_{0,e} = \begin{bmatrix} p_x \\ p_y \\ p_z \end{bmatrix} \quad (34)$$

$$\boldsymbol{r}_{1,e} = \begin{bmatrix} p_x \\ p_y \\ p_z - d_1 \end{bmatrix} \quad (35)$$

$$r_{2,e} = \begin{bmatrix} p_x - c_1c_2a_2 \\ p_y - c_2s_1a_2 \\ p_z - s_2a_2 - d_1 \end{bmatrix} \tag{36}$$

$$r_{3,e} = \begin{bmatrix} p_x - c_1c_{23}a_3 - c_1c_2a_2 \\ p_y - c_{23}s_1a_3 - c_2s_1a_2 \\ p_z - s_{23}a_3 - s_2a_2 - d_1 \end{bmatrix} \tag{37}$$

$$r_{4,e} = \begin{bmatrix} r_{13}d_6 + c_1s_{234}d_5 \\ r_{23}d_6 + s_1s_{234}d_5 \\ r_{33}d_6 - c_{234}d_5 \end{bmatrix} \tag{38}$$

$$r_{5,e} = \begin{bmatrix} r_{13}d_6 \\ r_{23}d_6 \\ r_{33}d_6 \end{bmatrix} \tag{39}$$

The previous results allow computing J_L as shown next:

$$J_{L_1} = \begin{bmatrix} -p_y \\ p_x \\ 0 \end{bmatrix} \tag{40}$$

$$J_{L_2} = \begin{bmatrix} -c_1(p_z - d_1) \\ -s_1(p_z - d_1) \\ s_1p_y + c_1p_x \end{bmatrix} \tag{41}$$

$$J_{L_3} = \begin{bmatrix} c_1(s_{234}s_5d_6 + c_{234}d_5 - s_{23}a_3) \\ s_1(s_{234}s_5d_6 + c_{234}d_5 - s_{23}a_3) \\ -c_{234}s_5d_6 + s_{234}d_5 + c_{23}a_3 \end{bmatrix} \tag{42}$$

$$J_{L_4} = \begin{bmatrix} c_1(s_{234}s_5d_6 + c_{234}d_5) \\ s_1(s_{234}s_5d_6 + c_{234}d_5) \\ -c_{234}s_5d_6 + s_{234}d_5 \end{bmatrix} \tag{43}$$

$$J_{L_5} = \begin{bmatrix} -d_6(s_1s_5 + c_1c_{234}c_5) \\ d_6(c_1s_5 - c_{234}c_5s_1) \\ -c_5s_{234}d_6 \end{bmatrix} \tag{44}$$

$$J_{L_6} = \begin{bmatrix} 0 \\ 0 \\ 0 \end{bmatrix} \tag{45}$$

As previously mentioned, to verify if the combination of joint angles results in a singularity, the determinant of the Jacobian matrix must be computed to determine if it is equal to or close to zero. This can be conducted with the following expression:

$$\det(J) = s_3s_5a_2a_3(c_2a_2 + c_{23}a_3 + s_{234}d_5) \tag{46}$$

Three types of singularities (shoulder, wrist, and elbow) exist for this robot. Information about them can be found in [20,21], and they are briefly described next.

A shoulder singularity happens when the last factor in Equation (46), which involves angles θ_2, θ_3, and θ_4, is equal to zero. One example can be seen in Figure 2, which shows that the end effector cannot be moved along z_6.

Wrist singularities exist when $s_5 = 0$, which mathematically happens when $\theta_5 = 0$ or $\theta_5 = \pm\pi$. This renders z_4 and z_6 parallel.

An elbow singularity is present when $s_3 = 0$, which happens when $\theta_3 = 0$ or $\theta_3 = \pm\pi$. This means that the arm is fully stretched or bent; however, only the former case is physically possible.

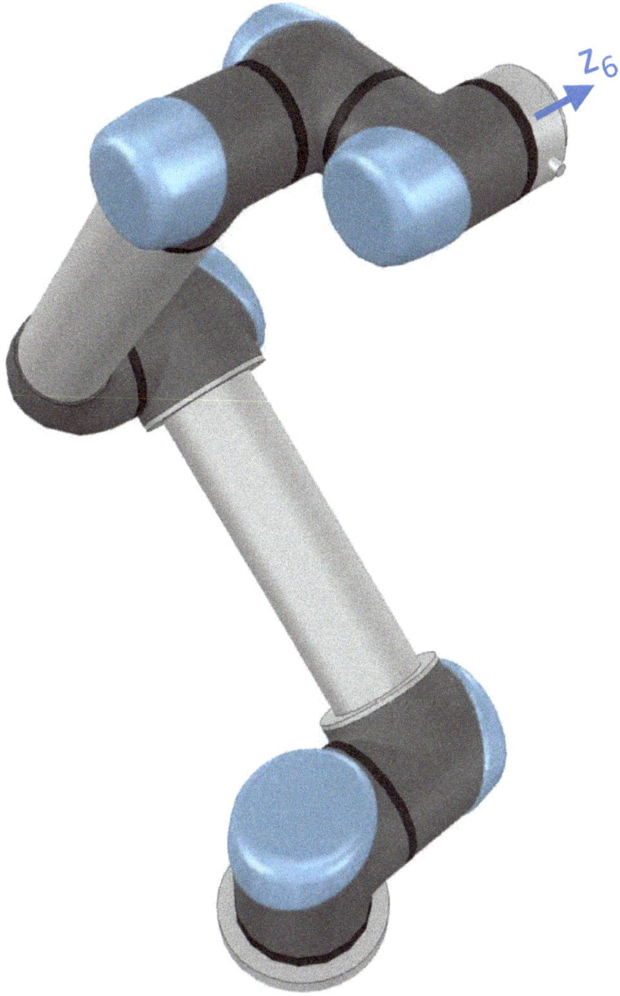

Figure 2. Example of shoulder singularity.

2.4. Algorithms to Select a Solution

This section presents two complete algorithms that can be used to select one set of angles that does not result in a singularity. In this work, angles close to 0 were considered to be 0, and angles close to $\pm\pi$ were defined as π; 2π was added or subtracted until the computed angle was within $[-\pi, \pi]$.

Not all values computed by the IK solutions are valid, i.e., some of them lead to computational errors. The cases in which this happens are described next:

- Computed θ_1 angles are not acceptable if they are complex and no valid set can be computed. For the used IK solution, it is enough to validate the result of the square root in Equation (20).
- θ_5 is not acceptable if it is complex or if $|s_5| \leq 1 \times 10^{-12}$; since the IK solution uses atan2, only the latter validation is necessary. The value of the limit for $|s_5|$ was chosen because if the other sines and cosines in Equation (46) are equal to 1, it results in $\det(\boldsymbol{J}) = 1.5190 \times 10^{-4}$, which can be considered equal to zero. The other reason was that although $\sin(0) = \sin(\pi) = \sin(-\pi) = 0$, the computational tools do not

always give the exact value, e.g., in MATLAB R2021a $\sin(\pi) = 1.2246 \times 10^{-16}$; this means that the determinant is not always exactly 0.

- The angles for θ_3 are not valid if $|s_3| \leq 1 \times 10^{-12}$, or when either θ_3 or s_3 is complex. For the used IK solution, it is enough to validate s_3. The value defined for θ_3 to be valid was obtained using the same considerations described previously for θ_5.
- Angles θ_2 and θ_4 are not considered to be acceptable if $|d_5 s_{234} + a_2 c_2 + a_3 c_{23}| \leq 1 \times 10^{-9}$; the value of the limit was chosen because if s_3 and s_5 are both equal to 1 and this term is 1×10^{-9}, then $\det(\boldsymbol{J}) = 1.6660 \times 10^{-4}$, which can be considered to be zero.
- Lastly, a complete set of angles is not considered to be valid when the manipulator reaches the outer limit of its workspace, which happens when $\theta_3 = 0$ and $\theta_4 = \pi/2$. However, this leads to $s_3 = 0$, which was already defined as not valid.

Modifications to the following algorithms may be necessary depending on the application and the order in which the IK solution that is used computes the angles. These algorithms do not consider the presence of obstacles.

2.4.1. Algorithm 1

The first algorithm computes all the sets of angles that take the end effector to a previously specified pose, and then selects the one that requires moving the joints the least overall.

To select a set of variables, some works maximize a cost function depending on the objective (avoiding singularities, joint limits, or obstacles) [29] or find trajectories that do not include the mutation of any joint angles at 180° [30]; however, this algorithm selects the set as in [31], where the total joint displacement is minimized. Although something similar was performed in [32], where the solution that minimized joint movement was selected, weights were used to prioritize moving smaller joints, which were assigned smaller weights.

Even though a specific selection criterion is used, depending on the objective, it is possible to change it without affecting the rest of the steps.

This algorithm consists of the following steps:

1. Both θ_1 solutions are computed, and complex angles are discarded.
2. The previously obtained values for θ_1 are used to compute θ_5. The sets containing values of θ_5 that are not considered valid are rejected.
3. θ_6 is computed for the remaining sets.
4. The values of θ_3 are computed and verified. Again, the solutions with angles that are not acceptable are discarded.
5. Lastly, θ_2 and θ_4 are computed, and the sets of angles that are not valid are rejected.
6. The algorithm selects the solution with the minimal difference with respect to the current joint positions, which is computed with the following equations:

$$\Delta\theta_i = \theta_{i,p} - \theta_{i,j} \tag{47}$$

and

$$diff_j = \sqrt{\sum_{i=1}^{6} \Delta\theta_i^2}, \tag{48}$$

where p refers to a previous value, i to a joint, and j to the computed set.

2.4.2. Algorithm 2

When there are two possible solutions for an angle, the second algorithm chooses the one that moves the specific joint less and verifies it does not result in a singularity. This is to compute only one complete set of angles that results in the desired pose.

This algorithm uses the FSM shown in Figure 3; this technique was used because it is a comprehensible computational model. The FSM technique defines the steps that have to be followed to complete a specified sequential task; a state is triggered by an event or completion signal activated in the previous state. In the proposed FSM, each step computes one or two angles, selects the closest option (if two possibilities exist), and verifies if the chosen angles are valid. If a value is not acceptable, it is modified, and a previous state is triggered. In this case, the task is to compute and select angles that do not result in singular configurations.

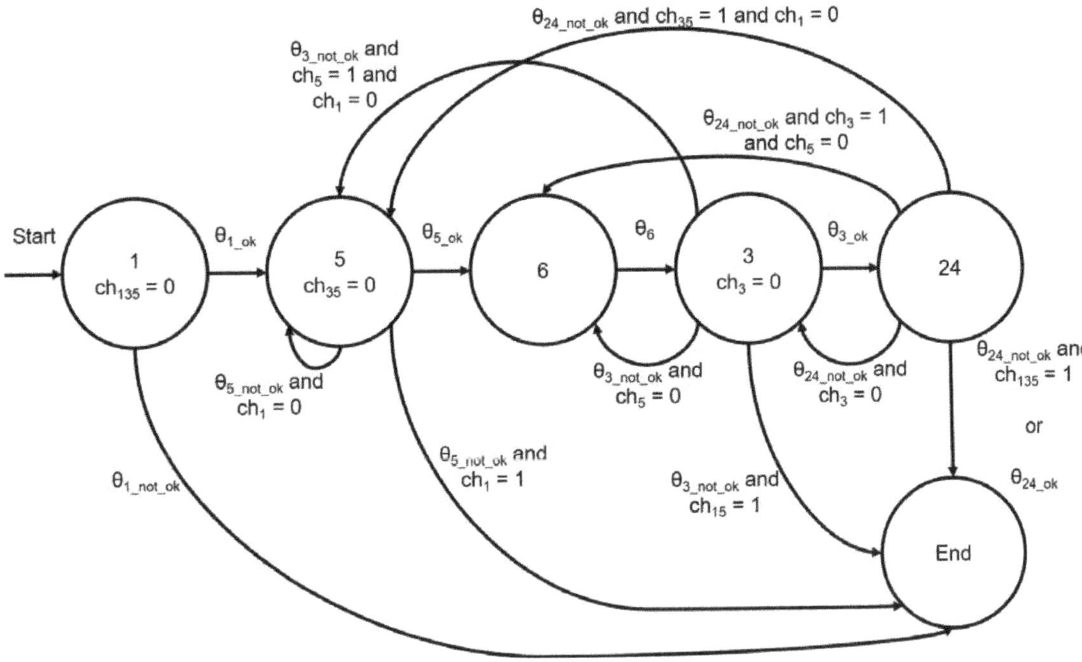

Figure 3. Finite state machine (FSM) for Algorithm 2; ch_{ijk} refers to the number of times that angles i, j, and k have been changed to try a different set of angles.

The states are described next:

- State 1: The values of θ_1 are computed and verified, and the one closer to the previous θ_1 is selected. If a valid θ_1 is found, State 5 is next.
- State 5: As for θ_1, the two possible angles for θ_5 are computed and verified, and the one closer to the previous value is selected. If an acceptable θ_5 is found, State 6 is next. Since the difference between the values of θ_5 is the sign, only one of them is necessary for the validation. If the computed angles are not valid, State 5 is repeated using the other θ_1; however, no solution exists if θ_1 has already been changed.
- State 6: Here, θ_6 is computed. This is followed by State 3.
- State 3: As for θ_1 and θ_5, both θ_3 values are computed and verified, and the one closer to the previous θ_3 is selected; State 24 is run if an acceptable θ_3 exists. Again, it is only necessary to verify one of the calculated angles; if θ_3 is not valid, one of the following states is next:
  - State 6: if θ_5 has not been modified, the other value for θ_5 is used.
  - State 5: if θ_5 has been changed and θ_1 has not, the other possible angle for θ_1 is tested.

    - End: no set of angles exists if, after changing θ_1, both possibilities for θ_5 result in unacceptable values for θ_3.

- State 24: θ_2 and θ_4 are computed in this state; for this reason, it is called State 24. If the resulting set of angles is valid, the algorithm has found a solution ("End" is the next state); otherwise, one of the following states is run:
    - State 24: if the value of θ_3 has not been changed, it is modified, and the state is repeated.
    - State 6: if both θ_3 angles have been used and only one θ_5 has been used, the other possible θ_5 is tested.
    - State 5: if θ_5 has been changed and both θ_3 options have been tested, but θ_1 has not been modified. The other angle θ_1 is used.
    - End: if the manipulator cannot be taken to the desired pose even after changing θ_1, θ_5, and θ_3.
- End: this last state is reached when a valid set of angles is found or if it is impossible to find one.

3. Results and Discussion

This section compares the results obtained with the algorithms presented in Section 2.4. For the comparison, the desired poses are shown in Table 2 (only sets 7 and 9 do not result in singularities). However, to choose a set of angles for said poses, the previous ones were assumed to be $\theta_1 = \theta_2 = \theta_3 = \theta_4 = \theta_5 = \theta_6 = 0$.

Table 2. Joint angles used to compute the desired poses (rad).

Test	θ_1	θ_2	θ_3	θ_4	θ_5	θ_6
1	π	$\pi/4$	$\pi/2$	$\pi/2$	0	$\pi/5$
2	π	$-\pi/2$	0	π	$\pi/3$	0
3	π	0	$\pi/2$	π	π	0
4	$\pi/2$	$\pi/2$	0	$-\pi/2$	$\pi/3$	π
5	$\pi/2$	0	$\pi/4$	π	$-\pi$	0
6	$-\pi/3$	$3\pi/4$	0	π	$-\pi$	0
7	$\pi/3$	$\pi/3$	$\pi/2$	$\pi/4$	$\pi/3$	0
8	$\pi/5$	$\pi/2$	$\pi/2$	$\pi/2$	π	$\pi/2$
9	$-\pi$	$\pi/3$	$-\pi/2$	$\pi/2$	$\pi/6$	$\pi/2$
10	$-\pi/2$	$\pi/2$	0	π	$-\pi$	$\pi/2$

Both algorithms were programmed in MATLAB R2021a. Other results can be obtained depending on how the algorithms and IK solutions are programmed (e.g., if values close to $-\pi$ are defined as $-\pi$ and not as π, or if the joint angles are not limited to be within $[-\pi, \pi]$).

Tables 3 and 4 present the sets of angles selected by Algorithms 1 and 2, respectively. Different results were chosen only in three tests (7, 8, and 9).

Table 3. Joint angles selected by Algorithm 1 (rad).

Test	θ_1	θ_2	θ_3	θ_4	θ_5	θ_6
1	2.3815	0.7054	1.6608	0.7755	0.7601	1.4137
2	−1.7132	−2.0438	0.5154	2.5800	1.4998	1.4473
3	−0.6392	1.4741	1.9941	−0.3266	0.6392	−1.5708
4	1.5708	0.9088	1.3914	0.8414	−1.0472	0
5	−1.9116	2.2113	1.1655	−0.2352	0.3408	−0.7854
6	2.4303	0.2367	0.7022	2.2027	−0.3359	−2.3562
7	1.0472	1.0472	1.5708	0.7854	1.0472	0
8	−2.0715	−0.1327	1.5357	1.7386	−0.4418	0
9	3.1416	−0.4429	1.5708	−0.0807	0.5236	1.5708
10	−2.9992	0.9905	0.9407	1.2104	−1.7132	0

Table 5 shows the computed determinants and proves that the final sets did not result in singular configurations.

Table 4. Joint angles selected by Algorithm 2 (rad).

Test	θ_1	θ_2	θ_3	θ_4	θ_5	θ_6
1	2.3815	0.7054	1.6608	0.7755	0.7601	1.4137
2	−1.7132	−2.0438	0.5154	2.5800	1.4998	1.4473
3	−0.6392	1.4741	1.9941	−0.3266	0.6392	−1.5708
4	1.5708	0.9088	1.3914	0.8414	−1.0472	0
5	−1.9116	2.2113	1.1655	−0.2352	0.3408	−0.7854
6	2.4303	0.2367	0.7022	2.2027	−0.3359	−2.3562
7	−0.8445	0.5569	1.3797	2.0214	2.8289	0.9248
8	−2.0715	0.3281	1.0257	−1.3537	0.4418	3.1416
9	−0.3414	2.5392	0.8718	−1.9651	2.6900	−2.2975
10	−2.9992	0.9905	0.9407	1.2104	−1.7132	0

Table 5. Determinants computed with the selected joint angles (rad).

<table>
<tr><th>Test</th><th>Algorithm 1</th><th>Algorithm 2</th></tr>
<tr><td>1</td><td colspan="2">4.9919×10^6</td></tr>
<tr><td>2</td><td colspan="2">-7.7601×10^6</td></tr>
<tr><td>3</td><td colspan="2">-2.9926×10^7</td></tr>
<tr><td>4</td><td colspan="2">−0.1915</td></tr>
<tr><td>5</td><td colspan="2">-3.2505×10^7</td></tr>
<tr><td>6</td><td colspan="2">-2.2865×10^7</td></tr>
<tr><td>7</td><td>-2.1859×10^7</td><td>0.7623×10^7</td></tr>
<tr><td>8</td><td>-3.4649×10^7</td><td>2.9645×10^7</td></tr>
<tr><td>9</td><td>5.2815×10^7</td><td>-3.5288×10^7</td></tr>
<tr><td>10</td><td colspan="2">-1.2625×10^7</td></tr>
</table>

Table 6 presents the differences between the selected angles and the previous ones; these were computed as in Algorithm 1 using Equation (48) and prove that Algorithm 2 sometimes chose a set that required moving the joints more than Algorithm 1.

Table 6. Differences computed with the selected joint angles (rad).

<table>
<tr><th>Test</th><th>Algorithm 1</th><th>Algorithm 2</th></tr>
<tr><td>1</td><td colspan="2">3.4792</td></tr>
<tr><td>2</td><td colspan="2">4.2870</td></tr>
<tr><td>3</td><td colspan="2">3.0888</td></tr>
<tr><td>4</td><td colspan="2">2.6521</td></tr>
<tr><td>5</td><td colspan="2">3.2697</td></tr>
<tr><td>6</td><td colspan="2">4.1197</td></tr>
<tr><td>7</td><td>2.5247</td><td>3.9838</td></tr>
<tr><td>8</td><td>3.1441</td><td>4.1651</td></tr>
<tr><td>9</td><td>3.9091</td><td>4.8683</td></tr>
<tr><td>10</td><td colspan="2">3.9066</td></tr>
</table>

3.1. Discussion

The results show that the algorithms did not always choose the same angles. However, selecting any of the computed sets that did not result in a singularity render it safer for the user to interact with the robot.

Although Algorithm 1 selected the angles that moved the robot the least, which means that the robot would reach the desired pose faster, Algorithm 2 would not lead to abrupt movements, which is even safer for the user.

It is recommended to use Algorithm 1 when it is necessary to choose the set of angles that moves the joints the least; however, the memory of the device used to compute the IK solution should be enough to store up to eight possibilities. Algorithm 2 is suggested for applications using devices with low computational resources and when the new configuration does not need to be as similar as possible to the previous one.

The derived expression and proposed methods to compute the determinant of the Jacobian require fewer parameters and calculations than the ones found in other works; this implies a computational advantage particularly useful for real-time applications. The proposed methods need modifications to use the solution in [20]. Even though it is possible to use the equation in [21] in the proposed algorithms, the determinant does not result in zero when either θ_3 or θ_5 is equal to π; these mathematical singularities exist and can be detected with Equation (46).

3.2. Applications

A complete IK solution that avoids singularities is helpful in different applications; one of them is robot design, particularly to implement the inverse kinematic control of a robot with the same geometry. Other more advanced applications include image-based visual servoing and the control of robotic digital twins.

Image-based visual servoing, as seen in Figure 4, refers to the use of features extracted from images to move the robot to a desired feature or pose. In the latter case, it is necessary to include the IK solution as a part of an external controller to choose one set of angles that can move the end effector as desired while avoiding singularities.

Using a digital twin as supervisory control of a physical robot requires the IK solution to be implemented in the digital twin controller to verify that the physical system is working as expected and under safe conditions, which is particularly important for cobots. The proposed validated IK solution that uses the FSM is suitable because it avoids abrupt changes in the joint motion, this also renders the system safer because joint velocities can be safely controlled if singularities can be detected and avoided.

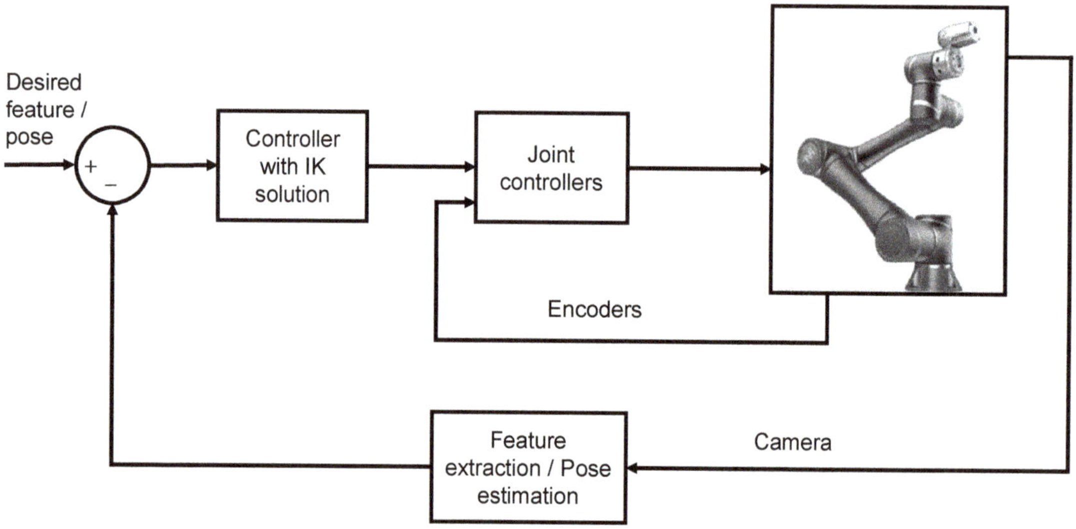

Figure 4. Image-based visual servoing control loop adapted from [33].

4. Conclusions

Inverse kinematic solutions can be used to control real and virtual robots. However, when an IK solution is used, it is also necessary to consider the robot's singular configurations to have complete and validated robot control algorithms. For that reason, this study used the Jacobian matrix of the UR5 robot and its determinant; the latter was used in two algorithms to select a set of angles that takes the robot to the desired pose without resulting in a singularity. In this way, the main contribution of this work is the definition of two alternative methods for the complete and validated computational implementation of IK solutions for the UR5 robot. Since reaching singularities leads to faster joint velocities, methods that help in avoiding singular configurations and in making it safer for people to interact with this kind of robot are essential additions to the IK solution and can be considered to be enhanced alternative inverse IK solutions for robotic arms with the UR geometry.

The results show that, although the two algorithms could lead to the same set of angles, this is not always true, which means that one of the presented algorithms can be more suitable depending on the application. For example, if one objective is to reach the desired pose faster, Algorithm 1 can be used to ensure that the set of angles that requires moving the joints the least overall is selected; however, by using an FSM and not computing all the possibilities, Algorithm 2 requires less storage space, which means that it can be used in devices with low computational resources; this algorithm also avoids abrupt joint movements.

As a final remark, the FSM-based method complements the IK solution procedure with advantages in the number of computations and performance by producing results that would not move the joints abruptly, which is desired for collaborative robots, and this method is helpful when using devices with low computational resources.

Future work will focus on using the IK solution and one of the selection algorithms to control a virtual UR5 that will later be used to control a physical UR5 robot. The chosen algorithm will be modified to select the set of angles that consumes the least power or to evaluate which one results in the trajectory that moves the robot as far as possible from singularities.

Author Contributions: Conceptualization, J.V., I.Y.S. and F.M.; Data curation, J.V.; Formal analysis, J.V.; Investigation, J.V.; Methodology, J.V.; Project administration, J.V. and F.M.; Resources, J.V., I.Y.S. and F.M.; Software, J.V.; Supervision, F.M.; Validation, J.V.; Visualization, J.V.; Writing—original draft, J.V.; Writing—review & editing, I.Y.S. and F.M. All authors have read and agreed to the published version of the manuscript.

Funding: this research was funded by Consejo Nacional de Ciencia y Tecnología grant number (doctoral scholarship) 752557.

Institutional Review Board Statement: Not applicable.

Informed Consent Statement: Not applicable.

Data Availability Statement: Not applicable.

Conflicts of Interest: The authors declare no conflict of interest.

References

1. Park, H.; Easwaran, A.; Andalam, S. TiLA: twin-in-the-loop architecture for cyber-physical production systems. In Proceedings of the 2019 IEEE 37th international conference on computer design (ICCD), Abu Dhabi, United Arab Emirates, 17–20 November 2019. [CrossRef]
2. Zhang, D.; Wei, B. A review on model reference adaptive control of robotic manipulators. *Annu. Rev. Control* **2017**, *43*, 188–198. [CrossRef]
3. Analytic Inverse Kinematics for the Universal Robots UR-5/UR-10 Arms. Available online: https://smartech.gatech.edu/handle/1853/50782 (accessed on 24 October 2022).
4. Supplementary Material: An Ultrasound Robotic System Using the Commercial Robot UR5. Available online: https://www.researchgate.net/publication/292987030_Supplementary_Material_An_Ultrasound_Robotic_System_Using_the_Commercial_Robot_UR5 (accessed on 24 October 2022).
5. Kinematics of a UR5. Available online: https://rasmusan.dk/wp-content/uploads/ur5_kinematics.pdf (accessed on 24 October 2022).
6. Liu, Q.; Yang, D.; Hao, W; Wei, Y. Research on kinematic modeling and analysis methods of UR robot. In Proceedings of the 2018 IEEE 4th Information Technology and Mechatronics Engineering Conference (ITOEC), Chongqing, China, 14–16 December 2018. [CrossRef]

7. Kebria, P.M.; Al-Wais, S.; Abdi, H.; Nahavandi, S. Kinematic and dynamic modelling of UR5 manipulator. In Proceedings of the 2016 IEEE International Conference on Systems, Man, and Cybernetics (SMC), Budapest, Hungary, 9–12 October 2016. [CrossRef]
8. Villalobos, J., Sanchez, I.Y., Martell, F. Alternative Inverse Kinematic Solution of the UR5 Robotic Arm. In *Advances in Automation and Robotics Research*; Moreno, H.A., Carrera, I.G., Ramírez-Mendoza, R.A., Baca, J., Banfield, I.A., Eds.; Springer: Cham, Switzerland, 2022; pp. 200–207. [CrossRef]
9. Zhao, R.; Shi, Z.; Guan, Y.; Shao, Z.; Zhang, Q.; Wang, G. Inverse kinematic solution of 6R robot manipulators based on screw theory and the Paden–Kahan subproblem. *Int. J. Adv. Robot. Syst.* **2018**, *15*, 1–11. [CrossRef]
10. Villalobos, J.; Sanchez, I.Y.; Martell, F. Statistical comparison of Denavit–Hartenberg based inverse kinematic solutions of the UR5 robotic manipulator. In Proceedings of the 2021 International Conference on Electrical, Computer, Communications and Mechatronics Engineering (ICECCME), Mauritius, Mauritius, 7–8 October 2021. [CrossRef]
11. Schreiber, L.-T.; Gosselin, C. Determination of the Inverse Kinematics Branches of Solution Based on Joint Coordinates for Universal Robots–Like Serial Robot Architecture. *J. Mech. Rob.* **2022**, *14*, 034501. [CrossRef]
12. Wang, H.; Qi, H.; Xu, M.; Tang, Y.; Yao, J.; Yan, X.; Li, M. Research on the relationship between classic Denavit–Hartenberg and modified Denavit–Hartenberg. In Proceedings of the 2014 seventh international symposium on computational intelligence and design, Hangzhou, China, 13–14 December 2014. [CrossRef]
13. Kohli, D.; Osvatic, M. Inverse kinematics of general 6R and 5R, P serial manipulators. *J. Mech. Des.* **1993**, *115*, 922–931. [CrossRef]
14. Raghavan, M.; Roth, B. Inverse kinematics of the general 6R manipulator and related linkages. *J. Mech. Des.* **1993**, *115*, 502–508. [CrossRef]
15. Manocha, D.; Canny, J.F. Efficient inverse kinematics for general 6R manipulators. *IEEE Trans. Rob. Autom.* **1994**, *10*, 648–657. [CrossRef]
16. Husty, M.L.; Pfurner, M.; Schröcker, H.-P. A new and efficient algorithm for the inverse kinematics of a general serial 6R manipulator. *Mech. Mach. Theory* **2007**, *42*, 66–81. [CrossRef]
17. Xin, S.Z.; Feng, L.Y.; Bing, H.L.; Li, Y.T. A simple method for inverse kinematic analysis of the general 6R serial robot. *J. Mech. Des.* **2007**, *129*, 793–798. [CrossRef]
18. Qiao, S.; Liao, Q.; Wei, S.; Su, H.-J. Inverse kinematic analysis of the general 6R serial manipulators based on double quaternions. *Mech. Mach. Theory* **2010**, *45*, 193–199. [CrossRef]
19. Wang, J.; Lu, C.; Zhang, Y.; Sun, Z.; Shen, Y. A numerically stable algorithm for analytic inverse kinematics of 7-degrees-of-freedom spherical-rotational-spherical manipulators with joint limit avoidance. *J. Mech. Rob.* **2022**, *14*, 051005. [CrossRef]
20. FarzanehKaloorazi, M.H.; Bonev, I.A. Singularities of the typical collaborative robot arm. In Proceedings of the International Design Engineering Technical Conferences and Computers and Information in Engineering Conference, Quebec, Canada, 26–29 August 2018. [CrossRef]
21. Weyrer, M.; Brandstötter, M.; Husty, M. Singularity avoidance control of a non–holonomic mobile manipulator for intuitive hand guidance. *Robotics* **2019**, *8*, 14. [CrossRef]
22. Geometric Jacobians Derivation and Kinematic Singularity Analysis for Smokie Robot Manipulator & the Barrett WAM. Available online: https://arxiv.org/abs/1707.04821 (accessed on 24 October 2022).
23. Development of Direct Kinematics and Workspace Representation for Smokie Robot Manipulator & the Barret WAM. Available online: https://arxiv.org/abs/1707.04820 (accessed on 24 October 2022).
24. Pires, F.; Melo, V.; Almeida, J.; Leitão, P. Digital twin experiments focusing virtualisation, connectivity and real-time monitoring. In Proceedings of the 2020 IEEE Conference on Industrial Cyberphysical Systems (ICPS), Tampere, Finland, 10–12 June 2020. [CrossRef]
25. Gao, C.; Piao, X.; Tong, W. Optimal motion control for IBVS of robot. In Proceedings of the 10th World Congress on Intelligent Control and Automation, Beijing, China, 6–8 July 2012. [CrossRef]
26. Xiao, Y.; Fan, Z.; Li, W.; Chen, S.; Zhao, L.; Xie, H. A manipulator design optimization based on constrained multi-objective evolutionary algorithms. In Proceedings of the 2016 International Conference on Industrial Informatics-Computing Technology, Intelligent Technology, Industrial Information Integration (ICIICII), Wuhan, China, 3–4 December 2016. [CrossRef]
27. Craig, J.J. *Introduction to Robotics: Mechanics and Control*, 3rd ed.; Pearson Prentice Hall: Hoboken, NJ, USA, 2005; pp. 135–159.
28. Asada, H.; Slotine, J.-J.E. *Robot Analysis and Control*, 1st ed.; John Wiley & Sons: New York, NY, USA, 1986; pp. 51–71.
29. Zaplana, I.; Basanez, L. A novel closed-form solution for the inverse kinematics of redundant manipulators through workspace analysis. *Mech. Mach. Theory* **2018**, *121*, 829–843. [CrossRef]
30. Gong, M.; Li, X.; Zhang, L. Analytical inverse kinematics and self-motion application for 7-DOF redundant manipulator. *IEEE Access* **2019**, *7*, 18662–18674. [CrossRef]
31. Kalra, P.; Mahapatra, P.B.; Aggarwal, D.K. An evolutionary approach for solving the multimodal inverse kinematics problem of industrial robots. *Mech. Mach. Theory* **2006**, *41*, 1213–1229. [CrossRef]
32. Tong, Y.; Liu, J.; Liu, Y.; Yuan, Y. Analytical inverse kinematic computation for 7-DOF redundant sliding manipulators. *Mech. Mach. Theory* **2021**, *155*, 104006. [CrossRef]
33. Corke, P.I. *Visual Control of Robots: High-Performance Visual Servoing*; Research Studies Press: Taunton, UK, 2005; pp. 152–154.

robotics

MDPI

Article

Mechanical Design of a Biped Robot FORREST and an Extended Capture-Point-Based Walking Pattern Generator

Hongxi Zhu * and Ulrike Thomas *

Robotics and Human-Machine Interaction Lab, Chemnitz University of Technology, Reichenhainer Str. 70, 09126 Chemnitz, Germany
* Correspondence: hongxi.zhu@etit.tu-chemnitz.de (H.Z.); ulrike.thomas@etit.tu-chemnitz.de (U.T.)

Abstract: In recent years, many studies have shown that soft robots with elastic actuators enable robust interaction with the environment. Compliant joints can protect mechanical systems and provide better dynamic performance, thus offering huge potential for further developments of humanoid robots. This paper proposes a new biped robot. The new robot combines a torque sensor-based active elastic hip and a spring-based passive elastic knee/ankle. In the first part, the mechanical design is introduced, and in the second part, the kinematics and dynamics capabilities are described. Furthermore, we introduce a new extended capture-point-based walking pattern generator that calculates footstep positions, which are used as input for the controller of our new biped robot. The main contribution of this article is the novel mechanical design and an extended walking pattern generator. The new design offers a unique solution for cable-driven bipeds to achieve both balancing and walking. Meanwhile, the new walking pattern generator can generate smooth desired curves, which is an improvement over traditional generators that use a constant zero-moment-point (ZMP). A simple cartesian controller is applied to test the performance of the walking pattern generator. Although the robot has been built, all experiments regarding the pattern generator are still simulated using MATLAB/Simulink. The focus of this work is to analyze the mechanical design and show the capabilities of the robot by applying a new pattern generator.

Keywords: biped; mechanical design; kinematics

Citation: Zhu, H.; Thomas, U. Mechanical Design of a Biped Robot FORREST and an Extended Capture-Point-Based Walking Pattern Generator. *Robotics* **2023**, *12*, 82. https://doi.org/10.3390/robotics12030082

Academic Editor: Raffaele Di Gregorio

Received: 16 December 2022
Revised: 30 April 2023
Accepted: 24 May 2023
Published: 7 June 2023

1. Introduction

While humans can easily walk with two legs, it remains very challenging to realize two-legged humanoid robots. Over the years, several biped robots have been developed. Compared to wheeled mobile robots, two-legged humanoid robots offer significant advantages, as they can move in difficult or uneven terrains, climb stairs, walk over obstacles, and reach spaces as humans do, even in challenging environments. Therefore, when designing biped robots, researchers need to ensure that they can move like humans with minimal energy consumption and that they are highly compliant when interacting with their environment or with humans.

In the past, the joints of most humanoid robots have been directly driven by actuators, which consist of motors and gears. One of the earliest famous humanoid robots was the ASIMO [1], developed by Honda and first introduced in 2000. It can walk, run, and jump. Additionally, AIST developed the humanoid robot series HRP-(1–5) [2–6]. Unlike the others, the HRP-4 is a smart version with a slim lightweight body and a female shape. The latest generation, HRP-5, is designed to perform heavy tasks or to operate in hazardous environments. Toyota developed the robot T-HR3 [7], which enables the entire robot body to be controlled via a wearable device that maps the movements of the user's hands, arms, and feet to the robot. Takanishi Laboratory developed the robot WABIAN-2R [8], which can walk with straight knees, heel contact, and toe-off. Another excellent humanoid robot is the robot TORO [9], developed by the German Aerospace Center (DLR). The robot TORO is built based on the legs of the former DLR biped [10] and is completely impedance-controlled based on the torque controller drive technology, similar to the DLR lightweight arms. It can adapt to an uneven environment similar to a foam mattress. Researchers at TU Munich developed two robots, Johnnie [11] and Lola [12]. The robot Johnnie can walk on flat and uneven ground and around curves. The next-generation robot, Lola, can recognize obstacles and achieve dynamic real-time movement through planning and control. Researchers at Oregon State University invented the biped robot Cassie [13], which has two bird-like legs. The researchers at PAL Robotics developed the biped robot REEM-C in 2013 [14]. In 2017, PAL built their second biped robot TALOS [15], which uses torque control. The robot can walk on uneven terrain. The newer biped robot, called Kangaroo, was released in 2022 [16]. All the robot's joints in Kangaroo are driven by ball-screw linear actuators. The company Tesla also released their biped robot called Optimus in 2022 [17].

All of the abovementioned robots use motor-gear drive systems and rigid joints. However, their dynamic behavior is limited by motor performance. Thus, researchers have developed bipeds that are driven by elastic actuators to improve the dynamic performance and safe interaction with environments. Atlas [18] is one of the world's most famous humanoid robots developed by Boston Dynamics. Atlas uses a hydraulic actuator system to drive all the joints. Compared to motor-gear drivers, the dynamic performance of a hydraulic driver system is more powerful. At Ritsumeikan University, researchers developed a torque-controlled hydraulic humanoid robot called TaeMu [19]. It is capable of full-body compliant balancing. In addition to hydraulic bipeds, another possibility to improve bipeds is to combine motor-gearbox systems with elastic springs. The robot cCub [20] is such a compliant robot that uses compact units based on series elastic actuators. A serial elastic actuator combines the common motor with an elastic spring and thus can provide an elasticity with constant stiffness [21]. Furthermore, the adjustable stiffness of a leg may improve the performance in walking on a variety of terrains [22]. Although variable stiffness actuators that can adjust their stiffness are available [23], they have not yet been applied to biped robots. The humanoid robot Valkyrie [24], which was developed by NASA, applies serial elastic actuators to drive every joint. The robot Valkyrie aims to solve challenging tasks in space. Another elastic biped robot is MABEL [25]. Its hip is constrained to revolute motion in the sagittal plane. It can only walk in a circle around a tower. In [26], the Technical University of Darmstadt presented an elastic biped robot called BioBiped. The robot BioBiped has two three degrees of freedom (DOF) legs and uses musculoskeletal technology. This robot can jump while its trunk is externally constrained to vertical translation. Another similar elastic biped robot, the C-Runner [27], was developed by the DLR. The elastic elements give the legs high impact robustness and are able to store energy in mechanical springs during dynamic movements. The bipeds MABEL, BioBiped, and C-Runner use elastic cable-driven systems. The springs can absorb shocks and store energy for running or jumping. They show good dynamic performance and huge potential for running and jumping. Due to the design difficulties of elastic cable-driven systems for multi-DoF joints (hips and ankles), such bipeds lack of the necessary number of DoFs for balancing. They can only walk around a central tower or in a sagittal plane. In order to achieve a fully compliant leg, we add a spring in the knee and design a new elastic cable-driven ankle with two DoFs to overcome these disadvantages. By applying the new ankle, higher velocities and torques can be reached. With our new leg, our robot is able to walk in a human-like way and save energy. Moreover, jumping and running will become possible in the near future.

In the past, many researchers have developed different control strategies to achieve stable balancing. A comprehensive robot control strategy for bipeds comprises a walking pattern generator and a balance controller. A walking pattern generator plans the desired trajectory for the biped. A simple generator based on the linear inverted pendulum (LIP) and ZMP concepts was developed by AIST [28]. The walking biped robot model is commonly simplified as a LIP, which assumes that the robot's center of mass (CoM) can be approximated as a point mass located above the stance foot. This simplification facilitates the development of control strategies for maintaining dynamic balance during walking, and this technique forms the basis of most walking pattern generators used today. Moreover, a walking pattern generator using preview control of ZMP was described in [29]. A virtual ZMP tracking control was utilized to generate the actual trajectory from the previewed reference ZMP in future steps. In [30], a third-order polynomial interpolation curve was employed for online gait generation, with the aim of facilitating its implementation in real systems. The real-time walking and running gait pattern generation for the bipedal robot ASIMO was presented in [31,32], where the generator utilized a model consisting of three point masses. One point mass was located at the end of the LIP, another was situated at the ankle of the supporting foot, and the third was positioned at the ankle of the swing foot. The compensation of the dynamics error between the approximate dynamic model and the real ASIMO robot was discussed in [33]. A walking pattern generator based on the capture point (CP) was described in [34]. The capture point was determined from the orbital energy of the LIP model. Specifically, the CP represents the point at which the velocity of the CoM is zero. This property enables the CP to be used for calculating a suitable foothold, thus preventing the robot from falling. The walking robot was stabilized by CP and ZMP controllers. Most generators for walking rely on a constant CoM height to simplify the planning process. However, this approach limits the robot to bending its knees during locomotion. In contrast, human-like walking involves a stretched knee, heel strike, and toe push-off, resulting in variations in the CoM height throughout the gait cycle. An enhanced centroidal moment pivot and virtual repellent point were presented in [35] to expand the 2D capture point concept into a 3D divergent component of motion (DCM). This approach facilitates the planning of the robot's CoM height trajectory, thereby enhancing the robot's ability to adapt to uneven terrain. The method, as presented in [36], accomplished the CoM height trajectory by adjusting the natural frequency of the DCM for locomotion on uneven terrain. A concept of spatially quantized dynamics (SQD) was proposed in [37] to achieve a more human-like walking pattern with a stretched knee. It involved discretizing the trajectory into constant distance intervals and transforming the spatial walking pattern into the time domain. In [38], a pattern generator for walking with variable height was presented and enabled 3D walking over uneven terrains based on capture inputs. The paper [39] proposed a planning algorithm capable of generating continuous-time walking patterns, including seamless transitions between flat-contact and heel-to-toe walking gaits. In addition to the conventional linear inverted pendulum (LIP) model, a spring-loaded inverted pendulum (SLIP) model was proposed in [40]. This model incorporated a virtual spring along the inverted pendulum, which can be compressed to absorb energy upon the landing of the swing foot and subsequently accelerate the CoM when the swing foot takes off. The SLIP model proved to be a valuable tool in controlling and analyzing the running and hopping of bipeds. In [41], a 3D SLIP model was employed to achieve high-speed running for a whole-body humanoid robot via simulation. Furthermore, in [42], a task-decomposed energy-exchange dynamics learning method was proposed, which combined model-based reinforcement learning to capture the simplified SLIP biped dynamics and utilize them for control. The difference between the LIP and SLIP is explained in Appendix A.

The balance controller is utilized to stabilize bipedal robots and prevent them from falling over. Controllers for ASIMO, including ground reaction force control and model ZMP control, were presented in [43]. In [44], an approach for balancing a humanoid robot with multiple contacts was presented and implemented in the robot TORO. A passivity-based controller was applied to the robot TORO in [45] to achieve balancing on soft terrain, such as a mattress. The controllers for multicontact and divergent components of motion were combined in [46]. In [47], the passivity controller was extended so that the robot TORO could balance itself on an unstable ground surface. An energy-efficient controller based on optimization was presented in [48]. The researcher minimized the energy cost of walking and calculated the target value of each step incrementally. An online foot position compensator was proposed in [49] to improve the robustness of walking. The control strategy of the biped robot DURUS was introduced in [50]. It divided walking into a sequence of distinct events and realized multicontact walking. The control strategy for the biped robot HRP-4 was presented in [51]. A 3D LIP method was implemented, and feedback linearization was used for joint tracking. A two-level variable horizon predictive controller was proposed in [52]. The two levels calculated the landing location and generated trajectories for landing in the desired time. A model-predictive-based control law, which uses extended centroidal dynamics, was presented in [53] to consider heavy limbs. An online nonlinear model predictive control approach was developed in [54] to realize the desired walking behaviors. In [55], the authors provided a robust controller using model predictive control to compensate for the gap between simulation and reality. The authors of [56] proposed a full-body predictive model control scheme based on differential dynamic programming that took into account the full dynamics of the system and determined the optimal actuation for the robot's lower body. A passivity-based inverse dynamics controller using a global energy tank was introduced in [57]. The approach used a task space inverse dynamics quadratic programming to calculate the desired torque for satisfying a set of tasks. Currently, most controllers are designed for rigid biped robots, and only a few studies have introduced control approaches for passive elastic bipeds [20,24]. The control strategy with passive elasticity is a challenge for our future work.

In this paper, a new design of a biped robot, called FORREST, is described, which was developed at the TU Chemnitz and is shown in Figure 1. The new biped robot combines a software-based elastic hip and spring-based passive elastic knees/ankles for walking and running. Each leg provides six DoFs. Currently, most cable-driven biped robots cannot maintain balance. We designed a novel elastic cable-driven ankle to fill this gap, by providing better dynamics. A CP-based extended walking pattern generator is introduced in this work. A simple CP controller and a torque-based Cartesian PD controller are used to test the walking pattern generator. This article is structured as follows: First, Section 2 provides an overview of the robot. In in Section 3, the mechanical design of each part of the robot is described. Section 4 analyzes the kinematic and dynamic performance of the knee and ankle, and the dynamic model is introduced. The extended walking pattern generator is presented in Section 5. Finally, Section 6 describes the control strategy and shows the experiments of locomotion with FORREST conducted in simulation.

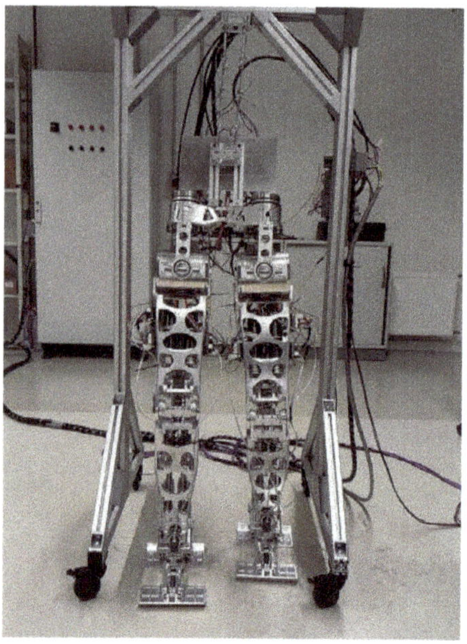

Figure 1. Biped robot FORREST.

2. Overview of FORREST's Design

This section provides an overview of the design of the biped robot FORREST. Figure 2 shows the kinematic scheme of the biped robot and the CAD model of one leg. Each leg consists of six joints: a three DoF hip, a one DoF knee, and a two DoF ankle. Most biped robots have six DoFs per leg. The total weight of the robot is 35 kg. Table 1 lists the weight and the size of each segment.

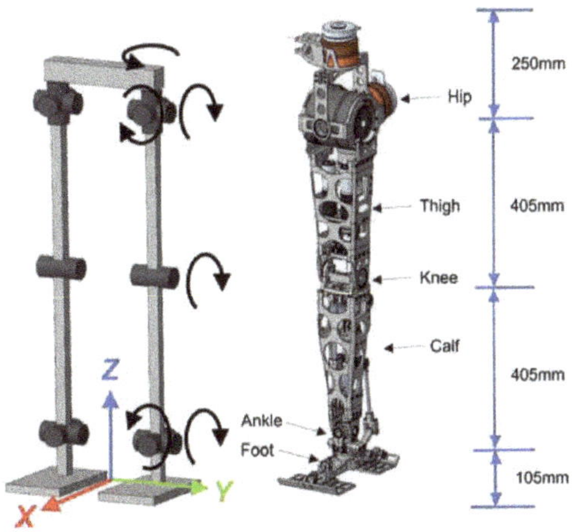

Figure 2. Kinematic scheme for the leg and its CAD model. The left picture shows that each leg consists of a three DoF hip, a one DoF knee, and a two DoF ankle. The right picture shows the height of each part.

Table 1. Overview of weight and height.

Segment	Weight [kg]	Height [mm]
base	17.14	250
thigh	5.83	405
calf	2.8	405
foot	0.35	105
total	35.1	1165

The joints of the hip are driven by actuators from SENSODRIVE, which consist of a BLDC motor, a harmonic gear, and an integrated torque sensor. This allows for software-based compliance at the hip. The knee has one degree of freedom and is driven by a motor with a ball screw spindle with a pitch of 4 mm. The ankle has two degrees of freedom and is driven by two MAXON motors using a parallel mechanism. The performance of the joints, considering the gear and drive system, is listed in Table 2.

Table 2. Torque and speed performance of joints.

Joint	Max Toque [Nm]	Max Speed [rpm]
hip 1	120	31.9
hip 2	107	19.9
hip 3	315	29
knee	195	15.8
ankle 1	212	35
ankle 2	138	47

Considering the human anatomy, the muscles in the thigh and calf regions play an important role while humans are running. These muscles can dampen impact, store energy, and release it during bursts, exhibiting behavior similar to that of springs, which we will utilize in our work. However, due to the design difficulty of elastic cable-drive systems for multi-DoF joints such as the hip and ankle, most cable-driven bipeds cannot balance [25–27]. Our goal is to overcome this disadvantage by designing a cable-driven ankle for our prototype. Only the knee and ankle joints use springs to store impact energy and achieve human-like running. FORREST's main purpose is to serve as an experimental platform for combining a torque-controlled software-based elastic hip with mechanical elastic knees/ankles. Further advantages of this design lie in the high reactivity of the new ankle joint, improving the possibility for balancing, and the combination of spring-based deformable joints. Compared to the other bipeds shown in Table 3, a typical biped can balance itself but only has active compliance, while most cable-driven bipeds have passive compliance but cannot balance themselves. However, our biped not only includes passive compliance but can also balance itself.

Table 3. Comparison with other bipeds.

	FORREST	Toro	Lola	TALOS	C-Runner	MABEL	BioBiped
balance	○	○	○	○	×	×	×
compliance	active + passive	active	active	active	passive	passive	passive

3. Design of the New Biped Robot

In this section, the mechanical design of each part of the leg is discussed. The hip joints are driven by a motor–gearbox system, while the knee is driven by an elastic ball screw spindle system. The ankle, on the other hand, is driven by an elastic parallel mechanism. The new biped robot, FORREST, aims to combine software-based compliance, which means joints where the output torque is measured at the link side and fed back into the control, and mechanical compliance, which enables it to be more responsive than previous solutions.

3.1. Hip

The structure of the hip is similar to that of most other biped robots to simplify the design. Figure 3 shows the front and back views of the hip. All the joints of the hip are rigid joints without springs and are driven by SENSORDRIVE actuators. The compliance of the joints is achieved by the software-based impedance controller, which offers compliance in one DOF. A high-precision inertia measurement unit is mounted on the hip to estimate the orientation and angular velocity.

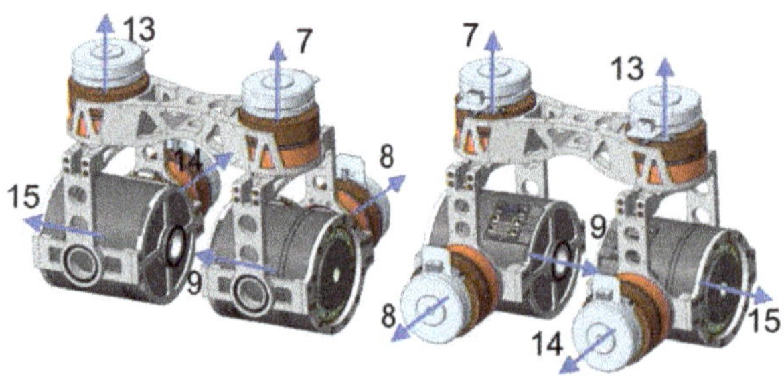

Figure 3. Front (**left**) and back (**right**) view of the hip. The blue arrow indicates the z-axis direction of the joint, and the number indicates the joint number, which corresponds to the link id in Table 4.

Table 4. Kinematic and dynamic parameters of FORREST.

Link Id	Child Id	Parent Id	T_i^p	m [kg]	r [m]	I_d [kg m²]
1	2	0	$T_{roty}(\pi/2)$			
2	3	1	$T_{rotx}(-\pi/2)$			
3	4	2	$T_{roty}(-\pi/2)$			
4	5	3	$T_{rotx}(\pi/2)$			
5	6	4	$T_{roty}(-\pi/2)$			
6	7, 13	5	$T_{rotx}(-\pi/2)T_{rotz}(\pi)$	4.45	$\begin{bmatrix}0 & 0 & 0.44\end{bmatrix}^T$	$\begin{bmatrix}0.23 & 0.18 & 0.06\end{bmatrix}$
7	8	6	$T_{trans}([0, 0.12, 0])$	2.26	$\begin{bmatrix}-0.13 & 0 & 0.02\end{bmatrix}^T$	$\begin{bmatrix}0.01 & 0.06 & 0.05\end{bmatrix}$
8	9	7	$T_{roty}(-\pi/2)$	4.22	$\begin{bmatrix}0 & 0 & 0\end{bmatrix}^T$	$\begin{bmatrix}0.01 & 0.01 & 0.01\end{bmatrix}$
9	10	8	$T_{rotx}(\pi/2)$	4.06	$\begin{bmatrix}-0.21 & 0 & 0\end{bmatrix}^T$	$\begin{bmatrix}0.01 & 0.22 & 0.22\end{bmatrix}$
10	11	9	$T_{trans}([0, 0, -0.405])$	1.95	$\begin{bmatrix}-0.17 & 0 & 0\end{bmatrix}^T$	$\begin{bmatrix}0 & 0.1 & 0.1\end{bmatrix}$
11	12	10	$T_{trans}([0, 0, -0.405])$	0.08	$\begin{bmatrix}0 & 0 & 0\end{bmatrix}^T$	$\begin{bmatrix}0 & 0 & 0\end{bmatrix}$
12		11	$T_{rotx}(\pi/2)$	1.46	$\begin{bmatrix}-0.07 & 0 & 0.01\end{bmatrix}^T$	$\begin{bmatrix}0.01 & 0.02 & 0.01\end{bmatrix}$
13	14	6	$T_{trans}([0, -0.12, 0])$	2.26	$\begin{bmatrix}-0.13 & 0 & 0.02\end{bmatrix}^T$	$\begin{bmatrix}0.01 & 0.06 & 0.05\end{bmatrix}$
14	15	13	$T_{roty}(-\pi/2)$	4.22	$\begin{bmatrix}0 & 0 & 0\end{bmatrix}^T$	$\begin{bmatrix}0.01 & 0.01 & 0.01\end{bmatrix}$
15	16	14	$T_{rotx}(\pi/2)$	4.06	$\begin{bmatrix}-0.21 & 0 & 0\end{bmatrix}^T$	$\begin{bmatrix}0.01 & 0.22 & 0.22\end{bmatrix}$
16	17	15	$T_{trans}([0, 0, -0.405])$	1.95	$\begin{bmatrix}-0.17 & 0 & 0\end{bmatrix}^T$	$\begin{bmatrix}0 & 0.01 & 0.01\end{bmatrix}$
17	18	16	$T_{trans}([0, 0, -0.405])$	0.08	$\begin{bmatrix}0 & 0 & 0\end{bmatrix}^T$	$\begin{bmatrix}0 & 0 & 0\end{bmatrix}$
18		17	$T_{rotx}(\pi/2)$	1.46	$\begin{bmatrix}-0.07 & 0 & 0.01\end{bmatrix}^T$	$\begin{bmatrix}0.01 & 0.02 & 0.01\end{bmatrix}$

3.2. Knee

As is often the case, the knee joint of the biped uses one actuator, which is mounted on the knee axis. To achieve an elastic knee joint, the knee of our robot is driven by a ball screw linear system, which is presented in Figure 4. The linear system consists of a Maxon EC Flat Motor, a ball screw spindle with a 4 mm pitch, a nut, and a compression spring. In the human body, muscles in the thigh can absorb shocks when jumping. The elastic linear actuator in our robot provides the same function as these muscles. Another advantage is that the motors, which drive the ankle, can also be mounted in the thigh when the knee is driven by a ball screw. Compared to other robots with actuators mounted on joint axes, our knee and ankle motors are mounted on the thigh to raise the CoM of the biped and reduce the weight of the calf. The stiffness of the spring in the knee drive system is 74 N/mm, and it can produce a maximum force of 1800 N. Assuming that the weight of the upper body is 30 kg and using the formula for kinetic energy, this spring has a maximum capacity to absorb 33 J of energy and can convert the required kinetic energy for the upper body to move at a velocity of approximately 1.5 m/s into elastic potential energy, which is sufficient for the purpose of studying walking. The choice of spring must also satisfy the limitations of the motor. The stiffness will be adjusted to correspond to the muscle stiffness during running and jumping in the future, but at this stage, we have chosen the highest possible stiffness to reduce the difficulty of control.

3.3. Ankle

A parallel mechanism is often used to drive the ankle of a biped robot, as it can provide more power and greater stability than serial kinematics. Our new biped robot uses a novel cable-driven system to drive the ankle, which is based on a parallel mechanism. Two motors, mounted in the thigh, as shown in Figure 4, are used to drive the two DoFs of the ankle. Each motor output connects two steel wires, where one wire drives the positive rotation, and the other drives the negative rotation, as shown in Figure 4. These wires are marked with red and black lines, respectively, and are transmitted through the knees by several fixed pulleys. One end of each wire is connected to a motor, while the other end is connected to an extension spring. The two wires drive a linear slider via a moving pulley. The stiffness of each extension spring in the ankle drive system is 64 N/mm, and it can produce a maximum force of 760 N. The selection of the ankle spring stiffness is based on the same methodology utilized for the selection of the knee spring stiffness.

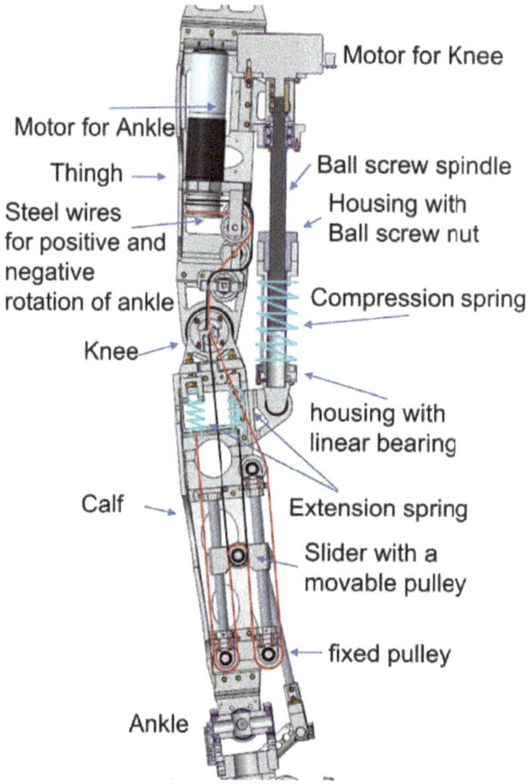

Figure 4. The section view of the thigh and calf and the drive system for the knee and ankle. The red and black lines are two wires, which are connected with one motor, to drive the slider on the calf.

Figure 5 shows the working principle of this cable-driven system. One wire connects the motor with the one side of the spring and passes through the movable pulley. The other side of the spring is fixed on the calf. The movable pulley is mounted on the slider. When the motor rotates in positive direction, the slider can move up. According to the characteristics of the moving pulley, the output force can be doubled. Because the wire can transmit only pulling forces, a second wire is necessary to drive the slider in a negative direction. The right hand side of Figure 5 shows the path of the second wire. This wire must pass through several fixed pulleys and a movable pulley of the slider. When the motor rotates in the negative direction, the slider moves down.

Two linear sliders on the calf are used to drive the ankle. Each linear slider is connected to a link by a rod end. The other end of the link connects to the foot via a universal joint. With the cooperation of two sliders, one ankle can be driven. This design provides a solution that combines a wire-driven system and a parallel mechanism to actuate the ankle. We can mount the motor that drives the ankle joint on the thigh using the wire-driven system. The output force of the motor on the parallel mechanism can be doubled by using movable pulleys, and furthermore, a single motor with a cable can drive the positive and negative directions of each input of the parallel mechanism.

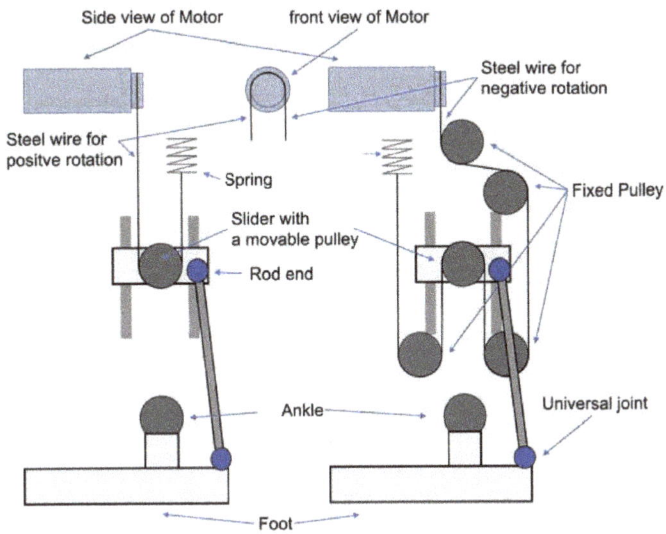

Figure 5. Functional principle of the parallel mechanism of the ankle. The picture on the left shows the wire that drives the slider upward. The right picture shows the wire that drives the slider down.

3.4. Foot

Most biped robots use flat feet. When such a biped robot walks, the floating foot must be parallel to the ground. Our proposed biped robot uses a movable toe and heel. With an additional degree of freedom of the toes and heels, the landing and lifting of the floating foot can be improved. Springs are used on the toe and heel to reduce the shock effect when the floating foot contacts the ground. Figure 6 shows the foot of the FORREST. In addition to the elastic toe and heel, rubber mats were installed under the toe and heel. The rubber mat can dampen the shock when a foot touches the ground. A 6-axis force/torque sensor will be installed between the foot and ankle. With the measurement of this sensor, the ZMP of the robot can be calculated.

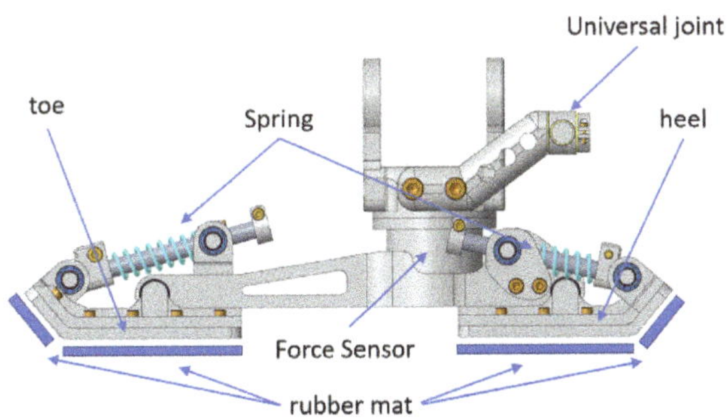

Figure 6. The side view of the foot with the additional DoFs from the toe and the heel.

4. Kinematic/Dynamic Analysis and Dynamic Model

In this section, we first introduce the kinematic parameters of FORREST. The hip joints are driven directly by actuators, and their kinematic and dynamic performance is equivalent to that of the motor. The knee and ankle joints are actuated by lead screws and parallel mechanisms, and we modeled and analyzed their kinematic and dynamic performance. Additionally, we introduce the dynamic model of FORREST.

4.1. Kinematic and Dynamic Parameter of FORREST

We used a floating-base tree-structure model to define the kinematic and dynamic parameters of FORREST. Table 4 describes these parameters of FORREST. Each link had its own ID, child ID, and parent ID to define the tree structure. The links of the driver system of the knee and ankle are not included. Links 1–6 are the translational (links 1–3) and rotational (links 4–6) joints of the floating base. Links 7–12 build the left leg, and links 13–18 build the right leg. $\boldsymbol{T}_i^p$ defines the homogeneous transformation matrix of link i with respect to the coordinate frame of the parent link p. Compared to the traditional Denavit–Hartenberg (DH) parameter, we can define any kinematics chain by a sequence of homogeneous transformation matrices. In addition, link i rotates about axis i, which is identical to the modified DH parameter. The vector $\boldsymbol{rc}_i{}^i$ is the CoM of the link expressed in its own coordinate frame. $\boldsymbol{I}_d$ contains the diagonal parameters of the inertia matrix. The off-diagonal elements of the inertia matrix are not described here. The kinematic and dynamic parameters were used for the simulation and control design, as described in the next section.

4.2. Knee

Figure 7 shows the dimensions of the knee drive system. Points A and B are two free joints that connect the ball screw unit with the thigh and calf, respectively. It was assumed that the origin of the coordinate system was fixed at the knee (point O), and the thigh was fixed in the coordinate system along the y-axis. When the knee joint q_4 rotated, the calf could move in the coordinate system. The picture shows the original position of points A and B.

$$\boldsymbol{x}_A = \begin{bmatrix} 0.235 \\ 0.065 \end{bmatrix} \quad , \quad \boldsymbol{x}_B = \begin{bmatrix} -0.08 \\ 0.065 \end{bmatrix}, \tag{1}$$

when $q_4 = 0$. When the knee joint rotated, the point B moved and was calculated by

$$\boldsymbol{x}_{B^*} = \boldsymbol{R}(q_4)\boldsymbol{x}_B = \begin{bmatrix} cos(q_4) & -sin(q_4) \\ sin(q_4) & cos(q_4) \end{bmatrix} \boldsymbol{x}_B. \tag{2}$$

According to the structure of the knee in Figure 7, we obtained the angle of the two passive joints q_{19} and q_{20} at both ends of the linear drive system, with

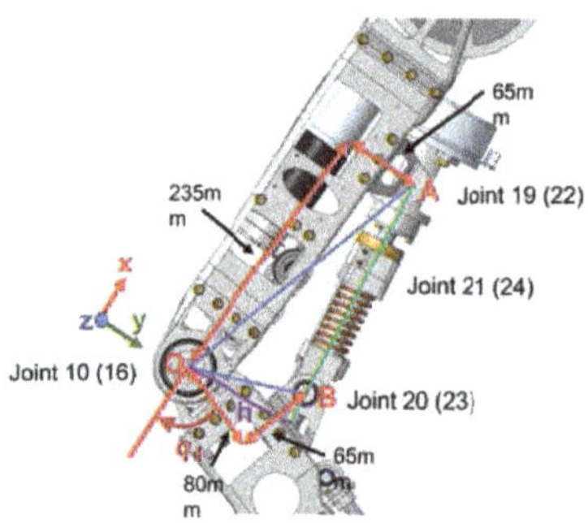

Figure 7. The dimension of the drive system of the knee and the joint numbers of left leg; the joint numbers of the right leg are notated in round brackets. A frame O is fixed on the knee. The thigh is fixed on the y-axis.

$$q_{19} = atan2(x_A - x_{B^*}, y_A - y_{B^*}), \tag{3}$$

$$q_{20} = q_{19} - q_4. \tag{4}$$

By using Heron's formula, the height h of the triangle AOB^* was computed with

$$\begin{aligned} h =& \frac{2\eta}{\|\overrightarrow{AB^*}\|_2}, \\ \eta =& \sqrt{s(s-\|\overrightarrow{OA}\|_2)(s-\|\overrightarrow{OB^*}\|_2)(s-\|\overrightarrow{AB^*}\|_2)}, \\ s =& \frac{\|\overrightarrow{OA}\|_2+\|\overrightarrow{OB^*}\|_2+\|\overrightarrow{AB^*}\|_2}{2}. \end{aligned} \tag{5}$$

The relationship of the speed $\dot{q}_k$ and the torque τ_k between the motor and the knee joint and the compliance k_k of the knee were:

$$\dot{q}_k = \frac{n_m p}{60h}, \tag{6}$$

$$\tau_k = \frac{2\pi\tau_m h}{p}. \tag{7}$$

$$k_k = \frac{\Delta\tau_{ks}}{\Delta q_4} = \frac{k_s \cdot \Delta\|\overrightarrow{AB^*}\|_2 h}{J_{AB}^{-1} \cdot \Delta\|\overrightarrow{AB^*}\|_2} = \frac{k_s \cdot h}{J_{AB}^{-1}}, \tag{8}$$

$$J_{AB} = \frac{d\|\overrightarrow{AB^*}\|_2}{dq_4}. \tag{9}$$

Herein, n_m and τ_m represent the motor speed and torque, respectively, and p is the screw pitch. By using these values, we determined the kinematic and dynamic performance of the knee joint. τ_{ks} denotes the torque caused by the spring, k_s is the spring stiffness, and J_{AB} is the Jacobian. Figure 8 illustrates the relationship between the maximum joint speed/torque/compliance and the knee joint angle. It is obvious that the joint torque and stiffness were highest at 60°, where the joint velocity was minimal.

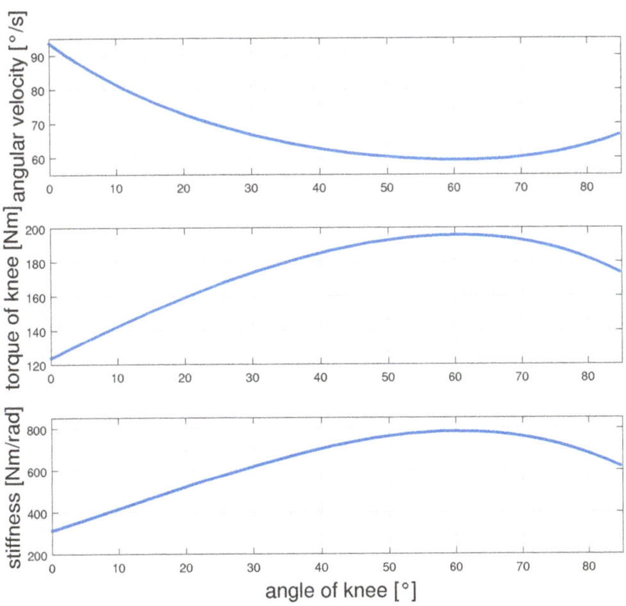

Figure 8. The velocity, torque, and compliance performance of the knee joint with respect to the joint angles.

4.3. Ankle

The ankle joint has two degrees of freedom and is driven by a parallel mechanism. Figure 9 illustrates the dimensions of the parallel mechanism. Points A and B represent the ends of a rod that connects a slider with point A on the calf and the foot with point B. A_1 and B_1 are the endpoints of the left rod, while A_2 and B_2 correspond to the right rod. The length of the rod was 0.185 m. We assumed that the origin of a coordinate system was fixed on the ankle (point O) to obtain the kinematic and dynamic relationship between the ankle and the motors. The calf was fixed in this coordinate system, while the foot was movable. The original positions of $A_{1/2}$ and $B_{1/2}$ were

$$A_1 = \begin{bmatrix} -0.015 \\ 0.04 \\ -0.06 \end{bmatrix}, \quad B_1 = \begin{bmatrix} x_{B_1} \\ 0.04 \\ -0.0345 \end{bmatrix}, \quad A_2 = \begin{bmatrix} -0.015 \\ -0.04 \\ -0.06 \end{bmatrix}, \quad B_2 = \begin{bmatrix} x_{B_2} \\ -0.04 \\ -0.0345 \end{bmatrix}. \tag{10}$$

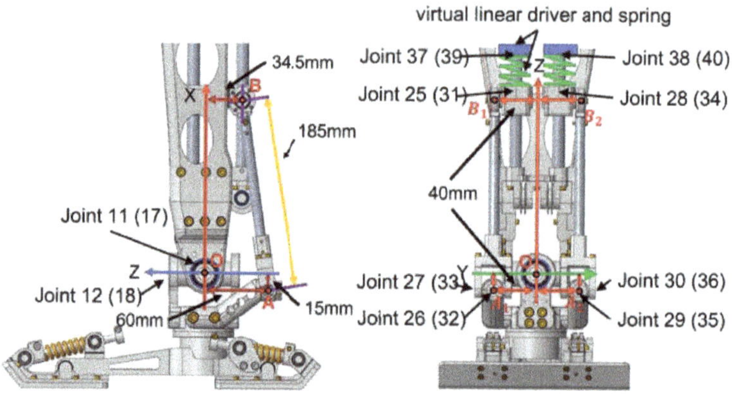

Figure 9. The dimensions of the parallel mechanism that drives the ankle joint together with the joint numbers of the left leg; the joint numbers of the right leg are in round brackets.

When the ankle joints q_{11} and q_{12} rotated, the new position of points A_1 were calculated by

$$A_1^* = \boldsymbol{R}_y(q_{11})\boldsymbol{R}_z(q_{12})\boldsymbol{A}_1 = \begin{bmatrix} cos(q_{11}) & 0 & sin(q_{11}) \\ 0 & 1 & 0 \\ -sin(q_{11}) & 0 & cos(q_{11}) \end{bmatrix} \begin{bmatrix} cos(q_{12}) & -sin(q_{12}) & 0 \\ sin(q_{12}) & cos(q_{12}) & 0 \\ 0 & 0 & 1 \end{bmatrix} \boldsymbol{A}_1. \tag{11}$$

As point B_1 can only slide along the x-axis, the y and z positions of B_1 are constant. Using the Euclidean norm, the new x position of B_1 was determined as follows:

$$x_{B_1^*} = x_{A_1^*} + \sqrt{L^2 - (y_{A_1^*} - y_{B_1^*})^2 - (z_{A_1^*} - z_{B_1^*})^2}. \tag{12}$$

The calculation of $z_{B_2^*}$ was identical. All the passive joints of the parallel mechanism were obtained with

$$\begin{aligned} q_{25} &= 0.405 - x_{b_1}^*, \quad q_{26} = asin\left(\frac{y_{b1}^* - y_{a1}}{L \cdot cos(q_{27})}\right), \quad q_{27} = asin\left(\frac{x_{b1}^* - x_{a1}}{L}\right), \\ q_{28} &= 0.405 - x_{b_2}^*, \quad q_{29} = asin\left(\frac{y_{b2}^* - y_{a2}}{L \cdot cos(q_{30})}\right), \quad q_{30} = asin\left(\frac{x_{b2}^* - x_{a2}}{L}\right). \end{aligned} \tag{13}$$

Then, we calculated the Jacobian matrix, which describes the relationship between the ankle joints and sliders.

$$J = \begin{bmatrix} \frac{\partial q_{25}}{\partial q_{11}} & \frac{\partial q_{25}}{\partial q_{12}} \\ \frac{\partial q_{28}}{\partial q_{11}} & \frac{\partial q_{28}}{\partial q_{12}} \end{bmatrix}. \tag{14}$$

The relationship of the speed and torque between the motor and knee joint was

$$\begin{bmatrix} \dot{q}_{11} \\ \dot{q}_{12} \end{bmatrix} = J^{-1} \begin{bmatrix} 2\pi r_1 n_{m_1} \\ 2\pi r_2 n_{m_2} \end{bmatrix}, \tag{15}$$

$$\begin{bmatrix} \tau_{11} \\ \tau_{12} \end{bmatrix} = J^{T} \begin{bmatrix} 2\tau_{m_1}/r_1 \\ 2\tau_{m_2}/r_2 \end{bmatrix}, \tag{16}$$

$$k_{11} = \frac{\Delta\tau_{11s}}{\Delta q_{11}} = \frac{J^T_{[1,-]} \boldsymbol{k}_s J \begin{bmatrix} \Delta q_{11} & 0 \end{bmatrix}^T}{\Delta q_{11}}. \tag{17}$$

Herein, n_{m_1}, n_{m_2}, τ_{m_1}, and τ_{m_2} are the speeds and torques of the motors, respectively. $\dot{q}_{11}$, $\dot{q}_{12}$, τ_{11}, and τ_{12} are the ankle joint speeds and torques, respectively. r_1 and r_2 are the radii of the cable drums, which are driven by motors. k_{11} is the stiffness of the ankle joint 11. The subscript of the Jacobi matrix $J^T_{[1,-]}$ represents the elements of the first row. k_s is the stiffness of the spring matrix. Assuming that joint 12 does not move when the stiffness of joint 11 is calculated, the method of calculating the joint 12 stiffness is the same. Figures 10–12 show the relationship between the maximum joint speed/torque and the angle of the ankle joints.

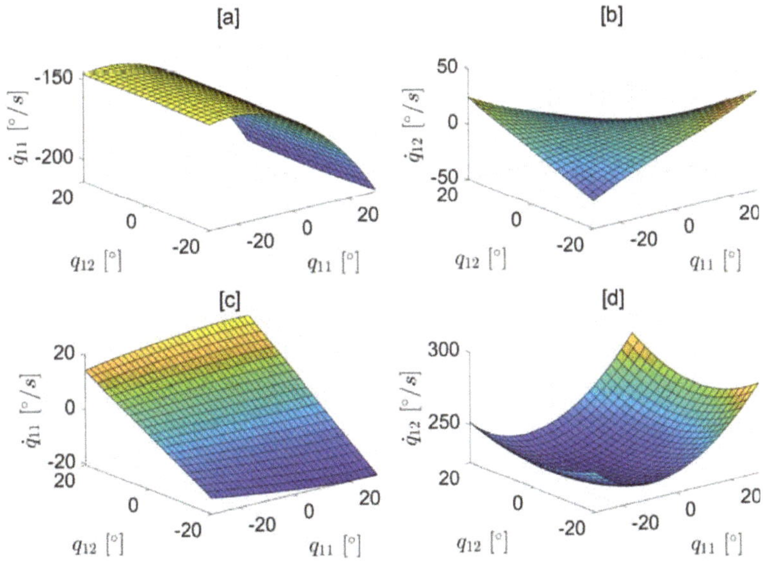

Figure 10. The velocity performance of the ankle joint with respect to the relevant joint angles. (**a**,**b**) show the velocity performance of the ankle joints q_{11} and q_{12}, when the two motors rotate in the same direction at maximum speed. (**c**,**d**) show the velocity performance of q_{11} and q_{12}, when the two motors rotate in the opposite direction to each other at maximum speed.

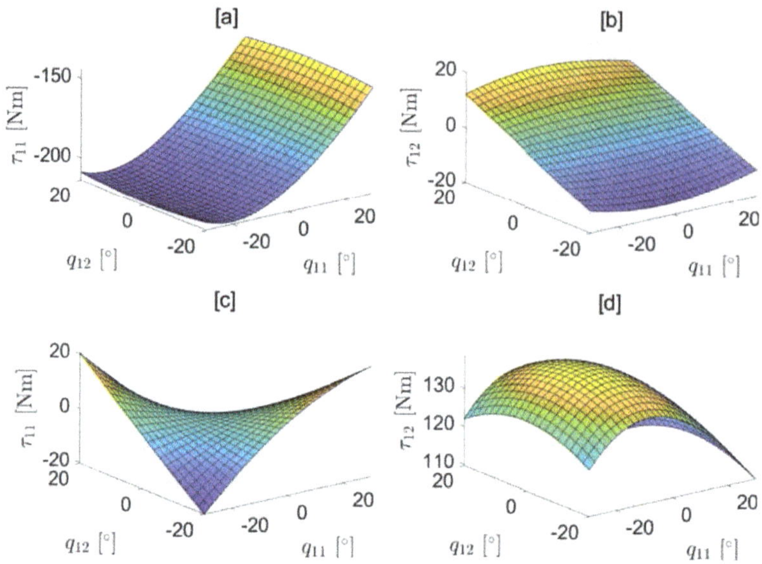

Figure 11. The torque performance of the ankle joint with respect to the relevant joint angles. (**a**,**b**) show the torque performance of ankle joints q_{11} and q_{12}, when the two motors' outputs generate the maximum torque in the same direction. (**c**,**d**) show the torque performance of q_{11} and q_{12}, when the two motors' outputs generate the maximum torque in the opposite direction to each other.

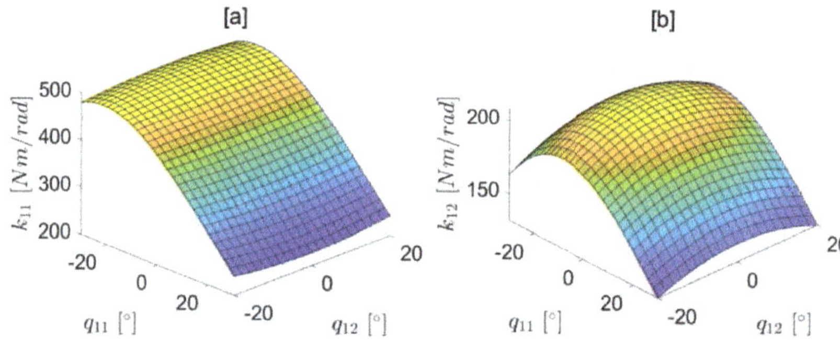

Figure 12. The stiffness of the ankle joint with respect to the relevant joint angles. (**a**) shows the stiffness performance of the ankle joint q_{11}. (**b**) shows the stiffness performance of q_{12}.

4.4. Dynamic Model

The dynamic model of FORREST is a floating base model with closed chains and elastic joints. We neglected the cables and pulleys of the cable-driven parallel mechanism to simplify the modeling. The drive system of the ankle joints was simplified as two virtual linear drivers mounted on the calf, and the motors and sliders were connected by virtual springs. The virtual drivers and springs can be found in Figure 9. We used the Newton–Euler method to calculate the dynamics of our biped. To solve the inverse dynamics, we first calculated the velocity and acceleration of all the passive joints that were not equipped with encoders. Using (18) and (19), we obtained the Jacobian matrices of the passive joints of the left leg. The following formulas with the joint numbers in this section were used to calculate the dynamics of the left leg. The calculation of the right leg was identical.

$$J_i = \frac{\partial q_i}{\partial q_{10}} \quad , \quad \dot{J}_i = \frac{\partial q_i}{\partial q_{10} \partial q_{10}} \dot{q}_{10} \quad and \quad i = 19, 20, \tag{18}$$

$$
\begin{aligned}
J_i &= \begin{bmatrix} \frac{\partial q_i}{\partial q_{11}} & \frac{\partial q_i}{\partial q_{12}} \end{bmatrix} &&, i = 25-30, \\
\dot{J}_i &= \begin{bmatrix} \frac{\partial q_i}{\partial q_{11}\partial q_{11}}\dot{q}_{11} + \frac{\partial q_i}{\partial q_{11}\partial q_{12}}\dot{q}_{12} & \frac{\partial q_i}{\partial q_{12}\partial q_{11}}\dot{q}_{11} + \frac{\partial q_i}{\partial q_{12}\partial q_{12}}\dot{q}_{12} \end{bmatrix} &&, i = 25-30.
\end{aligned} \tag{19}
$$

The velocity and acceleration of all the passive joints were calculated by

$$
\begin{aligned}
\dot{q}_i &= J_i(q_{10})\dot{q}_{10} &&, i = 19, 20, \\
\ddot{q}_i &= \dot{J}_i(q_{10}, \dot{q}_{10})\dot{q}_{10} + J_i(q_{10})\ddot{q}_{10} &&, i = 19, 20,
\end{aligned} \tag{20}
$$

$$
\begin{aligned}
\dot{q}_i &= J_i(q_{11}, q_{12})\begin{bmatrix} \dot{q}_{11} \\ \dot{q}_{12} \end{bmatrix} &&, i = 25-30, \\
\ddot{q}_i &= \dot{J}_i(q_{11}, q_{12})\begin{bmatrix} \dot{q}_{11} \\ \dot{q}_{12} \end{bmatrix} + J_i(q_{11}, q_{12})\begin{bmatrix} \ddot{q}_{11} \\ \ddot{q}_{12} \end{bmatrix} &&, i = 25-30.
\end{aligned} \tag{21}
$$

Then, we calculated the angular and linear velocity and acceleration of each link coordinate and CoM in the world coordinate system by (22)–(28). Herein, ω_i, $\dot{\omega}_i$, v_i, and a_i are the angular velocity/acceleration and linear velocity/acceleration of link coordinate i, respectively. $\dot{\omega}_{p_i}$ and a_{p_i} are the accelerations of the parent link of link i. For a floating base biped model, the base acceleration is related to the foot contact force/moment. The contact force/moment was measured by sensors that were mounted on the ankle. The acceleration of the base was still unknown. Therefore, the equations of the link acceleration needed to be divided into a term with base accelerations and a term without base accelerations ($\dot{\omega}_i^*$ and a_i^*). $\dot{\omega}_b$ and a_b are the accelerations of the floating base coordinate (link 6). The operator $S()$ converts a vector into a skew-symmetric matrix and simplifies the calculation of a cross product.

$$
\omega_i = \begin{cases} \omega_{p_i} + \dot{q}_i z_i & \text{, revolute,} \\ \omega_{p_i} & \text{, prismatic,} \end{cases} \tag{22}
$$

$$
\dot{\omega}_i = \dot{\omega}_b + \dot{\omega}_i^*, \tag{23}
$$

$$
\dot{\omega}_i^* = \begin{cases} \dot{\omega}_{p_i} + \ddot{q}_i z_i + S(\omega_{p_i})(\dot{q}_i z_i) & \text{, revolute,} \\ \dot{\omega}_{p_i}^* & \text{, prismatic,} \end{cases} \tag{24}
$$

$$
v_i = \begin{cases} v_{p_i} & \text{, revolute,} \\ v_{p_i} + \dot{q}_i z_i & \text{, prismatic,} \end{cases} \tag{25}
$$

$$
a_i = a_b - S(r_{i,b})\dot{\omega}_b + a_i^*, \tag{26}
$$

$$
a_i^* = \begin{cases} a_{p_i}^* - S(r_{i,p_i})\dot{\omega}_{p_i}^* + S(\omega_{p_i})S(\omega_{p_i})r_{i,p_i} & \text{, revolute,} \\ a_{p_i}^* - S(r_{i,p_i})\dot{\omega}_{p_i}^* + S(\omega_{p_i})S(\omega_{p_i})r_{i,p_i} + \ddot{q}_i z_i + 2S(\omega_{p_i})(\dot{q}_i z_i) & \text{, prismatic,} \end{cases} \tag{27}
$$

$$
\dot{\omega}_{c_i} = \dot{\omega}_i. \tag{28}
$$

We needed to calculate the acceleration of the CoM of all links to obtain the Newton–Euler equations. The angular acceleration of the CoM was the same as that of the link coordinate. a_{c_i} represents the linear acceleration of the CoM of link i.

$$
\begin{aligned}
a_{c_i} &= a_i - S(r_{c_i,i})\dot{\omega}_i + S(\omega_i)S(\omega_i)r_{c_i,i} \\
&= a_b - S(r_{c_i,b})\dot{\omega}_b + a_i^* - S(r_{c_i,i})\dot{\omega}_i^* + S(\omega_i)S(\omega_i)r_{c_i,i}.
\end{aligned} \tag{29}
$$

When the accelerations were calculated, we obtained the Newton–Euler equations of each link. F_i^k and M_i^k are the forces and moments exerted on link i by the parent and child of link i, respectively. Here, because of the elastic drive system, the springs connected link 2, 0 and 2, 1, link 2, 5 and 3, 7, and link 2, 8 and 3, 8. The spring force can be considered an interaction force between two links and was included in the Newton–Euler equations. Usually, when the robot kinematics are calculated, the deformation of the spring can be obtained, and the spring force can be calculated. On the other hand, we can also compute it with the Newton–Euler method.

$$m_i(\boldsymbol{a}_b - \boldsymbol{S}(\boldsymbol{r}_{c_i,b})\dot{\boldsymbol{\omega}}_b + \boldsymbol{a}_i^* - \boldsymbol{S}(\boldsymbol{r}_{c_i,i})\dot{\boldsymbol{\omega}}_i^* + \boldsymbol{S}(\boldsymbol{\omega}_i)\boldsymbol{S}(\boldsymbol{\omega}_i)\boldsymbol{r}_{c_i,i}) = \sum_k \boldsymbol{F}_i^k + m_i\boldsymbol{g}, \tag{30}$$

$$\boldsymbol{I}_i\dot{\boldsymbol{\omega}}_b + \boldsymbol{I}_i\dot{\boldsymbol{\omega}}_i^* + \boldsymbol{S}(\boldsymbol{\omega}_i)\boldsymbol{I}_i\boldsymbol{\omega}_i = -\sum_k \boldsymbol{S}(\boldsymbol{r}_{c_i,k})\boldsymbol{F}_i^k + \sum_k \boldsymbol{M}_i^k. \tag{31}$$

The Newton–Euler Equation of each link was transformed into a matrix form (32). The acceleration of the floating base, forces, and torques on the link were unknown variables. $\boldsymbol{E}$ is a 3×3 unit matrix. $\boldsymbol{F}_i$ and $\boldsymbol{M}_i$ represent the vector of all applied forces $\boldsymbol{F}_i^k$ and moments $\boldsymbol{M}_i^k$ by the parent and child k. $\boldsymbol{S}_{\boldsymbol{r}_{c_i,k}}$ is a matrix, which contains all the skew-symmetric matrices $\boldsymbol{S}(\boldsymbol{r}_{c_i,k})$ for all the applied forces $\boldsymbol{F}_i^k$ in (31).

$$\begin{bmatrix} m_i\boldsymbol{E} & -m_i\boldsymbol{S}(\boldsymbol{r}_{c_i,b}) & -\boldsymbol{E} & 0 \\ 0 & \boldsymbol{I}_i & \boldsymbol{S}_{\boldsymbol{r}_{c_i,k}} & -\boldsymbol{E} \end{bmatrix} \begin{bmatrix} \boldsymbol{a}_b \\ \boldsymbol{\omega}_b \\ \boldsymbol{F}_i \\ \boldsymbol{M}_i \end{bmatrix} = \begin{bmatrix} m_i(-\boldsymbol{a}_i^* + \boldsymbol{S}(\boldsymbol{r}_{c_i,i})\dot{\boldsymbol{\omega}}_i^* - \boldsymbol{S}(\boldsymbol{\omega}_i)\boldsymbol{S}(\boldsymbol{\omega}_i)\boldsymbol{r}_{c_i,i} + \boldsymbol{g}) \\ -\boldsymbol{I}_i\dot{\boldsymbol{\omega}}_i^* - \boldsymbol{S}(\boldsymbol{\omega}_i)\boldsymbol{I}_i\boldsymbol{\omega}_i \end{bmatrix}. \tag{32}$$

Furthermore, the constraints of the passive joints needed to be considered. The moment along the passive joint axis was zero. It can also be described that the interaction moment between link i and its parent link p_i along the joint axis was zero. All moments on the two spherical joints, which connected link 25 (28) and 27 (30), were zero. The force on the two sliders (link 25, 28) along the linear axis can only be produced by the spring, and the other force source (link 10) along the linear axis must be zero according to (35).

$$\boldsymbol{z}_i^{p_i,T}\boldsymbol{M}_i^{p_i} = 0, \quad i = 10, 11, 12, 19, 20, 25\text{–}30 \tag{33}$$

$$\boldsymbol{M}_{25}^{27} = \begin{bmatrix} 0 & 0 & 0 \end{bmatrix}^T, \quad \boldsymbol{M}_{28}^{30} = \begin{bmatrix} 0 & 0 & 0 \end{bmatrix}^T \tag{34}$$

$$\boldsymbol{z}_i^{10,T}\boldsymbol{F}_i^{10} = 0, \quad i = 25, 28. \tag{35}$$

By combining the Newton–Euler Equation (32) and the constraints (33)–(35) of all links, we obtained the matrix form of the system of linear equations for the biped (36). Here, $\boldsymbol{F}$ and $\boldsymbol{M}$ are the vectors of all the interaction forces and moments, including the spring forces. The acceleration of the base and all forces/moments were calculated by solving (36). However, due to the presence of the elastic knee and ankle joints, these joints were under-actuated and could not be directly driven by motors. The angular acceleration of the joint was related to the spring force, not the motor output force. Therefore, this inverse dynamics method could not be directly used for inverse dynamics-based control. When setting the desired joint acceleration, the spring force did not match. To solve this problem, it was necessary to extend the dynamics to a fourth order system (including jerk and snap), which will be covered in further work.

$$\boldsymbol{H}\begin{bmatrix} \boldsymbol{a}_b \\ \boldsymbol{\omega}_b \\ \boldsymbol{F} \\ \boldsymbol{M} \end{bmatrix} = \boldsymbol{Y} \tag{36}$$

5. Extended CP-Based Walking Pattern Generator

We used an LIP-based walking pattern generator to generate the walking gait trajectory. A common LIP model has a resultant external force by the ground $\boldsymbol{F}_{ext}$, which must be along the LIP. The vertical component of the external force is cancelled out by gravity, which keeps the height z_c of the CoM constant. In our work, we aimed to extend the walking pattern generator by adding a variable height z_c to the CoM. A similar result using a different method was introduced in [35]. We assumed that the LIP was subject to an external force F_{az} that could produce vertical acceleration in addition to the external forces F_{ax}, F_{ay}, and F_{ag}, which were the components along the LIP of the resultant external force. The new dynamic equation of the LIP was

$$
\begin{aligned}
m\ddot{x}_c &= \frac{x_c - p_x}{r} f, \\
m\ddot{y}_c &= \frac{y_c - p_y}{r} f, \\
m\ddot{z}_c &= \frac{z_c - p_z}{r} f + \frac{z_c}{r} f - mg,
\end{aligned}
\tag{37}
$$

and the scalar of resultant force f satisfied

$$
\frac{z_c}{r} f = mg. \tag{38}
$$

Herein, the three components of $x = [x_c \quad y_c \quad z_c]$ were the position of the CoM. $p = [p_x \quad p_y \quad p_z]^T$ was the position of the extended ZMP (eZMP), which had the same x–y position as the common 2D ZMP of the LIP and an additional z position. Its position on the z-axis was symmetrical to the endpoint of the vector F_{az} about a plane passing through the CoM and perpendicular to the z-axis. Figure 13 shows the forces on the LIP Model. The scalar f represents the resultant force of F_{ax}, F_{ay}, and F_{ag} and should be along the LIP. The scalar r represents the length of the vector from ZMP to the CoM. With the additional force F_{az}, the resultant force on the LIP moves the CoM upward.

By combining (37) and (38), we obtained the new dynamic equation of the LIP

$$
\ddot{x} = \frac{1}{m}(F_{ext} + F_g) = \frac{1}{m}\left(\begin{bmatrix} F_{ax} \\ F_{ay} \\ F_{az} + F_{ag} \end{bmatrix} + \begin{bmatrix} 0 \\ 0 \\ -mg \end{bmatrix} \right) = \omega^2 (x - p) = \frac{g}{z_c} \begin{bmatrix} x_c - p_x \\ y_c - p_y \\ z_c - p_z \end{bmatrix}, \tag{39}
$$

and the components of external force satisfied

$$
\frac{F_{ax}}{x_c - p_x} = \frac{F_{ay}}{y_c - p_y} = \frac{F_{ag}}{z_c} = \frac{mg}{z_c}. \tag{40}
$$

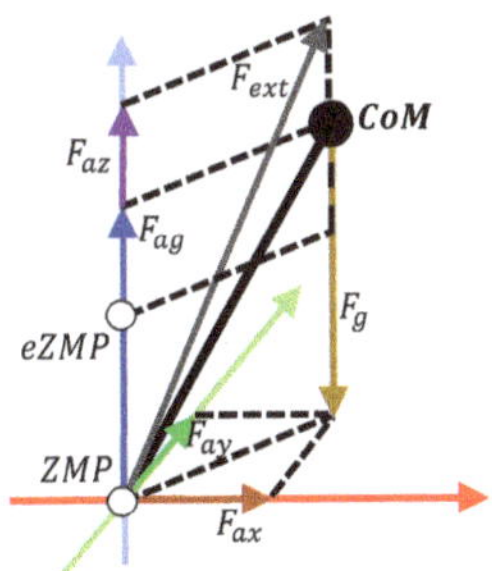

Figure 13. Three-dimensional linear inverted pendulum model.

The time constant ω was $\sqrt{g/z_c}$. In the following, we describe the use of the 3D capture point ξ dynamics to generate our desired trajectory. The 3D CP is also called the divergent component of motion (DCM). The introduction of the CP and DCM can be found in [34,35].

$$
\dot{x} = -\omega(x - \xi) \tag{41}
$$

$$
\dot{\xi} = \omega(\xi - p). \tag{42}
$$

The dynamics of the CoM (41) constitute a stable first-order open loop system, which ensures that the position of the CoM will always converge to the CP. On the other hand, the dynamics of the CP (42) constitute an unstable first-order open loop system. Based on the behavior of the CoM, CP, and ZMP, it is sufficient to control the ZMP to track the desired CP trajectory. However, a common walking pattern generator utilizes a constant ZMP position, leading to instantaneous ZMP exchanges and sudden changes in CoM acceleration, which can result in an unsmooth velocity trajectory. To overcome this disadvantage, we introduced three additional phases between two ZMPs. Figure 14 illustrates an example of the curve of the ZMP and CP. Each step comprised three phases: two exchange phases, lasting T_1 and T_3, and one constant ZMP phase, lasting T_2. We assumed that the initial and final ZMPs, P_0 and P_2, were the same as the initial and final CPs, ξ_0 and ξ_e, in each step. Our objective was to search for a desired ZMP P_1 to satisfy the above assumptions.

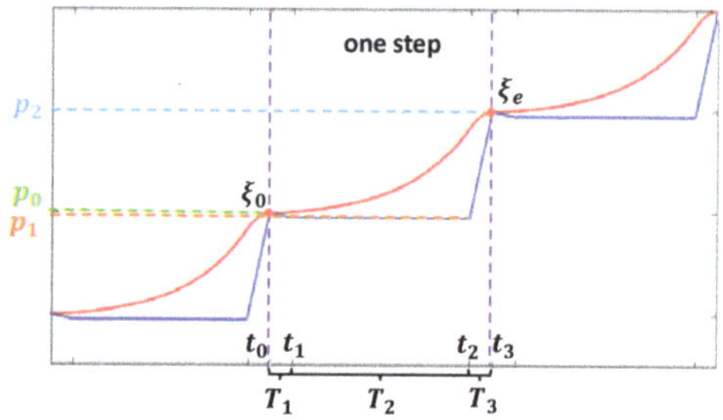

Figure 14. Curves of the ZMP and the CP. The blue line represents the ZMP, which is composed of three parts during each step with respective durations of T_1, T_2, and T_3. Conversely, the red line signifies the CP that coincides with the ZMP at the start and end of each step.

The function of the ZMP is:

$$P(t) = \begin{cases} k_{p_1}t + p_0 & t_0 \le t < t_1 \\ p_1 & t_1 \le t \le t_2 \\ k_{p_2}t + p_1 & t_2 < t \le t_3 \end{cases} \tag{43}$$

$$k_{p_1} = \frac{p_1 - p_0}{T_1} \quad , \quad k_{p_2} = \frac{p_2 - p_1}{T_3}. \tag{44}$$

By solving the ordinary differential equation, we obtained the position equation of the CoM about time

$$x(t) = \begin{cases} (x_{0_1} - p_0)cosh(\omega t) + \dfrac{\dot{x}_{0_1} - k_{p_1}}{\omega} sinh(\omega t) + k_{p_1}t + P_0 & t_0 \le t \le t_1 \\ (x_{0_2} - p_1)cosh(\omega t) + \dfrac{\dot{x}_{0_2}}{\omega} sinh(\omega t) + p_1 \quad t_1 < t \le t_2 \\ (x_{0_3} - p_1)cosh(\omega t) + \dfrac{\dot{x}_{0_3} - k_{p_2}}{\omega} sinh(\omega t) + k_{p_2}t + p_1 & t_2 < t \le t_3, \end{cases} \tag{45}$$

and the CP equation

$$\xi(t) = \begin{cases} \xi_{0_1} e^{\omega t} + \left(P_0 + \dfrac{P_1 - P_0}{\omega T_1}\right)(1 - e^{\omega t}) + \dfrac{P_1 - P_0}{T_1} t & t_0 \le t \le t_1 \\ \xi_{0_2} e^{\omega t} + P_1(1 - e^{\omega t}) \quad t_1 < t \le t_2 \\ \xi_{0_3} e^{\omega t} + \left(P_1 + \dfrac{P_2 - P_1}{\omega T_3}\right)(1 - e^{\omega t}) + \dfrac{P_2 - P_1}{T_3} t & t_2 < t \le t_3. \end{cases} \tag{46}$$

Herein, x_{0_1}, x_{0_2}, and x_{0_3} are the initial positions of the CoM for each phase. $\dot{x}_{0_1}$, $\dot{x}_{0_2}$ and $\dot{x}_{0_3}$ are its initial velocity. ξ_{0_1}, ξ_{0_2} and ξ_{0_3} are the initial position of the CP. It is known that $\xi_{0_1} = \xi_0$ and $\xi(t_3) = \xi_e = P_2$. Because the final CP of first phase is the initial CP of the second phase, and the final CP of the second phase is the initial CP of the third phase, we combined the three formulas of (46) and obtained

$$\begin{aligned}\xi_e(t_3) = P_2 = \xi_0 e^{\omega(T_1+T_2+T_3)} + \frac{P_0}{\omega T_1}\left(e^{\omega(T_1+T_2+T_3)} - e^{\omega(T_2+T_3)} - \omega T_1 e^{\omega(T_1+T_2+T_3)}\right) \\ + \frac{P_1}{\omega T_1 T_3}\left(T_1 e^{\omega T_3} + T_3 e^{\omega(T_2+T_3)} - T_3 e^{\omega(T_1+T_2+T_3)} - T_1\right) + \frac{P_2}{\omega T_3}\left(1 - e^{\omega T_3} + \omega T_3\right).\end{aligned} \tag{47}$$

Once P_0 and P_2 are known, we can use (48) to calculate the desired P_1.

$$\begin{aligned}P_1 = -\frac{\xi_0 \omega T_1 T_3 e^{\omega(T_1+T_2+T_3)} + P_0 T_3\left(e^{\omega(T_1+T_2+T_3)} - e^{\omega(T_2+T_3)} - \omega T_1 e^{\omega(T_1+T_2+T_3)}\right)}{T_1 e^{\omega T_3} + T_3 e^{\omega(T_2+T_3)} - T_3 e^{\omega(T_1+T_2+T_3)} - T_1} \\ - \frac{P_2 T_1 (1 - e^{\omega T_3})}{T_1 e^{\omega T_3} + T_3 e^{\omega(T_2+T_3)} - T_3 e^{\omega(T_1+T_2+T_3)} - T_1}.\end{aligned} \tag{48}$$

Then, we can obtain the trajectory of the CoM and CP by (45) and (46). Figure 15 shows the curves of the ZMP, CP, CoM, and the velocity curve of the CoM. By using this new method, we obtained a position curve of the CoM with a smooth velocity curve. The variable height of the CoM also became possible. However, for this article, the CoM height only decreased from its initial position to a constant value. We will discuss in future articles the application of variable height in efficient walking, such as walking with extended knees. Furthermore, the ZMP and the CP were the same at the end of each step, ensuring CP stability. Figure 16 shows the ZMP, CP, and CoM curve in the xy plane. It is evident that the trajectories of the CP and ZMP overlapped because the CP and ZMP shared the same starting and ending points for each footstep. At the beginning of each step, the ZMP moved away from the CP towards the P_1, generating a suitable repulsive force to accelerate the CP away from ZMP. As the end of each step neared, the ZMP quickly approached the CP and coincided with it at the conclusion.

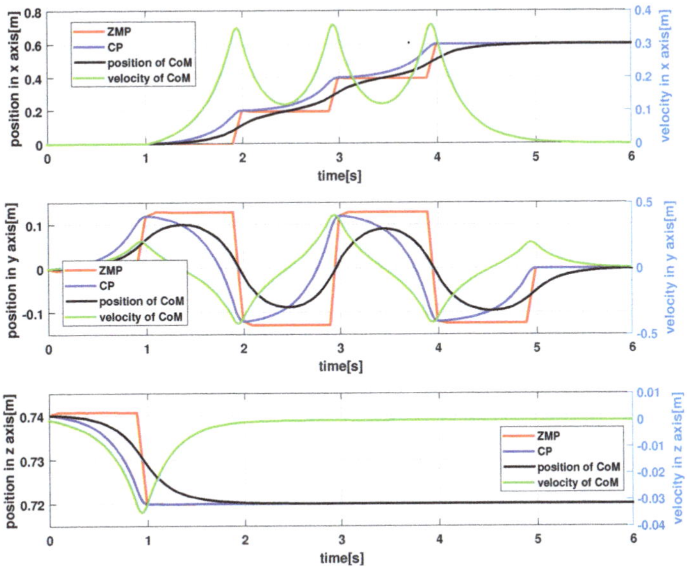

Figure 15. Curves of the ZMP, CP, and the position and velocity of the CoM.

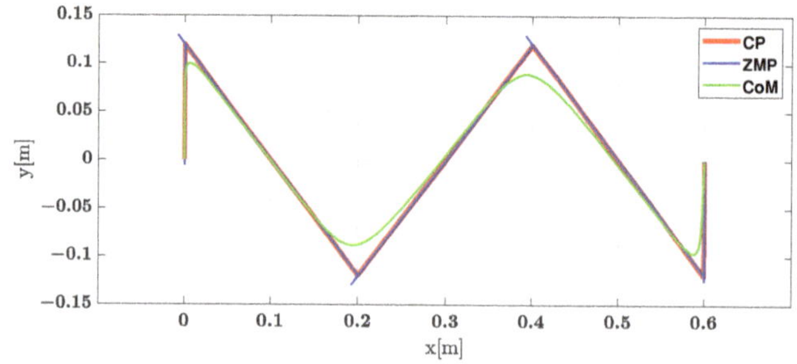

Figure 16. Curve of the ZMP, CP, and the position and velocity of the CoM in the x-y plane.

6. Control Strategy

In this section, we introduce the control strategy for the biped FORREST. Currently, the control strategy for balancing the robot is under development. We used MATLAB/Simulink to test and validate the performance of the controller, which can be implemented in our biped in the future. Figure 17 shows the control scheme for our biped, which consists of an LIP-based walking pattern generator, a capture point controller, a cartesian PD controller, an optimizer for contact force distribution, and an inverse dynamics module. We defined the step distance and step time, and the walking pattern generator calculated the desired trajectories of the CoM, CP, and feet. The capture point controller calculated the feedforward acceleration of the CoM, while the PD controller obtained the desired force on the CoM according to the desired trajectories. An optimization algorithm was implemented to obtain the optimized force distribution from the desired force. Finally, the inverse dynamics module calculated the desired torque of each joint and sent it to the robot. Firstly, we introduce the walking pattern generator of the robot.

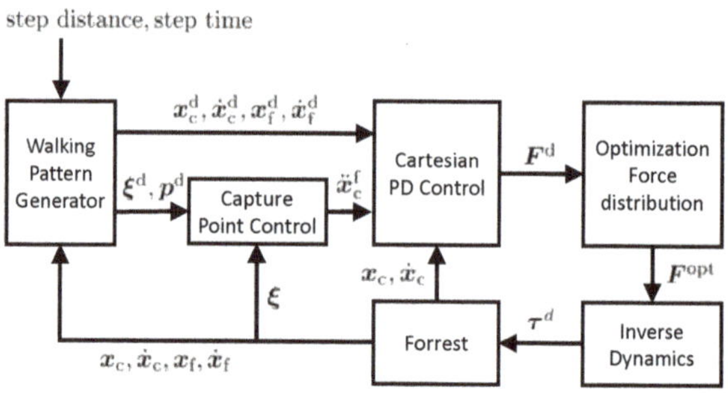

Figure 17. Control schema for the biped.

6.1. Capture Point Control

We used the method from [35] to track the CP trajectory. We defined a stable dynamic equation of the error of CP for $k > 0$.

$$\dot{\xi}_d - \dot{\xi} = -k(\xi_d - \xi). \tag{49}$$

By substituting (42) into (49), we obtained the controlling ZMP $\boldsymbol{p}^c$

$$\boldsymbol{p}^c = \boldsymbol{p}^d + (1 + k/\omega)(\xi - \xi_d). \tag{50}$$

Finally, we used $\boldsymbol{p}^c$ to calculate the desired acceleration of the CoM. The desired acceleration was used as a feedforward input in the Cartesian PD control.

$$\ddot{x}_c^f = \omega^2(x_c^d - \boldsymbol{p}^c). \tag{51}$$

6.2. Cartesian PD Controller

In this section, we introduce the Cartesian PD controller, which is based on [44], the optimization of contact force distribution, and the inverse dynamics that we implemented in our simulation. We used the Cartesian PD controller with feed forward from (51) to keep the robot in balance. The control law is

$$\begin{aligned}\begin{bmatrix}\boldsymbol{a}_c^d\\ \boldsymbol{\omega}_c^d\end{bmatrix} &= \boldsymbol{K}_{pc}\tilde{\boldsymbol{x}}_c + \boldsymbol{K}_{dc}\dot{\tilde{\boldsymbol{x}}}_c + \ddot{\boldsymbol{x}}_c^f\\ \begin{bmatrix}\boldsymbol{a}_f^d\\ \boldsymbol{\omega}_f^d\end{bmatrix} &= \boldsymbol{K}_{pf}\tilde{\boldsymbol{x}}_f + \boldsymbol{K}_{di}\dot{\tilde{\boldsymbol{x}}}_f.\end{aligned} \tag{52}$$

$\tilde{\boldsymbol{x}}_c$ contains the vector of the position and orientation error of the CoM. $\tilde{\boldsymbol{x}}_f$ are the vectors of the errors of the two feet. By using (53), the desired wrench $\boldsymbol{F}_c^d$ on the CoM was obtained.

$$\boldsymbol{F}_c^d = \begin{bmatrix} m_c\boldsymbol{a}_c^d \\ \boldsymbol{I}_c\dot{\omega}_c^d + \omega_c \times (\boldsymbol{I}_c\omega_c)\end{bmatrix} \tag{53}$$

In order to calculate the force distribution, we defined the optimized wrench on the feet, $\boldsymbol{F}f^{opt}$, as shown in (54). This wrench consists of the vertical and horizontal contact forces, $\boldsymbol{f}l^{opt}$ and $\boldsymbol{f}r^{opt}$, of the left and right feet. The horizontal elements of the contact moments, $\boldsymbol{M}l^{opt}$ and $\boldsymbol{M}r^{opt}$, were zero. Additionally, $\boldsymbol{p}l^{opt}$ and $\boldsymbol{p}_r^{opt}$ represent the center of pressure (CoP) of the left and right feet, respectively.

$$\boldsymbol{F}_f^{opt} = \begin{bmatrix}\boldsymbol{F}_l^{opt} & \boldsymbol{M}_l^{opt} & \boldsymbol{p}_l^{opt}\\ \boldsymbol{F}_r^{opt} & \boldsymbol{M}_r^{opt} & \boldsymbol{p}_r^{opt}\end{bmatrix} = \begin{bmatrix} f_{l,x}^{opt} & f_{l,y}^{opt} & f_{l,z}^{opt} & 0 & 0 & \tau_{l,z}^{opt} & p_{l,x}^{opt} & p_{l,y}^{opt}\\ f_{r,x}^{opt} & f_{r,y}^{opt} & f_{r,z}^{opt} & 0 & 0 & \tau_{r,z}^{opt} & p_{r,x}^{opt} & p_{r,y}^{opt}\end{bmatrix} \tag{54}$$

We used the following constrained quadratic optimization problem:

$$\min_{\boldsymbol{F}_f^{opt}}(\boldsymbol{F}_c^d - \boldsymbol{F}_c^{opt})\boldsymbol{W}(\boldsymbol{F}_c^d - \boldsymbol{F}_c^{opt}) \tag{55}$$

with

$$\boldsymbol{F}_c^{opt} = \begin{bmatrix}\boldsymbol{F}_l^{opt} + \boldsymbol{F}_r^{opt}\\ \boldsymbol{F}_l^{opt} \times \boldsymbol{p}_{l,c} + \boldsymbol{F}_r^{opt} \times \boldsymbol{p}_{r,c} + \boldsymbol{M}_l^{opt} + \boldsymbol{M}_r^{opt}\end{bmatrix} \tag{56}$$

and with the constraints for the support foot. The optimized vertical force must be larger than the minimum vertical force. The horizontal forces must be smaller than the friction force, where μ is the friction factor. The position of the CoP must be within the range of the support polygon S of the feet. The optimized wrench was substituted into (53) to calculate the optimized acceleration of the CoM. By combining the optimized acceleration of the CoM and the desired acceleration of the two feet, we calculated the desired acceleration of the joints using (57).

$$\begin{bmatrix}\boldsymbol{a}_c^{opt}\\ \boldsymbol{a}_l^d\\ \boldsymbol{a}_r^d\end{bmatrix} = \dot{\boldsymbol{J}}\dot{\boldsymbol{q}} + \boldsymbol{J}\ddot{\boldsymbol{q}}^d. \tag{57}$$

Finally, we substituted the current joint position, velocity, desired acceleration, and optimized wrench of the feet into the floating base inverse dynamics of the biped, which was described in the previous section, to calculate the desired torque $\boldsymbol{\tau}^d$ of the joints.

$$\boldsymbol{\tau}^d = invdyn(\boldsymbol{q}, \dot{\boldsymbol{q}}, \ddot{\boldsymbol{q}}^d, \boldsymbol{F}_l^{opt}, \boldsymbol{F}_r^{opt}) \tag{58}$$

6.3. Results

We implemented our new walking pattern generator in a simulation using the Simscape Multibody Toolbox of Simulink to validate its performance. The dynamic model of the robot was exported from the CAD model. Four contact points were set to the four corners of the foot bottom plate to simulate the contact between the robot and the ground. The step length was set to 20 cm, and the time per step was 1 s. The sample time of the simulation was 0.1 ms. The cycle time of the controller was 1 ms. The process of walking is displayed in Figure 18. The robot took two steps from 1 s to 3 s. The left foot was the support foot between 1 s and 2 s, and the right foot was the support foot between 2 s and 3 s. The red circle is the CoM of the biped. The green circle is the CoP. Figure 19 shows the trajectory of the CoM and CoP in the xy-plane. The red curve is the desired trajectory, and the blue curve is the measured trajectory. The biped tracked the trajectory accurately in simulation. We compared the position error of the extended generator with the traditional CP-based generator using the same controller to verify the performance of the extended CP-based walking pattern generator. Figure 20 shows the position error of the extended CP and traditional CP. As seen in Figure 20, the difference was not very significant. By using both walking pattern generators, the biped could walk and keep balance. In order to confirm the performance, we compared the absolute value of both errors by (59). If f is larger than zero, it means that the error of extended CP is smaller. The result was that 67.5% of the time, the x position error of the extended CP was smaller, and 57.4% of the time the y position error of the extended CP was also smaller.

$$f = |\tilde{x}_{trad}| - |\tilde{x}_{ext}| \tag{59}$$

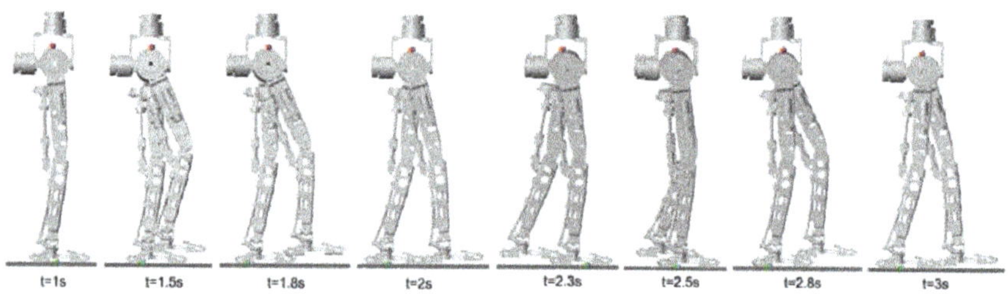

Figure 18. Animation of locomotion from 1 s to 3 s. The left foot is the support foot from 1 s to 2 s. The right foot is the support foot from 2 s to 3 s.

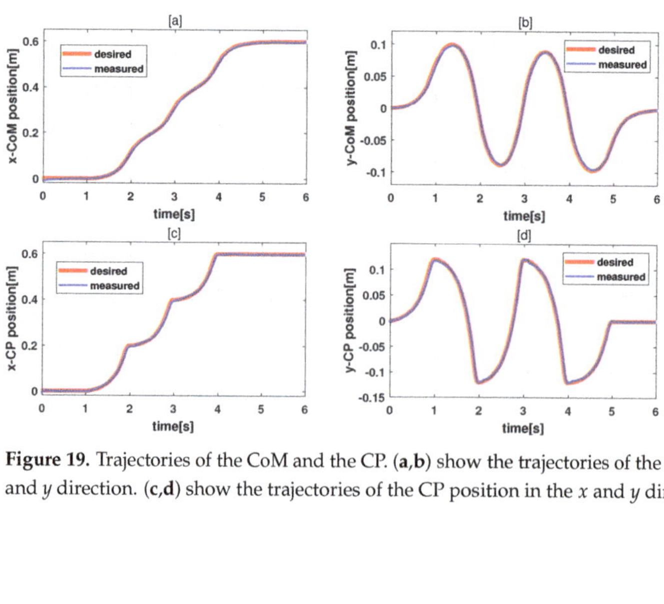

Figure 19. Trajectories of the CoM and the CP. (**a**,**b**) show the trajectories of the CoM position in the x and y direction. (**c**,**d**) show the trajectories of the CP position in the x and y direction.

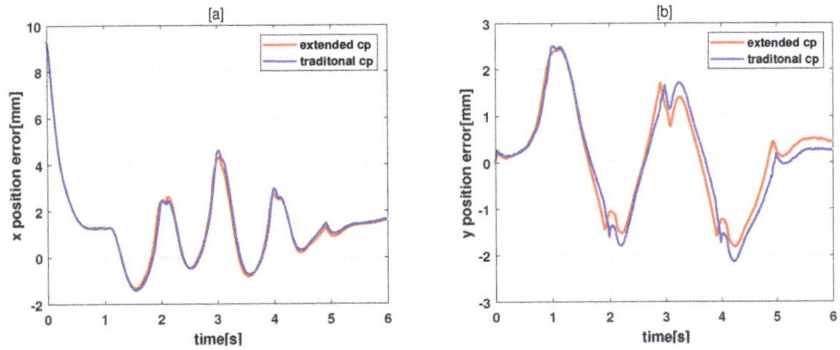

Figure 20. Trajectories of the CoM and the CP. (**a**) Error of the CoM position in the x-direction. (**b**) Error in the y-direction.

7. Conclusions

In this paper, we proposed a new biped robot called FORREST. The robot consists of a hip with torque-controlled joints and mechanical elastic knee/ankle joints. The massive compliance in the robot will enable human–robot cooperation and bring humanoid robots into daily life. FORREST serves as an experimental platform for combining active elastic joints and passive elastic joints. The knee is actuated by an elastic ball screw system, and we use a novel elastic cable-driven parallel mechanism to drive the ankle joint. Unlike most cable-driven bipeds, our robot can maintain balance. Thus, in the future we will show how to use the robot in our daily surroundings. In this paper, we first described the structure and working principles of the knee and ankle, analyzed the kinematic and dynamic performance of the joints, and introduced the dynamic model of the entire robot. We proposed an extended CP-based walking pattern generator for our robot to achieve a smoother walking pattern including smooth trajectories. We validated the performance of the new generator using a simple control strategy in simulation, and the results show that the new walking pattern generator performs better than traditional generators. In the near future, we plan to realize the real walking of our biped and develop a new control strategy that takes into account its elasticity.

Author Contributions: The conceptual design of the robot and the methodology was done by both authors equally. H.Z. built the robot, implemented and modified the algorithms and prepared the original draft, where U.T. supervised the work, reviewed and edited the draft. All authors have read and agreed to the published version of the manuscript.

Funding: Parts of the work were funded by the Deutsche Forschungsgemeinschaft (DFG, German Research Foundation)—Project-ID 416228727, 416228728 and 491193532 the Chemnitz University of Technology.

Data Availability Statement: The data generated for this article are confidend and stored for ten years at Chemnitz University of Technology.

Acknowledgments: We would like to express our gratitude to Lars Gebhardt, Tino Scheibe, and Annett Stark from the technical staff for their support during the production.

Conflicts of Interest: No conflict of interest.

Appendix A. Comparison between the LIP and SLIP Models

The difference between an LIP (linear inverted pendulum) and a SLIP (spring-loaded inverted pendulum) is that the height of the CoM can be changed in the LIP, allowing for convex (Figure A1) shapes to facilitate efficient stretched-knee walking. On the other hand, in the SLIP, the height of the CoM must be concave due to the compression of the spring. The SLIP model proves to be highly valuable for controlling and analyzing the locomotion patterns of bipeds, particularly in running and hopping. Its characteristics make it particularly suitable for studying these dynamic activities. However, when it comes to efficient stretched-knee walking, the LIP model is more appropriate and advantageous.

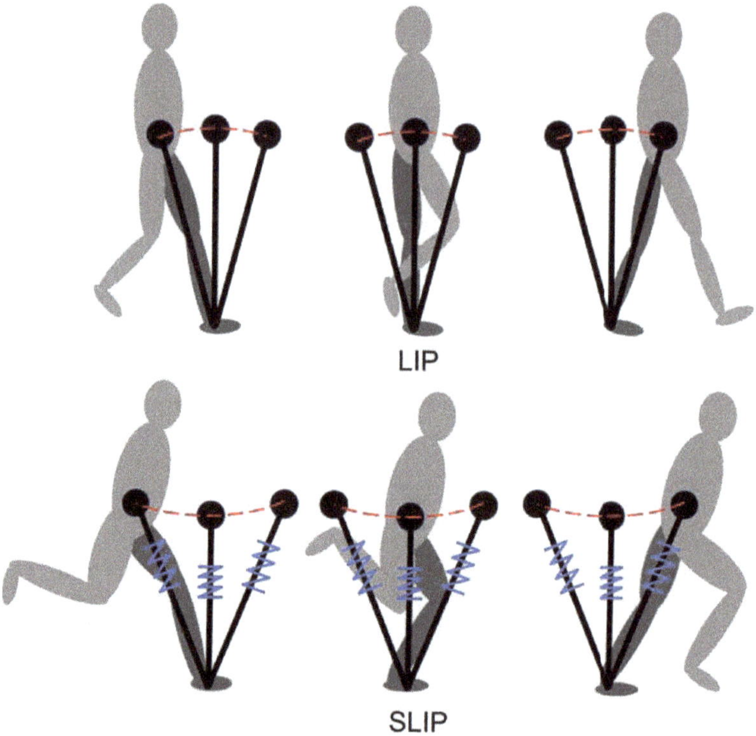

Figure A1. Comparison between the LIP and SLIP.

References

1. Sakagami, Y.; Watanabe, R.; Aoyama, C.; Matsunaga, S.; Higaki, N.; Fujimura, K. The intelligent ASIMO: System overview and integration. In Proceedings of the IEEE/RSJ International Conference on Intelligent Robots and Systems, Lausanne, Switzerland, 30 September–4 October 2002; Volume 3, pp. 2478–2483. [CrossRef]
2. Hirukawa, H.; Kanehiro, F.; Kaneko, K.; Kajita, S.; Fujiwara, K.; Kawai, Y.; Tomita, F.; Hirai, S.; Tanie, K.; Isozumi, T.; et al. Humanoid robotics platforms developed in HRP. *Robot. Auton. Syst.* **2004**, *48*, 165–175. [CrossRef]
3. Kaneko, K.; Kanehiro, F.; Kajita, S.; Hirukawa, H.; Kawasaki, T.; Hirata, M.; Akachi, K.; Isozumi, T. Humanoid robot HRP-2. In Proceedings of the IEEE International Conference on Robotics and Automation, ICRA'04, New Orleans, LA, USA, 26 April–1 May 2004; Volume 2, pp. 1083–1090. [CrossRef]
4. Kaneko, K.; Harada, K.; Kanehiro, F.; Miyamori, G.; Akachi, K. Humanoid robot HRP-3. In Proceedings of the 2008 IEEE/RSJ International Conference on Intelligent Robots and Systems, Nice, France, 22–26 September 2008; pp. 2471–2478. [CrossRef]
5. Kaneko, K.; Kanehiro, F.; Morisawa, M.; Akachi, K.; Miyamori, G.; Hayashi, A.; Kanehira, N. Humanoid robot HRP-4—Humanoid robotics platform with lightweight and slim body. In Proceedings of the 2011 IEEE/RSJ International Conference on Intelligent Robots and Systems, San Francisco, CA, USA, 25–30 September 2011; pp. 4400–4407. [CrossRef]
6. Kaneko, K.; Kaminaga, H.; Sakaguchi, T.; Kajita, S.; Morisawa, M.; Kumagai, I.; Kanehiro, F. Humanoid Robot HRP-5P: An Electrically Actuated Humanoid Robot with High-Power and Wide-Range Joints. *IEEE Robot. Autom. Lett.* **2019**, *4*, 1431–1438. [CrossRef]
7. Toyota Global Newsroom. Toyota Unveils Third Generation Humanoid Robot T-HR3. November 2017. Available online: https://newsroom.toyota.co.jp/en/download/20110424 (accessed on 31 December 2022).
8. WABIAN-2R Biped Robot. Available online: http://www.takanishi.mech.waseda.ac.jp/top/research/wabian/ (accessed on 31 December 2022).
9. Englsberger, J.; Werner, A.; Ott, C.; Henze, B.; Roa, M.A.; Garofalo, G.; Burger, R.; Beyer, A.; Eiberger, O.; Schmid, K.; et al. Overview of the torque-controlled humanoid robot TORO. In Proceedings of the 2014 IEEE-RAS International Conference on Humanoid Robots, Madrid, Spain, 18–20 November 2014; pp. 916–923. [CrossRef]

10. Ott, C.; Baumgärtner, C.; Mayr, J.; Fuchs, M.; Burger, R.; Lee, D.; Eiberger, O.; Albu-Schäffer, A.; Grebenstein, M.; Hirzinger, G. Development of a biped robot with torque controlled joints. In Proceedings of the 2010 10th IEEE-RAS International Conference on Humanoid Robots, Nashville, TN, USA, 6–8 December 2010; pp. 167–173. [CrossRef]
11. Lohmeier, S.; Loffler, K.; Gienger, M.; Ulbrich, H.; Pfeiffer, F. Computer system and control of biped "Johnnie". In Proceedings of the IEEE International Conference on Robotics and Automation, ICRA'04, New Orleans, LA, USA, 26 April–1 May 2004; Volume 4, pp. 4222–4227. [CrossRef]
12. Lohmeier, S. Design and Realization of a Humanoid Robot for Fast and Autonomous Bipedal Locomotion. Ph.D. Thesis, Technische Universität München, Munich, Germany, 2010.
13. Reher, J.; Ma, W.L.; Ames, A.D. Dynamic walking with compliance on a cassie bipedal robot. In Proceedings of the 2019 18th European Control Conference (ECC), Naples, Italy, 25–28 June 2019.
14. PAL Biped Robot Reem. 2013. Available online: https://pal-robotics.com/robots/reem-c/ (accessed on 31 December 2022).
15. PAL Biped Robot Talos. 2017. Available online: https://pal-robotics.com/robots/talos/ (accessed on 31 December 2022).
16. Roig, A.; Kothakota, S.K.; Miguel, N.; Fernbach, P.; Hoffman, E.M.; Marchionni, L. On the hardware design and control architecture of the humanoid robot kangaroo. In Proceedings of the 6th Workshop on Legged Robots during the International Conference on Robotics and Automation (ICRA 2022), Philadelphia, PA, USA, 27 May 2022.
17. Tesla Biped Robot Optimus. 2022. Available online: https://spectrum.ieee.org/robotics-experts-tesla-bot-optimus (accessed on 31 December 2022).
18. Feng, S.; Xinjilefu, X.; Atkeson, C.G.; Kim, J. Optimization based controller design and implementation for the atlas robot in the darpa robotics challenge finals. In Proceedings of the 2015 IEEE-RAS 15th International Conference on Humanoid Robots (Humanoids), Seoul, Republic of Korea, 3–5 November 2015.
19. Hyon, S.H.; Suewaka, D.; Torii, Y.; Oku, N. Design and experimental evaluation of a fast torque-controlled hydraulic humanoid robot. *IEEE/ASME Trans. Mechatron.* **2016**, *22*, 623–634. [CrossRef]
20. Ugurlu, B.; Tsagarakis, N.G.; Spyrakos-Papastavridis, E.; Caldwell, D.G. Compliant joint modification and real-time dynamic walking implementation on bipedal robot cCub. In Proceedings of the 2011 IEEE International Conference on Mechatronics, Istanbul, Turkey, 13–15 April 2011.
21. Negrello, F.; Garabini, M.; Catalano, M.G.; Malzahn, J.; Caldwell, D.G.; Bicchi, A.; Tsagarakis, N.G. A modular compliant actuator for emerging high performance and fall-resilient humanoids. In Proceedings of the 2015 IEEE-RAS 15th International Conference on Humanoid Robots (Humanoids), Seoul, Republic of Korea, 3–5 November 2015.
22. Farley, C.T.; Gonzalez, O. Leg stiffness and stride frequency in human running. *J. Biomech.* **1996**, *29*, 181–186. [CrossRef] [PubMed]
23. Wolf, S.; Eiberger, O.; Hirzinger, G. The DLR FSJ: Energy based design of a variable stiffness joint. In Proceedings of the 2011 IEEE international conference on robotics and automation, Shanghai, China, 9–13 May 2011.
24. Radford, N.A.; Strawser, P.; Hambuchen, K.; Mehling, J.S.; Verdeyen, W.K.; Donnan, A.S.; Holley, J.; Sanchez, J.; Nguyen, V.; Bridgwater, L.; et al. Valkyrie: Nasa's first bipedal humanoid robot. *J. Field Robot.* **2015**, *32*, 397–419. [CrossRef]
25. Grizzle, J.W.; Hurst, J.; Morris, B.; Park, H.W.; Sreenath, K. MABEL, a new robotic bipedal walker and runner. In Proceedings of the 2009 American Control Conference, St. Louis, MO, USA, 10–12 June 2009.
26. Radkhah, K.; Lens, T.; von Stryk, O. Detailed dynamics modeling of BioBiped's monoarticular and biarticular tendon-driven actuation system. In Proceedings of the 2012 IEEE/RSJ International Conference on Intelligent Robots and Systems, Vilamoura-Algarve, Portugal, 7–12 October 2012.
27. Loeffl, F.; Werner, A.; Lakatos, D.; Reinecke, J.; Wolf, S.; Burger, R.; Gumpert, T.; Schmidt, F.; Ott, C.; Grebenstein, M.; et al. The dlr c-runner: Concept, design and experiments. In Proceedings of the 2016 IEEE-RAS 16th International Conference on Humanoid Robots (Humanoids), Cancun, Mexico, 15–17 November 2016.
28. Kajita, S.; Kanehiro, F.; Kaneko, K.; Yokoi, K.; Hirukawa, H. The 3D linear inverted pendulum mode: A simple modeling for a biped walking pattern generation. In Proceedings of the 2001 IEEE/RSJ International Conference on Intelligent Robots and Systems. Expanding the Societal Role of Robotics in the the Next Millennium (Cat. No. 01CH37180), Maui, HI, USA, 29 October–3 November 2001; Volume 1.
29. Kajita, S.; Kanehiro, F.; Kaneko, K.; Fujiwara, K.; Harada, K.; Yokoi, K.; Hirukawa, H. Biped walking pattern generation by using preview control of zero-moment point. In Proceedings of the 2003 IEEE international conference on robotics and automation (Cat. No. 03CH37422), Taipei, Taiwan, 14–19 September 2003; Volume 2.
30. Park, I.-W.; Kim, J.-Y.; Oh, J.-H. Online biped walking pattern generation for humanoid robot khr-3(kaist humanoid robot-3: Hubo). In Proceedings of the 2006 6th IEEE-RAS International Conference on Humanoid Robots, Genova, Italy, 4–6 December 2006; pp. 398–403.
31. Takenaka, T.; Matsumoto, T.; Yoshiike, T. Real time motion generation and control for biped robot-1st report: Walking gait pattern generation. In Proceedings of the 2009 IEEE/RSJ International Conference on Intelligent Robots and Systems, St. Louis, MO, USA, 10–15 October 2009.
32. Takenaka, T.; Matsumoto, T.; Yoshiike, T.; Shirokura, S. Real time motion generation and control for biped robot-2nd report: Running gait pattern generation. In Proceedings of the 2009 IEEE/RSJ International Conference on Intelligent Robots and Systems, St. Louis, MO, USA, 10–15 October 2009.

33. Takenaka, T.; Matsumoto, T.; Yoshiike, T. Real time motion generation and control for biped robot-3rd report: Dynamics error compensation. In Proceedings of the 2009 IEEE/RSJ International Conference on Intelligent Robots and Systems, St. Louis, MO, USA, 10–15 October 2009.
34. Englsberger, J.; Ott, C.; Roa, M.A.; Albu-Schäffer, A.; Hirzinger, G. Bipedal walking control based on capture point dynamics. In Proceedings of the 2011 IEEE/RSJ International Conference on Intelligent Robots and Systems, St. Louis, MO, USA, 10–15 October 2011.
35. Englsberger, J.; Ott, C.; Albu-Schäffer, A. Three-dimensional bipedal walking control using divergent component of motion. In Proceedings of the 2013 IEEE/RSJ International Conference on Intelligent Robots and Systems, Tokyo, Japan, 3–7 November 2013; pp. 2600–2607.
36. Hopkins, M.A.; Hong, D.W.; Leonessa, A. Humanoid locomotion on uneven terrain using the time-varying divergent component of motion. In Proceedings of the 2014 IEEE-RAS International Conference on Humanoid Robots, Madrid, Spain, 18–20 November 2014; pp. 266–272.
37. Kajita, S.; Benallegue, M.; Cisneros, R.; Sakaguchi, T.; Nakaoka, S.; Morisawa, M.; Kaneko, K.; Kanehiro, F. Biped walking pattern generation based on spatially quantized dynamics. In Proceedings of the 2017 IEEE-RAS 17th International Conference on Humanoid Robotics (Humanoids), Birmingham, UK, 15–17 November 2017; pp. 599–605.
38. Caron, S.; Escande, A.; Lanari, L.; Mallein, B. Capturability based pattern generation for walking with variable height. *IEEE Trans. Robot.* **2020**, *36*, 517–536. [CrossRef]
39. Tazaki, Y.; Hanasaki, S.; Yukizaki, S.; Mitazono, Y.; Nagano, H.; Yokokohji, Y. A continuous-time walking pattern generator for realizing seamless transition between flat-contact and heel-to-toe walking. *Adv. Robot.* **2023**, *37*, 316–328. [CrossRef]
40. Geyer, H.; Seyfarth, A.; Blickhan, R. Spring-mass running: Simple approximate solution and application to gait stability. *J. Theor. Biol.* **2005**, *232*, 315–328. [CrossRef] [PubMed]
41. Wensing, P.M.; Orin, D.E. High-speed humanoid running through control with a 3D-SLIP model. In Proceedings of the 2013 IEEE/RSJ International Conference on Intelligent Robots and Systems, Tokyo, Japan, 3–7 November 2013.
42. Kuo, C.-Y.; Shin, H.; Kamioka, T.; Matsubara, T. TDE2-MBRL: Energy-exchange Dynamics Learning with Task Decomposition for Spring-loaded Bipedal Robot Locomotion. In Proceedings of the 2022 IEEE-RAS 21st International Conference on Humanoid Robots (Humanoids), Ginowan, Japan, 28–30 November 2022; pp. 550–557. [CrossRef]
43. Takenaka, T.; Matsumoto, T.; Yoshiike, T.; Hasegawa, T.; Shirokura, S.; Kaneko, H.; Orita, A. Real time motion generation and control for biped robot-4th report: Integrated balance control. In Proceedings of the 2009 IEEE/RSJ International Conference on Intelligent Robots and Systems, St. Louis, MO, USA, 10–15 October 2009.
44. Henze, B.; Dietrich, A.; Ott, C. An approach to combine balancing with hierarchical whole-body control for legged humanoid robots. *IEEE Robot. Autom. Lett.* **2015**, *1*, 700–707. [CrossRef]
45. Henze, B.; Roa, M.A.; Ott, C. Passivity-based whole-body balancing for torque-controlled humanoid robots in multi-contact scenarios. *Int. J. Robot. Res.* **2016**, *35*, 1522–1543. [CrossRef]
46. Mesesan, G.; Englsberger, J.; Henze, B.; Ott, C. Dynamic multi-contact transitions for humanoid robots using divergent component of motion. In Proceedings of the 2017 IEEE International Conference on Robotics and Automation (ICRA), Singapore, 29 May–3 June 2017.
47. Henze, B.; Balachandran, R.; Roa-Garzon, M.A.; Ott, C.; Albu-Schäffer, A. Passivity analysis and control of humanoid robots on movable ground. *IEEE Robot. Autom. Lett.* **2018**, *3*, 3457–3464. [CrossRef]
48. Ding, J.; Xiao, X. Two-stage optimization for energy-efficient bipedal walking. *J. Mech. Sci. Technol.* **2020**, *34*, 3833–3844. [CrossRef]
49. Liu, C.; Zhang, T.; Liu, M.; Chen, Q. Active balance control of humanoid locomotion based on foot position compensation. *J. Bionic Eng.* **2020**, *17*, 134–147. [CrossRef]
50. Reher, J.P.; Hereid, A.; Kolathaya, S.; Hubicki, C.M.; Ames, A.D. Algorithmic foundations of realizing multi-contact locomotion on the humanoid robot DURUS. In *Algorithmic Foundations of Robotics XII: Proceedings of the Twelfth Workshop on the Algorithmic Foundations of Robotics*; Springer International Publishing: Cham, Switzerland, 2020.
51. Orozco-Soto, S.M.; Ibarra-Zannatha, J.M.; Kheddar, A. Gait Synthesis and Biped Locomotion Control of the HRP-4 Humanoid. In Proceedings of the 2021 XXIII Robotics Mexican Congress (ComRob), Tijuana, Mexico, 27–29 October 2021.
52. Daneshmand, E.; Khadiv, M.; Grimminger, F.; Righetti, L. Variable horizon mpc with swing foot dynamics for bipedal walking control. *IEEE Robot. Autom. Lett.* **2021**, *6*, 2349–2356. [CrossRef]
53. García, G.; Griffin, R.; Pratt, J. MPC-based locomotion control of bipedal robots with line-feet contact using centroidal dynamics. In Proceedings of the 2020 IEEE-RAS 20th International Conference on Humanoid Robots (Humanoids), Munich, Germany, 19–21 July 2021.
54. Galliker, M.Y.; Csomay-Shanklin, N.; Grandia, R.; Taylor, A.J.; Farshidian, F.; Hutter, M.; Ames, A.D. Planar Bipedal Locomotion with Nonlinear Model Predictive Control: Online Gait Generation using Whole-Body Dynamics. In Proceedings of the 2022 IEEE-RAS 21st International Conference on Humanoid Robots (Humanoids), Ginowan, Japan, 28–30 November 2022.
55. Villa, N.A.; Fernbach, P.; Naveau, M.; Saurel, G.; Dantec, E.; Mansard, N.; Stasse, O. Torque Controlled Locomotion of a Biped Robot with Link Flexibility. In Proceedings of the 2022 IEEE-RAS 21st International Conference on Humanoid Robots (Humanoids), Ginowan, Japan, 28–30 November 2022.

56. Dantec, E.; Naveau, M.; Fernbach, P.; Villa, N.; Saurel, G.; Stasse, O.; Taïx, M.; Mansard, N. Whole-Body Model Predictive Control for Biped Locomotion on a Torque-Controlled Humanoid Robot. In Proceedings of the 2022 IEEE-RAS 21st International Conference on Humanoid Robots (Humanoids), Ginowan, Japan, 28–30 November 2022.
57. Ramuzat, N.; Boria, S.; Stasse, O. Passive inverse dynamics control using a global energy tank for torque-controlled humanoid robots in multi-contact. *IEEE Robot. Autom. Lett.* **2022**, *7*, 2787–2794. [CrossRef]

robotics

MDPI

Article

A Dynamic Approach to Low-Cost Design, Development, and Computational Simulation of a 12DoF Quadruped Robot

Md. Hasibur Rahman [1,2], Saadia Binte Alam [1,2,*], Trisha Das Mou [1,2], Mohammad Faisal Uddin [1,2] and Mahady Hasan [1,2]

1 Department of Computer Science Engineering, IUB, Dhaka 1229, Bangladesh
2 RIoT Research Center, IUB, Dhaka 1229, Bangladesh
* Correspondence: saadiabinte@iub.edu.bd

Abstract: Robots equipped with legs have significant potential for real-world applications. Many industries, including those concerned with instruction, aid, security, and surveillance, have shown interest in legged robots. However, these robots are typically incredibly complicated and expensive to purchase. Iron Dog Mini is a low-cost, easily replicated, and modular quadruped robot built for training, security, and surveillance. To keep the price low and its upkeep simple, we designed our quadruped robot in a modular manner. We provide a comparative study of robotic manufacturing cost between our proposed robot and previously established robots. We were able to create a compact femur and tibia structure with sufficient load-bearing capacity. To improve stability and motion efficiency, we considered the novel Watt six-bar linkage mechanism. Using the SolidWorks modeling software, we analyzed the structural integrity of the robot's components, considering their respective material properties. Furthermore, our research involved developing URDF data for our quadruped robot based on its CAD model. Its gait trajectory is planned using a 14-point Bezier curve. We demonstrate the operation of the simulation model and briefly discuss the robot's kinematics. Computational methods are emphasized in this research, coupled with the simulation of kinematic and dynamic performances and analytical/numerical modeling.

Keywords: kinematics; simulations; quadruped robot; watt six-bar linkage; PyBullet; URDF; 12DoF

Citation: Rahman, M.H.; Alam, S.B.; Mou, T.D.; Uddin, M.F.; Hasan, M. A Dynamic Approach to Low-Cost Design, Development, and Computational Simulation of a 12DoF Quadruped Robot. *Robotics* **2023**, *12*, 28. https://doi.org/10.3390/robotics12010028

Academic Editor: Raffaele Di Gregorio

Received: 16 December 2022
Revised: 29 January 2023
Accepted: 30 January 2023
Published: 17 February 2023

1. Introduction

Numerous quadrupedal robots have been developed in recent decades. The potential of mobile robots to replace people in high-risk situations has made this a fascinating area of robotics study [1], including first response, flammable and toxic substance disposal, and lidar field navigation. Most remotely operated robots fall into three categories: Those with wheels, those with a crawler-type under-carriage, and those with legs [2]. Although mobile robots that move on wheels or crawlers may be effective on the ground level, their mechanisms are severely limited by obstacles such as hills and mountains. Legged robots have a wider variety of potential applications, as they can traverse difficult terrain [3]. The number of feet may be used to categorize robots into three groups: Bipeds, quadrupeds, and multi-legged. More people are interested in quadruped robots than in the past, as they are more stable, can carry more weight than biped robots, have better mobility performance than multi-legged robots, and are more efficient at moving than the aforementioned robots [4]. Rapid and accurate evaluation of dynamic characteristics is essential for precise modeling, estimation, and control of robots. Scientists in the field of robotics can greatly benefit from having access to techniques that can reduce error and speed up the development process. In this study, we provide a computationally based, organized simulation model run in the Pybullet physics engine, in order to address a wide variety of issues that arise when constructing quadruped robots. Our research here describes the connection between a kinematics equation and the characteristics of the joints of a four-legged robot in depth. The robot's design features legs with three degrees of freedom. in this study, we emphasize the use of computational methods, such as modeling and simulation of kinematic and practical implementations.

Bipedal and quadrupedal robots are incredibly well-known for their complex, sophisticated mechanics and mathematical techniques, although robotics researchers deal with a wide variety of robots at present. Parts of the quadruped robot are relevant to creatures with four legs. Simulating a high-quality, mechanically constructed, and hydraulically operated quadruped robot requires the employment of at least twelve degrees of freedom [5]. Scientists have previously developed robot engine controls for a quadruped robot [6]. Implementing a four-legged robot relies heavily on simulations run in software. This research suggested a simulation model with outlined components and parameters, such as kinematics, designs, and a standard robot description format. To simulate the motion of a robot's 2DoF leg, kinematics solutions have been implemented in a purpose-built 3D program [7]. One research highlight was demonstrating a novel robot design using electrically operated motors [8]. Additionally, most quadruped robots, such as wildcat [9], use a two-joint construction for their primary leg component. This framework is uncomplicated, understandable, and straightforward to manage. However, there are significant biological advantages to the three-joint limb structure of toed animals, such as cats, dogs, and lions, regarding walking velocity. MIT Cheetah utilized a three-segment construction that allowed for a running velocity of 6 m/s and the efficient passage of obstacles [10]. Additionally, the Cheetah-cub was created with a pantograph leg arrangement, in order to simplify controlling three links with only two joints and allowing it to achieve a trot [11]. Additionally, Pneupard—a genuine "cat-sized" robot—has utilized the same pantograph technique [12]. It has been shown that placing motors at the shoulder works well for high-velocity locomotion. Ming Lu has highlighted a 2DoF-based parallel leg formation [13]. A hybrid-legged wheeled robot, which uses a wheel and leg for walking, has been described in [14]. Kinematic analysis can be used as the determining element for quadruped robots, and recent studies [15,16] have detailed exquisite mathematical kinematics techniques. A motion observer study for a four-legged robot capable of traversing rugged terrain has been carried out [17]. Gait pattern creation is critical for quadruped robots, as they require precise trajectory adjustment. Sooyeong Yi has described the two-phase discontinuous gaits of quadruped walking robots [18]. Thanhtam Ho has created a biomimetic self-contained quadruped bounding robot [19].

The researcher gave an overview of a quadruped robot model developed in a dynamic simulator with an alternate gait creation mechanism [20]. Through structural simulation, one of the essential load test analyses of a robot leg has been detailed [21]. Knowing how much weight it can withstand and how long it can withstand pressure is beneficial. The researcher demonstrated a variable-based design for a quadruped robot using the equivalent motion, a validation approach, and a non-programmable method [22–24]. Previously, researchers have employed computer modeling and analysis before constructing a massive robot, thus reducing instrument losses and making the model construction process faster. MIT researchers have used 3D design control to demonstrate improved design and advancement [25].

In SolidWorks, we created a working prototype of a four-legged robot with precise joint measurements, specifications, and dimensions. Initially, we intended to construct our prototype using standard methods, where the servomotors are usually in joint areas like the coxa–femur joint and femur–tibia joint; however, the size and weight of our servomotors became assembly limiting considerations. Due to the poor weight distribution in the body and joints of our quadruped robot, the design was precarious. To address this problem, we built a Watt six-bar linkage mechanism with all servomotors in the coxa joint region and linkages connecting the femur and tibia to their respective servomotors. This structure allowed us to improve the weight distribution. Figure 1 shows the quadruped robot iron dog mini. Table 1 shows some feature comparisons between the existing and proposed robots.

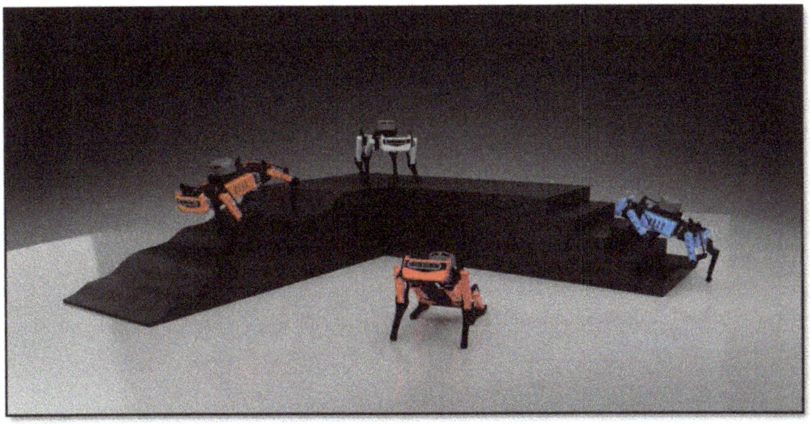

Figure 1. Iron Dog mini, a low-cost quadruped robot with Watt six-bar linkage mechanism.

Table 1. Comparison of existing models with the proposed model. The table includes the technology adoption, features, and mechanisms of previous and current research based on quadruped robots.

Robot	Year	State-of-the-Art Technology Adoption							
		Light Weight	**Heavy Weight**	**Carry Load**	**Low Cost**	**Modular**	**Power Efficient**	**Mechanism**	**Agile**
Sony Aibo	1999	✓	-	-	-	-	✓	Electric	-
BigDog [26]	2005	-	✓	✓	-	-		Hydraulic	✓
Scalf1 [6]	2011	-	✓	✓	-	-	-	Hydraulic	-
Frog [27]	2013	-	✓	✓	-	-	-	-	-
Alpha Dog [28]	2012	-	✓	✓	-	-	-	Hydraulic	✓
HyQ [29]	2011	-	✓	✓	-	-	-	Hydraulic and Electric	✓
Baby Elephant [30]	2013	-	✓	✓	-	-	-	Serial–parallel Hybrid	✓
AnyMal [31]	2016	-	✓	✓	-	-	✓	Electric	✓
Spot [32]	2017	-	✓	✓	-	✓	✓	Electric	✓
MIT Cheetah 3 [25]	2018	-	✓	✓	-	-	✓	Electric	✓
Unitree Laikago	2017	-	✓	✓	-	✓	✓	Electric	✓
Stoch 2 [33]	2019	-	✓	-	-	-	-	Electric-five bar linkage	✓
Proposed	2022	✓	-	✓	✓	✓	✓	Electric-Six bar Linkage	✓

The contributions and motivation of this work are described below:

- We introduce the novel Watt six-bar linkage mechanism for better walking motion.
- The motivation of this research was to develop a low-cost and modular quadruped assistive robot platform for use in security and surveillance operations.
- We innovatively designed robot parts to make it modular, such that users can quickly assemble and disassemble the robot.
- We explain the kinematic equations, demonstrate the URDF process, and test several commands in the PyBullet physics engine.
- We discuss the material characteristics and structural analysis of the robot's parts.

2. Design Principle

Most quadruped robots have 8–16 degrees of freedom (DoF). While these are the most common, there are various other viable architectural options. The axis of all eight joints (four hips, four knees) is parallel to itself, making up all eight degrees of freedom. Although an 8-DoF robot lacks the hip joint's transverse swing flexibility, they are nonetheless easy to control and capable of fast forward and backward motion. However, the motion performance of 8-DoF quadruped robots is hindered by their poor steering capabilities and inability to carry out transverse motion. A quadruped robot with 16 DoF has more joints for agile maneuvering, but is more complicated and, hence, more difficult to control. To allow for three rotatable joints in each leg, we developed a 12 DoF design. The notion of a four-legged robot was mostly inspired by domesticated feline and canine animals, which are distinguished from other domestic species by the presence of an endoskeleton, allowing for greater movement. The development of a quadruped robot with 12 degrees of freedom is a key part of our study. Our intricate quadruped robot is made up of many different elements with rotating joints. The coxa, femur, and tibia are the most vital components of the proposed quadruped robot.

A rod end, servo arm, rod end linker, and servo horn are also included in the design. Three typical bones make up each leg: The coxa (hip bone), femur (thigh bone), and tibia (shin bone). Twelve servo actuator motors are employed to translate the links and modify the joints. The different parts of robot are shown in Figure 2.

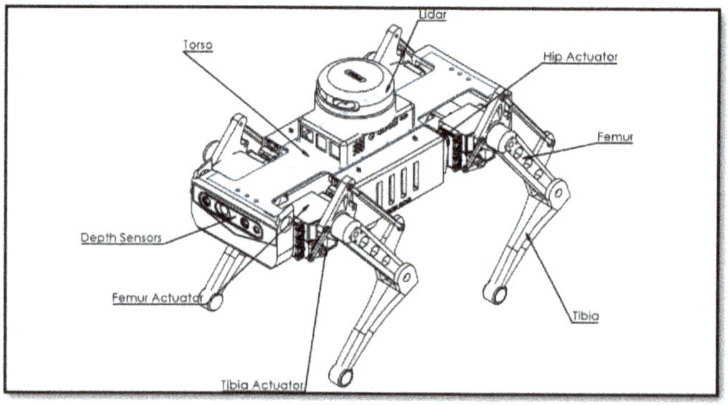

Figure 2. Different essential parts of the quadruped robot equipped with six-bar linkage mechanism.

The locomotion of the quadruped robot depends on the translations and rotations of the coxa joint, femur, and tibia. The kinematic analysis defines the motion of the quadruped robot's legs, where we can correlate the forward and inverse kinematics. The objective of this study is to create a quadruped robot with a body made entirely of 3D-printed parts and reasonably priced servo motors. Given its size and capabilities, the robot has been built to be able to carry a reasonable payload. As such, we designed the robot's torso in such a way that we additional components can be mounted on our robot. Our design principle allows the quadruped robot to be modular, such that the user can easily replace any broken or defective components without replacing the whole robot, massively reducing the affordability and maintenance costs of our quadruped robot. Our modular-based design is shown in Figure 3. Table 2 shows essential parameters of Iron dog mini.

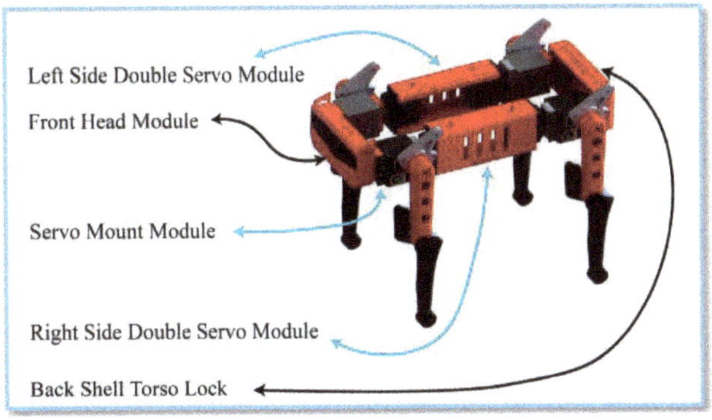

Figure 3. Different modules of Iron dog mini.

Table 2. Essential Parameters of Iron Dog mini, including the height, weight, and mass values of different links.

Parameter	Value				Total Mass (g)	Parameter	Value				Total Mass (g)
	Height (cm)	Width (cm)	Number of Uses	Single Part Mass (g)			Height (cm)	Width (cm)	Number of Uses	Single Part Mass (g)	
Base	30	6	2	92	184	Servo Arm	2.4	0.63	4	5	20
Femur	12	2.5	4	55	220	Long rod end	4.5	0.25	4	7	28
Tibia	17	2.5	4	48	192	Short Rod end	1.2	0.2	4	1.8	7.2
Actuator	5.6	2	12	80	960	Front Head Module	7	12	1	55	55
Rolling Servo mount	4	2	4	10	40	Back Shell For torso lock	7	12	1	48	48
Side Servo Mount	4	2	4	8	32	Side pitch mount (double Servo Module)	12	5.6	2	120	240
Servo Horn	0.14	0.025	4	6	24	Screws	0.3	0.05	74	1	74
Free linker	6.32	4.60	4	12	48	Battery	-	-	1	250	250
Other Parts	-	-	-	150	150	Total Weight of Robot After Fully Mounted				2572.2 g	

3. Working Mechanism

Iron dog mini consists of 12 servo motors that control the joint angles at the coxa, femur, and tibia. Figure 2 shows detailed information on the locations and orientations of the servo motors. Our design provides better weight balance, as we concentrated the weight close to the four corners of the robot's main body by locating the three servo motors of a single leg at the hip joint area. Actuator 1 controls the joint angle of the coxa. The hip of the robot is directly coupled with the servo motor, where the hip joint helps to stabilize the robot during motion.

Actuator 2 controls the Femur, and the femur link is directly coupled with the servo motor. Actuator 3 controls the Tibia, but the Tibia joint is not directly coupled with the servo motor. To control the Tibia link, we introduced the Watt six-bar linkage mechanism into our design, which is a unique approach. Conventional mechanisms such as four-bar linkage mechanisms have some motion constraints due to their limited design variables. We have pointed out some advantages of the six-bar linkage mechanism over the four-bar linkage mechanism.

- The watt six-bar linkage mechanism provides a greater range of motion for leg actuation than the four-bar linkage mechanism.

- The watt six-bar linkage mechanism produces leg motion during gait generation, which is very close to the leg motion of a four-legged animal compared to the four-bar linkage mechanism.
- The four-bar linkage mechanism has many motion constraints. Therefore, the four-bar linkage mechanism robot has a limited range of motion for its leg. The four-bar linkage has a total of eight design variables.
- The watt six-bar linkage mechanism has fourteen design variables.
- The six-bar linkage has more motion parameters than the four-bar linkage, increasing the range of motion.

Additionally, the gait generation produced by the watt six-bar linkage mechanism is found to be highly comparable to that of a four-legged animal, making it an ideal candidate for use in quadrupedal robots. Therefore, the six-bar linkage is a more convenient mechanism for the leg actuation of quadruped robots.

We designed and optimized the six-bar linkage mechanism to manipulate the tibia link effectively, and the bar linkages are shown in Figure 4. The Watt six-bar linkage mechanism has more stability, adequate movement, and better motion efficiency than conventional four- and five-bar linkage mechanisms. The actuator that is used has a significant impact on the performance of a quadruped robot's movement. Considering the robot's weight, the actuator must provide significant torque while maintaining a rapid response time and a compact footprint. Brushless motors are widely used, due to their excellent dynamic qualities; however, they are typically more costly and more extensive in size. Thus, we used a high-voltage servo motor which has a built-in gear reducer. Moreover, we used a metal gear servo motor, providing 35 kg torque per centimeter. The single-leg configuration is shown in Figure 5.

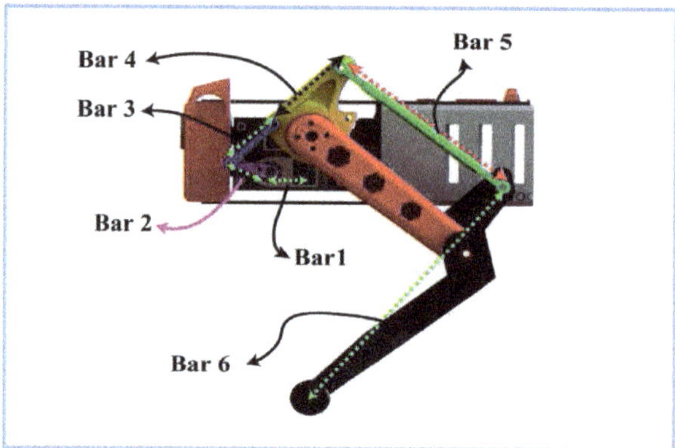

Figure 4. Watt Six-Bar linkage mechanism.

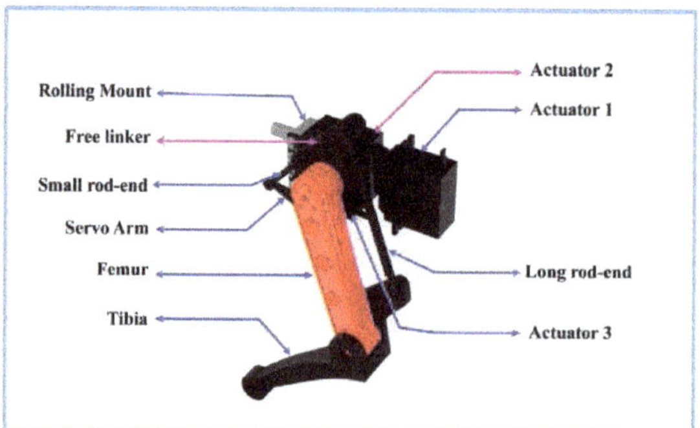

Figure 5. Single-leg configuration.

3.1. Mathematical Analysis

Quadruped robots use two types of kinematics: Forward and inverse kinematics. Robot end-effector joint positions are determined using forward kinematics, whereas manipulator joint values can be derived from inverse kinematics. In order to calculate the forward kinematics, we multiply the number of matrices. To better understand how forward kinematics operate in quadruped robots, we constructed a block-based dummy quadruped robot and figured out all the frames. In our previous work, we have demonstrated a comprehensive derivation of the true nature of forward kinematics [15]. Figure 6 depicts all the various frames and coordinate systems in detail. Additionally, Figure 6 was used to reach the foot frame, which serves as the illustration's final effector. The zeroth frame and hip joints represent the coxa coordinate X_0, Y_0, Z_0. To find the forward kinematics equation, we need to multiply each frame from zeroth to the fourth. If we want to reach the zeroth frame from the first frame, there will be a translation, denoted by L_1. When reaching the first frame from the second frame, there is no translation but, instead, a rotation, denoted by θ_1. Passing from frame two to frame three involves L_2 and θ_2 for translation and rotation, respectively. Finally, we have L_3 translation and θ_3 rotation between the third and fourth frames. After multiple translations and rotations, we obtain the final transformation matrix, according to Denavit Hartenberg's convention. We can determine the matrix representing all possible transformations between frames 0 and 4 [15] as $T_0^4 = T_0^1 * T_1^2 * T_2^3 * T_3^4$:

$$T_0^4 = \begin{bmatrix} \phi_{11} & \phi_{12} & \phi_{13} & \phi_{14} \\ \phi_{21} & \phi_{22} & \phi_{23} & \phi_{24} \\ \phi_{31} & \phi_{32} & \phi_{33} & \phi_{34} \\ \phi_{41} & \phi_{42} & \phi_{43} & \phi_{44} \end{bmatrix}. \quad (1)$$

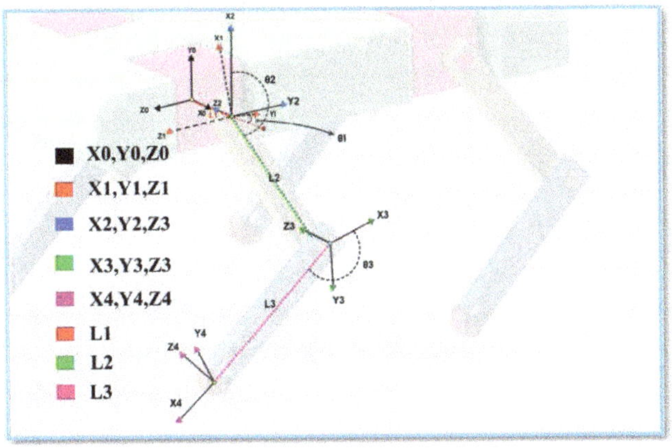

Figure 6. Frame-by-frame coordinate system of a quadruped robot using 3D design.

From this transformation matrix, our end effector value can be determined from the fourth column (i.e., ϕ_{14}, ϕ_{24}, and ϕ_{34}). After the forward kinematics are established, inverse kinematics are required to regulate the joint settings. We sketched through the actual model for better understanding.

Based on Figure 7,

$$\theta_1 = \alpha_3 - \alpha_1 \,. \quad (2)$$

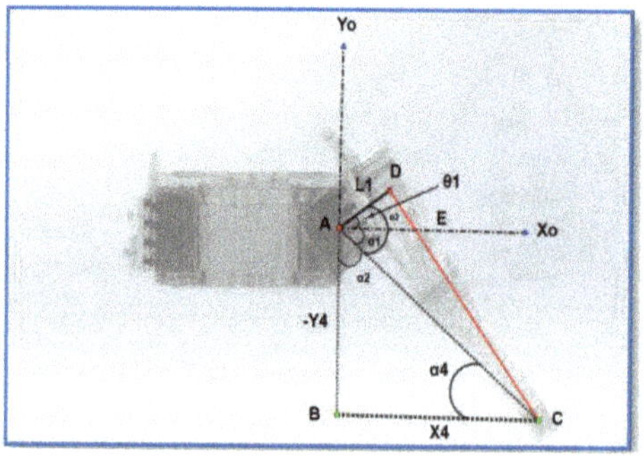

Figure 7. Determining θ_1 utilizing the front view of the robot leg.

In $\triangle$ADC,

$$\alpha_3 = \arctan\left(\frac{\sqrt{x_4^2 + y_4^2 - L_1^2}}{L_1}\right). \tag{3}$$

In $\square$ABCE,

$$\alpha_1 + \alpha_2 = 90^\circ \tag{4}$$

$$\Rightarrow \alpha_1 = 90^\circ - \alpha_2. \tag{5}$$

Furthermore, in $\triangle$ABC,

$$\alpha_2 + \alpha_4 + \angle B = 180^\circ, \tag{6}$$

$$\alpha_2 + \alpha_4 = 180^\circ - 90^\circ\ [\angle B = 90^\circ], \tag{7}$$

$$\alpha_2 = 90^\circ - \alpha_4. \tag{8}$$

Substituting this value into Equation (5) :

$$\alpha_1 = 90^\circ - (90^\circ - \alpha_4) \tag{9}$$

$$\Rightarrow \alpha_1 = \alpha_4. \tag{10}$$

In $\triangle$ABC,

$$\alpha_4 = arctan\left(\frac{-y_4}{x_4}\right). \tag{11}$$

Now, substituting the values of α_1 and α_3 into Equation (2),

$$\theta_1 = \alpha_3 - \alpha_1 \tag{12}$$

$$\Rightarrow \theta_1 = arctan\left(\frac{\sqrt{x_4^2 + y_4^2 - L_1^2}}{11}\right) - arctan\left(\frac{-y_4}{x_4}\right). \tag{13}$$

Based on Figure 8,

$$\theta_2 = -90^\circ + \alpha_1. \tag{14}$$

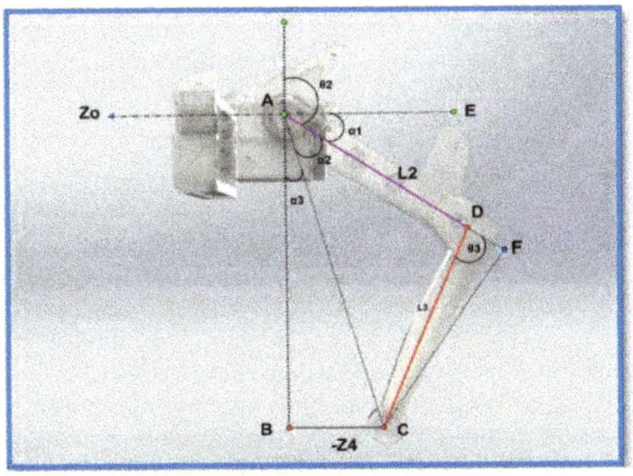

Figure 8. Determining θ_2 utilizing the side view.

From □ABFE,

$$\alpha_1 + \alpha_2 + \alpha_3 = 90^\circ \tag{15}$$

$$\therefore \alpha_1 = 90^\circ - \alpha_2 - \alpha_3. \tag{16}$$

From △ABC,

$$\alpha_3 = \arctan\left(\frac{-Z_4}{\sqrt{x_4^2 + y_4^2 - L1^2}}\right). \tag{17}$$

From $\triangle ACF$,

$$\alpha_2 = \arctan\left(\frac{CF}{AF}\right). \tag{18}$$

Furthermore, from $\triangle CDF$,

$$\sin\theta_3 = \frac{CF}{CD}, \tag{19}$$

$$CF = CD\sin\theta_3, \tag{20}$$

$$CF = L_3\sin\theta_3\ [CD = L_3], \tag{21}$$

$$\cos\theta_3 = \frac{DF}{CD}, \tag{22}$$

$$DF = L_3\cos\theta_3\ [CD = L_3], \tag{23}$$

$$AF = AD + DF, \tag{24}$$

$$AF = L_2 + L_3\cos\theta_3[AD = L_2]. \tag{25}$$

Similarly,

$$\therefore\ \alpha_2 = \arctan\left(\frac{L_3\sin\theta_3}{L_2 + L_3\cos\theta_3}\right). \tag{26}$$

Now, substituting the values of α_2 and α_3 into Equation (16),

$$\alpha_1 = 90^\circ - \arctan\left(\frac{L_3\sin\theta_3}{L_2 + L_3\cos\theta_3}\right) - \arctan\left(\frac{-Z_4}{\sqrt{x_4 + y_4{}^2 - L_1^2}}\right), \tag{27}$$

$$\theta_2 = -90^\circ + \alpha_1 \tag{28}$$

$$\Rightarrow \theta_2 = -\arctan\left(\frac{L_3\sin\theta_3}{L_2 + L_3\cos\theta_3}\right) - \arctan\left(\frac{-z_4}{\sqrt{x_4^2 + y_4^2 - L_1}}\right). \tag{29}$$

Based on Figure 9,

$$\theta_3 = 180° - \alpha. \tag{30}$$

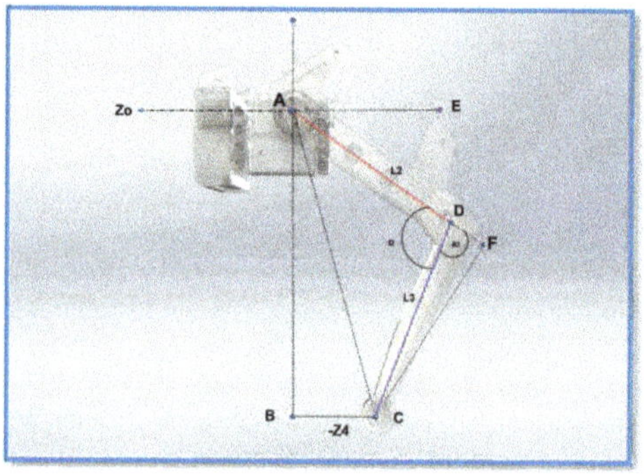

Figure 9. Determining θ_3.

In $\triangle$ACD,

$$AC^2 = AD^2 + CD^2 - 2 \cdot AD \cdot CD \cdot \cos\alpha \tag{31}$$

$$\Rightarrow \alpha = \arccos\left(\frac{AD^2 + CD^2 - AC^2}{2 \cdot AD \cdot CD}\right). \tag{32}$$

In $\triangle$ABC,

$$AC^2 = AB^2 + BC^2, \tag{33}$$

$$AC^2 = \left(\sqrt{X_4^2 + y_4^2 + L_1^2}\right)^2 + (-z_4)^2, \tag{34}$$

$$AC^2 = x_4^2 + y_4^2 + L_1^2 + z_4^2. \tag{35}$$

Now, substituting the value of AC^2 into Equation (32),

$$\alpha = \arccos\left(\frac{L_2^2 + L_3^2 - X_4^2 - Y_4^2 - L_1^2 - z_4^2}{2L_2L_3}\right)\left[\begin{array}{c} AD = L_2 \\ CD = L_3 \end{array}\right], \tag{36}$$

$$\theta_3 = 180° - \arccos\left(\frac{L_2^2 + L_3^2 - x_4^2 - y_4^2 - L_1^2 - z_4}{2L_2L_3}\right). \tag{37}$$

In summary, in the derivation process of θ_1, θ_2 and θ_3 in the inverse kinematic analysis, we took the left front leg of our quadruped robot into consideration. We positioned the robot's leg to mimic its normal standing posture (side view). Using the front view of our leg, we positioned the coxa angle, which mirrors the roll position of the robot, allowing us to identify θ_1. From the two right-angle triangles ABC and ADC in Figure 7, we were able to obtain the equation for θ_1 that we required. Using the profile view of our leg in Figure 8, we determined θ_2. Two right-angle triangles, ABC and CDF, and one obtuse triangle, and ADC, are shown in Figure 8, and from these triangles we derived the desired equation for θ_2. Using the profile of our leg, we were able to compute θ_3 from Figure 9. Figure 9 depicts two triangles, one with a right-angle (ABC) and another with an acute angle (ADC), from which our desired equation for θ_3 was derived.

$$\theta_1 = \text{arctan2}\left(\left(\sqrt{x_4^2 + y_4^2 - L_1^2}\right), L_1\right) - \text{arctan2}(-y_4, x_4), \tag{38}$$

$$\theta_2 = -\text{arctan2}(L_3\sin(\theta_3), (L_2 + L_3\cos(\theta_3))) - arctan2\left(-z_4, \sqrt{x_4^2 + y_4^2 - L_1^2}\right), \tag{39}$$

$$\theta_3 = \pi - \arccos\left(\frac{L_2^2 + L_3^2 - x_4^2 - y_4^2 - L_1^2 - Z_4^2}{2L_2L_3}\right). \quad (40)$$

3.2. Simulation Model Workflow

Workflow Procedure

Next, we describe the computational method used to simulate our quadruped robots. We used the Python-based Pybullet physics engine in the Ubuntu operating system for this scientific simulation following multiple stages, as presented in Figure 10. Parts were created in the SolidWorks software using precise measurements.

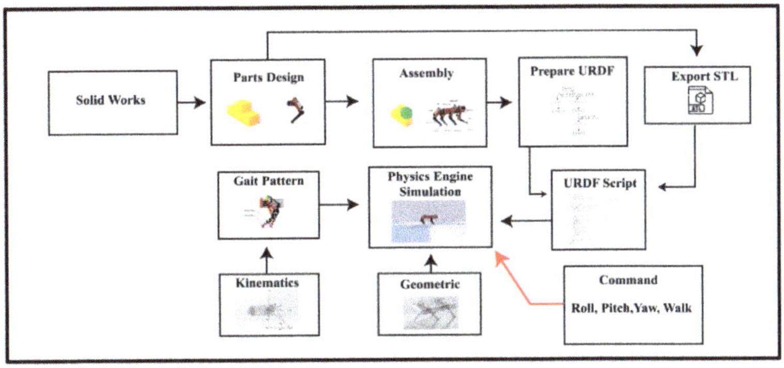

Figure 10. Workflow diagram for the proposed simulation model.

A great deal of thought and care must go into the design of each component before it can be assembled into a functioning robot. There are several essential parts for robots. About 64 unique parts were conceived for this study. Each of the robot's legs consists of three separate parts (the coxa, femur, and tibia), and the design is based on the principle of three degrees of freedom. Throughout the entirety of the design phase, there was a continuous process of considering the design's framework in addition to the materials that were to be utilized.

Certain tests, such as finite element modeling (FEM), finite element analysis (FoS), and strain, must be performed on the robot's core components to ascertain their durability and the maximum payload the design can withstand. In Section 4.1, we summarize relevant information in this aspect.

After making all the pieces, they must be put together in a way that allows the robot to operate well. As there are three movable and one fixed joint, each joint must be mated with extreme precision. The six-bar linkages were created to generate exterior revolute joints in addition to the leg joints. A rod-end and actuator were utilized to rotate the desired joint in the six-bar linkages.

For the simulation process, creating a URDF script was very much necessary. URDF contains all of the robot data, such as link length, link mass, material, inertia, geometry, and the path of the robot STL file. STL files are exported for every part of the robot, which contain the 3-dimensional surface geometry of the parts. This format is suitable for use with the PyBullet physics engine.

After that, we had to link our robot's gait pattern to the kinematics algorithm. The various link parameters and joint angles were contained in the Kinematics Algorithm, which provides 3D foot positions for us. We tested several gait patterns. To develop a walking motion for a quadruped robot, kinematics is insufficient. Gait generation is required to operate the robot in both real and virtual simulations. There are various gait patterns available, including the creep, trot, pace, cantor, and gallop gaits. We created a creep gait, trot gait, and pace gait for our simulation investigation, through investigating the walking patterns of genuine dogs and cats. Manually adjusting the arms and joints of the robot to the appropriate angle is challenging, although it is plausible to adjust the motor inclination by issuing repeated instructions, this is considered inappropriate

A kinematics-based system for controlling both virtual and actual robots was created as part of this research project. Using the simulation model, complete control can be exerted over the angle of each joint, as well as the foot locations and the end effectors. This method incorporates both forward and inverse kinematics. The values of the end effectors are provided as output after all the robot's connection parameters, and the manipulator's required angle is considered.

We can also set the end effector value to achieve the appropriate location. One of the most difficult challenges is taking control of a quadruped robot from its center of gravity. If the four-legged equations are not linked to the center of gravity point, the robot will never be able to stand on its own. For example, if the robot's rear side is heavier than its front side, then the device will always be unable to walk, as it will always fall on its back. To solve this problem, we first located the center points of the robot and then directly connected them to each leg. After completing all the stages (i.e., Part design, Assembly, URDF creation, Gait pattern, and kinematics setting), we executed all the scripts for simulation using the PyBullet physics engine. We connected our implemented robot with the simulation model during the simulation process. The hardware sensors transmitted sensory data every 20 milliseconds. We can control the simulation remotely, either using a controller or giving a command.

3.3. Step Trajectory and Gait Generation

There are various different gait patterns used by quadruped robots, including creep, trot, and gallop, among others. Creep and trot are often employed and swapped, based on the surroundings and the desired pace of mobility. Trot serves as the most frequent gait. The two diagonal legs move in tandem, with the left front leg and right back leg moving first and the right front leg and left back leg moving second. According to the trot diagram in Figure 11, the trot is a type of gait in which two legs are activated in tandem. At one moment, the front right leg and the back left leg swing, while the right back leg and left front leg form a stance. Two motions are present: Swing and stance. The stance phase means that the toe has contact with ground; during the swinging phase, the toe is in the air.

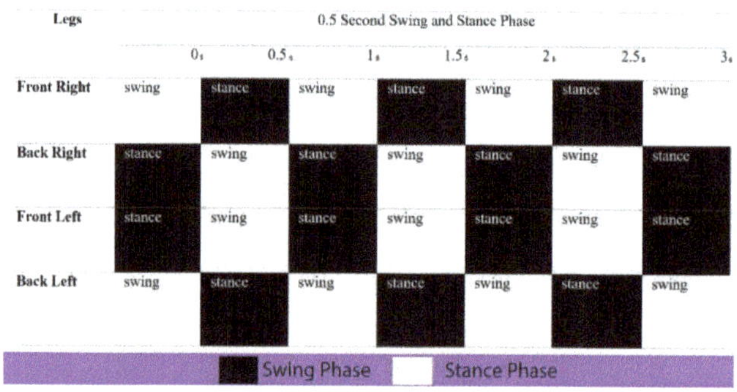

Figure 11. Trot Gait Patterns.

We used the Bernstein polynomial and Bezier curve for our robot's trajectory generation [34]. We created a 14-point Bezier curve to control the footstep trajectory. The control points that make up the Bezier curve were used to build the swing phase. Equation (41) shows the method for calculation of the binomial factor (d, l), and Equation (42) is used to simplify the Bezier curve calculation, where K is the current point index; t is the time; and c denotes the coordinate. Here, we have $P_0(a_0, d_0)$, $P_1(a_1, d_1)$, $P_2(a_2, d_2)$, $P_3(a_3, d_3)$, $P_4(a_4, d_4)$, $P_5(a_5, d_5)$, $P_6(a_6, d_6)$, $P_7(a_7, d_7)$, $P_8(a_8, d_8)$, $P_9(a_9, d_9)$, $P_{10}(a_{10}, d_{10})$, $P_{11}(a_{11}, d_{11})$, $P_{12}(a_{12}, d_{12})$, $P_{13}(a_{13}, d_{13})$ as the 14 control points. Equations (43) and (44) are used for curve construction. In Figure 12a, we show the trajectory steps of swing and stance, where t = 0.5 seconds; in Figure 12b, the parametric Bezier curve is shown (the curve was plotted using MATLAB). We show all the control point values in Table 3.

$$f(d,l) = \frac{d!}{K!(d-l)!}, \tag{41}$$

$$b(t,K,c) = f(N,K)\cdot(1-t)^{(N-K)\cdot t^{K}\cdot c}, \tag{42}$$

$$\begin{aligned} a = (b(t,0,a_0) + \; & b(t,1,a_1) + b(t,2,a_2) + b(t,3,a_3) + b(t,4,a_4) + b(t,5,a_5) \\ & + b(t,6,a_6) + b(t,7,a_7) + b(t,8,a_8) + b(t,9,a_9) \\ & + b(t,10,a_{10}) + b(t,11,a_{11}) + b(t,12,a_{12}) + b(t,13,a_{13})), \end{aligned} \tag{43}$$

$$\begin{aligned} d = (b(t,0,d_0) + \; & b(t,1,d_1) + b(t,2,d_2) + b(t,3,d_3) + b(t,4,d_4) + b(t,5,d_5) \\ & + b(t,6,d_6) + b(t,7,d_7) + b(t,8,d_8) + b(t,9,d_9) \\ & + b(t,10,d_{10}) + b(t,11,d_{11}) + b(t,12,d_{12}) + b(t,13,d_{13})). \end{aligned} \tag{44}$$

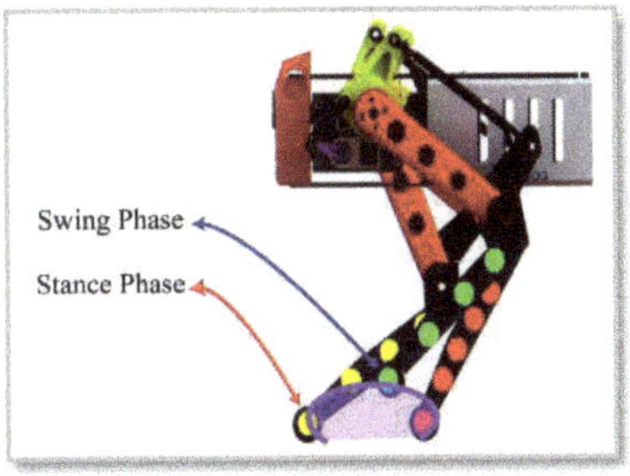

(a)

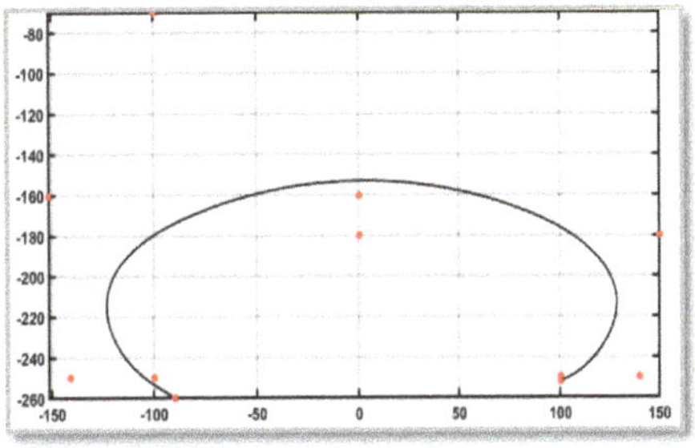

(b)

Figure 12. (**a**) Step trajectory stance and swing phase; (**b**) trajectory generation using 14-point Bezier curve.

Table 3. Control Points for Bezier curve.

Control Points		(mm)		(mm)
p_0	a_0	100	d_0	−252.4
p_1	a_1	100	d_1	−250
p_2	a_2	140	d_2	−250
p_3	a_3	150	d_3	−180
p_4	a_4	150	d_4	−180
p_5	a_5	150	d_5	−180
p_6	a_6	−100	d_6	−70
p_7	a_7	0	d_7	−180
p_8	a_8	0	d_8	−160.5
p_9	a_9	−151.5	d_9	−160.5
p_{10}	a_{10}	−151.5	d_{10}	−160.5
p_{11}	a_{11}	−141	d_{11}	−250
p_{12}	a_{12}	−100	d_{12}	−250
p_{13}	a_{13}	−90	d_{13}	−260

4. Results and Discussion

4.1. Material Analysis

As our quadruped robot was designed with a low cost in mind, we had to choose a material for 3D printing that is both cheap and strong for fabrication. Therefore, we chose ABS plastic (Acrylonitrile butadiene styrene) as our printing material.While designing and developing our low-cost quadruped robot, we used SolidWorks to create around 64 distinct components. We conducted load simulation in SolidWorks for various portions (i.e., femur and tibia) of our quadruped robot, in order to better understand the strength and stiffness of our design and the material used. We obtained analytical data, such as the strain, displacement, and Factor of Safety (FoS) for the femur and tibia of our four-legged robot throughout the simulation process. By looking at these data, we could make changes to how the components were made, thus improving their performance in the long run. Table 4 shows the mechanical properties of abs plastic.

Table 4. Mechanical Properties of ABS Plastic.

Property	Value	Unit
Tensile strength	30	MPa
Mass density	1020	kg/m^3
Elastic modulus	2000	MPa
Shear modulus	318.9	MPa
Poisson's ratio	0.394	N/A

In SolidWorks, we have utilized our femur and tibia designs to perform load analysis simulation. During load simulation, we considered approximately 7.5 kg of payload on our quadruped robot. The weight of our quadruped robot is around 2.752 kg. So, with the added payload, the total weight would be close to 10.25 kg, equivalent to a 100 N load. During the simulation process, we have obtained diagrams showing the performance of our designed parts under load. In Figure 13a,b, the displacement diagrams for the femur and tibia parts are shown, with the scale on the right indicating the degree of deformation of the components (in mm). We can see that, under a 100 N load on the femur, the maximum deformation was 6.596×10^{-3} mm. When the 100 N load was applied on the Tibia, the maximum deformation was 5.254×10^{-1} mm. For visual representation of the deformation, the components are highlighted with multiple colors, indicating the most deformed area (red color) and the least deformed area (blue color), according to the scale. Note that the deformation scale was adjusted for clearer visual analysis.

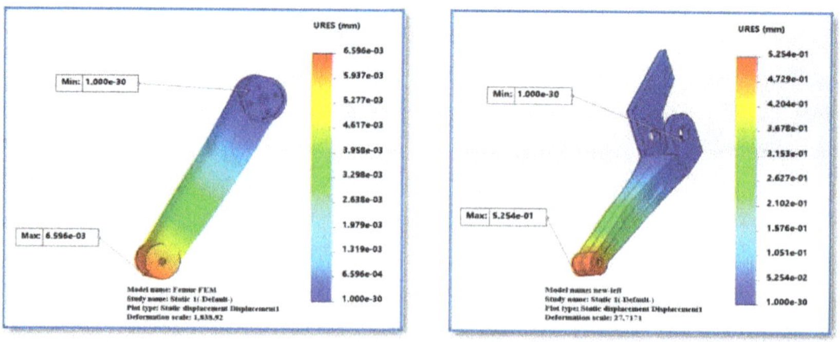

(a) (b)

Figure 13. (**a**) Displacement analysis diagram for femur; and (**b**) Displacement analysis diagram for tibia.

In the simulation process, we also obtained Factor of Safety diagrams, as shown in Figure 14a,b for the femur and tibia, respectively. The Factor of Safety indicates the safety of any mechanical component during operations; therefore, we can determine how safe the component is under certain conditions, according to the factor of safety scale.

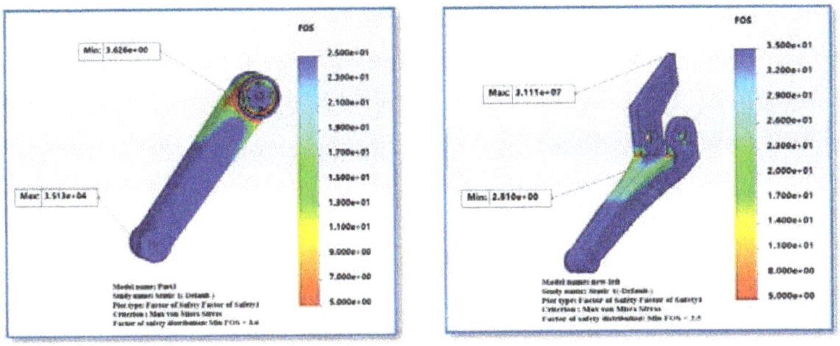

(a) (b)

Figure 14. (**a**) Factor of Safety analysis diagram for femur; and (**b**) Factor of Safety analysis diagram for tibia.

A base number indicates the minimum factor of safety where the component is safe. A value lower than the base number will indicate the failing point of the component. The factor of safety can vary for different materials, and the process of finding the factor of safety varies for ductile and brittle materials. For our ABS Plastic, which is a brittle material, the factor of safety is determined as the ratio of ultimate tensile stress to working stress.

The diagrams show the Factor of Safety scale on the right, which indicates the safety factor of components. Here, the factor of safety for the femur ranged from 3.6 up to 3.5×10^4, depending on the working stress in different parts of the femur. For the tibia, the factor of safety ranged from 2.510 to 3.111×10^7. The components are highlighted with multiple colors, indicates the areas where the components are safe (blue area) and where the components are at the risk of failing (red area).

From the simulation process, we also obtained strain analysis diagrams, as shown in Figure 15a,b for the Femur and Tibia, respectively. Strain is associated with the ratio of deformation under load to the original state. The strain is directly proportional to the applied stress. We can analyze the strain condition of our femur and tibia from the load analysis. For the femur, we can see that, with a deformation scale of 1838.92, the strain ranged from 4.004×10^{-9} to 4.358×10^{-5}. For the tibia, with a deformation scale of 27.71, the strain ranged from 3.457×10^{-10} up to 3.257×10^{-3}. The components are highlighted with multiple colors, indicates the areas where strain is more significant. Note that the deformation scale was adjusted for clear visual analysis.

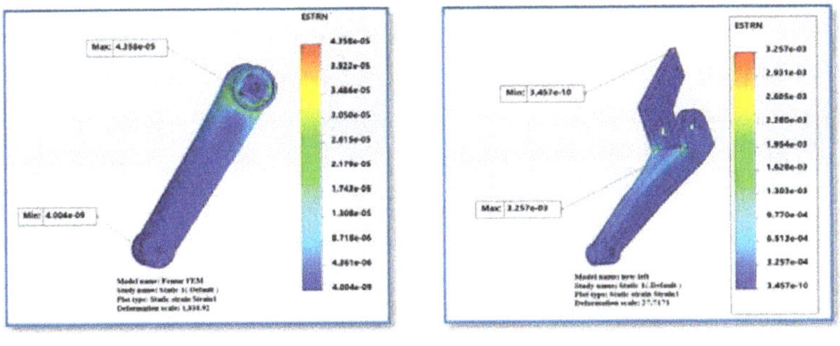

(a) (b)

Figure 15. (**a**) Strain analysis diagram for femur; and (**b**) strain analysis diagram for tibia.

4.2. Unified Robot Description Format Verification

The URDF refers to a robot as a tree of links connected by joints. The joints define how one link moves concerning another, thereby defining the location of the links in space. The links represent the moving components of the robot. URDF contains every link's inertia, mass, and rotation axis information. To create a scientific simulation of kinematics after finishing the assembly of the robot, we created a URDF file for the robot, which contains the joint parameter information such as translation and rotation along the x, y, and z axes, as well as roll, pitch, and yaw movements. Table 5 shows the urdf verification parameters.

Table 5. Verification at various angles with URDF.

Robot Positions	Hips Joint Angle (Degree)		Femur Joint Angle (Degree)		Tibia Joint Angle (Degree)	
Ideal State Figure 16a	Coxa FR	0	Femur FR	0	Tibia FR	0
	Coxa FL	0	Femur FL	0	Tibia FL	0
	Coxa BR	0	Femur BR	0	Tibia BR	0
	Coxa BL	0	Femur BL	0	Tibia BL	0
Standing State Figure 16b	Coxa FR	0	Femur FR	−45	Tibia FR	90
	Coxa FL	0	Femur FL	45	Tibia FL	90
	Coxa BR	0	Femur BR	−45	Tibia BR	90
	Coxa BL	0	Femur BL	45	Tibia BL	90

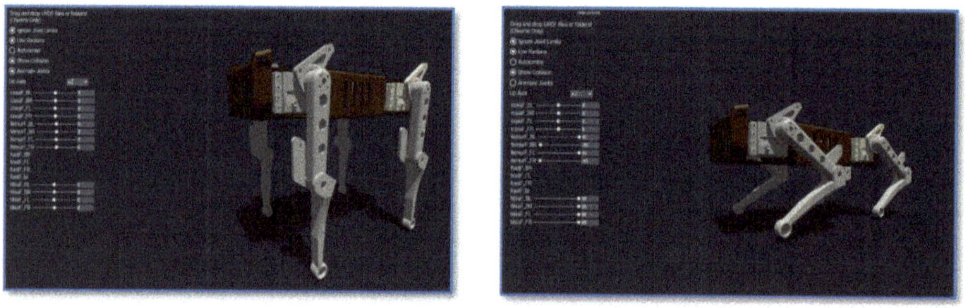

(a) (b)

Figure 16. (**a**) Ideal state position; and (**b**) standing state position.

In the URDF file, we can utilize visualization and material information to create the simulation effectively. If the URDF file has a problem, the simulation results will be poor. For this issue, after creating the URDF file with meshes, we verified the design using a Web-based URDF visualization tool. Scenario 1 shows the ideal state of our quadruped robot. We created the URDF file in SolidWorks, where we set the positions of all joint angles and links, as shown in Figure 16a. All of the links were set to zero position, relative to the base link (black body). For each leg, the coxa was the parent link, the femur was the child link relative to the coxa, and the tibia was the child link relative to the Femur. Figure 16b shows the standing state of our quadruped robot, where all the links and joint angles were manipulated within their respective parameter ranges; in particular, the coxa remains at zero position, while the femur link is set to 45° relative to the coxa and the tibia link is set to 90° relative to the Femur. The main idea of testing the URDF is to verify the angles of the coxa associated with the other links (i.e., the femur and tibia). Figure 17 shows the URDF tree.

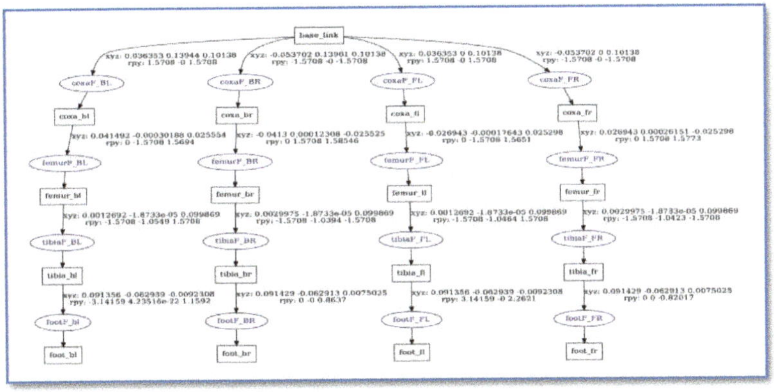

Figure 17. Unified robot description format tree. This diagram shows every joint detail with x, y, z and r, p, y values. Each leg has four joints and four links; three joints are revolute, and one is fixed.

4.3. Dynamic Simulation Results

We created a quadruped robot using a novel mobility strategy for the leg called the Watt six-bar linkage mechanism. The Pybullet physics engine platform was used for testing of the developed robot, and the robot's dynamic model was established through simulation. The robot was divided into multiple parts, according to its components, and each piece was handled as a separate entity. The maximum torque and speed of each joint, as well as the mass, inertia, and collision geometry of each component, were precisely entered into the simulation environment. In Figure 18, we show our quadruped robot from various angles. We also provide detailed explanations of the mechanism and the kinematic equations with pertinent figures below.

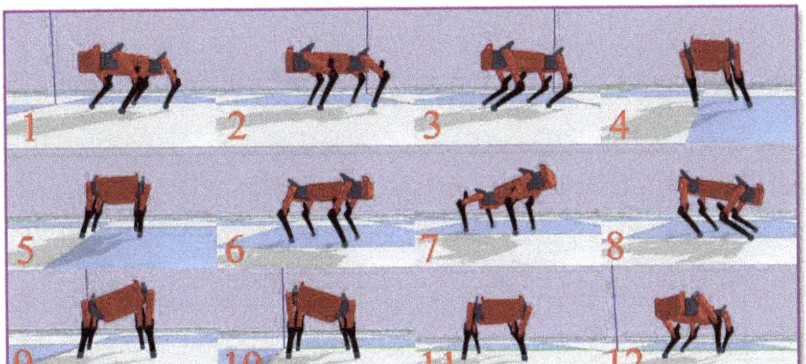

Figure 18. Computational simulation results.

Before the simulation process, we connected gait patterns (e.g., trot and creep gait) in our simulation, in order to observe the walking motion of the quadruped robot. In Figure 18, Scenario 1 shows the natural standing position of the quadruped robot. Scenarios 2–6 demonstrate translation of the body of our quadruped robot along its x, y, and z axes. We can see the pitch of the body along its y-axis in Scenarios 7 and 8, while Scenarios 9 and 10 show the roll of the body along its x-axis. Finally, we can see the yaw movement of our robot in Scenarios 11 and 12. Throughout the simulation, we captured several alternative orientations of the robot. Specifically, we demonstrated our quadruped robot's ideal state, roll, pitch, and yaw positions. Following that, we built a real robot that was coupled with a simulation engine. We put our quadruped robot through specific tests, in order to evaluate its pitch position.

Having matched the hardware configuration to the software simulation, we could adjust the pitch position of the actual robot to influence the model in the simulation using the Pybullet Physics Engine. Figure 19a displays the ideal functioning of our simulation model. The sensor's roll and pitch angle data were successfully acquired during our hardware orientation test. We constructed a roll–pitch vs. time graph (Figure 19b), based on the information gathered by our sensors. Our simulation model's pitch orientation is depicted in Figure 19c. We set the robot's pitch orientation on the hardware test-bench. The pitch position roll–pitch vs. time graph is depicted in Figure 19d.

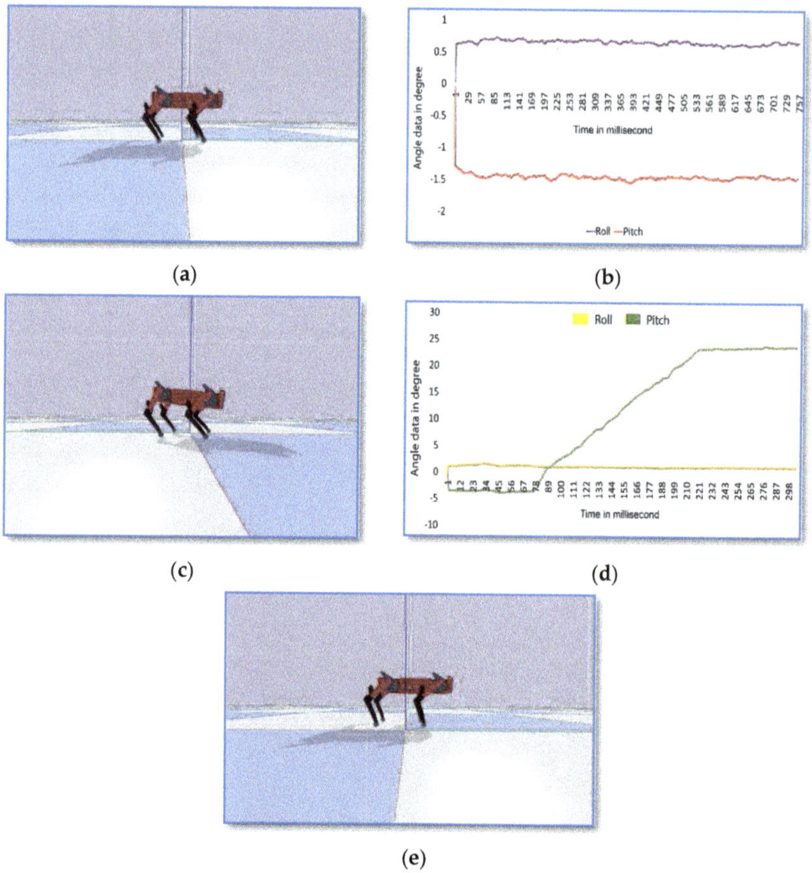

Figure 19. (**a**) Ideal state position of robot in side view; (**b**) Ideal state graph; (**c**) Pitch 9 of robot; (**d**) Roll–Pitch vs. time graph; and (**e**) Self-Balanced position.

With hardware and software synchronization, we not only could test how the hardware setup manipulates the simulation model, but also vice versa. We used Proportional Integral Derivative with our kinematics equations. While configuring our robot into a pitch position, the simulated model moved such that its center of mass took precedence over the initial pitch. Figure 19e demonstrates how the simulated model adjusted its femur and tibia to counteract the pitch position. The central body always lies flat against the reference plane in the virtual world. Viewed from the front, our simulation model in its ideal state is depicted in Figure 20a. The sensor's roll and pitch angle data were successfully acquired during our hardware orientation test. Figure 20b shows a graph depicting the roll–pitch angle over time, as measured by our sensors. The roll orientation of our simulation model is displayed in Figure 20c. The related roll–pitch vs. time graph for roll position is shown in Figure 20d.

Figure 20e shows how the simulation model stabilized when we positioned the gyroscope sensor to induce roll. The core coordinate system was maintained in its initial location, while the rest of the model counteracted the roll by shifting the coxa angle. In this way, the simulated model can maintain a level stance relative to the reference plane.

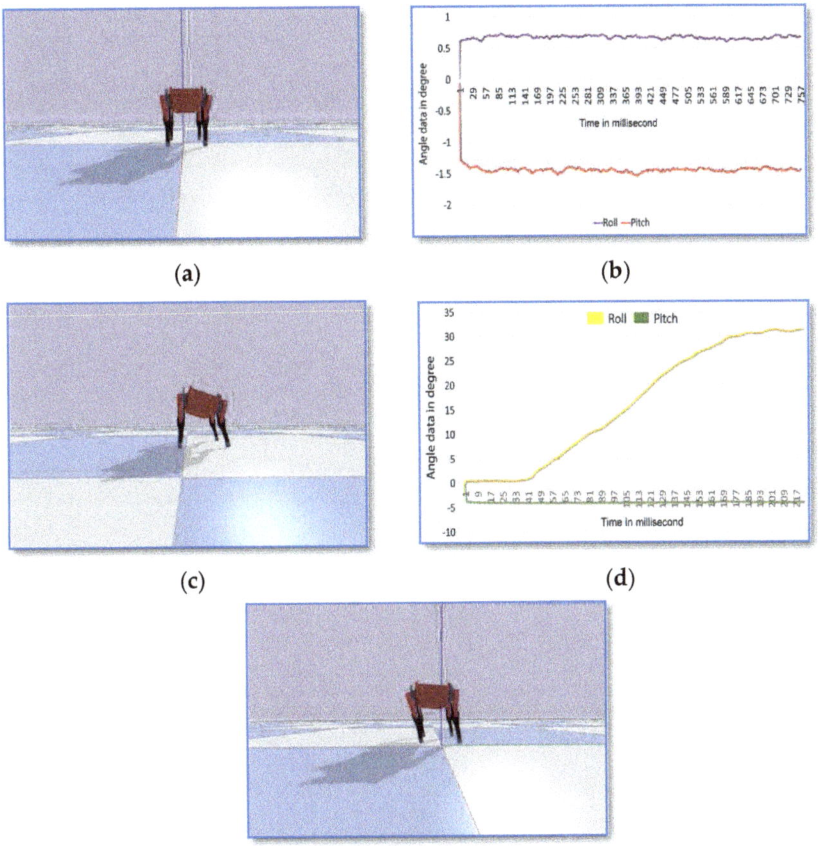

Figure 20. (**a**) Ideal state position of the robot in front view; (**b**) Ideal state graph; (**c**) Roll orientations of the robot; (**d**) Roll–Pitch vs. time graph; and (**e**) Self-Balanced position.

We constructed a real robot using 3D-printed components and the six-bar connection system. When we assessed the robot's trot gait pattern, its walking was essentially flawless. Figure 21 clearly depicts the motion of the trot gait pattern. According to the flat terrain test, we simultaneously recorded the roll vs. pitch sensor data as a graph. During the trotting period, the robot recorded a minimum roll angle of $-5°$ and a maximum roll angle of $16°$. The minimum pitch angle was $-11°$, while the highest pitch angle was $9°$. The angle data are displayed in Figure 22. The gyroscope results of the standing robot were $-1.2°$ pitch and $0.5°$ roll. Some photographs of our robot are shown in Figure 23.

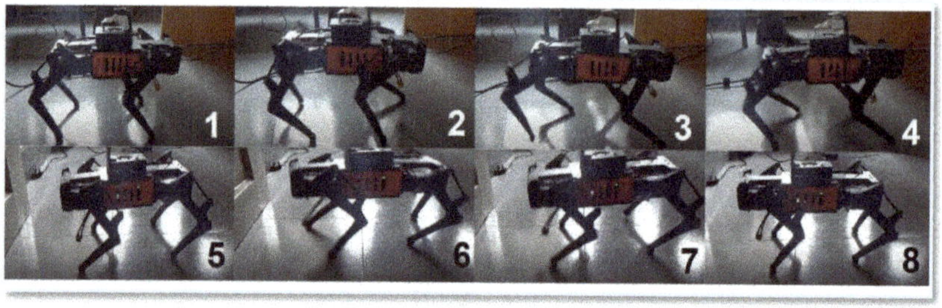

Figure 21. Still pictures of adopted watt six-bar linkage mechanism-based quadruped robot. Sequence 1–8 shows the trot gait transition on flat terrain.

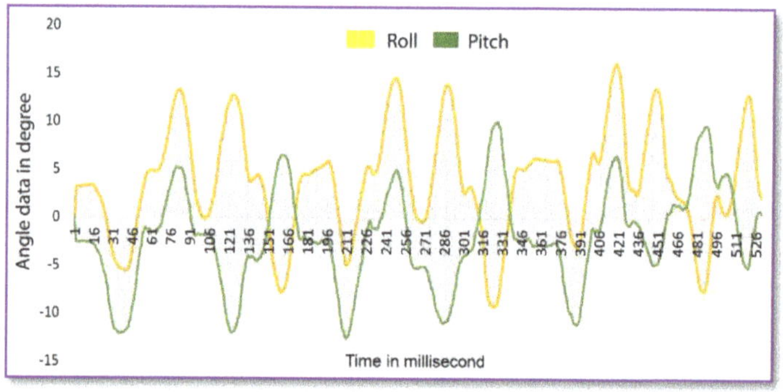

Figure 22. Roll pitch data extracted from running trot gait, according to Figure 21 (real robot flat terrain test).

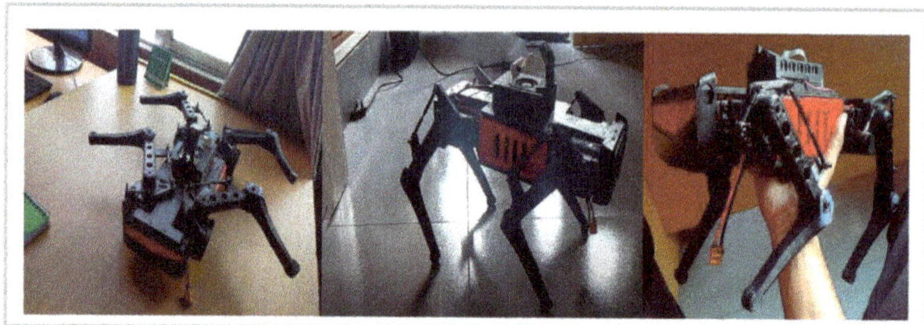

Figure 23. Still pictures of the implemented version of Iron Dog mini.

Using the Pybullet physics engine simulation, we achieved dynamic simulation results. Several translational and rotational movements were demonstrated. Kinematics (in particular, Forward and Inverse) aided us in achieving this motion. Forward kinematics provide the positions of the end effectors, while inverse kinematics provide the locations of the joints. Therefore, x, y, and z translation, as well as roll, pitch, and yaw movement can be easily achieved using these two mathematical approaches. We have, thus, briefly described the kinematics portion of the mathematical analysis. During the simulation time, robot walking was achieved through the use of a trot gait pattern and a 14-point Bezier curve. According to the mathematical analysis, gait pattern, and trajectory generation, the outcome of the simulation was satisfactory. Details of the practical implementation and a simulation video can be found elsewhere [35].

5. Conclusions and Future Work

This study on the low-cost development and simulation of a quadruped robot, considering a six-bar linkage mechanism, is expected to play an important role in the vigorous development of quadruped robots in the future. Within the scope of this investigation, we presented a process for computational simulation and its underlying mathematical rationale. We developed an innovative concept, providing thorough explanations of the steps followed to generate the simulation model. A cutting-edge Watt six-bar linkage system was added to redistribute the weight better. Overall, we have developed a relatively affordable quadruped robot, allowing researchers to more easily pursue careers in this field. Our four-legged robot costs only USD 350, which is significantly less than other quadruped robots such as the Spot (USD 74,500), Unitree a1 (USD 15,000), and Xiaomi Cyberdog (USD 1599). Compared with existing robots, the proposed robot is lightweight, can go through rough terrain, and can carry loads. We also used relatively cheap servo motors and 3D-printed components.

In comparison to more traditional methods, including the belt-pulley approach utilized in previous quadruped robots, we opted to use the Watt six-bar linkage mechanism, due to its intuitive design. We provided in-depth kinematic analysis data. We may consider training the robot's gait using this model as a template. With any luck, our upcoming study will demonstrate the real-time software and hardware synchronization system, including a reinforcement learning technique. One of the goals of this research is to use AI in a simulation environment, then take the results and use them to fine-tune a physical robot. In Table 6 we exhibited the cost breakdown table.

Table 6. Cost Breakdown table.

Component	Per Unit Cost (USD)	Quantity	Total Cost (USD)
SPT5535LV Actuator-Servo	15	12 Pieces	180
Metal Servo Arm	0.50	4 Pieces	2
Metal Servo Horn	0.40	4 Pieces	1.6
Hex spacer	0.10	30 Pieces	3
M3 Screws	0.05	85 Pieces	4.3
Brass Inserts	0.02	85 pieces	1.7
16Awg Silicon wire	1.5 (per yard)	5 yards	7.5
3D printing Abs material	21.99 (per kg)	2 kg	43.98
Raspberry Pi 4b	40	1 Pieces	40
Bluetooth Joystick Controller	16	1 Pieces	16
Arduino Pro Mega	8	1 Pieces	8
41 A Buck Converter	5	1 Pieces	5
7.4 V 2200 mAh Battery	18	1 Pieces	12
IMU 6050	0.6	1 Pieces	0.6
	Total Cost		332.18$

Throughout this research, we designed and built our quadruped robot and developed a kinematic method for its simulation in the Pybullet Physics Engine. We used 3D printing technology to test the model in the actual world when the simulation was complete. With the help of our simulated model, we could correct most issues, including improper scaling and changes that would have harmed the robot. The use of a simulation model makes it straightforward to examine the kinematic model and create a gait. Using the simulation, we can refine our robot model to keep up with the real robot. Our long-term goal is to advance robotics by creating a low-cost, highly developed quadruped robot that the next generation of scientists and engineers can easily use.

Author Contributions: Methodology, M.H.R.; Formal analysis, S.B.A., T.D.M., M.F.U. and M.H.; Investigation, M.H.R. and T.D.M.; Writing—original draft, M.H.R.; Visualization, M.H.R.; Supervision, S.B.A.; Project administration, M.H.R.; Funding acquisition, S.B.A. and M.H. All authors have read and agreed to the published version of the manuscript.

Funding: This research received no external funding.

Institutional Review Board Statement: Not applicable.

Informed Consent Statement: Not applicable.

Data Availability Statement: Data is available upon request from the authors.

Acknowledgments: We would like to express our appreciation and gratitude to Fayed Al Monir's help with the mathematics and (FEM), (FoS) testing of our research work.

Conflicts of Interest: The authors declare no conflict of interest.

Abbreviations

The Following abbreviations are used in this research article.

URDF	Unified Robot Description Format
DoF	Degrees Of Freedom
FEM	Finite Element Method
FoS	Factor of Safety
MIT	Massachusetts Institute of Technology
CAD	Computer Aided Design
STL	Stereolithography
PID	Proportional Integral Derivative
IMU	Inertial Measurement Unit
FR	Front Right
FL	Front Left
BR	Back Right
BL	Back Left
g	gram
Cm	Centimeter

References

1. Yao, L.; Hao, Y.; Lu, Z. Design and driving model for the quadruped robot: An elucidating draft. *Adv. Mech. Eng.* **2021**, *13*, 16878140211009035. [CrossRef]
2. Meng, X.; Wang, S.; Cao, Z.; Zhang, L. A review of quadruped robots and environment perception. In Proceedings of the 2016 35th Chinese Control Conference (CCC), Chengdu, China, 27–29 July 2016; IEEE: New York, NY, USA, 2016.
3. Biswal, P.; Mohanty, P.K. Mohanty. Development of quadruped walking robots: A review. *Ain Shams Eng. J.* **2021**, *12*, 2017–2031. [CrossRef]
4. Todd, D.J. *Walking Machines: An Introduction to Legged Robots*; Springer Science & Business Media: Berlin/Heidelberg, Germany, 2013.
5. Rong, X.; Li, Y.; Ruan, J.; Li, B. Design, and simulation for a hydraulic actuated quadruped robot. *J. Mech. Sci. Technol.* **2012**, *26*, 1171–1177. [CrossRef]
6. Yang, K.; Zhou, L.; Rong, X.; Li, Y. Onboard hydraulic system controller design for quadruped robot driven by gasoline engine. *Mechatronics* **2018**, *52*, 36–48. [CrossRef]
7. Atique, M.U.; Sarker, R.I.; Ahad, A.R. Development of an 8DOF quadruped robot and implementation of Inverse Kinematics using Denavit-Hartenberg convention. *Heliyon* **2018**, *4*, e01053. [CrossRef]
8. Gor, M.M.; Pathak, P.M.; Samantaray, A.K.; Alam, K.; Kumar, P.; Anand, D.; Vijay, P.; Sarkar, R.; Yang, J.-M.; Kwak, S.W. Development of a compliant legged quadruped robot. *Sādhanā* **2018**, *43*, 102. [CrossRef]
9. WildCat-The World's Fastest Quadruped Robot. Available online: https://www.bostondynamics.com/wildcat (accessed on 14 March 2019).
10. Ananthanarayanan, A.; Azadi, M.; Kim, S. Towards a bio-inspired leg design for high-speed running. *Bioinspir. Biomim.* **2012**, *7*, 046005. [CrossRef]
11. Spröwitz, A.; Tuleu, A.; Vespignani, M.; Ajallooeian, M.; Badri, E.; Ijspeert, A.J. Towards Dynamic Trot Gait Locomotion-Design, Control, and Experiments with Cheetah-cub, a Compliant Quadruped Robot. *Int. J. Robot. Res.* **2013**, *32*, 932–950. [CrossRef]

12. Rosendo, A.; Liu, X.; Nakatsu, S.; Shimizu, M. A combined CPG-stretch reflex study on a musculoskeletal pneumatic quadruped. In *Biomimetic and Biohybrid Systems, Proceedings of the Third International Conference, Living Machines 2014, Milan, Italy, 30 July–1 August 2014*; Springer International Publishing: Cham, Switzerland, 2014.
13. Lu, M.; Jing, B.; Duan, H.; Gao, G. Design of a Small Quadruped Robot with Parallel Legs. *Complexity* **2022**, *2022*, 9663746. [CrossRef]
14. Pinto, V.; Soares, I.; Rocha, M.; Lima, J.; Gonçalves, J.; Costa, P. Design, modeling, and control of an autonomous legged-wheeled hybrid robotic vehicle with non-rigid joints. *Appl. Sci.* **2021**, *11*, 6116. [CrossRef]
15. Rahman, M.H.; Islam, M.M.; Al Monir, M.F.; Alam, S.B.; Rahman, M.M.; Shidujaman, M.; Islam, R. Kinematics analysis of a quadruped robot: Simulation and Evaluation. In Proceedings of the 2022 2nd International Conference on Image Processing and Robotics (ICIPRob), Colombo, Sri Lanka, 12 March 2022; IEEE: New York, NY, USA; pp. 1–6.
16. Biswal, P.; Prases, K. *Mohanty. Kinematic and Dynamic Modeling of a Quadruped Robot. Machines, Mechanism and Robotics*; Springer: Singapore, 2022; pp. 369–378.
17. Morlando, V.; Teimoorzadeh, A.; Ruggiero, F. Whole-body control with disturbance rejection through a momentum-based observer for quadruped robots. *Mech. Mach. Theory* **2021**, *164*, 104412. [CrossRef]
18. Yi, S. Reliable gait planning and control for miniaturized quadruped robot pet. *Mechatronics* **2010**, *20*, 485–495. [CrossRef]
19. Ho, T.; Lee, S. Piezoelectrically actuated biomimetic self-contained quadruped bounding robot. *J. Bionic Eng.* **2009**, *6*, 29–36. [CrossRef]
20. Sun, Z.; Zhu, Z.; Zhang, G.; Li, Y.; Rong, X. Design and Control of a Quadruped Robot with Changeable Configuration. In Proceedings of the International Conference on Intelligent Robotics and Applications, Harbin, China, 1–3 August 2022; Springer: Cham, Switzerland, 2021.
21. Shi, Y.; Li, S.; Guo, M.; Yang, Y.; Xia, D.; Luo, X. Structural design, simulation and experiment of quadruped robot. *Appl. Sci.* **2021**, *11*, 10705. [CrossRef]
22. Cong, Q.; Shi, X.; Wang, J.; Xiong, Y.; Su, B.; Xu, W.; Liu, H.; Zhou, K.; Jiang, L.; Tian, W. Stability Study and Simulation of Quadruped Robots with Variable Parameters. *Appl. Bionics Biomech.* **2022**, *2022*, 9968042. [CrossRef]
23. Rodinò, S.; Matteo Curcio, E.; di Bella, A.; Persampieri, M.; Funaro, M.; Carbone, G. Design, simulation, and preliminary validation of a four-legged robot. *Machines* **2020**, *8*, 82. [CrossRef]
24. Takei, Y.; Tazawa, R.; Kaimai, T.; Morishita, K.; Saito, K. Dynamic simulation of non-programmed gait generation of quadruped robot. *Artif. Life Robot.* **2022**, *27*, 480–486. [CrossRef]
25. Bledt, G.; Powell, M.J.; Katz, B.; Di Carlo, J.; Wensing, P.M.; Kim, S. MIT Cheetah 3: Design and control of a robust, dynamic quadruped robot. In Proceedings of the 2018 IEEE/RSJ International Conference on Intelligent Robots and Systems (IROS), Madrid, Spain, 1–5 October 2018; IEEE: New York, NY, USA, 2018.
26. Raibert, M.; Blankespoor, K.; Nelson, G.; Playter, R. BigDog, the rough-terrain quadruped robot. *IFAC Proc. Vol.* **2008**, *41*, 10822–10825. [CrossRef]
27. Zong, M.; Yang, Y.; Fu, M.; Wang, S.; Deng, Z. Dynamic modeling and kinematic analysis of Frog robot. In Proceedings of the 32nd Chinese Control Conference, Xi'an, China, 26–28 July 2013; IEEE: New York, NY, USA; pp. 5848–5853.
28. Raibert, M. Alphadog, the rough-terrain robot. In *Adaptive Mobile Robotics*; World Scientific: Singapore, 2012; p. 7.
29. Semini, C.; Tsagarakis, N.G.; Guglielmino, E.; Focchi, M.; Cannella, F.; Caldwell, D.G. Design of HyQ-a hydraulically and electrically actuated quadruped robot. *Proc. Inst. Mech. Eng. Part I J. Syst. Control Eng.* **2011**, *225*, 831–849. [CrossRef]
30. Chen, X.; Gao, F.; Qi, C.; Zhao, X. Spring parameters design to increase the loading capability of a hydraulic quadruped robot. In Proceedings of the 2013 International Conference on Advanced Mechatronic Systems, Luoyang, China, 25–27 September 2013; IEEE: New York, NY, USA, 2013.
31. Hutter, M.; Gehring, C.; Jud, D.; Lauber, A.; Bellicoso, C.D.; Tsounis, V.; Hwangbo, J.; Bodie, K.; Fankhauser, P.; Bloesch, M.; et al. Anymal-a highly mobile and dynamic quadrupedal robot. In Proceedings of the 2016 IEEE/RSJ International Conference on Intelligent Robots and Systems (IROS), Daejeon, Republic of Korea, 9–14 October 2016; IEEE: New York, NY, USA, 2016.
32. Ackerman, E. *Boston Dynamics' SpotMini Is all Electric, Agile, and has a Capable Face-Arm*; IEEE Spectrum: New York, NY, USA, 2016.
33. Tirumala, S.; Sagi, A.; Paigwar, K.; Joglekar, A.; Bhatnagar, S.; Ghosal, A.; Bharadwaj, A.; Kolathaya, S. Gait library synthesis for quadruped robots via augmented random search. *arXiv* **2019**, arXiv:1912.12907.
34. Oruç, H.; George, M. Phillips. q-Bernstein polynomials and Bézier curves. *J. Comput. Appl. Math.* **2003**, *151*, 1–12. [CrossRef]
35. Rahman, M.H. Practical Implementation and Simulation Video of Iron Dog Mini. Available online: https://github.com/irondogmini (accessed on 7 January 2023).

Author Index

MDPI
St. Alban-Anlage 66
4052 Basel
Switzerland
Tel. +41 61 683 77 34
Fax +41 61 302 89 18
www.mdpi.com

Robotics Editorial Office
E-mail: robotics@mdpi.com
www.mdpi.com/journal/robotics

www.ingramcontent.com/pod-product-compliance
Lightning Source LLC
Chambersburg PA
CBHW061815210926
51850CB00006B/710
9783036580760